AQA Science

Science B: Science in Context

Teacher's Book

New GCSE

James Hayward
Jo Locke
Nicky Thomas
Louise Burt
Andrea Johnson

Series Editor
Lawrie Ryan

GCSE Science B: Science in Context

Contents

Welcome to AQA GCSE Science!	1

Unit 1 – My world

Theme 1 – My wider world — 2

Chapter 1 Our changing planet and universe	2
Summary answers	18
Examination-style answers	19
Chapter 2 Using materials our planet provides	20
Summary answers	40
Examination-style answers	41

Theme 2 – Life on our planet — 42

Chapter 3 Life on our planet	44
Summary answers	56
Examination-style answers	57
Chapter 4 Biomass, energy flow and the importance of carbon	58
Summary answers	70
Examination-style answers	71
End of unit examination-style answers	72

Unit 2 – My family and home

Theme 1 – My family — 74

Chapter 5 Body systems	76
Summary answers	92
Examination-style answers	93
Chapter 6 Human inheritance and genetic disorders	94
Summary answers	104
Examination-style answers	105

Theme 2 – My home — 106

Chapter 7 Materials used to construct our homes	108
Summary answers	122
Examination-style answers	123
Chapter 8 Fuels and electricity	124
Summary answers	140
Examination-style answers	141

Theme 3 – My property — 142

Chapter 9 Using energy and radiation	144
Summary answers	158
Examination-style answers	159
End of unit examination-style answers	160

Unit 3 – Making my world a better place

Theme 1 – Improving health and wellbeing — 162

Chapter 10 The use (and misuse) of drugs	164
Summary answers	176
Examination-style answers	177
Chapter 11 Modern medicine	178
Summary answers	194
Examination-style answers	195

Theme 2 – Making and improving products — 196

Chapter 12 Improving materials	196
Summary answers	210
Examination-style answers	210
Chapter 13 Selective breeding and genetic engineering	212
Summary answers	220
Examination-style answers	220

Theme 3 – Improving our environment — 222

Chapter 14 Environmental concerns when making and using products	222
Summary answers	232
Examination-style answers	233
Chapter 15 Our environment at home	234
Summary answers	252
Examination-style answers	253
End of unit examination-style answers	254

Controlled Assessment	**256**

Welcome to AQA GCSE Science!

New AQA GCSE Science remains the only series to be endorsed and approved by AQA. This Teacher's Book is written and reviewed by experienced teachers who have worked closely with AQA on their specifications. This book is structured around the Student Book and offers guidance, advice, support for differentiation and lots of practical teaching ideas to give you what you need to teach the AQA specifications.

Learning objectives
These tell you what your students should know by the end of the lessons and relate to the learning objectives in the corresponding Student Book topic, although extra detail is provided for teachers.

Learning outcomes
These tell you what your students should be able to do to demonstrate that they have achieved against the learning objectives. These are differentiated where appropriate to provide suitable expectations for all your students. Higher Tier outcomes are labelled.

Specification link-up: Science B
These open every spread so you can see the AQA specification references covered in your lessons, at a glance.

Lesson structure
This provides you with guidance and ideas for tackling topics in your lessons. There are short and long starter and plenary activities so you can decide how to structure your lesson. Explicit **support** and **extension** guidance is given for some starters and plenaries.

Support
These help you to give extra support to students who need it during the main part of your lesson.

Extend
These provide ideas for how to extend the learning for students aiming for higher grades.

Further teaching suggestions
These provide you with ideas for how you might extend the lesson or offer alternative activities. These may also include extra activities or suggestions for homework.

Summary answers
All answers to questions within the Student Book are found in the Teacher's Book.

Practical support
For every practical in the Student Book you will find the corresponding feature which gives you a list of the equipment you will need to provide, safety references and additional teaching notes. There are also additional practicals given that are not found in the Student Book.

The following features are found in the Student Book. You may also find additional guidance in the Teacher's Book:

 Did you know … ?

 Maths skills

This feature indicates where there are opportunities to introduce Science in context concepts:

 Science in context

Activity

Controlled Assessment
There are chapters dedicated to Controlled Assessment in the Student Book. The teacher notes within this book give you detailed guidance on how to deliver this part of the course.

End of chapter pages
And at the end of each chapter you will find Summary answers and AQA Examinations-style answers. You will also find:

Kerboodle resources
Kerboodle is our online service that holds all of the electronic resources for the series. All of the resources that support the chapter that are provided on Kerboodle are listed in these boxes.

Where you see in the Student Book, you will know that there is an electronic resource on Kerboodle to support that aspect.

Just log on to www.kerboodle.com to find out more.

Practical support
These list the suggested practicals from AQA that you need to be aware of. Support for these practicals can be found on Kerboodle, or are covered within the practical support section of the Teacher's Book. The indicates that there is a practical in Kerboodle. The 📖 indicates that the practical is covered in this Teacher's Book.

Bump up your grades
These are written by AQA examiners giving advice on how students can pick up additional marks to improve their grades.

Examiner's tips
These are written by AQA examiners giving advice on what students should remember for their exams and highlighting common errors.

Unit 1, Theme 1

My wider world

AQA Specification link-up: Science B 3.3.1

Scientists have used different types of evidence to prove the many changes that have taken place in the universe and on our own planet over time. Living organisms have also adapted to the changing environmental conditions.

In this theme there are four contexts:

3.3.1.1 Our changing universe
3.3.1.2 Our changing planet
3.3.1.3 Materials our planet provides
3.3.1.4 Using materials from our planet to make products

Our changing universe

There are regular missions taking place to explore our neighbouring planets, with NASA's Mars Lander landing on Mars in 2010. The NASA website www.nasa.gov and the European Space Agency website www.esa.int are excellent resources with up-to-date information about space missions. Both sites contain live data.

How do we learn about other parts of the universe?

Activity:
- Write down what you know about light and waves.
- Prepare a poster showing the structure of one telescope, how it works and some of its discoveries.
- Find out why more distant stars tell us about the early universe.

Why do we think the universe is changing?

Activity:
- Sketch a diagram showing the parts of the solar system.
- Prepare a poster showing what happened in the Big Bang and the evidence that backs up this model.

Possible misconceptions

Students need to be aware of the electromagnetic spectrum before studies of telescopes and the Big Bang make sense. They need to know that the longer wavelength waves have less energy and are emitted by cooler objects. They also need to know the order of the members of the spectrum.

Students also have problems comprehending the size and scale of the solar system and universe. They often question what was present before the Big Bang and what is outside the universe – answers to these questions tend to be unsatisfactory.

What do we know about the Earth's place in the universe?

Keep an eye open for events such as partial eclipses of the Moon or Sun, comets or meteoroid showers. These are not as rare as you may think, although it can be inconvenient to view many of these phenomena. www.seasky.org/index.html contains a calendar of events.

Activity:
- Prepare a leaflet describing one model of the solar system, as if it was written at that time. Include the evidence you used.
- Write a postcard from another planet.
- Build a scale model of the solar system.

Unit 1, Theme 1

My wider world

In Unit 1, Theme 1 you will work in the following contexts, covered in Chapters 1 and 2:

Our changing planet and universe

How do we learn about other parts of the universe?

Light from distant stars has taken millions of years to reach us. So, when we see a star, we are actually seeing it as it was a long time ago. Telescopes based on Earth or in space help scientists discover what the stars and universe were like in the past. They detect different forms of electromagnetic radiation so we can study distant stars, black holes and other objects even further back in time.

Why do we think the universe is changing?

Scientists study light from distant stars. They can say what elements are present in stars billions of miles away and how fast stars and galaxies are moving. Studying the light from distant galaxies helps astronomers explain changes in the universe that are happening now, as well as modelling how the universe began.

What do we know about the Earth's place in the universe?

The Sun seems to move through the sky as the day goes on. Over many centuries, astronomers have studied how the Earth and its neighbouring planets orbit our star, the Sun.

Our changing planet

How has the Earth changed since it formed?

The landscape around us does not seem to change. If we look a bit closer we can tell that things are not as they seem. Rocks have fossils in them. These rocks have been crushed and bent into odd shapes. The Earth has cooled and continues to change since it formed billions of years ago. Its structure is layered and cracked near the surface. Tectonic plates move constantly so that continents and oceans are now dramatically different from when they first formed.

What causes the changes on Earth?

Convection currents in the molten rock under the Earth's surface force the tectonic plates to move. This is the cause of earthquakes and volcanoes that can devastate large areas.

Has our atmosphere always been the same?

We take the air for granted but did you know that we are only here today because of microorganisms? When the Earth formed, gases were released during the many violent eruptions from volcanoes that were active at that time. Water vapour condensed to form oceans. The remaining gases somehow allowed single-celled organisms to grow. Some of these produced the oxygen now in the atmosphere.

Why can the Earth support life?

Our atmosphere surrounds the Earth and the layer of gases helps to sustain life. It plays a vital part in keeping temperatures on Earth stable. It allows some energy from the Sun through. It also absorbs some radiation going out from the Earth.

Materials our planet provides

Raw materials

Mining and quarrying extract millions of tonnes of material from the Earth every year. Some of these materials are elements, some are compounds and some are mixtures. Mixtures like crude oil or rock salt are quite easy to separate. This is because the substances in them are not chemically joined. Compounds, found in mineral ores, are harder to separate, because the elements they are made of are chemically joined. Some materials such as gold or sulfur exist as elements in the Earth's crust. Not all materials we use come from the Earth's crust. For example, gases such as nitrogen are extracted from the atmosphere.

Using materials from our planet to make products

Making products

Most of the time, raw materials need to be chemically changed before they can be used. This means that knowledge of chemical reactions is vital when making new products. To make it easier to communicate about chemical reactions, they can be written as equations. Chemists can work out exactly how much of each raw material they need from balanced chemical equations. If they get it wrong, money is wasted and costs go up for us.

Check that the basic concepts and definitions are clear for all students – planets, moons and stars get muddled up, as do galaxies, solar systems and the universe. Students may not realise that our nearest star is the Sun – it is a very ordinary, small to medium size sun in the stable phase of its life cycle.

Our changing planet

There are regular newspaper stories about earthquakes and volcanoes. Check the National Earthquake Information Centre website: www.earthquake.usgs.gov. Students should be aware that certain regions of the world are at much greater risk than others. Earthquake-proof buildings and advance warning systems are issues to consider when thinking about the damage earthquakes can cause. Seismographs and ways of monitoring the position and intensity of earthquakes are of interest too.

How has the Earth changed since it formed?

Activity:
- Prepare a timeline showing the main events that have occurred since the Earth was formed.

What causes the changes on Earth?

Preparation:
- Find out how we use convection currents in our daily lives.

My wider world

How has space exploration helped us? New materials used for space missions have everyday uses too. For example, prosthetics (artificial limbs) are lighter and tougher. Our homes can be insulated using much thinner but more effective materials. Fibreglass roofing fabric is being used as a stunning new building material and improved thermal protection on racing cars keeps the drivers safer.

Volcanoes and earthquakes can have huge effects on our everyday life. In 2010 the ash from a volcano in Iceland halted flights all over the world.

Weather forecasters, use information from satellites, as well as taking readings from weather stations to predict the weather. We can use satellites to monitor the weather locally and different satellites to track weather systems such as cyclones. When you hear the weather forecast in the morning it will be based on information from satellites.

As well as the constant need for raw materials, getting new materials from the Earth's crust keeps many people employed. Without quarries there wouldn't be much employment in some communities. There are many skilled jobs in a quarry, such as blasting, monitoring the environment, running machinery and transporting the raw materials.

Activity:
- Draw a diagram showing how tectonic plates cause earthquakes and volcanoes.
- Find out how people living in countries on fault lines prepare for earthquakes and volcanoes.

Possible misconceptions

Remind students that volcanoes are not randomly located – they are found along the edges of tectonic plates. They are found in many countries around the globe as well as under oceans.

Remind students that tectonic plates slide, rather than split. The way the plates slide varies: they can move apart, e.g. in the mid-Atlantic ridge, slide past one another or collide. The plates are often not linked to political boundaries (e.g. continental plates).

Has our atmosphere always been the same?

The emission of greenhouse gases is a very topical issue, which students will come across in other subjects (e.g. geography) as well as in the media.

This topic links to the ways we produce our electricity as well as how we use resources for transport and heating.

Students need to understand that all greenhouse gases are not the same, with different causes and effects.

They should be aware that greenhouse gases are an international issue. The UK is generally considered to be responsible for about two per cent of the world's emissions.

Activity
- Prepare diagrams showing the proportions of gases in our atmosphere, now and in the past.
- Draw a flow chart showing how the proportions of gases in our atmosphere have changed.

Why can the Earth support life?

Activity:
- Prepare a poster describing how the conditions on one other planet are affected by its atmosphere. Include details of the gases.
- Prepare a pie chart showing the different greenhouse gases and their relative proportions in our atmosphere.
- Find out which gases have the greatest warming effects.

Possible misconceptions

These can be wide-ranging as the topic is very widely reported by non-scientists and pressure groups. The greenhouse gases are not all as effective as each other: as well as carbon dioxide, water vapour and methane are also responsible for global warming.

The greenhouse effect is not all bad. Without it the Earth would be much colder. The public's concern is about the apparent effect of human activity on the extent of warming.

Students may be surprised how much conditions on Earth have fluctuated over millions of years.

Materials our planet provides

Activity:
- People do not often consider where the products they use actually come from. A good introduction to this unit would be to analyse the materials a particular product was made from and the processes involved in obtaining each raw material. This could be broken down in terms of geographical location, type of process (i.e. refining, extracting, transporting, selling) to assess the individual components of a product. This also provides an opportunity to categorise materials into elements/compounds/mixtures.
- Chocolate bars are excellent for this kind of activity because they provide metals, plastics, materials from living organisms, and processing aspects of the product's manufacture.

Possible misconceptions

'Compound' and 'molecule' are not the same thing. Use examples and models to address this.

Gases are not all lighter than air. To demonstrate this, limestone and acid can be reacted in a conical flask to fill the flask with CO_2, which can then be 'poured' into a beaker containing a lit tea-light candle, extinguishing it.

Using materials from our planet to make products

Activity:
- Despite being a good example of these principles, the Haber process is a very abstract concept for students. To break down and introduce the idea about understanding reactions affecting raw material use (and effectively cost), students can use the analogy of a toy factory producing stuffed animals. If the factory is making octopuses, what would the ratio of heads to legs be when they were making body parts? How would this ratio change for pandas? If they had 120 gorilla heads, how many legs would they need? How many tails?
- This approach also helps with the idea of balancing equations and the conservation of mass.

Possible misconceptions

The symbol of an element does not always start with its initial letter. Discuss the etymology of elements such as potassium (K was for the Latin kalium).

Unit 1, Theme 1 – My wider world

1.1 Observing our solar system

Learning objectives

Students should learn:
- that observations of the solar system and the galaxies in the universe can be carried out on Earth or from space
- that observations are made with telescopes that may detect visible light or other electromagnetic radiation (EM), such as radio waves or X-rays.

Learning outcomes

Most students should be able to:
- describe how telescopes are used to observe the universe
- describe the difference between space-based and Earth-based telescopes
- state some advantages and disadvantages of each.

Some students should also be able to:
- explain that different celestial objects emit different types of EM radiation
- explain why different telescopes detect different celestial objects.

AQA Specification link-up: Science B 3.3.1.1
- Know that observations of the solar system and the galaxies in the universe can be carried out on the Earth or from space.
- Know that observations are made with telescopes that may detect visible light or other electromagnetic radiation such as radio waves or X-rays from space, and that these observations provide evidence for changes taking place in the universe.

Lesson structure

Starters

Images from space – Show images of celestial objects, including nebulae (clouds where stars are born), planets (which orbit stars) and moons (which orbit planets), stars (e.g. our Sun), supernovas (explosions caused by massive dying stars), neutron stars and pulsars (left over after a supernova). Explain different telescopes are needed for each image due to the different distances and types of EM radiation involved. *(5 minutes)*

What can you see? – All students list the different objects they can see from Earth without a telescope. To extend students, ask them to rank a selection of objects in order of distance, explaining reasons for their apparent sizes [e.g. planets are less bright but much closer than stars, so seem the same distance away; the Sun and Moon appear to be the same size because the Sun is much further away but much larger]. Students may not realise that some objects they think are stars are actually planets or the International Space Station (ISS). The NASA site states when the ISS is visible from your location. To support students, ask them to rank the most obvious objects in order of distance [Sun, Moon, clouds, planets, stars]. *(10 minutes)*

Main

- Run through a very brief history of telescopes. Consider why astronomers who had such limited information available to them before the first telescopes were invented were still able to identify wandering stars (planets) and predict eclipses.
- Hand round a selection of convex lenses for students to use to magnify images. Explain how two lenses in a tube made the first simple telescopes. The main benefit is that the image is brighter (light is focused) as well as magnified. Limitations are that these were only used with visible light.
- Explain how developments now mean that other forms of EM radiation are detected using mirrors, lenses and metal dishes.
- Use the Hubble website http://hubblesite.org to show a slideshow of images. Discuss features of these images, such as the type of EM radiation used and the type of telescope used.
- Students use the gallery found on this site for them to prepare a PowerPoint presentation, including one image for each type of celestial object, or each type of electromagnetic wave. There is detailed information about the telescope too.

Plenaries

What have I learned? – Students list three main points that they remember from the lesson. *(5 minutes)*

Telescopic information – Students write down three things telescopes have told us about objects in space – this can be linked to earlier work to support students or to the less familiar objects such as pulsars or black holes to extend students. *(10 minutes)*

Support
- Include practical activities for students or this can become a very dry topic.
- Details of how a telescope works are not needed, but students can look at pictures of an optical telescope and a radio telescope and spot differences in their structure – for example radio waves are so much longer than light waves that the receiving dish must be much larger for the radio telescope.

Extend
- The Hubble website and NASA websites have very detailed information about telescopes as well as superb photographs. Students could research one space probe or telescope, identifying its mission and including an update on the results received on Earth, including images.

Further teaching suggestions

Making telescopes
Students can make a simple telescope. You will need two convex lenses (focal length, approx. 5 cm and 30 cm) and an adjustable cardboard tube (in two sections, with one section easily able to slide in and out of the other. It needs to be able to extend to a total of approx. 50 cm). Slot the lenses securely to each end of the adjustable tube; look through the lens with the shortest focal length at a distant object and adjust the length of the tube until the image is focused.

Chapter 1 – Our changing planet and universe

Science in context

Students can see objects in the night sky such as the Moon and some stars. Students in city centres will find this hard as light pollution is significant – this is why telescopes are situated in remote places. Other students will see stars at night and the Moon easily especially if it is a clear night. If clouds are present, they absorb different EM radiation blocking our view – another disadvantage of ground-based telescopes.

Suggest students use binoculars or telescopes to examine the night sky in more detail.

Did you know … ?

The more massive an object, the more gravity it creates (mass is the amount of matter contained in it rather than the volume). Neutron stars are small but incredibly dense, so they are very massive. Black holes are the most massive objects we know of. They have such strong gravitational fields that light passing near them is distorted and cannot escape the gravity of the black hole. X-rays are detected from the regions surrounding the black hole.

(Imagine the surface of a trampoline distorted by a person standing on it – a ball pushed across the trampoline changes direction to end up at the person's feet. This is in effect how light waves are affected by very strong gravitational fields. This means we detect a black hole by its effect on radiation in neighbouring regions.)

Unit 1, Theme 1 – My wider world

1.1 Observing our solar system

Learning objectives
- How can we observe the solar system and the galaxies in the universe?
- What types of radiation can be detected by telescopes?

How observing the universe has developed

For centuries, people could only look at **stars** using their eyes. Astronomers drew star charts, attempting to show the movements of the stars and their patterns. They could not explain exactly what they saw as the stars seemed too small to see any details.

When Galileo started using the **telescope** in the 1600s, he could see moons circling around nearby planets. He could also see Saturn's rings and the stars in our galaxy, the Milky Way. Since then, other scientists have improved the design of telescopes. Now we can observe stars that were formed early in the history of the universe. Their light has taken billions of years to reach us.

The problem with looking at a distant object is that very little light from it reaches us. Simply making an object's image larger can make it too dim to see. Telescopes use lenses and mirrors. They collect more light than our eyes can. They focus this light to make a brighter image. They also make the image bigger.

Our telescopes now are more powerful than ever before. For example, the Hubble Space Telescope shows galaxies and stars as they were 13 billion years ago.

a How do telescopes help us to see distant objects?

Telescopes can be based on the ground or in space. Telescopes on the ground are easier to use. They can be updated, maintained and visited easily. However, our atmosphere can spoil the images. For example, clouds can block our view of the sky.

Telescopes based in space take much clearer images because there is no light pollution or atmosphere present. However, they are expensive to run and very hard to mend or visit. You will see how the choice of telescope depends on different factors.

Figure 1 Telescopes are needed to study distant stars in more detail

b Write down one advantage and one disadvantage of ground- and space-based telescopes.

AQA Examiner's tip
A satellite telescope is only about 60 miles up in space. Our nearest star is 3.97×10^{13} km away. So, even though a satellite telescope is nearer the stars than one on the Earth's surface, it is not the shorter distance that makes the difference to the quality of the image. Satellite telescopes are put into space so that they are above clouds, light pollution and atmospheric pollution.

Different types of electromagnetic radiation

We can feel the Sun's warmth and it tans our skin. The Sun emits (gives out) light. It also, like other stars, emits infrared radiation, which warms us, and ultraviolet radiation, which tans our skin.

Modern telescopes detect light as well as other forms of radiation from stars. The images they make help astronomers find out even more information about distant objects.

Different stars emit different types of electromagnetic radiation. The type of radiation emitted by stars depends on their temperature. This affects the choice of telescope as some forms of radiation are absorbed by our atmosphere.

Radio waves are caused by the coolest sources – such as some stars, and planets like the Earth. Since radio waves can pass through our atmosphere, large **ground-based telescopes** detect these signals.

Newly formed stars and cool dust clouds emit low-energy **microwaves**. The dust clouds contain tiny solid particles, and are seen as dark patches between stars. Ground-based telescopes also detect a background of microwaves coming from all directions throughout the universe. These microwaves were created at the Big Bang, the very start of the universe itself.

Some planets, stars such as the Sun and dust clouds emit **infrared radiation** – very hot stars emit **ultraviolet radiation**. Very little infrared or ultraviolet radiation from distant stars and galaxies travels through our atmosphere. This means **space-based telescopes** are needed to see objects giving out infrared and ultraviolet radiation. The Hubble Space Telescope is mainly designed to look at visible light, but can also detect some infrared and ultraviolet radiation.

X-rays and gamma rays are only detected by space-based telescopes as our atmosphere absorbs this radiation. **X-rays** are given out during supernovas – the huge explosions caused by the collapse of very massive stars as their lifecycle ends. X-rays are also given out by the uppermost level of the Sun's atmosphere. **Gamma rays** are the most energetic rays. We can see the remains of supernovas, neutron stars and black holes, by using telescopes to see the gamma rays they emit.

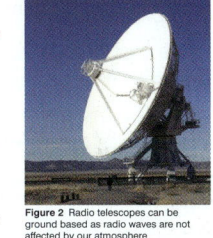
Figure 2 Radio telescopes can be ground based as radio waves are not affected by our atmosphere

Figure 3 Galaxies like this one emit gamma rays as they produce stars at their core

c Explain what affects the different types of electromagnetic radiation given out by different objects in space.

Gathering data

Space probes are spacecraft carrying scientific instruments. They are designed to gather data and send it back to Earth. The probes travel through the solar system and may land on other planets. They are not designed to return to Earth and so they can explore much further than other telescopes.

Summary questions
1. Describe the main differences between space- and ground-based telescopes.
2. Explain why we use space probes and space-based telescopes, as well as ground-based telescopes.
3. Make a list of objects in space so that each emits a different type of electromagnetic radiation.

Did you know … ?
Anyone can apply to use the Hubble Space Telescope to view the universe. Hubble sees light from the most distant stars. Their light has taken so long to reach us that we see these stars as they appeared billions of years ago.

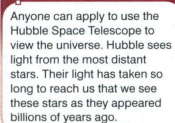 **links**
For more information on the electromagnetic spectrum see 9.6 Electromagnetic waves.

Key points
- Telescopes are used to see light and other radiation coming from distant objects.
- Telescopes can be based on the ground or in space.
- Different telescopes see different types of electromagnetic radiation, allowing us to see different objects in space.

Summary answers

1. Telescopes based on the ground can be small and portable or very large and based in observatories; if they are space-based telescopes they tend to be large and based on satellites. They can detect different types of radiation. Ground-based telescopes may use lenses and mirrors or large dishes to receive and process the radiation.

2. The space probes travel far into the solar system and can send back data from very distance places. Space-based telescopes orbit Earth and provide detailed information from where they are. Earth-based telescopes can be cheap and portable, so can be used to look at many places in the sky. They are limited to the wavelengths of EM radiation that can penetrate the Earth's atmosphere.

3. radio wave – cool sources, such as some stars and planets like Earth
   microwave – newly formed stars, cool dust clouds
   infrared – stars such as the Sun
   visible – stars
   ultraviolet – very hot stars
   X-ray – supernova
   gamma rays – neutron stars, black holes, supernovas

Answers to in-text questions

a They make the images brighter and larger.

b Ground-based: advantages – cheaper/easier to service or upgrade; disadvantages – images less clear, effects of light pollution.
Space-based: advantages – clearer images, can examine all types of EM radiation; disadvantages – more expensive, very hard to upgrade or perform maintenance/repairs.

c Hotter objects, such as neutron stars, emit higher energy EM radiation, such as gamma rays.

Unit 1, Theme 1 – My wider world

1.2 How the universe began

Learning objectives

Students should learn:

- that a moving wave source has a change in the observed wavelength and frequency
- that there is a red-shift in light observed from most distant galaxies
- that the red-shift is evidence that the universe is expanding and supports the Big Bang theory.

Learning outcomes

Most students should be able to:

- describe the red-shift in terms of the Doppler effect
- state that the universe is changing and that it began as a Big Bang
- describe the main evidence that the universe is changing.

Some students should also be able to:

- explain why the red-shift provides evidence that the universe is expanding
- explain how cosmic background radiation supports the Big Bang theory.

AQA Specification link-up: Science B 3.3.1.1

- Understand that if a wave source is moving relative to an observer there will be a change in the observed wavelength and frequency (Doppler effect).
- Explain why there is a red-shift in light observed from most distant stars and galaxies. The further away stars or galaxies are, the more their light is red-shifted. This indicates that distant galaxies are moving away from us, and that the further away a galaxy is the faster it is moving away.
- Explain how the observed red-shift provides evidence that the universe is expanding and supports the 'Big Bang' theory (that the universe began from a very small initial point).

Within this context, candidates should be able to use scientific data and evidence to discuss, evaluate or suggest implications of the following:

- the evidence for the origin, structure and continuing evolution of the universe.

Lesson structure

Starters

Start of the universe – Show an animation or video clip showing the start of the universe. Ask students to identify the main changes in the temperature, size and structure within the universe over the first few minutes. Students should understand that as the temperature falls, the universe gets larger and that some energy changes into matter (or that particles such as subatomic particles start to appear). Extend students by asking them to sketch an approximate timeline showing that as time increases and the temperature falls, the particles start to join up forming subatomic particles, atoms, nebulae and galaxies. To have an idea of the scale of the developments, ask students what scale they would use if asked to plot their timeline. *(5 minutes)*

I wasn't there – As we were not present at the start of the universe, we have to base our theories on evidence. Link this to a crime scene – police find evidence to support their theory when a crime takes place. The courts reach a verdict based on that evidence. The theory of the Big Bang is backed by several pieces of evidence, and in general the scientific community accepts this. *(10 minutes)*

Main

- Explain the main stages in the theory, and then include as many demonstrations as practical. Keep linking back to the main ideas.
- Demonstrate the Doppler effect using water waves. Place a small floating object in water in a ripple tank. By illuminating the ripple tank, the waves can be seen more easily by students. As the object is pushed through the water, the waves bunch up or spread out.
- Show a continuous spectrum by shining light through a prism. Link this to a diagram of the EM spectrum, showing the relative positions of red light and blue light.
- Demonstrate line spectra using spectroscopes to look at lights in the classroom, or use them in a darkened room to view the spectra when doing a flame test on different metal compounds (e.g. sodium chloride, calcium chloride). A line spectrum will be visible if students look directly at the coloured flame through the spectroscope.
- Almost all spectra from galaxies are red-shifted, so they are all moving apart. Use a balloon with spots drawn on it to demonstrate that when a balloon is inflated all the spots move apart.

Plenaries

Big Bang – Students sort cards into a sequence showing the stages in the Big Bang. *(5 minutes)*

The big tweet – Ask students to prepare a tweet (140 character statement) describing the evidence for the Big Bang or their responses to the lesson. Students share their tweets and try to improve them to produce a group tweet. For lower ability, hand out a prepared tweet and ask them if they agree or disagree with it. What else would they include if they were writing it? For example, 'Universe began as a Big Bang billions of years ago, is still expanding, red-shift proves stars moving away as their wavelengths are stretched.' *(10 minutes)*

Support

- Although the demonstrations in the main part of the lesson structure help students understand the red-shift, it is a very challenging concept. Concentrate on helping students to recognise and remember what the evidence shows instead.

Extend

- Students investigate other models of the origins of the universe, including the Steady State theory.
- Students summarise the evidence for and against each theory, using reputable websites such as www.nasa.gov.

Chapter 1 – Our changing planet and universe

Further teaching suggestions

The expanding universe

Discuss the implications of an expanding universe with a finite age – the night sky is dark because light from distant stars has not reached us yet. The universe is cooling because the energy created at the Big Bang, which caused such massively hot temperatures, is spreading throughout larger and larger spaces. The rate of expansion is speeding up but will it stop? The universe may keep expanding; it may stop expanding or may reach a maximum size and then contract in a 'Big Crunch'.

Did you know …?

There are three main types of spectra:
- line spectra – several bright coloured lines in fixed positions given out from a heated element
- continuous spectra – no lines visible, but a continuous range of colours like a rainbow
- absorption spectrum – a continuous spectrum with dark lines corresponding to the positions seen on the line spectrum; these are seen commonly from stars.

Unit 1, Theme 1 – My wider world

Chapter 1 – Our changing planet and universe

1.2 How the universe began

Learning objectives
- What happens to the observed wavelength and frequency of a moving wave source?
- What do we mean by 'the red shift' in light observed from most distant galaxies?
- What evidence do we have that the universe is expanding and how does this support the Big Bang theory?

Wavelengths and wave source

Last century, Edwin Hubble used telescopes to measure the light from distant stars and galaxies. Light from stars is a blend of different colours. Different colours of light have a different blend of colours present. **Spectroscopy** shows these different colours of light as lines on a coloured spectrum. The blend of colours of light changes depending on which elements are present in the star.

a How can you tell which elements are present in a star?

Hubble used the pattern of lines to identify the elements in galaxies. The pattern of lines matched the elements Hubble expected to see. However, all lines were moved along the spectrum towards the redder light. More distant galaxies had a larger **red-shift** than closer ones.

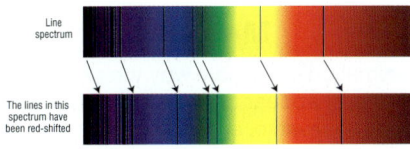

Figure 1 A line spectrum contrasted with lines that have been red-shifted

To explain this, imagine that a train travels past, moving very fast. As it comes closer, the pitch of the noise from its engines rises. As it moves away again, the pitch falls. This is the **Doppler effect**.

The Doppler effect happens because a wave source moving towards us will squash the waves. When the wave source moves away, it stretches the waves. If the wavelength is stretched, then the frequency is lower. It happens with light waves as well as sound waves.

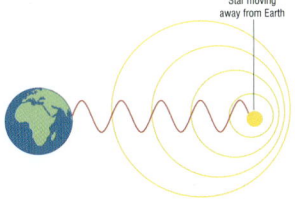

Figure 2 The Doppler effect means that waves are stretched apart if a star moves away from the Earth

links
For more information on the frequency and wavelength of waves see 9.5 How fast do waves travel?

b How can we tell that the stars are moving?

Red light has a longer wavelength than blue light. When a star or galaxy moves away from us, its light appears redder than expected because its wavelength is stretched. Hubble's results were evidence that galaxies move away from us. The galaxies further away have a bigger red shift because they are moving away faster than galaxies closer to us.

c How could Hubble tell that galaxies further away are moving faster than closer galaxies?

The universe is expanding

One puzzle was that galaxies in all directions are moving away from the Earth. Does this mean the Earth is at the centre of the universe? Scientists believe that this is not the case. It is like blowing up a balloon with spots marked all over it. As it gets bigger, all the spots move away from each other. Similarly, if the whole universe is expanding, then galaxies will move away from each other.

Figure 3 As the balloon gets bigger, the spots spread apart

If the universe is getting bigger, then it must have been smaller in the past. In fact, many scientists believe that the universe was once so small that all matter and energy were squashed in one tiny place. This is called the 'cosmic egg'. At some point there was a huge explosion that scientists called the **Big Bang**. As the matter was flung outwards, the universe started expanding. It has continued to expand for 14 billion years and is still expanding today.

There is more evidence that the universe began as an explosion. Scientists using a radio telescope in the 1960s detected a weak signal coming from all directions, which they called **cosmic background radiation**. This was caused by microwave radiation left over from the Big Bang. It is often called the 'echo of the Big Bang' as it is what remains of the radiation created at that time.

d What evidence is there that the Big Bang took place?

Summary questions

1 Copy and complete using the words below:
 Big Bang   Doppler effect   red-shift
 a We believe the universe began as a big explosion called the ....
 b The .... is evidence that the universe is still expanding.
 c The .... means that the wavelength of light stretches when a galaxy moves away from us.

2 Explain how the red-shift and the Doppler effect are linked.

3 Write down **one** piece of evidence that:
 a the universe began billions of years ago
 b the universe is still expanding.

Examiner's tip
When stars appear red, this does not mean that the stars emit red light. A star may appear red because it is moving rapidly away from the Earth. Use the word 'appear' in your description or you may lose a mark.

Did you know …?
The researchers who discovered the cosmic background radiation were investigating something completely different. They thought the noise was caused by pigeon droppings in the receiver of the telescope.

Key points
- The universe began as a huge explosion called the Big Bang.
- The red-shift is evidence that the universe is still expanding.
- The cosmic background radiation is evidence that the Big Bang took place.
- The wavelength and frequency of a moving light source change.

Summary answers

1 a Big Bang
  b red-shift
  c Doppler effect

2 The red-shift occurs because of the Doppler effect. Light waves are stretched when stars move away. Their light appears at a lower frequency and longer wavelength.

3 a cosmic background radiation
  b red-shift

Science in context

The Doppler effect can be heard on Earth – the sirens of passing emergency vehicles and trains rise in pitch as they approach, and fall as they recede.

Answers to in-text questions

a When it is heated, each element has its own line spectrum, so examine the line spectra for different stars.
b the red-shift
c Their red-shift is greater.
d the red-shift, cosmic background radiation

Unit 1, Theme 1 – My wider world

1.3 Our place in the universe

Learning objectives

Students should learn:
- that our solar system is a star with planets orbiting it
- that the solar system is part of the galaxy, the Milky Way
- the implications of the Earth's position in the solar system.

Learning outcomes

Most students should be able to:
- describe how ideas about the Earth's position in the solar system has changed over time
- state why the Earth's position in the solar system is vital for life on our planet
- explain the link between galaxies, solar systems, stars, moons and planets.

Some students should also be able to:
- suggest reasons why we have life on Earth but think it unlikely there is life in other parts of the solar system.

AQA Specification link-up: Science B 3.3.1.1

Within this context, candidates should be able to use scientific data and evidence to discuss, evaluate or suggest implications of the following:
- the position of the Earth in the solar system.

Lesson structure

Starters

Parts of the solar system – Review earlier work briefly – Students list the components of the solar system, put them in order and describe how they are linked. Remind students of terms such as 'year', 'day', etc. *(5 minutes)*

Orbiting planets – Show an animation of the planets orbiting the Sun, such as www.kidsastronomy.com. Students list the points they notice. More able students should notice the shorter year for closer planets; the arrangement of smaller rocky planets close to the Sun and larger ones further out; the rotation of the planets on their axes causing days of varying lengths and the strange orbit of the comet. They should also spot that the animation is not to scale and that the orbit is shown as a circle where it is actually an ellipse. Less able students should notice that the planets all orbit the Sun, at different speeds with smaller planets closer to the Sun. *(10 minutes)*

Main

- Explain with a computer animation if possible why the planets appear to wander in the night sky. Planets (such as Mars) appear to move forwards across the constellations (fixed pattern of stars), and then reverse, appearing to move backwards. This occurs because the Earth moves faster than Mars and has a shorter year. As both planets orbit the Sun, the Earth catches up with, and overtakes, Mars.
- Students can research attempts to find life away from Earth and the difficulties connected with this search. The SETI institute is a good starting point. There are several strands to investigate – for example are there places with conditions suitable for us to inhabit? Is there is intelligent life elsewhere? Has there been any form of life anywhere? The BBC series *Wonders of the Solar System* is very relevant here. Although these are hour-long programmes, suitable clips bring the topic up to date.
- As extension work, there are many resources online to facilitate research. Students can research and prepare a PowerPoint presentation showing the changes in thinking about the solar system over time, or concentrate on the work, life and times of one particular astronomer (Galileo, Kepler, Copernicus, Ptolemy, Aristotle).

Plenaries

Message to an alien – Students consider what to include on a plaque being sent out to alien life forms. Students should decide what information is important as well as how to communicate it. Students should suggest the information that is important. *(5 minutes)*

Three key facts – Students present three key facts about their research to the class. *(10 minutes)*

Support

- This topic is very visual. Students can label diagrams of different models of the solar system, or make their own models. There are many animations and diagrams online too.
- Card-sorting exercises to place different objects in their correct place, or to put objects in order of size would be appropriate.

Extend

- Students could be introduced to light years (the standard astronomical way of measuring distance – the distance light travels in one year).
- Many students do not appreciate the vastness of space, and the practical difficulties of exploring it. Consider the time factor too – in terms of mankind's existence, we have only been able to communicate outside Earth for a tiny proportion of time.

Further teaching suggestions

Analysing planets

Provide data about the different planets for analysis (enter 'the planets data' into your internet search engine). Include some of these features: day length, year length, mass and diameter of planets, gravitational field, and temperature. Students look for patterns (e.g. moving away from the Sun, the year length increases and temperature falls; gravitational field strength depends on wide mass (not diameter) of a planet). More able students can calculate the density of the different planets, which shows the four rocky planets closer to the Sun and the four gas giants further out. They can also calculate the speed of the planets in orbit – this will show that planets further out orbit at a slower speed as well as travelling further.

Chapter 1 – Our changing planet and universe

Did you know …?

The temperature on different planets does not just depend on the distance away from the Sun. Venus is hotter than Mercury because of its atmosphere.

The moons orbiting planets have unique conditions that may mean they can sustain life. Europa, orbiting Jupiter, is considered one of the most likely places to find extraterrestrial life.

Students may be interested to find out that light from the Sun, which we see, left the Sun some 8 minutes ago. If the Sun went out now it would be 8 minutes before we were plunged into darkness.

Science in context

Students know the Sun rises and sets, they see changes in the phases of the Moon, and some may have noticed our neighbouring planets in the night sky.

Summary answers

1 a Milky Way
  b Sun
  c Moon
  d Sun

2 450 BCE – Sun at centre with Earth rotating round it; 140 BCE – Earth at centre with Sun orbiting it; 1543 – Sun at centre with Earth and planets orbiting it; later, elliptical orbits; 1609 – other planets had moons; modern day – Sun orbited by eight planets plus dwarf planets.

3 Yes, to look for resources or in the interests of scientific research and understanding, etc.; no, in terms of cost and time, risk of finding unfriendly aliens, etc.

Answers to in-text questions

a A star gives out its own light, a planet reflects the star's light. Planets orbit stars.

b Ancient Greek's: Earth in centre, Sun orbits Earth and stars surround it on crystal spheres; Copernicus: Sun at centre, Earth orbits the Sun.

c We are close enough to the Sun for water to be liquid, but not too close, so water does not evaporate.

d There are billions of other galaxies and solar systems – there will be conditions suitable for life on millions of these.

Unit 1, Theme 1 – My wider world

1.4 The Earth's structure

Learning objectives

Students should learn:
- how the structure of the Earth has changed since it was formed
- the Earth's structure.

Learning outcomes

Most students should be able to:
- state that the Earth consists of a core, mantle and crust, surrounded by its atmosphere.
- describe how the crust was formed by cooling.

Some students should also be able to:
- describe the composition of each layer of the Earth.

Specification link-up: Science B 3.3.1.2

- Know that the surface of the Earth has changed over time as a result of cooling.
- Know that the Earth consists of a mantle, core and crust, surrounded by the atmosphere.

Lesson structure

Starters

Unscramble – Key word anagrams for **core, mantle, crust, nickel, iron, earth, melting, cooling, geology**. Students identify the keywords they have heard before and those they have not. *(5 minutes)*

Sections – Show students photographs of sections through everyday objects, and discuss what the photographs show. Start with easy objects such as an apple, then move on to more complex items, such as a leaf, skin, a tree, a television, a house, an engine, etc. Use these to start a discussion about the different ways we can represent things with pictures. To extend students, get groups to critique a particular cross-sectional diagram, listing the strengths of using a cross-section as a model, and the potential weaknesses. To support students, perform the above starter but only identify the objects and the parts represented in the cross-section. *(10 minutes)*

Main

- Show a schematic of the structure of the Earth, talking students through the composition of the layers. Students can label their own diagrams. Alternatively, provide students with text that they must use to produce their own schematic.
- Provide students with modelling clay (at least 300 g per group) and balances. Give them data on the percentage volume of each layer of the Earth (inner core 0.7%; outer core 15.7%; mantle 82%; crust 1.6%). Their task is to divide their clay up into masses that represent the relative volume they make up of the Earth.
- They could also try to construct the Earth using these masses, but it will be difficult. One way would be to roll each piece into a sphere, divide it in two, and then press out into a hemisphere that will encompass the next smallest. Students will soon notice they would literally have to paint the crust layer on to do this accurately. This will help reinforce the idea that the crust is very thin indeed, which in turn will help the next lesson when considering plate tectonics.

Plenaries

What have I learned? – Students sort statements into groups, depending on which part of the Earth they refer to (i.e. made of nickel and iron, flows very slowly, largest by volume, etc.). *(5 minutes)*

Composition of the Earth – To support students, prepare true or false questions about the composition of the Earth that the class can vote on. To extend students, use a text distillation activity. In pairs, students come up with five sentences to describe the topic, which they then summarise in five words, and then further summarise their five words into one. *(10 minutes)*

Support

- This topic lends itself well to modelling and analogy. A small, well supported group could handle, describe and cut apart hard-boiled eggs. Learning could be reinforced by discussing the merits and disadvantages of using an egg as a model, that is no liquid part of core, wrong shape, albumen does not move in the same way as the mantle.

Extend

- Students could learn about how core sampling works, and interpret diagrams and pictures of core samples.

Chapter 1 – Our changing planet and universe

Further teaching suggestions

Dr James Hutton

Students can research the work of James Hutton, finding out the evidence upon which he based his claims about deep time. This could lead to further debate or role play on the topic of the cultural significance of science and the significance of Hutton's theories at a time most of Europe believed the Earth to be 6000 years old.

Although this will not be tested in an examination, it may help students to put the ideas in this chapter into a wider historical as well as scientific context.

Unit 1, Theme 1 – My wider world

1.4 The Earth's structure

Learning objectives
- How has the Earth changed over time?
- What is the Earth's structure?

The age of the Earth

Three hundred years ago, scientists believed the Earth was only 6000 years old. There was good reason for this, of course. There were few historical records dating back this far, and an age of 6000 years agreed with religious texts. In the 1780s, a Scottish doctor called James Hutton challenged this belief.

Hutton spent a lot of time looking at rocks and identifying what they were made of. He started to see evidence that natural recycling was taking place, and wrote:

> A vast proportion of present rocks are composed of materials produced by the destruction of animal, vegetable and mineral substances of more ancient origin.

a About 300 years ago, how old did most scientists think the Earth was?

The Earth's composition

Geologists study the Earth, measuring its composition and structure. They try to explain the processes that change the Earth. This is very valuable information to companies that mine the Earth for materials such as **metal ores** or **fossil fuels**. Knowing the types of rock formations that might contain precious metals or crude oil is worth a lot of money. A career in geology can involve a lot of travel around the world and work outdoors.

By measuring levels of **radioactivity** in rocks, we now know the Earth to be around 4.5 billion years old. Some major changes have happened during that time. The Earth looks very different from the way it did 4.5 billion years ago.

When the Earth had just formed, it was a ball of **molten** rock in constant motion. The dense metals iron and nickel started to sink into the centre of this ball. They are still hot billions of years later. Lighter elements around the outside of the Earth then cooled down, forming a solid crust. Its surface was very unstable and covered in **volcanoes**.

Figure 1 Dr James Hutton was the first to suggest the Earth changes very slowly

Figure 2 Geologists at work taking a core sample – this helps build up a picture of the structure of the Earth's crust

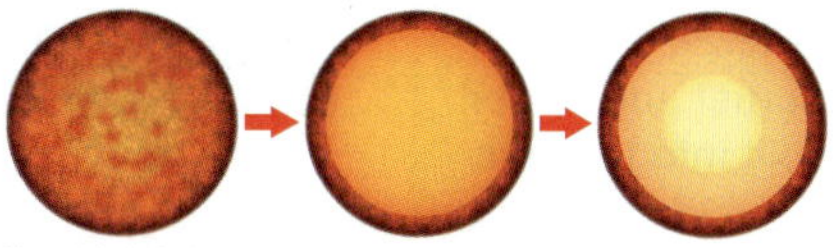

Figure 3 As the Earth cooled, the materials in it started to form layers

b Why does the core of the Earth contain iron and nickel?

We now know a lot about the structure of the Earth. Firstly, we know its **core** is a mixture of iron and nickel, which is solid in the middle, but liquid in the outer part of the core. Surrounding the core is a layer called the **mantle**. The mantle has many of the properties of a solid, but it can flow slowly in parts. This movement in the mantle can have dramatic and devastating consequences for us.

On top of the mantle is a very thin **crust** of solid rock, upon which we live. This solid crust is surrounded by a layer of gases we call our atmosphere.

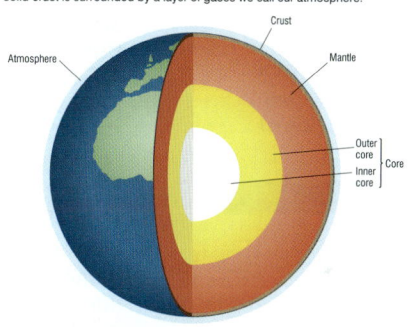

Figure 4 The structure of the Earth

Did you know ...?
The Earth's crust is very thin. If the world was the size of a football, the crust would be as thick as a postage stamp on its surface.

Summary questions

1 Copy and complete using the words below:
   core crust atmosphere rock mantle flows
   The surface of the Earth is called the .... Underneath this is a layer of .... which we call the .... This layer .... very slowly over time. The centre of the Earth is called the .... The Earth is surrounded by its ....

2 What skills and qualities do you think a geologist needs?

Key points
- The Earth was originally a ball of molten rock, which slowly cooled down to form a crust.
- Today, the Earth consists of a metallic core surrounded by a mantle, a thin crust and gaseous atmosphere.

Summary answers

1 crust, rock, mantle, flows, core, atmosphere

2 Possible qualities:
- Interest in Earth science
- Organised
- Methodical
- Patient
- Good at either team or independent work

Possible skills:
- Able to work with precise equipment
- Good ICT skills
- Good communication skills

Answers to in-text questions

a about 6000 years old

b Iron and nickel are dense metals so they sank a long way down when the Earth was molten.

Unit 1, Theme 1 – My wider world

1.5 Changes in the Earth's surface

Learning objectives

Students should learn:
- that the Earth's crust and upper mantle is split into tectonic plates
- that convection currents in the mantle cause the movement of tectonic plates
- that movement of the Earth's tectonic plates causes earthquakes and the eruption of volcanoes.

Learning outcomes

Most students should be able to:
- state that the Earth's crust and upper part of its mantle is split into tectonic plates
- describe how convection currents in the mantle cause plates to move
- state that the movement of tectonic plates causes earthquakes and the eruption of volcanoes.

Some students should also be able to:
- explain the processes taking place at plate boundaries and relate them to earthquakes and the eruption of volcanoes.

Answers to in-text questions

a Soup at the bottom of the saucepan rises to the surface when it is heated and then sinks back down when it cools upon reaching the top. The same is true of the mantle – heat from the mantle causes material to rise and circulate in convection currents in the same way.
b There are 8 major plates.
c Plates moving toward each other: Eurasian/Pacific; Nazca/South American. Plates sliding past each other: Pacific/North American.
d It is less dense than surrounding material in the crust.

Support

- This topic lends itself well to pictures and models. Attention can be given to the behaviour of solids, liquids and gases when describing the behaviour of the mantle. Model volcanoes can be made.

Extend

- Students can research in depth the various methods of predicting volcanic eruptions and earthquakes.

AQA Specification link-up: Science B 3.3.1.2

- Know that the Earth's crust and the upper part of the mantle are cracked into a number of large pieces (tectonic plates).
- Explain how convection currents within the mantle cause the movement of tectonic plates.
- Describe how movement of tectonic plates can have disastrous consequences such as earthquakes and the eruption of volcanoes.

Within this context, candidates should be able to use scientific data and evidence to discuss, evaluate or suggest implications of the following:
- the accurate prediction of earthquakes and volcanic eruptions.

Lesson structure

Starters

- **Label cross-section** – Students can recap their learning from the previous lesson by labelling a cross-section through the Earth. *(5 minutes)*
- **Earthquakes and volcanoes** – Show students video footage of earthquakes or volcanoes and assess their initial ideas of their origins. Small groups can discuss the following and feed back to the rest of the class: In which parts of the world might an earthquake or volcano occur? Why are these phenomena unlikely in the UK? When was the last earthquake or eruption you remember seeing in the news? Locate news articles about some of the volcanoes or earthquakes and give to another group. Use text comprehension strategies to enable this group to make notes on the news articles. The two groups can then share information. To extend students, provide them with world maps and information on where earthquakes or eruptions occur. Students find the locations on the map and start to form conclusions about hot spots. *(10 minutes)*

Main

- Demonstrate convection in liquids and discuss students' observations. If demonstrating convection by heating water with potassium manganate(VII), carefully sprinkle some glitter on to the meniscus of the water. Students can describe in words or sketch the movement they see. Students can then relate the convection they have observed in water to the movement of the Earth's mantle. It would also be useful to compare the effectiveness of the model with the way the mantle actually behaves (i.e. the mantle is much slower, hotter, more viscous, etc.).
- Present students with three maps, showing plate boundaries and vectors, incidences of volcanoes and earthquake epicentres. Students use the direction of movement of the plates and the locations of earthquakes and volcanoes to suggest the type of plate movement responsible for each.
- Expert group activity. The class forms two groups: volcanologists and seismologists. Each group is provided with stimulus material about the prediction of volcanoes or earthquakes, together with information about the types of equipment used by each specialist. They must use their information to create learning activities for the other group. The two groups are then assessed with a short quiz.
- Students make simple seismometers using a stand, slotted masses, a spring and a small paintbrush. Details are given in 'Practical support'.

Plenaries

- **Summary tables** – Students can produce summary tables for volcanoes and earthquakes, with headings such as 'recent examples', 'hot spots', 'type of plate movement', 'diagram of plate movement'. *(5 minutes)*
- **Revision of key words** – Students can write definitions for the lesson's key words (i.e. seismometer, tectonic, convection). To extend students, students can write revision questions for the key words (i.e. 'what piece of equipment measures vibrations?'). *(10 minutes)*

Practical support

Demonstrating convection

Equipment and materials required:
Eye protection, large pyrex beaker (2 litre preferably), sticky-tac, glitter, tripod, wire gauze, heatproof mat, tweezers, Bunsen burner, potassium manganate(VII). (CAUTION: OXIDISING AGENT)

Details
Fill the beaker $\frac{3}{4}$ with water, using tweezers push two or three potassium manganate(VII) crystals into the surface of the sticky-tac and sink it into the water near the side of the beaker. Sprinkle glitter on the surface so that it sits on the meniscus.

Heat the edge of the beaker near the crystals using a 'roaring flame' in order to see localised convection before the heat is conducted to the whole base of the beaker. Plumes of purple potassium manganate(VII) will rise to the surface and to sink back down once they reach the top. The glitter will be visibly propelled around the surface of the water.

Safety: CLEAPSS Hazcard 81 Potassium manganate(VII) – oxidising/harmful. Will stain clothing and skin – use tweezers to handle crystals. Wear eye protection.

Making a seismometer

Equipment and materials required: Clamp stand, small spring, slotted masses and hanger, elastic bands, paintbrush, ink, paper/card.

Details
Hang the masses from the clamp stand using the spring. There should be enough masses for them to resist movement if the apparatus is wobbled slightly. Use the elastic band to attach the paintbrush to the mass assembly. Attach paper or card to the base of the stand, so that the paintbrush (dipped in ink) just touches it.

When the base of the seismometer is moved, the masses should remain in roughly the same place, so that the brush draws a pattern corresponding to the movement of the 'ground'.

Safety: Take care with falling masses.

Unit 1, Theme 1 – My wider world

1.5 Changes in the Earth's surface

Learning objectives
- What are tectonic plates?
- What causes tectonic plates to move?
- What effects do movement of the Earth's tectonic plates cause?
- Can scientists predict earthquakes and volcanic eruptions?

Have you ever heated soup in a saucepan? If you leave it for a few minutes, you will see a skin starts to form on the surface. If you keep watching, you may see cracks or wrinkles start to appear, and very slight movements in the skin. The interior structure of the Earth behaves in a very similar way to your saucepan of soup.

Just like the soup in your saucepan, the Earth's mantle is being heated strongly from below. The heat is coming mainly from nuclear reactions in the core, as unstable elements go through radioactive decay. The heat causes **convection currents** in the mantle, which in turn cause areas of crust to move about.

a Explain how the circulation of soup in the saucepan above can be used to represent convection in the Earth's mantle.

Tectonic plates

In fact, the Earth's crust isn't a single continuous shell like an egg shell. It has been cracked and split into several huge pieces. We call these pieces **tectonic plates**. They are made up of the Earth's crust and the upper part of the mantle. These plates are slowly moved around by the convection currents within the mantle. This has been happening for billions of years.

Figure 1 Convection currents are caused when a fluid is heated. The warmer fluid is less dense than its cooler surroundings, so it rises.

Figure 2 The Earth's crust is composed of several tectonic plates. The arrows on the diagram show their direction of movement

?? Did you know ...?
Tectonic plates also move away from each other, with material from the mantle rising up to fill the gap. In this way, the UK and US move about 2.5 cm further apart every year.

b How many tectonic plates are shown in the diagram above?

c Which plates are moving toward each other, and which are slipping past each other?

Earthquakes and volcanoes

Seismologists and **volcanologists** study the places where these plates meet (**plate boundaries**). They do this because these places are often the sites of **earthquakes** and **volcanoes**.

Figure 3 Earthquakes are caused when plates grind against each other

Seismologists study earthquakes. Enormous strain builds up at the tectonic plate boundaries. When the force pushing the plates gets high enough, the plates can suddenly slip a short way past each other. This causes them to shudder as they grind against each other. Buildings nearby can collapse or be severely damaged. For example, in 2009, an estimated 230 000 people died when an earthquake hit Haiti.

Volcanologists study volcanoes. These occur when one plate slides under another. The plate being forced downward then starts to melt as it is pushed deeper into the mantle. The molten crust material is less dense than the solid crust above it. This allows it to rise back upwards, breaking through the surface to form a volcano.

d Why does molten crust material return to the Earth's surface?

Seismologists and volcanologists cannot stop earthquakes and volcanoes. However, they can help save lives by predicting when they will strike. A tool used by seismologists is the **seismometer**, which detects vibrations in the Earth's crust. By placing seismometers in different locations, they can build up a picture of how the crust is behaving. They can also detect tell-tale vibrations just before an earthquake. Volcanologists can use **GPS** (Global Positioning System) devices to detect whether a volcano is swelling up, which happens just before an eruption. This again gives some advance warning but accurate predictions are still very difficult to make.

Figure 4 Vibrations in the Earth's crust cause this simple seismometer to move. The weight and spring hold the pen still, so a pattern is drawn when the rotating drum moves.

Summary questions

1. Copy and complete using the words below:
   toward   core   past   plates   mantle

   The heat of the Earth's .... causes convection currents in the ....
   These currents move the tectonic .... of the Earth's crust around.
   When tectonic plates move .... each other, earthquakes can occur.
   When tectonic plates move .... each other, volcanoes can be formed.

2. Describe how the movement of tectonic plates can cause earthquakes and volcanoes.

3. Evaluate the usefulness of the methods scientists use to predict earthquakes and volcano eruptions.

Key points
- Tectonic plates are made up of the Earth's crust and upper part of its mantle.
- Heat from the Earth's core causes convection currents in the mantle.
- Convection currents in the mantle cause tectonic plates to move.
- Earthquakes and volcanoes occur mainly at boundaries between tectonic plates. Accurate predictions of events are very difficult to make.

Summary answers

1. core, mantle, plates, past, towards

2. - **Earthquakes**
     When plates slide past each other, strain forces build up. When the forces are strong enough, plates quickly slip along a short distance. The vibrations caused as they grind past each other cause the earthquake.
   - **Volcanoes**
     When one plate is forced under another, the edge being forced down will begin to melt as it gets deeper into the mantle. The molten rock can then rise back up to the surface and escape from the crust by forming a volcano.

3. Seismometers can only give a few seconds' notice of an earthquake, which isn't long enough to evacuate areas. However, even short notice may save lives. GPS devices can detect small movements in tectonic plates, or volcanic craters, so will become increasingly important. However, judging the timing of an event remains an inexact science.

Unit 1, Theme 1 – My wider world

1.6 The Earth's changing atmosphere

Learning objectives

Students should learn:
- what the Earth's early atmosphere was like
- which gases are now present in the Earth's atmosphere
- how and why the Earth's atmosphere has changed since it first formed.

Learning outcomes

Most students should be able to:
- state that the Earth's atmosphere is mainly nitrogen and oxygen, with some water vapour and carbon dioxide
- explain that the Earth's early atmosphere came from volcanic activity
- state that the Earth's early atmosphere was probably mainly carbon dioxide with little or no oxygen
- state that microorganisms converted much of the early atmosphere's carbon dioxide into oxygen by photosynthesis.

Some students should also be able to:
- state that there may also have been water vapour, with small proportions of methane, hydrogen and ammonia **[HT only]**
- explain how the gases in the early atmosphere could have made the first molecules needed for life **[HT only]**

Support
- Follow the pictures from the spread in the Student Book, with students describing the differences they can see at each stage. Students draw their own representations of the stages in the Student Book. Use coloured dot stickers to represent different gases, sticking appropriate proportions of stickers onto the sky of each picture as a more visual representation of the stages of atmospheric development.

Extend
- Students can research the atmosphere of Venus, drawing comparisons with the Earth's current and early atmosphere. Suggest means of terraforming Venus (hypothetically transforming a planet into one sufficiently similar to the Earth to support terrestrial life), making its atmosphere closer in composition to the Earth's.

AQA Specification link-up: Science B 3.3.1.2
- Know that during the first billion years of the Earth's existence there was intense volcanic activity.
- Know that volcanic activity released the gases that formed the early atmosphere and water vapour that condensed to form the oceans.
- Understand that some theories suggest that, during this period, the Earth's atmosphere was mainly carbon dioxide and there would have been little or no oxygen gas.
- There may also have been water vapour and small proportions of methane, hydrogen and ammonia. **[HT only]**
- Describe how plants and algae produced the oxygen that is now in the atmosphere by photosynthesis.

Lesson structure

Starters

Pose the question – Which came first, plants or animals? Pairs of students have two minutes to come up with as many arguments as they can for each answer (i.e. plants need carbon dioxide to photosynthesise and animals produce carbon dioxide when they respire; animals need glucose to be able to respire, and plants produce glucose when they photosynthesise, and so on). *(5 minutes)*

Main

- Students read about the development of the Earth's atmosphere from the Student Book and use it to produce a series of cartoon illustrations of how we believe it evolved.
- Students can use a card matching activity to connect evidence with theory (i.e. there was intense volcanic activity on the early Earth / volcanoes emit carbon dioxide and water vapour / the early atmosphere probably contained a lot of carbon dioxide and water; nitrogen is unreactive / ammonia is converted into nitrogen and water by oxygen / nitrogen probably built up in the atmosphere because ammonia reacted with oxygen).
- Review the word equation for photosynthesis (students may need to revise earlier work here), explaining the theory that early microorganisms helped reduce the concentration of carbon dioxide in the early atmosphere. Use this to introduce the idea of terraforming Mars by introducing algae to its surface. There are many video clips found on the internet which describe this process (just enter 'terraforming' and 'Mars' into your search engine).
- Groups of students can brainstorm the types of measurements they would need to take in order to provide evidence for the way Earth's atmosphere is believed to have developed. Ideas could include sampling gases emitted by volcanoes, trying to grow plants in sealed containers of carbon dioxide and testing for oxygen. This activity could provide an opportunity to develop ideas about reliable methodology and data collection.
- Introduce higher tier students to the Miller Urey experiment by showing them diagrams of the experimental apparatus and discussing how it worked. This could be adapted to an apparatus labelling activity to involve less able students. **[HT only]**
- Give small groups of students molecular models of the gases used in the Miller Urey experiment and ball-and-stick diagrams of some simple amino acids such as glycine, serine or alanine. Students try to make amino acids from the gases of the early atmosphere. **[HT only]**

Plenaries

Summarising learning objectives – Students write a short 'blurb' for a book jacket of a book entitled *How We Got Our Atmosphere*. To extend students, they can review each others' blurbs with reference to the learning objectives. To support students, provide key words to scaffold the blurb. *(5 minutes)*

Students as teachers – In pairs or threes, students decide which questions they would ask the class to assess their learning, and why. *(10 minutes)*

Chapter 1 – Our changing planet and universe

Further teaching suggestions

Investigating where water came from

Students can research the theory that comets provided much of the Earth's early water. What evidence is there for this?

Did life begin like this?

Students could research the Miller Urey experiment independently, focusing on the arguments for and against the repeatability and reproducibility of their findings. The class could debate whether the experiment provides a valid hypothesis about how life began on Earth. **[HT only]**

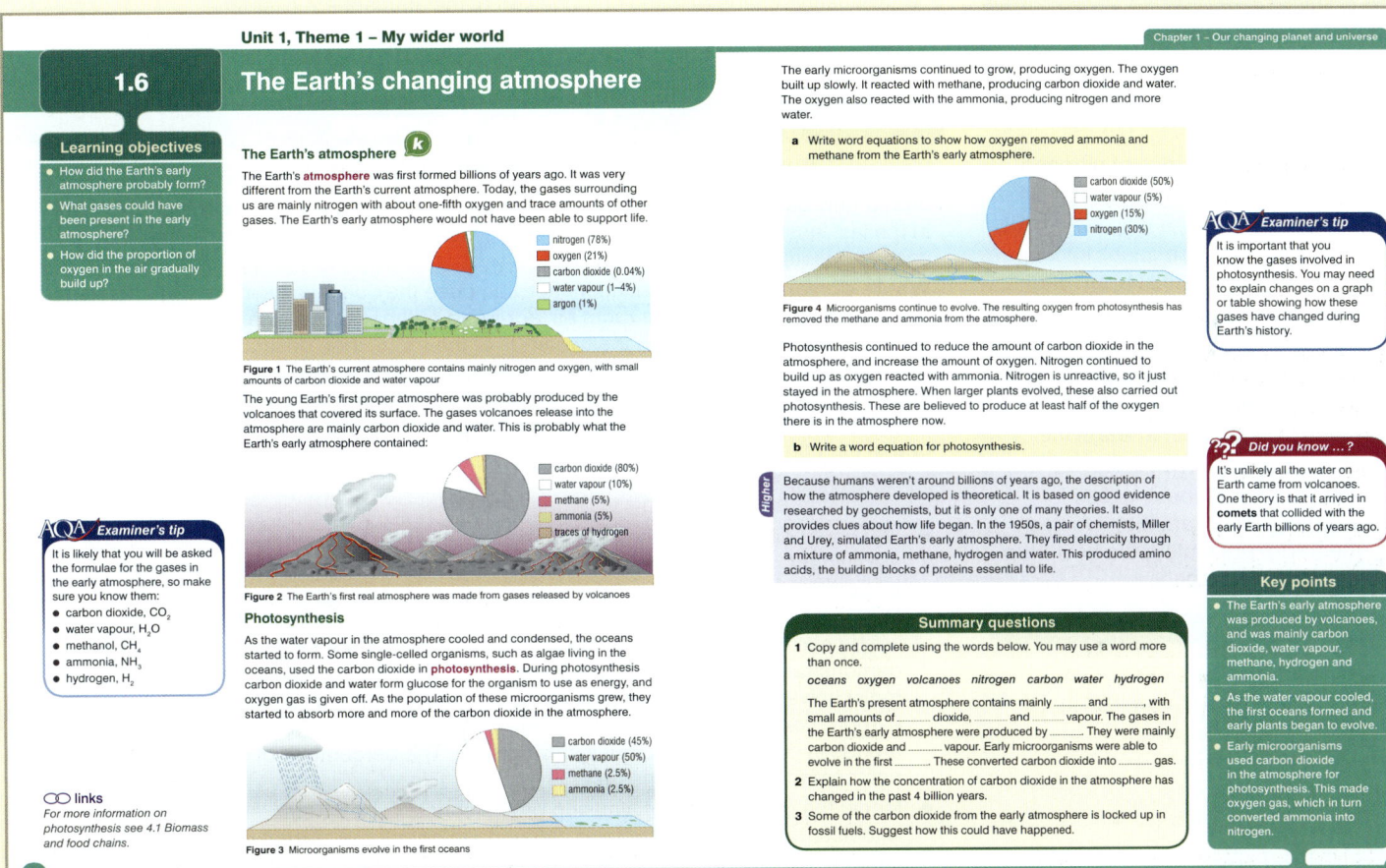

Summary answers

1. nitrogen, oxygen, carbon, hydrogen, water, volcanoes, water, oceans, oxygen

2. The concentration of carbon dioxide increased, due to intense volcanic activity. Then it slowly decreased as microorganisms evolved and started using it in photosynthesis.

3. Carbon from carbon dioxide in the early atmosphere could be incorporated into microorganisms and later into plants, during photosynthesis. If these (or animals that ate them) died under the right conditions, so that their decay slowed, and heat and pressure were applied over millions of years, they would become fossil fuels.

Answers to in-text questions

a. oxygen + ammonia $\longrightarrow$ nitrogen + water; oxygen + methane $\longrightarrow$ carbon dioxide + water

b. carbon dioxide + water $\xrightarrow{light}$ glucose + oxygen

Unit 1, Theme 1 – My wider world

1.7 Maintaining our atmosphere

Learning objectives

Students should learn:
- the roles of the gases in the Earth's atmosphere
- how our atmosphere allows the Sun to heat the Earth
- how greenhouse gases keep the Earth warm enough to support life.

Learning outcomes

Most students should be able to:
- state the composition of the Earth's atmosphere
- describe the importance of nitrogen, oxygen, water and carbon dioxide in maintaining life
- state that greenhouse gases keep the Earth warm enough to support life.

Some students should also be able to:
- explain how the gases in the atmosphere allow short-wave radiation from the Sun to warm the Earth
- explain that greenhouse gases absorb long-wave radiation emitted by the Earth, preventing the heat from radiating out into space.

AQA Specification link-up: Science B 3.3.1.2

- Describe how the atmosphere surrounding the Earth allows light energy radiated from the sun to pass through.
- Explain how greenhouse gases in the atmosphere keep temperatures on Earth stable and warm enough to support life by allowing short-wave radiation to pass through the atmosphere to the Earth's surface but absorbing the outgoing long-wave and short-wave radiation from the Earth.

Within this context, candidates should be able to use scientific data and evidence to discuss, evaluate or suggest implications of the following:
- evaluating changes to the composition of the atmosphere over time.

Lesson structure

Starters

Just a minute – Students talk for a minute on what they learned about the atmosphere in the last lesson, without pause or hesitation, or they are 'out' and another student takes over. *(5 minutes)*

Where do you stand? – Continuum activity: pose the question 'Is the greenhouse effect bad?' Denote one end of the room as 'agree', the other end as 'disagree'. Students arrange themselves between the two ends of the room, then describe the position they have taken. To extend students, ask them to justify their choice and explain why they agree or disagree with the choices their classmates have made. To support students, give pre-made statements about the greenhouse effect and ask them to decide to which end of the continuum the statement belongs. *(10 minutes)*

Main

- You could discuss and demonstrate the gases in air. Oxygen and carbon dioxide are evident – plants photosynthesise, all organisms respire and materials can be combusted or oxidised. Pass air from syringe to syringe over heated copper to remove the oxygen, forming copper(II) oxide (CuO). The resulting gas mixture will be nitrogen and argon. A lighted splint will be put out by this mixture. You can demonstrate it is not carbon dioxide by bubbling it through limewater or adding a moist universal indicator strip (limewater would become cloudy if the gas contained carbon dioxide, but it does not; universal indicator would change from green to orange in carbon dioxide, but it stays green). The percentage of oxygen can also be demonstrated by these means.
- Model the composition of air using grains of rice. Fill a 500 cm³ soft drink bottle with rice, weigh it, then use food colouring to colour the correct percentages of it differently to indicate the different gases in air. This can help students appreciate that air is mostly nitrogen, since a pie chart can be quite abstract.
- Investigate the effect of greenhouse gases. Compare the effects of air, carbon dioxide and methane on temperature in a sealed bottle (see 'Practical support').

Plenaries

Where do you stand? – Repeat the activity from the start of the lesson. Ask students if they have changed their minds during the lesson, and if so why. *(5 minutes)*

Revisit the learning objectives – Students write the answers on sticky notes and share their ideas. To extend students, they could suggest further questions to deepen their understanding. To support students, they could focus on writing single sentences describing things they have learned, and sharing them with each other. *(10 minutes)*

Support

- Thinking that greenhouse gases 'reflect' heat back to Earth is a common misconception, because a greenhouse is not a perfect analogy for global warming – make sure students are clear that greenhouse gases *absorb* energy (long-wave radiation) radiated by the Earth.

Extend

- Students can carry out their own research to find greenhouse gases other than carbon dioxide and suggest ways of incorporating them into the investigation activity.

Further teaching suggestions

Climate change?

Not everyone agrees the enhanced greenhouse effect is a real problem. Research different opinions, coming up with a list of arguments for and against the enhanced greenhouse effect as a real issue.

Practical support

Removing oxygen from air / producing a sample of nitrogen

Equipment and materials required: Two 100 cm³ glass gas syringes, combustion tube, copper turnings, Bunsen burner, safety mat, eye protection.

Details

Start with one syringe depressed and one extended. Heat the turnings in the glass tube until they are glowing, passing the air slowly back and forth across them. (CAUTION: HEAT.) Allow the apparatus to cool and then collect the gas in a syringe for testing.

Safety: Use eye protection. CLEAPSS Hazcard 26 Copper(II) oxide – harmful.

Comparing greenhouse gases

Equipment and materials required: Three 2 litre plastic bottles, Alka Seltzer tablets, water, three bungs with thermometers/temperature probes, gas tube, bright light.

Details

Add 200 cm³ water to the bottom of each bottle. Bung the first bottle and label 'air'. Add two Alka Seltzer tablets to the second bottle, this fills the bottle with carbon dioxide. Allow to settle before replacing the bung (increased pressure will skew results) and label carbon dioxide. Add methane to the third bottle via a gas tap. (CAUTION: FLAMMABLE) Replace the bung immediately (methane is lighter than air) and label methane. Place all three bottles in equal sunlight. Either monitor with a data logger or measure the temperature manually every two minutes. Plot the temperature of each bottle in a line graph.

Safety: CLEAPSS Hazcard 45A Methane – extremely flammable.

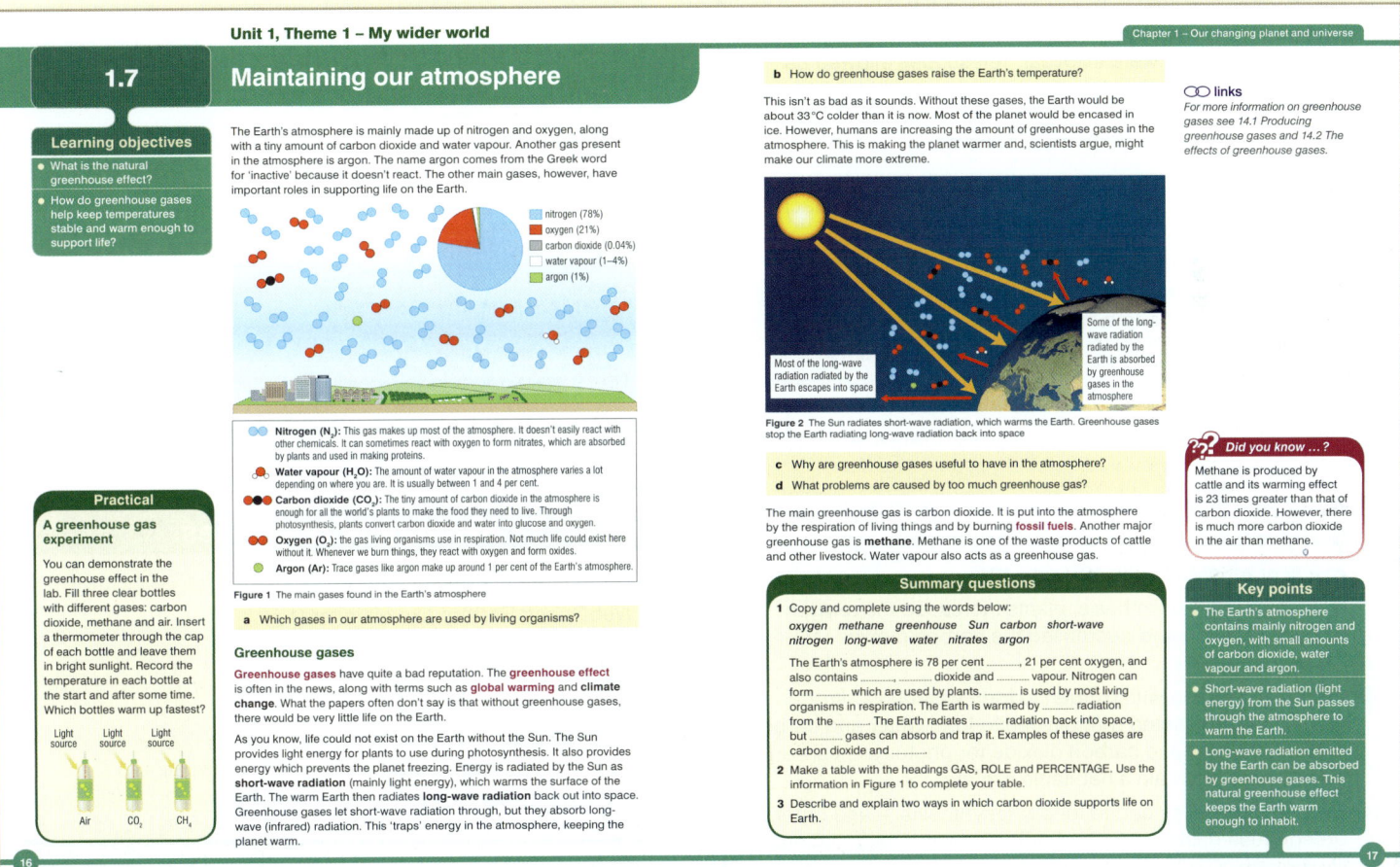

Summary answers

1. nitrogen, argon, carbon, water, nitrates, oxygen, short-wave, Sun, long-wave, greenhouse, methane

2.

Gas	Role	Percentage
Oxygen	Respiration, forms oxides	21
Carbon dioxide	Photosynthesis, maintains temperature	0.4
Nitrogen	Forms nitrates used by plants	78
Water vapour	Condenses to liquid water, photosynthesis	1–4
Argon	No major role; it is inert	1

3. Carbon dioxide is essential to photosynthesis – photosynthesis converts solar energy to chemical energy, enabling life on Earth. Photosynthesis also produces oxygen as a waste product. All living things need oxygen to respire.
Carbon dioxide keeps the planet at an inhabitable temperature by absorbing longwave radiation emitted by the Earth, preventing it being radiated away into space.

Answers to in-text questions

a oxygen, water vapour, carbon dioxide

b Greenhouse gases 'trap' energy in the atmosphere by absorbing the longwave radiated emitted by the Earth when the sun warms it.

c Greenhouse gases are useful because they raise the Earth's temperature enough to support life.

d Too much greenhouse gas causes global warming, when the Earth's temperature is raised too much, which may lead to climate change.

Unit 1, Theme 1 – My wider world

Summary answers

1

Inner core	Solid metal
Outer core	Liquid metal
Mantle	Slowly flowing rocks
Crust	Solid rock

2 Different celestial bodies emit different types of electromagnetic radiation. Also our atmosphere absorbs different types of electromagnetic radiation.

3 Red-shift tells us the universe is expanding, and appears to be expanding more rapidly at greater distances. Evidence for the Big Bang includes the background microwave radiation and although some light from very distant stars has started to reach us, it does not fill the night sky with light.

4 We thought that the Earth was central, orbited by the Sun, moon and planets. We then realised the Sun was central, and orbits were elliptical. Telescopes allowed us to see moons around other planets so we could tell how these orbits worked and see much fainter planets. Modern telescopes let us see other solar systems.

5 It would take too long (four years for a light signal to arrive so many more for a rocket); stars are too hot to approach closely.

6 Mars is too far from the Sun for there to be liquid water. There are so many other solar systems in our galaxy though, it is very likely that some have liquid water, which we think is vital for life.

7 longer year; cooler temperature; dimmer light (do not accept longer day)

8 Seismologists measure vibrations in the Earth's crust. They use seismometers.

9 a any year between 270 and 290 million years ago
  b any year between 380 and 395 million years ago
  c any number between 10% and 12%
  d Through photosynthesis, plants increased the oxygen concentration and decreased the carbon dioxide concentration.

10 Ammonia reacted with oxygen to form water and nitrogen gas. Methane reacted with oxygen to form water and carbon dioxide. Plants need carbon dioxide for photosynthesis and it also helps maintain a habitable climate.

11 Carbon dioxide is needed for photosynthesis. It and other greenhouse gases keep the Earth at a temperature where life thrives.

12 Energy from the core heats the mantle. This causes convection currents in the mantle. The convection currents drive tectonic plates around the surface of the Earth. When the plates rub past each other, they can shudder, causing earthquakes.

13 The Sun emits short-wave radiation, which heats the Earth's surface. The long-wave radiation then emitted by the Earth's warm surface is absorbed by CO_2 and other greenhouse gases. This causes the atmosphere to become warmer.

Unit 1, Theme 1 – My wider world

Summary questions

1 Match up the following words with their descriptions:
  a Inner core   Liquid metal
  b Outer core   Slowly flowing rocks
  c Mantle   Solid rock
  d Crust   Solid metal

2 Explain why we need to use different telescopes to gain a full idea of what is in the universe, including space-based telescopes, land-based telescopes and telescopes sensitive to different types of electromagnetic radiation.

3 Explain what the red shift tells us about the universe. Describe the evidence for the Big Bang.

4 Describe how our model of the solar system has changed. How did telescopes help us develop new models?

5 Explain why we cannot send a space probe to the star nearest the Sun.

6 Explain why we are less likely to find life on Mars than to find life in another solar system.

7 Mars is our closest planet but it is further away from the Sun than we are. Suggest three features of Mars that would be different as a result of this.

8 Seismologists make measurements in the Earth's crust. What sort of measurements do they make, and what equipment do they use?

9 Look at the graph and then answer the questions.

Change in oxygen

(graph: Proportion of oxygen in the atmosphere (%) vs Time before present (millions of years), 500 400 300 200 100 0, showing Present day level)

a How many years ago was the oxygen level the highest?
b Suggest the year that plants first started growing on the Earth.
c The present day level of oxygen is 21 per cent. Suggest the proportion of oxygen 200 million years ago.
d How did plants affect the concentrations of oxygen and carbon dioxide in the Earth's early atmosphere?

10 What happened to the ammonia and methane believed to be in the Earth's early atmosphere?

11 Give **two** reasons why it is important for there to be carbon dioxide in our atmosphere.

12 Explain in steps how energy from the Earth's core can end up causing an earthquake. Use diagrams to help explain each step.

13 Global warming is sometimes described as carbon dioxide 'trapping' energy in the Earth's atmosphere. Explain in detail how this happens.

Kerboodle resources

- Theme map: My wider world
- WebQuest: Earthquake disaster (1.5)
- Extension: The power of nature (1.5)
- WebQuest: Climate (1.6)
- Practical: Volcano! (1.5)
- Interactive activity: The changing universe
- Examination-style questions
- Answers to examination-style questions

AQA Practical suggestions

Practicals	AQA	k	📖
Collect gas produced by aquatic plants and test for oxygen.	✓		
Measure the amount of carbon dioxide in inhaled and exhaled air.	✓		
Pass air over heated copper using gas syringes and measure the percentage of oxygen. Then burn magnesium in the nitrogen to form Mg_3N_2. Add water to produce ammonia (nitrogen must have come from the air).	✓		✓
Make model volcanoes.	✓	✓	

Chapter 1 – Our changing planet and universe

Examination-style questions

1 Put the following in order of size starting with the smallest.
 A Moon
 B Sun
 C Milky Way
 D Earth
 E Solar system (4)

2 Choose words from the box to label the diagram of the Earth.

| Core | Crust | Atmosphere |
| Mantle | Ozone | |

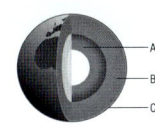

(3)

3 Complete the table.

Part of the early atmosphere	Formula
	H_2O
Carbon dioxide	
	CH_4
Ammonia	

(4)

4 The order of the planets in our solar system is given in the table below.
For each planet, choose the correct length of its year.

Planet	Year length in Earth years
Mercury	164.81
Venus	11.86
Earth	247.70
Mars	1.88
Jupiter	0.62
Saturn	29.456
Uranus	84.07
Neptune	0.24
Pluto (a dwarf planet)	1.00

(3)

5 Match the type of wave each telescope uses to how the telescope may be used by astronomers.

Type of wave received by telescope	Measuring temperature	Seeing through clouds and smoke	Looking at distant neutron stars
Gamma			
Infrared			
Microwave			

(2)

6 Choose the correct word from each box to finish these sentences.
 a .... is a greenhouse gas.

| Argon | Carbon monoxide | Methane |

(1)

 b Greenhouse gases absorb more .... radiation.

| short-wave | medium-wave | long-wave |

(1)

 c Therefore more heat is retained in the ....

| atmosphere | ground | sea |

(1)

7 In this question you will be assessed on using good English, organising information clearly and using specialist terms where appropriate.
Explain how scientists used line spectra to explain what is happening to the universe. (6)

8 Telescopes can be used on Earth and in space.
 a Suggest **two** benefits to astronomers who use telescopes on the surface of the Earth. (2)
 b Suggest **two** reasons why other astronomers would prefer to use a telescope on a satellite in space. (2)

9 The diagram shows the plate boundary between North America and Mexico.

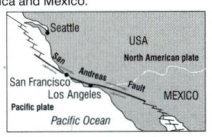

 a Explain what happens inside the Earth so that earthquakes occur. (3)
 b Suggest two reasons why there are a lot of earthquakes in Los Angeles. (1)
 c Suggest another town that might suffer from a lot of earthquakes (1)

Examination-style answers

1 ADBEC
(All correct = 4 marks, 4 correct = 3 marks, 3 correct = 2 marks, 1 or 2 correct = 1 mark)

2 A – core (1 mark)
 B – mantle (1 mark)
 C – crust (1 mark)

3

Part of the early atmosphere	Formula
Water	H_2O
Carbon dioxide	CO_2
Methane	CH_3
Ammonia	NH_4

(4 marks)

4

Planet	Year length in Earth years
Mercury	0.24
Venus	0.62
Earth	1.00
Mars	1.88
Jupiter	11.86
Saturn	29.456
Uranus	84.07
Neptune	164.81
Pluto (a dwarf planet)	247.70

(8 or 9 correct = 3 marks, 5, 6 or 7 correct = 2 marks, 2, 3 or 4 correct = 1 mark)

5

Wave type	Measuring temperature	Seeing through clouds and smoke	Looking at distant neutron stars
Gamma			✓
Infrared	✓		
Microwave			✓

(3 marks)

6 a Methane (1 mark)
 b long-wave (1 mark)
 c atmosphere (1 mark)

7 Marks awarded for this answer will be determined by the Quality of Written Communication (QWC) as well as the standard of the scientific response.

There is a full and detailed explanation of how scientists used the absorption spectra to explain what is happening to the universe. It is well structured with minimal repetition or irrelevant points. There is an accurate, fluent and clear expression of ideas with only minor errors in the use of technical terms, spelling, punctuation and grammar.
(5–6 marks)

There is some explanation of how scientists used the absorption spectra to explain what is happening to the universe with some omissions. The answer shows some attempt at structuring and the ideas are expressed with reasonable fluency and clarity. There are some errors in the use of technical terms, spelling, punctuation and grammar.
(3–4 marks)

There is an attempt to explain what is happening to the universe with some omissions. The answer is largely incomplete and may contain some valid points which are not clearly structured. It lacks fluency and/or clarity. It contains errors in the use of technical terms, spelling, punctuation and grammar. (1–2 marks)
No relevant content (0 marks)

Examples of points made in the response:
- When star is moving away, the absorption spectrum moves further towards the red end
- This is known as red-shift
- The further away the star, the more the red-shift
- This means that the star is moving faster
- Suggesting that there was some kind of explosion known as the Big Bang
- So scientists believe that the universe is expanding.

8 a Any **two** from: easier to fix, cheaper, not relying on computers. (2 marks)
 b Clearer picture OR more reliable (1 mark)
 Above clouds/atmosphere OR no light pollution (1 mark)

9 a Plates moving. (1 mark)
 Due to convection currents. (1 mark)
 When plates rub together, earthquakes occur. (1 mark)
 b It is on a plate boundary. (1 mark)
 The plates are rubbing against each other. (1 mark)
 c San Francisco (1 mark)

Unit 1, Theme 1 – My wider world

2.1 Building blocks of new products

Learning objectives

Students should learn:
- that all materials are made of atoms
- that compounds are combinations of different types of atoms that are chemically joined
- that mixtures are combinations of different materials that are not chemically joined.

Learning outcomes

Most students should be able to:
- classify materials as elements, compounds and mixtures
- locate elements on the periodic table.

Some students should also be able to:
- explain that atoms combine chemically to form compounds.

AQA Specification link-up: Science B 3.3.1.3
- Be able to classify materials as elements, compounds or mixtures.

Lesson structure

Starters

Spot the symbol – Allow students to use a periodic table to find the symbols for some common elements. *(5 minutes)*

Periodic table bingo – Provide students with periodic tables (showing symbols only) with some of the entries removed at random. Ask students to cross off the names of the elements as you call them out. To support students, use a limited section of the periodic table with symbols and names present, calling out symbols. To extend students, call out other facts about the elements they may know (i.e. needed in respiration, used in balloons, found in proteins and fertilisers). *(10 minutes)*

Main

- Start by reviewing students' knowledge of the periodic table. Check students know that elements are all made up of one type of atom.
- Use a presentation including particle diagrams to explain to the class how elements, compounds and mixtures differ. This could include photographs of examples (i.e. sodium, chlorine, sodium chloride and rock salt). Use the photographs to illustrate the homogeneity of elements and compounds, compared to mixtures.
- Demonstrate the formation of iron sulfide to the class.
- Present groups with a range of named substances in sealed containers. They use the names and appearance of the substances to suggest whether they are elements, compounds or mixtures.
- Students use the information presented to them in the lesson to produce an informational poster explaining the differences between elements, compounds and mixtures.

Plenaries

Q and A – Each student writes two revision questions and then asks another member of the class to answer them. *(5 minutes)*

Elements, mixtures and compounds – Show students a presentation with a number of unlabelled particle diagrams (like the three on the corresponding page of the Student Book). They decide whether they are elements, compounds or mixtures. To extend students, use photographs of materials and ask students to explain their reasoning. To support students, use tactile examples, such as molecular models, marbles, or Lego™ bricks. *(10 minutes)*

Support
- Molecular model kits can be used to give more tactile experience of elements, compounds and mixtures. Alternatively, just model the ideas using Lego™ bricks of different colours.

Extend
- Students can use extension material about Mendeleev's work on the periodic table to develop their ideas on the properties of elements.

Further teaching suggestions

The work of Lavoisier
- Look at the work of Lavoisier, either through texts, internet research or a teacher-led presentation. Discuss early ideas about elements and how/why they have changed in modern science.

Literacy in science
- Develop literacy skills by discussing all the terms that use the word 'compound', i.e. compound interest, compound fractures, compound eyes, compound words, etc. Use this discussion to reinforce the idea that compounds are multiple items that are connected to each other.

Chapter 2 – Using materials our planet provides

Practical support
Preparation of iron sulfide

Equipment and materials required: A range of named substances in sealed containers, iron filings, sulfur flowers, Bunsen burner, magnet, heatproof surface, eye protection.

Details

Students describe the physical appearance of sulfur and iron, and locate them on the periodic table. Demonstrate that iron is magnetic.

Mix roughly equal quantities of each together. Highlight that this is now a mixture. Prompt the students to suggest that they could separate it again with the magnet.

Heat the mixture on a suitable surface (i.e. a metal tray or heatproof mat) with the hottest flame of the Bunsen burner. This works best if the mixture is piled into a fat line about 5 to 10 cm long. (Safety – heating sulfur must be performed in a fume cupboard. Heating can be done in the laboratory if using ignition tubes plugged with mineral wool.)

Play the flame along the length of the mixture until it has all heated to the point of glowing. Students describe any changes they see occurring. Allow it to cool.

Draw the students' attention to the fact that a new substance has been made. It is a different colour from both of the original elements and is non-magnetic. Explain how the iron sulfide is different from the iron and sulfur mixture.

Students record their observations and use particle diagrams to show the differences between the elements, mixture and compound.

Safety: Use fume cupboard and eye protection. Warn asthmatics about fumes given off by the reaction. CLEAPSS Hazcard 96A Sulfur. Use CLEAPSS Guide L195 (sulfur/iron).

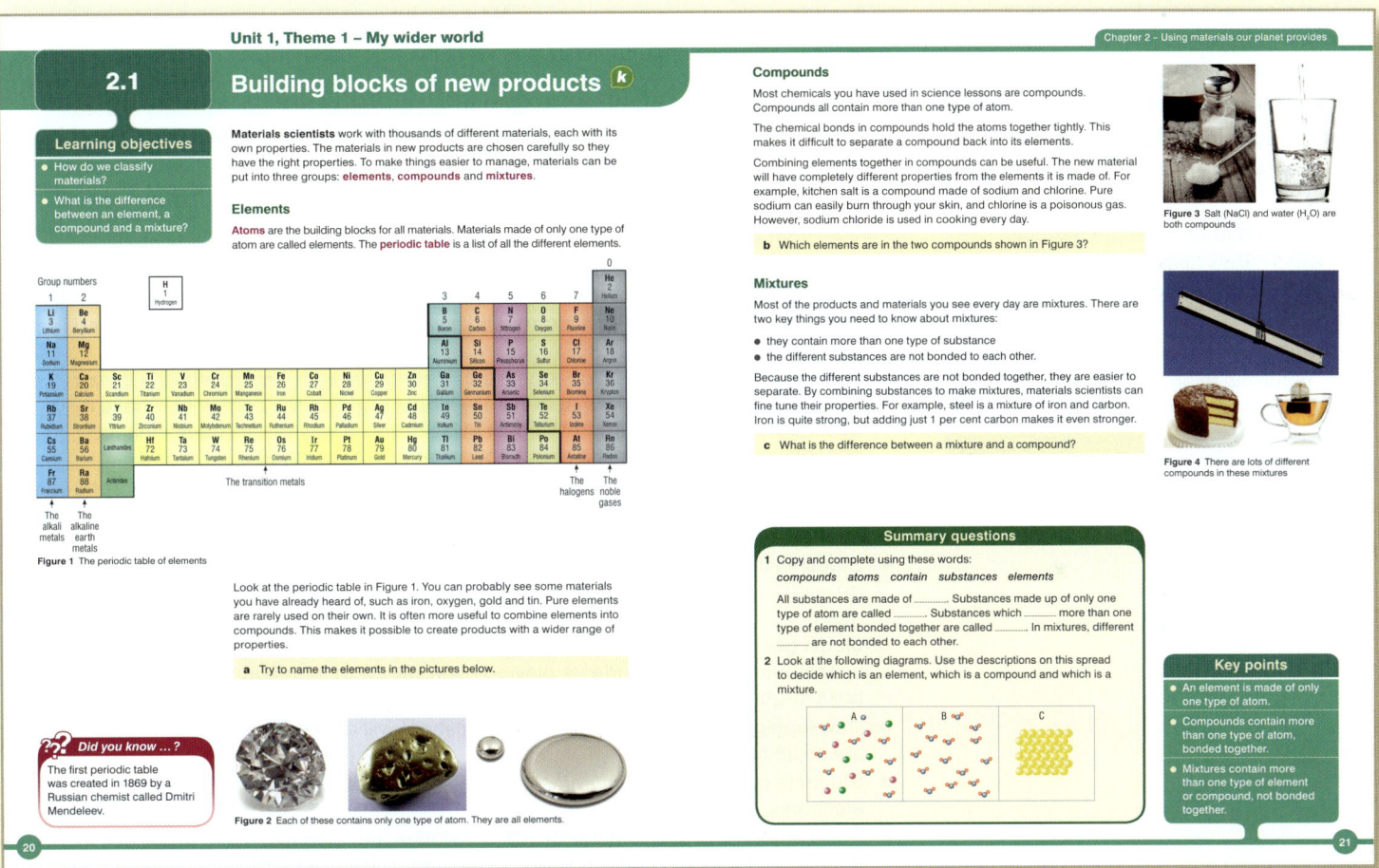

Summary answers

1 atoms, elements, contain, compounds, substances
2 a mixture
  b compound
  c element

Answers to in-text questions

a carbon, gold, mercury
b salt – sodium, chlorine; water – oxygen, hydrogen
c Mixtures can be separated more easily than compounds by physical methods. Different elements are bonded together in compounds but not in mixtures.

Unit 1, Theme 1 – My wider world

2.2 Inside atoms

Learning objectives

Students should learn:
- that atoms consist of protons, neutrons and electrons
- that protons are positive, neutrons are neutral and electrons are negative
- that protons and neutrons have a relative mass of one
- that atomic nuclei consist of protons and neutrons
- that electrons orbit the nucleus in shells.

Learning outcomes

Most students should be able to:
- describe the composition of a simple atom
- use information from the periodic table to draw models of atoms.

Some students should also be able to:
- explain how to work out the number of neutrons in an atom from its atomic number and mass number.

AQA Specification link-up: Science B 3.3.1.3

- Describe the structure of an atom in terms of number of protons, neutrons and electrons and their arrangement. Atoms contain the same number of protons (positive charge) as electrons (negative charge). The protons and neutrons (no charge) are at the centre, in the nucleus and the electrons are positioned around the outside of the atom.
- Define the terms atomic number and mass number.

Lesson structure

Starters

Chemical symbol dominos – Give groups of three or four students domino cards with element names on one half and symbols on the other. Students take turns putting cards down to match up with the previous card. To support students, use two sets of cards as a memory card game using pairs. To extend students, use triangular dominos which include a fact about an element. *(5 minutes)*

Snowballing – Get students to list what they already know about atoms and share their ideas with a neighbour. Pairs then discuss ideas in a small group, and a spokesperson is chosen from each small group to feed back to the rest of the class. *(10 minutes)*

Main

- Check students' ideas about atoms with a short quiz to review what they learned from previous work.
- Show the students a small, opaque, sealed container with a marble inside it. Ask them how we can find out what is inside without opening the container. Shake it? Use magnets? Take an X-ray? Weigh it? Use this as a way to introduce indirect observation, leading to ideas about how scientists have studied the atom.
- Discuss the etymology (the history of a word and which language it comes from) of the word 'atom'. Explain how the ancient Greeks decided there must be an indivisible unit of matter ('a' means 'not' and *'tomos'* means 'cut').
- Students can use the internet or library to research the history of the atom. This can be used to produce a timeline showing the development of different models of the atom.
- Model the structure of an atom using students (preferably in a fairly large open space). Start with two students playing hydrogen, and then add more students to build larger atoms.
- Developing an understanding of electron shells will aid understanding of other chemical concepts.
- Students can complete a summary table of the properties of protons, neutrons and electrons.
- Show students pictures of the CERN installation and discuss recent advances in atomic theory. What different scientific jobs are there in a facility like CERN? (Their website is quite comprehensive and has a job search facility.)

Plenaries

Sticky atoms – Give groups sheets of circular coloured stickers (three colours) and ask them to compete with other groups to make various atoms. *(5 minutes)*

Modelling atoms – Organise students to form two teams and then model various atoms (by playing the parts of protons, neutrons and electrons). The other team has to work out which atom is being modelled. To support students, provide teams with diagrams to help them. To extend students, ask them to look up atomic and mass numbers and then give the numbers of each subatomic particle. *(10 minutes)*

Support
- Use objects such as table-tennis balls stuck together to show nuclei. Make sure the objects are different colours to differentiate between protons and neutrons.

Extend
- Develop students' understanding of indirect observation by examining more examples, such as monitoring animal populations through stools or tracks; studying distant planets, the structure of the Earth, dark matter.

Chapter 2 – Using materials our planet provides

Further teaching suggestions

Rutherford's experiment

Discuss Rutherford's experiment. Then ask students to pretend to be the scientist explaining his findings to different people, i.e. one student can role-play Rutherford telling an eight-year-old nephew what he had discovered; another student can role-play Rutherford explaining the experiment to another scientist.

Unit 1, Theme 1 – My wider world

2.2 Inside atoms

Learning objectives
- What are protons, neutrons and electrons?
- How are protons, neutrons and electrons arranged in atoms?

Scientists need to understand how the atoms in a substance behave. This makes it easier to control the properties of a new material. To understand atoms, scientists needed to explain what they are made of.

Atoms are very, very small. In fact, if you laid about 1 billion in a line they would be as long as the full stop at the end of this sentence.

About 100 years ago, scientists discovered that the atom is made up of even smaller particles. These are called subatomic particles. There are three main types of subatomic particle you need to learn about: protons, neutrons and electrons.

Atomic structure

By knowing the structure of its atoms, you can predict how an element will behave. For example, will it conduct energy? Or, will it react with other substances? Look at the picture below, it shows two simple atoms. You can see they are both made from the same three types of particle.

Figure 1 The Large Hadron Collider at CERN will help answer questions about the universe

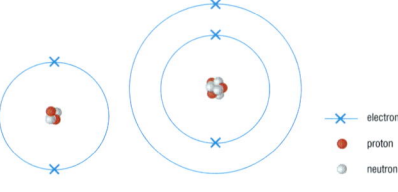

Figure 2 Helium and lithium atoms

There are two types of particle in the centre or **nucleus** of the atom. These are called **protons** and **neutrons**. The particles orbiting around the outside are called **electrons**.

Did you know …?
Particle physicists and quantum physicists study these tiny particles for clues about the universe. The best way to do this is to smash them together at enormous speeds. By doing this, physicists hope to answer big questions, such as what causes mass and gravity.

The particles have different charges and masses. The protons are positive. The neutrons are neutral and the electrons are negative.

The nucleus is small and very dense. The protons and neutrons are the heavy particles in an atom. We give them a relative mass of 1.

The smaller electrons have hardly any mass and orbit the nucleus. We call the area that the electrons orbit an electron shell. Because electrons have hardly any mass they are given a relative mass of 0.

a How many protons are in the helium atom shown in Figure 2?
b How many neutrons are in the lithium atom shown in Figure 2?

Atomic number and mass number

The helium and lithium atoms are different elements because they have different numbers of protons. The number of protons an atom has tells you what element that atom is. So if an atom has three protons, it has to be a lithium atom.

An atom contains the same number of protons (positive charge) and electrons (negative charge), so lithium has three electrons.

If you look at a periodic table, you will see that next to every element there are often two numbers. The smaller number tells you the number of protons in an atom of that element. This is called the **atomic number**, for example:

$_6$C $_2$He $_{10}$Ne

c Use the periodic table to find the atomic numbers of iron, chlorine, sodium and gold.

The top number is called the **mass number**. This tells you the total number of protons plus neutrons in an atom. A lithium atom has three protons and four neutrons. So its mass number is 7.

Particle	Where is it?	Charge	Mass
Proton	In the nucleus	Positive	1
Neutron	In the nucleus	Neutral	1
Electron	In shells outside the nucleus	Negative	Almost zero

So to recap:
- All atoms of the **same** element have the **same** number of protons.
- The **atomic number** of an element is the number of **protons** its atoms contain.
- The number of electrons in an atom is **equal** to the number of protons.
- The **mass number** tells us the number of **protons plus neutrons**.
- If you want to work out the number of neutrons in an atom, just **subtract** the atomic number from the mass number.

d Use the periodic table to calculate the number of neutrons in a potassium atom with a mass number of 39.

Summary questions

1 Copy and complete using these words:
nucleus   three   positive   one   negative
outside   protons   atomic

Atoms contain .... types of particles. Neutrons are neutral and are found in the .... Protons, which have a .... charge, are also found in the nucleus. Protons and neutrons have a relative mass of .... Electrons have a .... charge and are located .... the nucleus in shells. All atoms of the same element have the same number of .... This is shown on the periodic table by the .... number.

2 a Why is the nucleus the densest part of an atom?
b An atom contains nine protons. What type of atom is it?
c All atoms are neutral in charge. A particular atom contains five protons, how many electrons will it have?

Figure 3 The atomic and mass number of lithium. We show this as $^7_3$Li.

AQA Examiner's tip
You may be asked to draw and label a diagram of the atom when you are given just the atomic number and mass number, so make sure you know how to do this.

Did you know …?
Atoms may be small, but they can release huge amounts of energy. Some elements, such as uranium, have unstable nuclei. When a uranium nucleus is hit by a neutron it breaks apart, releasing energy and several neutrons. Each neutron then hits another uranium atom. This process is called a chain reaction. If this is controlled, you can generate nuclear power. An uncontrolled nuclear chain reaction takes place in a nuclear bomb.

Key points
- Protons are positive, neutrons are neutral and electrons are negative.
- The number of protons in an atom equals the number of electrons in that atom.
- Protons and neutrons are in the nucleus of an atom and have a relative mass of 1.
- Electrons have a very tiny mass. They orbit the nucleus.

Summary answers

1 three, nucleus, positive, one, negative, outside, protons, atomic

2 a It is dense because it contains the protons and neutrons.
  b fluorine
  c 5 electrons

Answers to in-text questions

a 2 protons
b 4 neutrons
c iron 26, chlorine 17, sodium 11, gold 79
d 39 – 19 = 20 neutrons

Unit 1, Theme 1 – My wider world

2.3 Different types of particles

Learning objectives

Students should learn:
- that atoms, molecules and ions are different types of particle
- that molecules are groups of atoms that have chemical bonds between their atoms
- that ions are charged particles.

Learning outcomes

Most students should be able to:
- identify atoms, molecules and ions from diagrams
- give examples of atoms, molecules and ions.

Some students should also be able to:
- relate the charge on an ion to loss and gain of electrons.

Answers to in-text questions

a oxygen, hydrogen, carbon
b element (oxygen), compound (methane), compound (carbon dioxide), element (nitrogen)
c 2–
d 3+

Support
- Use the see-saw analogy to help explain the charges resulting from ionisation.

Extend
- Remind students of the behaviour of the alkali metals. Ask them to try to explain why it is 'easier' to lose an electron as the atoms get larger down the group. Then ask students to predict the corresponding trend they might see in Group 7.

AQA Specification link-up: Science B 3.3.1.3
- Explain the difference between atoms, molecules and ions.

Lesson structure

Starters

Revision – Review students' understanding of the differences between elements, compounds and mixtures. Show the class a series of particle diagrams and ask students to decide which is which. This could also be done as a card-match activity. *(5 minutes)*

Atomic structure – Review students' understanding of atomic structure. Ask students to draw out the structures of various atoms from the first 20 elements in the periodic table. To support students, provide diagrams and periodic table information for them to pair together. To extend students ask them to consider elements beyond atomic number 20. *(10 minutes)*

Main

- Use molecular modelling kits to demonstrate the difference between atoms and molecules. Students can make models of various simple molecules, given their chemical formula. A common student misconception is that molecules are always compounds. Use examples of molecules of diatomic elements such as O_2, H_2 and N_2.
- Split the class into teams and give each team enough molecular modelling parts to make a more complicated molecule such as ethanol or ethanoic acid. Show them the empirical formula and challenge them to work out how the atoms could fit together. Reinforce the differences between atoms and molecules by asking questions such as 'How many atoms does the molecule have in total?', 'How many types of atoms are there in the molecule?'
- Lead students to consider the relative sizes of molecules with the nanometre activity described in the 'Practical support' section of this spread.
- Show students a presentation on nanotechnology. Explain how scientists working in this field need to be able to manipulate matter on the molecular level. How can we move individual molecules about? What are the benefits of this kind of technology?
- Ask the class to draw the atomic structure of lithium. (Introduce a basic knowledge of the electronic structures of the first 20 elements. This will help the explanation of ionisation.) Explain that lithium is more stable if it loses one electron. Ask the class to consider the balance of positive and negative charges – a see-saw diagram may help some students.
- A simple worksheet with statements such as 'If .... loses .... electron(s), its charge will be ....' will reinforce ideas about ionisation.
- Explain how different metal ions produce different-coloured flames.
- It may be appropriate to perform a class practical to investigate the different flame colours.

Plenaries

True or false – Read the class 10 true or false questions reviewing the content of this lesson. Mark out different sides of the room for true and false. Students move to the correct side of the room for each question. *(5 minutes)*

Three facts – Students write three facts they have learned in this lesson, under appropriate headings such as 'What are atoms and molecules?', 'What is nanotechnology?', 'What are ions?'. To support students, provide them with facts that they categorise under headings. To extend students, they could consider the social, economic and environmental impact of exploiting flame colours in fireworks. *(10 minutes)*

Summary answers

1 atoms, two, joined, ion
2 Because it is made of both oxygen and hydrogen.
3 positively charged

Chapter 2 – Using materials our planet provides

Practical support

Flame tests

Equipment and materials required: 0.1 mol/dm³ solutions of copper sulfate, copper chloride, strontium chloride, potassium chloride, calcium chloride, sodium chloride, Bunsen burners, heatproof mats, wire loops (preferably platinum), eye protection.

Details

Students first heat the wire loop over a blue Bunsen flame. The loop is then dipped in the salt solution and returned to the flame. The flame colour is recorded. Compare the colours from copper sulfate and copper chloride to prove the flame colour depends upon the metal in the compound. Good classroom management and student discipline are required during this practical.

Safety: Wear eye protection. No hazards for all chemicals used at the concentrations used – CLEAPSS Hazcard 27A; 27C; 19A

Making slime

Equipment and materials required: Borax powder, PVA glue, plastic cups or beaker, disposable pipettes, stirring rod (wooden or glass), glitter or food colouring are optional.

Details

Before the lesson, make up a Borax solution by slowly adding the Borax powder to water (25–30 cm³ per student) and stir until no more dissolves. You can also dilute the PVA glue if you think it is too thick. 2 plastic cups will be needed per student, fill with 25 cm³ of each solution.

Students can add glitter and food colouring to the PVA to make it look interesting, then they should add the borax solution drop by drop and stir. They should see slime being formed and that it loses the ability to stick to the inside of the cup!

Safety: Wear eye protection. The solutions of PVA and borax do not need hazard classification and the slime is safe to handle, although students should wash their hands when they have finished handling it. CLEAPSS Hazcard 14 Borax – toxic.

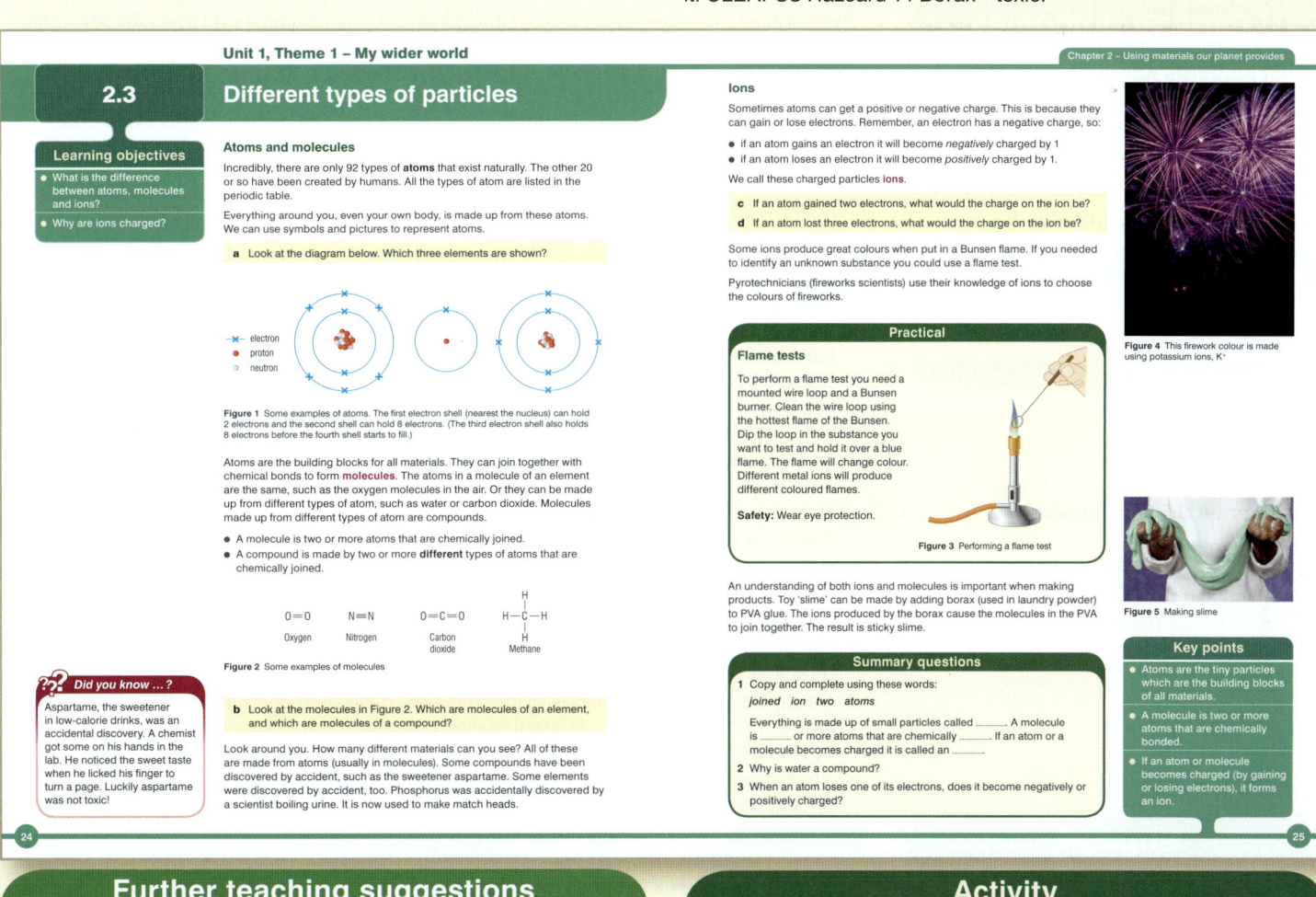

Further teaching suggestions

Nanotechnology

- Students can research a particular aspect of nanotechnology. Areas to consider might be: molecular computers, moving atoms and molecules around, medical nanotechnology.

Fireworks and pyrotechnics

- Relate the flame test practical to the pyrotechnics industry. How do we produce different-coloured fireworks? Show students pictures or videos of firework displays and ask them which metals were used to produce each explosion.

Activity

Nanometre activity

Equipment and materials required: A large piece of paper, rulers.

Details

Before the lesson, get a large piece of paper and draw a 1 × 1 m square on each side. Give students metre rulers and ask them to compare the relative sizes of a metre, a millimetre and a human hair. Now, show the class your 1 m square and ask a student to stand in it. Ask the class to imagine that the student is standing in a square the width of a human hair. Then mark out a millimetre square in the 1 m square. Explain to the class that this is the relative size of a red blood cell. Now turn the sheet over to reveal the second 1 m square. Tell the class this square is now equivalent to the size of a red blood cell. Draw out another 1 mm square in this box, and explain to the class that this is now approaching the size of a large molecule.

Unit 1, Theme 1 – My wider world

2.4 Making products with materials from the Earth

Learning objectives

Students should learn:

- that gold, sulfur, limestone and marble are used straight from the ground
- that rock salt and crude oil are both separated into useful components before they can be used
- that crude oil is separated by fractional distillation and rock salt is separated using filtration and evaporation.

Learning outcomes

Most students should be able to:

- name some materials that are used straight from the ground, and some that need to be separated
- describe some separation techniques.

Some students should also be able to:

- link separation techniques to differences in the mixed materials' properties.

AQA Specification link-up: Science B 3.3.1.3

- Give examples of substances used straight from the ground (gold, sulfur, limestone and marble).
- Describe how salt is separated from rock salt before use.
- Describe how fuels (hydrocarbons) are separated from crude oil (fractional distillation).

Lesson structure

Starters

Wordsearch – Provide students with a word search including the names of the materials to be discussed in the lesson. *(5 minutes)*

Raw material – Present students with a range of everyday objects and materials. Ask them to discuss, in small groups, what raw materials might have been needed to manufacture them. To support students, supply a list of raw materials to choose from. To extend students, include maps and discuss where each raw material comes from for each object. *(10 minutes)*

Main

- Brainstorm students' ideas on different materials. Ask students to produce spider diagrams showing materials that can be used as they are found, versus materials that need separation and processing before use.
- Discuss ideas about the uses of gold. How long has it been in use? Why is it so valuable? Why does it not rust? Where is it found? Why is it used in microelectronics? Explain how gold's unreactivity makes it a useful material.
- Students can consolidate their learning by writing short summaries with the above headings.
- Show students some pictures of marble or limestone buildings/statues. Draw out the important contribution these two materials have made to society.
- Show the students a sample of rock salt. Use their experience at Key Stage 3 of separating mixtures to suggest a way to get the salt out of the mixture. Students could write a brief method for a procedure to obtain pure salt.
- Play the class a video or animation showing the formation and extraction of crude oil. Highlight the range of products and materials that rely on crude oil.
- Demonstrate the distillation of a synthetic crude oil mixture in the laboratory. (See 'Practical support' for more details.)
- Relate students' observations to the fractional distillation of crude oil. Check they understand how the different boiling points of the fractions allow them to be separated.
- Students should be given information about the different fractions in crude oil. They should relate the size of molecule to its boiling point, viscosity, colour and volatility.

Answers to in-text questions

a element

b Sulfur is used to make sulfuric acid which can be made into fertilisers and used for car batteries.

Plenaries

True or false – Check students' understanding with true/false statements. As a whole class activity this could be done by standing up for true and sitting down for false. Rock salt is distilled [F], Gold is unreactive [T], Crude oil is a mixture [T], Crude oil is mined in quarries [F], Marble can be used straight from the ground [T]. *(5 minutes)*

Extended writing – Students write a paragraph about one of the materials they have learned about, explaining what elements of their lifestyle it supports and suggesting alternatives to it. To support students, give them a list of materials to which they must assign 'slight impact on my life', 'some impact on my life', and 'large impact on my life', writing a sentence next to each to explain their choice. To extend students, they write about how the material is obtained and the impact of that activity on the environment. *(10 minutes)*

Support

- Students could be given summary sheets with the headings 'name of material', 'uses of material' and 'where does it come from?'
- Students could carry out a rock salt separation experiment.

Extend

- When students are given information about the different fractions of crude oil they should relate the size of molecule to its boiling point, viscosity, colour and volatility.
- You can then take this a step further by linking this information about the fraction to its uses.

Chapter 2 – Using materials our planet provides

Further teaching suggestions

Use of sulfuric acid
Sulfur is used to make sulfuric acid, one of the most important manufactured chemical in the world. Students can use the internet to research the different uses of this chemical. Learning can be demonstrated by producing concept maps, posters, and leaflets or by listing the uses.

Determining salt content
Students can design a procedure to determine the percentage salt content of a sample of rock salt. Does rock salt from different areas have a different salt content?

Summary answers

1. **a** Straight from the ground – sulfur, gold. Separated – salt, crude oil.
   b Any **two** valid uses for the substance chosen. Sulfur – sulfuric/battery acid and fertilisers; gold – jewellery, electronics; salt – flavouring, preserving; crude oil – petrol, plastics.
2. Gold is an excellent electrical conductor and does not corrode.
3. They have different boiling points.
4. Add water to dissolve salt, filter out insoluble rock, evaporate water from the solution.

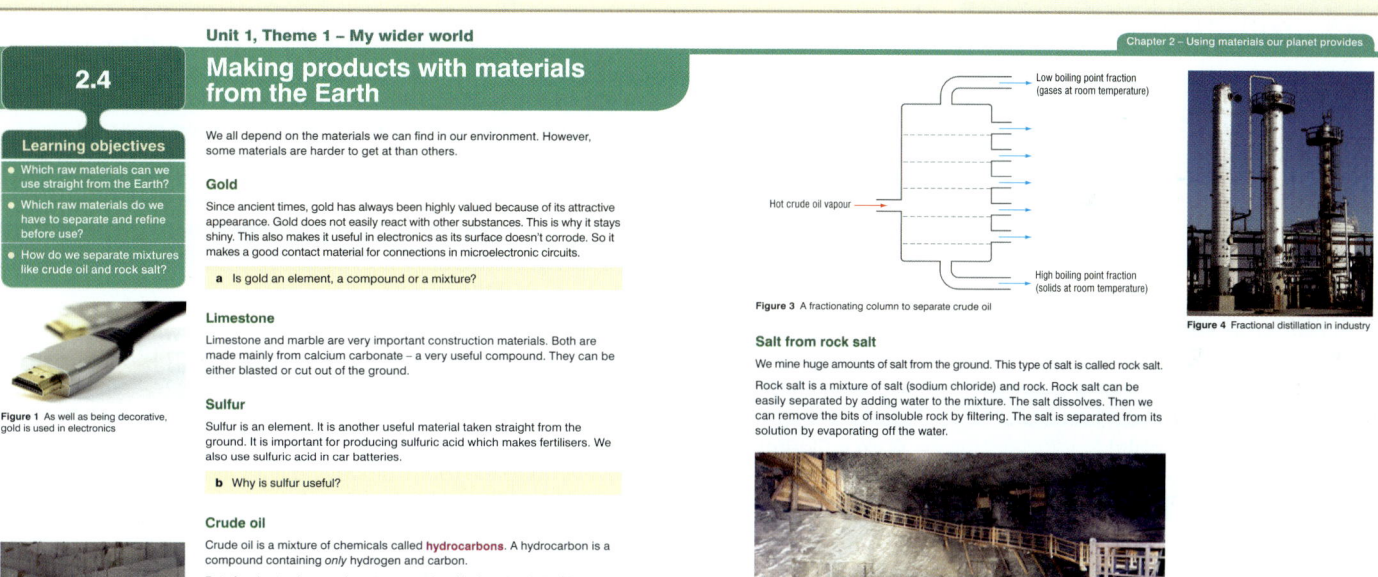

Practical support

Fractional distillation in the lab

Equipment and materials required:
5–10 cm³ synthetic crude oil (see *CLEAPSS Hazcard 45A and recipe card 20*), side-arm boiling tube, clamp and stand, electric mantle, delivery tube, 4 small test tubes, 4 watch glasses, mineral wool, pipette, 0–360 °C thermometer with bung to fit boiling tube, wooden splints, heatproof mat, beaker of cold water.

Details
Set up the equipment as shown in the diagram. Put 2–3 cm mineral wool in the bottom of the side-arm tube and add the synthetic crude oil to this with a pipette. Slowly and carefully, heat the crude oil using the electric mantle. The first fraction should start appearing between 80 °C and 100 °C.

Replace the collection tube with a fresh one at 100 °C, 150 °C, 200 °C and 250 °C.

Draw students' attention to the difference in colour between each fraction, use the watch glasses to demonstrate each fraction's viscosity and the wooden splints to show how easy each fraction is to ignite. You should observe that the fractions with higher boiling points are darker in colour, thicker, harder to ignite and will burn with a smoky flame.

This can be performed as a demonstration or as a class practical, limiting the amount of synthetic crude oil to 2 cm³. If a class practical is used, it may be possible to pour all the fractions back together and observe a mixture similar to the original sample.

Safety: Wear eye protection. CLEAPSS Guide L195 has further information.

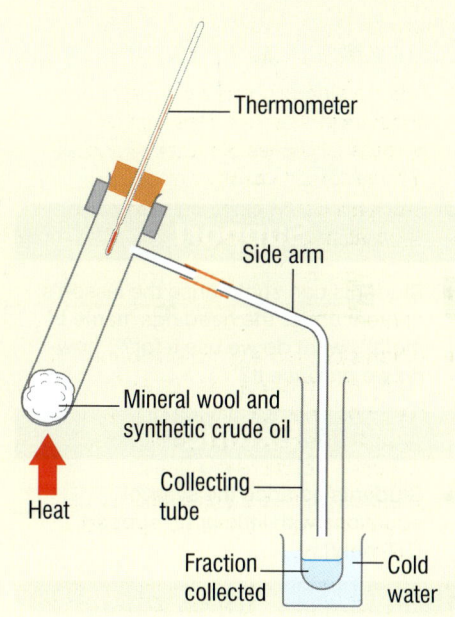

Unit 1, Theme 1 – My wider world

2.5 Extracting metals for construction

Learning objectives

Students should learn:

- that metals are extracted from their ores
- that reducing agents remove oxygen from oxides
- that iron and lead oxides are both produced using reducing agents.

Learning outcomes

Most students should be able to:

- state that reducing agents remove oxygen from metal oxides
- describe the production of lead and iron including word equations
- use a reactivity series to determine if an element can be used as a reducing agent.

Some students should also be able to:

- produce balanced symbol equations for the production of lead and iron from their ores. [HT only]

AQA Specification link-up: Science B 3.3.1.3

- Describe how metals are separated from their ores:
  - **a** metals more reactive than carbon, such as aluminium, are extracted by electrolysis of molten compounds. The use of large amounts of energy in the extraction of these metals makes them expensive
  - **b** metals less reactive than carbon are extracted from their ores using carbon and carbon monoxide as reducing agents
  - **c** lead and iron may be made from their oxides by reduction:
    - extraction of lead: carbon and carbon monoxide can act as reducing agents ($2PbO + C \longrightarrow 2Pb + CO_2$ and $PbO + CO \longrightarrow Pb + CO_2$).
    - extraction of iron: iron oxide (Fe_2O_3) and coke (carbon) are heated to produce iron. The coke burns to produce carbon dioxide ($C + O_2 \longrightarrow CO_2$). The carbon dioxide reacts with the coke to produce carbon monoxide ($C + CO_2 \longrightarrow 2CO$). When heated, the iron oxide reacts with the carbon monoxide to produce iron. Iron oxide is reduced and carbon monoxide is oxidised ($Fe_2O_3 + 3CO \longrightarrow 2Fe + 3CO_2$).

Lesson structure

Starters

Metallic properties key word recap – This is a card-match activity in which students match up the characteristic properties of metals with their definitions. This could also be done as a card-sorting activity of metallic and non-metallic properties. To help students, provide diagrams or pictures to support each property. To extend students, ask them to draw a diagram depicting each property. *(5 minutes)*

Metal survey – Students move around the room, surveying it for items made from metal. Each item is recorded, and small groups discuss why a metal was used for that particular item. *(10 minutes)*

Main

- Discuss the importance of metals to society. What are the key properties of metals that we use? What uses are they best suited for? What metals do we use the most and what for? How do metals such as gold and platinum differ from metals such as iron and copper? Where do metals come from? What is the definition of an ore?
- Use examples or practical results to review students' knowledge of displacement and the reactivity series.
- Demonstrate how copper can be obtained from a copper salt, by adding steel wool to copper(II) sulfate (safety: copper(II) sulfate is harmful). Draw students' attention to the colour changes.
- Alternatively, this could be performed as a class practical, first reducing copper oxide with sulfuric acid and displacing the copper with another metal. (See 'Practical support'.)
- Reduce lead oxide as a class practical or demonstration. (See 'Practical support'.)
- After each practical activity, talk through the word equations for the reactions and introduce the balanced symbol equations for Higher Tier students.
- Show students an animation of the electrolysis of aluminium oxide. Describe the process in words, or use a diagram.
- Make a table of the similarities (both involve stripping off oxygen, both are very hot and produce molten metal, etc.) and differences (one uses electricity, one uses heat and displacement, electrolysis uses more energy so is more expensive, etc.) between using CO and electricity to obtain metals from their ores. To support students, this could be approached as a card sorting activity.
- Use an animation/video to help explain the workings of the blast furnace, although the details of the actual blast furnace will not be examined. Draw attention to the reducing agent being carbon monoxide, rather than the coke itself.

Plenaries

Equations – Students rearrange mixed-up word equations for the production of lead and iron from their ores. To support students, use word equations and a prompt sheet describing the reaction in prose. To extend students, use balanced symbol equations for Higher Tier students. *(5 minutes)*

Formulae – Students test each other on the formulae of the chemicals used in this lesson. This could also be approached as a card-match activity. *(10 minutes)*

Support

- Students can summarise the lesson's content under the headings 'name of metal', 'what do we use it for?', 'how do we produce it?'

Extend

- Students balance the symbol equations with little or no support. [HT only]

Practical support

Extracting copper from copper oxide

Equipment and materials required: Copper(II) oxide, 1 mol/dm³ sulfuric acid, iron filings (or magnesium powder), Bunsen burner, safety mat, tripod, gauze, 250 cm³ beaker, stirring rod, filter paper, funnel, eye protection.

Details

Add about three spatulas of copper(II) oxide to 100 cm³ 1 mol/dm³ sulfuric acid. Stir, and heat gently for 10 minutes. Allow to cool and filter the mixture to obtain copper sulfate solution. Add iron filings to displace the copper, until the solution is colourless. Remove unreacted iron filings with a little acid. Filter and wash the sample of copper.

Safety: Wear eye protection. CLEAPSS Hazcard 26 Copper oxide – harmful. CLEAPSS Hazcard 98A Sulfuric acid – corrosive. CLEAPSS Hazcard 27C Copper sulfate – harmful. CLEAPSS Hazcard 59A Magnesium powder – highly flammable. CLEAPSS Hazcard 55A Iron filings.

Extracting lead from lead oxide

Equipment and materials required: Lead oxide, carbon powder, crucible, tripod, clay triangle, eye protection, Bunsen burner, tongs, small spatula.

Details

Add two spatulas each of lead oxide and carbon powder to the crucible. Mix the powders together well. Heat very strongly for 10 minutes. Allow to cool and examine for droplets of lead. Keep laboratory well ventilated, and/or use a fume cupboard. You can repeat this process with other metal oxides and compare their reactivity.

Safety: Wear eye protection. CLEAPSS Hazcard 56 Lead oxide – toxic. Take care with hot equipment.

Unit 1, Theme 1 – My wider world

2.5 Extracting metals for construction

Learning objectives
- What is an ore?
- Which common metals do we get from ores?
- How can we use reducing agents to get metals from ores?

Metals are vital to make products. Even products that don't contain metals are often made by machines that do.

Most metals exist in the Earth's crust as materials called **ores**. An ore is a rock containing a useful amount of metal. Ores contain enough metal to make it worth spending the money to extract it. We need to separate the metal from the other elements found in its compound in the ore.

Many ores contain oxides of the metal. These are compounds of the metal plus oxygen. We can remove oxygen from a compound with a **reducing agent**.

a What is an ore?

An element used as a reducing agent has to be **more reactive** than the metal in the oxide. Being more reactive means it is able to 'take' the oxygen away from the metal. If this happens, we say that the metal oxide has been **reduced**.

What will reduce what?

The **reactivity series** is used to work out which reducing agent could be used with which ore. A reactivity series is just a list of elements in order of how reactive they are (see Figure 2). Carbon is more reactive than iron or lead. This means we can use carbon or carbon monoxide as a reducing agent for iron and lead ores. In fact, any metal less reactive than carbon can be extracted in this way.

b What is a reducing agent?

c Name **two** metals whose ores could **not** be reduced by carbon. (Hint: look at the reactivity series.)

Producing iron from iron ore

Iron is a very important metal. We use it to make steel. Most iron comes from an ore called haematite. This contains iron oxide (Fe_2O_3).

Haematite is crushed and put into a huge tower called a **blast furnace**. Carbon (in the form of coke made from coal) is also added. Air is blasted through the furnace at high temperature. The following chemical reaction happens:

$$C + O_2 \rightarrow CO_2$$
carbon + oxygen → carbon dioxide

Then the carbon dioxide reacts with more carbon (coke) to produce carbon monoxide:

$$C + CO_2 \rightarrow 2CO$$
carbon + carbon dioxide → carbon monoxide

The carbon monoxide then reduces the iron oxide:

$$3CO + Fe_2O_3 \rightarrow 3CO_2 + 2Fe$$
carbon monoxide + iron oxide → carbon dioxide + iron

Figure 1 Mining iron ore

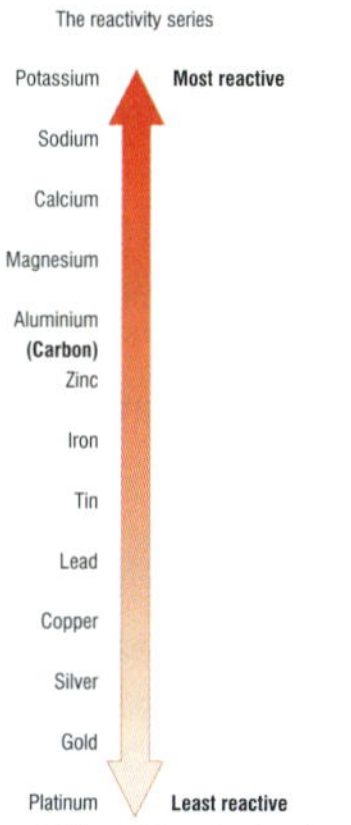

The reactivity series

Potassium — Most reactive
Sodium
Calcium
Magnesium
Aluminium
(Carbon)
Zinc
Iron
Tin
Lead
Copper
Silver
Gold
Platinum — Least reactive

Figure 2 This reactivity series shows how reactive each element is compared to the other elements.

In the process of reducing iron oxide, carbon monoxide is itself **oxidised**. It has oxygen added to it to form carbon dioxide.

Extracting lead from its ore

Lead is used in car batteries and roofing for buildings. Because lead is less reactive than iron, it is easier to extract than iron. It is easier to reduce lead oxide than iron oxide.

Carbon (again as coke) is added to lead oxide. This reduces the lead and makes carbon monoxide:

$$PbO + C \rightarrow Pb + CO$$
lead oxide + carbon → lead + carbon monoxide

The carbon monoxide also reduces some of the lead oxide:

$$PbO + CO \rightarrow Pb + CO_2$$
lead oxide + carbon monoxide → lead + carbon dioxide

Extracting more reactive metals

Metals that are more reactive than carbon need to be extracted another way. These are extracted using a process called **electrolysis**. Electrolysis uses electricity to separate compounds. Aluminium is extracted from its ore (bauxite) in this way.

First of all aluminium oxide is separated from the bauxite and melted in a large electrolytic cell. Then an electric current is passed through it. The aluminium oxide is melted to allow the movement of ions in the molten mixture. Because aluminium forms positive ions, it is attracted to the negative part of the cell. The pure molten aluminium can then be poured out of the cell.

Figure 3 Working at a blast furnace

Figure 4 Electrolytic cells at an aluminium plant

links
For more information on symbol equations see 2.9 Using equations.

AQA Examiner's tip
At Higher Tier you are expected to be able to write balanced symbol equations for the reactions involved in separating metals from their ores.

links
For more information on electrolysis see 12.1 Electrolysis and 12.2 Electroplating.

Key points
- Reducing agents remove oxygen from ores.
- An element used as a reducing agent has to be more reactive than the metal in the oxide.
- Iron oxide and lead oxide are reduced by carbon and carbon monoxide.

Summary questions
1. What is iron's main ore called?
2. What are the **two** reducing agents mentioned on these two pages?
3. Why could you not use carbon to reduce aluminium oxide into aluminium?

Further teaching suggestions

Finding and using ores

Students could research the global uses and sources of metals such as iron and aluminium. How have these changed in the last hundred years? Internet search terms could include: haematite, bauxite, mining, blast furnace.

Answers to in-text questions

a a mineral containing enough metal compounds to make it worth extracting the metal
b a substance that can remove the oxygen from a metal oxide
c magnesium and aluminium

Summary answers

1. haematite
2. carbon and carbon monoxide
3. Because carbon is not reactive enough.

Unit 1, Theme 1 – My wider world

2.6 Products from the atmosphere

Learning objectives

Students should learn:

- that nitrogen, argon and helium are extracted from the atmosphere for use in industry [HT only]
- that nitrogen is used for making ammonia, freezing and preserving [HT only]
- that helium is used for balloons, airships and cooling [HT only]
- that argon is used in incandescent light bulbs and other lighting. [HT only]

Learning outcomes

Most students should be able to:

- state uses for nitrogen, argon and helium [HT only]
- describe simply how nitrogen, argon and helium are obtained. [HT only]

Some students should also be able to:

- evaluate the usefulness of materials taken from Earth's atmosphere. [HT only]

AQA Specification link-up: Science B 3.3.1.3

- Describe air (the atmosphere) as a mixture of gases with different boiling points that can be fractionally distilled to provide new materials for industrial processes (helium for balloons, argon for filament lamps and electrical discharge tubes and nitrogen for ammonia – which is used for making fertilisers). Helium, argon and nitrogen can either be used directly or used to make another product. **[HT only]**

Lesson structure

Starters

Extinguisher – Demonstrate the nitrogen used in crisp packets by showing a freshly opened packet putting out a lit splint. *(5 minutes)*

What's in the atmosphere? – Model the composition of the atmosphere using different coloured beads in a large transparent container. Choose a suitable mass to represent the whole atmosphere. Given the percentage composition of the atmosphere, students can work out the mass of each colour of different colour beads to use in order to model the atmosphere. To support students, work on a smaller scale with small groups using 100 g so the calculations are easier. To extend students, for illustrative purposes, use beads that are proportionally different in mass (i.e. marbles or ball bearings for N_2 or O_2, plastic beads for He). Invite them to suggest ways to separate the beads afterwards, starting a discussion about relative properties such as boiling point. *(10 minutes)*

Main

- Fill a beaker with iced water. When droplets of water start to condense on its surface, invite students to explain what is happening. Why did the water condense? Why can't we normally see it? How much water is in the air? Can we remove all the water? What other gases are in the air? Could we condense them in a similar way?
- Revise fractional distillation of crude oil and compare with liquid air, emphasising the low temperatures and relatively simple mixture in liquid air compared with crude oil.
- Use photographs and diagrams to explain how a condenser extracts gases from the air. Start with a diagram of a dehumidifier to make it more accessible.
- Show students video clips of experiments and demonstrations with liquid nitrogen.
- Show pictures of airships and play the newsreel from the Hindenburg disaster to explain the importance of helium as an inert gas.
- Play video clips of argon in use: lasers, plasma lights, light bulb production lines.
- During video clips, students can take notes about how the gas is used.

Plenaries

Writing prose – Students write a paragraph about one of the gases they have learned about, explaining what elements of their lifestyle it supports and suggesting how their life would be different without it. To support students, produce a writing frame to organise this information. Use prompts such as '(gas) is important to society because ....' and 'without (gas) our lives would be different because ...'. To extend students, they write about how it is obtained and the impact of that activity on the environment. *(5 minutes)*

Leaflets – Students, either individually or in pairs, make summary informational leaflets about the gases we extract from the atmosphere, using their notes from the lesson. *(10 minutes)*

Support

- Use lots of photographs and diagrams to focus on listing and describing the uses of each gas.

Extend

- Students can research the industrial uses of nitrogen generation. What industrial processes use nitrogen? What kinds of businesses have nitrogen generators on site? Why is it sometimes cheaper for a factory to produce its own nitrogen than to buy it?

Chapter 2 – Using materials our planet provides

Further teaching suggestions

Diving
Students can find out what gas is used in scuba tanks. Why is it not pure oxygen?

Absorption spectra
Gain a better idea of how gases and light interact by observing absorption spectra. Use a hand spectrometer to look at a fluorescent tube, an incandescent light bulb and sunlight (reflected off clouds or a white surface). Compare the spectra and relate them to the different gases involved. For example, the bright peaks seen in the spectrum of a fluorescent tube come from gaseous mercury, whilst the dimmer continuous spectrum is from the phosphorus coating the inside of the tube.

This could be extended to explain how we know the sun is mainly helium.

Liquid nitrogen
A local college or university may have a nitrogen generator. A visit could be arranged to see it in action. Also, there may be outreach programmes run by local science education centres offering in-house demonstrations of liquid nitrogen.

Higher

2.6 Products from the atmosphere

Unit 1, Theme 1 – My wider world

Learning objectives
- How can we separate and collect gases in the atmosphere? [H]
- How do we use gases we obtain from the atmosphere? [H]

Did you know …?
Cryonics is the science of deep freezing organisms using liquid nitrogen and other coolants. Hundreds of people have had their body frozen after death. They hope to be revived in the distant future. They believe that by then medical science will be advanced enough to cure the cause of their death.

Most of the products we use from day to day are solids and liquids. However, without the gases we take from the atmosphere, many products couldn't be made at all. These gases are used when making light bulbs, neon signs, fertilisers, or when freezing things. Some of the gases can be used directly, for instance argon. Some gases, such as nitrogen, are used to make other products.

a Name **two** gases obtained from the Earth's atmosphere.

Extracting gases from the air
Air is a mixture of mainly nitrogen and oxygen. Air also contains some 'trace' gases, such as argon, carbon dioxide and helium. To use gases from the air, they must first be separated. To do this, air is compressed and cooled. This changes it into a liquid at a very low temperature. The liquefied gases can be separated by fractional distillation in the same way as crude oil. Figure 1 shows how nitrogen and oxygen can be fractionally distilled using this method. The bottom part of the column is not as cold as the top part. Because nitrogen has a lower boiling point than oxygen, it rises to the top of the column as a gas.

Figure 1 A simple fractionating column separating oxygen and nitrogen

Using nitrogen
Nitrogen (N_2) is the most used gas from the atmosphere. Most of it is used to produce **ammonia** (NH_3). Ammonia is then used to make **fertilisers**, cleaning fluids and nitric acid.

Pure nitrogen is useful by itself, either as a gas or a liquid. Because nitrogen is usually unreactive, it is used in the food industry as a **preservative**. For example, nitrogen is pumped into crisp packets before they are sealed. This replaces any oxygen and prevents **microbes** from growing. Liquid nitrogen is incredibly cold at –196 °C, and can be used for freezing things very quickly. This method is used in medicine to keep samples of cells (such as sperms or eggs) for a long time.

b Describe **two** ways nitrogen can be used.

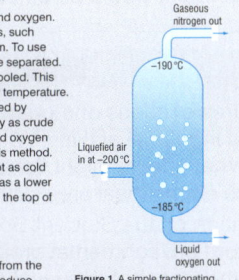

Figure 2 A container of liquid nitrogen

Using helium
Helium is almost totally unreactive and much lighter than air. This makes it very useful. It can be extracted from air in the same way as nitrogen, but most is obtained from natural gas from oil fields. Helium is used in balloons and airships because of its very low density. Like liquid nitrogen, liquid helium is very cold, existing at –269 °C. It is used to cool **superconductors** used in **MRI scanners**.

Figure 3 Liquid helium cools the superconducting magnets in this MRI scanner

c Describe how liquid helium can be used.

Chapter 2 – Using materials our planet provides

Using argon
Like helium, argon is almost totally unreactive. It also makes up nearly 1 per cent of the Earth's atmosphere; much more than helium. The largest domestic use for argon is lighting. **Filament (incandescent) light bulbs** are filled with argon so that the thin metal filament glows but doesn't burn and snap.

Argon produces light when an electrical current is passed through it. Because of this, it is used in filament and electric discharge tubes. It is also used in medical **lasers** for eye surgery and to make 'plasma' lighting.

Figure 5 Argon glows when electricity passes through it

Did you know …?
Helium is actually the second most common element in the universe, despite being quite rare on the Earth.

Figure 4 Filament (incandescent) light bulbs are filled with argon so the filament doesn't burn out

d Why is argon used in filament light bulbs?

Here is a summary of the different gases extracted from the atmosphere, and some of their uses:

Gas	Uses
Nitrogen	Making ammonia, freezing agent, preservative
Helium	Balloons, airships, coolant for superconductors
Argon	Filament light bulbs, electric discharge tubes, lasers

Summary questions
1 Copy and complete using these words:
nitrogen   airships   coolant   bulbs   discharge
fertilisers   preservative   dense   atmosphere

We use gases taken from the Earth's ……… to make new products. ……… is used to make ammonia, which is used in making ………. Nitrogen is also used as a ……… and a ………. Helium is used in balloons and ……… because it is much less ……… than air. Argon is used in filament light ………, electric ……… tubes and lasers.

2 Why is liquid helium more expensive than liquid argon?

3 Why can nitrogen be used to produce other chemicals, but not argon?

links
There is more on the gases in the atmosphere in 1.6 The Earth's changing atmosphere and 1.7 Maintaining our atmosphere.

Key points
- Gases are extracted from the atmosphere by liquefying air and separating the different liquids by fractional distillation. [H]
- Argon is used mainly in lighting. [H]
- Helium is used for cooling and for filling balloons and airships. [H]
- Nitrogen is mainly used to make ammonia. [H]

Summary answers
1 atmosphere, nitrogen, fertilisers, coolant, preservative, airships, dense, bulbs, discharge

2 There is much less helium in the atmosphere than argon.

3 Argon is unreactive.

Answers to in-text questions
a any **two** from: nitrogen, argon, oxygen, neon or helium
b freezing, preservative, making ammonia
c cooling superconductors, airships
d It is unreactive.

Unit 1, Theme 1 – My wider world

2.7 Exploiting the Earth's resources

Learning objectives

Students should learn:

- that extracting materials from the Earth has social, economic and environmental implications
- that there are legal and scientific safeguards in existence to reduce the harm caused by such activities.

Learning outcomes

Most students should be able to:

- describe some costs and benefits of extracting materials from the Earth
- explain ways in which the environmental and social costs of extraction can be reduced, including sustainable development and phytomining.

Some students should also be able to:

- relate various costs and benefits of these activities to specific stakeholder groups
- explain the benefits and drawbacks of using phytomining to absorb minerals from soil.

AQA Specification link-up: Science B 3.3.1.3

Candidates should be able to use scientific data and evidence to discuss, evaluate or suggest implications of the following:

- the social, economic and environmental impacts of exploiting the Earth's crust, sea and atmosphere, and living organisms
- methods of cleansing coal and metal mines such as phytomining.

Lesson structure

Starters

Materials from mining – Ask students to read the first paragraph of the Student Book spread. Ask them to make a list of materials we use that are obtained by mining and quarrying. *(5 minutes)*

What if mining did not happen? – Give students a list of materials produced by mining. Small groups discuss how society would be different if the materials were not available. To support students, adapt this as a card-match activity where students match the names of materials to reasons for their importance. To extend students, consider how the economy and the environment would be different if the mined materials were not available. *(10 minutes)*

Main

- Show the students pictures of mines and quarries. If possible, find some pictures of similar areas without any industrial activity. What are the obvious visual differences between the pictures? What effects might there be on local wildlife? What effect does the removal of trees have on the local area? What about the noise and dust?
- Talk through the information on the Student Book pages. Use the internet to find additional information from environmental groups. What is meant by the term 'stakeholder'? Who are the stakeholders involved when a mine is opened? Who are the stakeholders when a mine is closed? Distinguish between local, national and global implications of mining and quarrying. Balance this information against the great benefits to society provided by mining and quarrying. Show images of architecture and examples of important products made from metals.
- Have a class debate on the topic 'Mines are a menace to the Earth'. Students can use information from the Student Book, and their own research, to build arguments for and against mining and quarrying.
- Find definitions of sustainable development on the Internet. How could these principles be applied to mining or quarrying?
- Show students pictures of plants used for phytomining. Invite them to evaluate the pros and cons of the practice, such as the reduction of soil pollution versus the slow rate of uptake.
- If you have time you could look to set up a role-play of a town meeting about a new quarry being opened nearby (see 'Activity' box).

Plenaries

Costs and benefits connectives activity – In pairs, one student writes down a benefit of the mining industry. The other student continues '… but,' and writes down a drawback of mining. *(5 minutes)*

Concept maps – Students produce concept maps on 'Mining and the environment'. To support students, provide words for them to link together under the headings 'costs', 'benefits', 'stakeholders'. To extend students, ask them to write about the connections, describing why the connections were made. *(10 minutes)*

Support

- Students can divide a sheet of paper down the middle and write down the costs and benefits of mining on either side during the lesson. This list should be reviewed at the end of the lesson to clear up any misconceptions.

Extend

- Students can research an actual mine or quarry. What impact has it had on the local area? What goods and products were produced from the mined materials? The Pennines are rich in these quarries. You may need to put together some resources to help them.
- Alternately, the students could design a questionnaire to ask people living locally to a quarry what their opinions are of the quarry, its benefits to the community and any negative impact it has on the community.

Answers to in-text questions

a crude oil – plastics and fuels
metals – electronics, construction
rock – construction

b Land: landscape spoilt, habitats destroyed, area covered with dust. Air: carbon dioxide, carbon monoxide, sulfur dioxide, nitrogen oxides, particulates emitted. Water: poisonous salts washing into rivers, streams.

Chapter 2 – Using materials our planet provides

Activity

Role play
Role-play a town meeting about a new quarry being opened nearby. Characters to use could be:

Quarry foreman: wants to co-exist pleasantly with the town, but needs the quarry to go ahead.

Quarrying company environmental officer: needs to convince the townspeople that the quarry will be managed sustainably.

Quarrying company executive: mainly concerned with making money from the quarry. Needs to convince the townspeople of the importance of the quarry's products.

Government representative: needs the quarry to be opened. The nation's economy will be boosted by the materials it produces.

Local unemployed person: the quarry provides a chance of employment.

Local activist group: very worried and angry about the possible environmental damage the quarry could cause.

Local mother: concerned her children's health could suffer as a result of the quarry.

Local farmer: worried that pollutants from the quarry will damage his crop yield.

Unit 1, Theme 1 – My wider world

2.7 Exploiting the Earth's resources

Learning objectives
- How do materials taken from the Earth benefit society?
- What is the environmental impact of mining and quarrying?
- How can mining be managed sustainably?
- How can phytomining help clean up the mining process?

Mining and quarrying affect all of us. Every day we rely on products of materials taken from the Earth. We use electronics, travel by cars, buses and trains, use plastic products and build buildings of stone and rock.

We are constantly using other mined materials too. Our homes are made from brick, concrete and steel. We put salt from salt mines on our food. We use plastics made from oil that was extracted from the ground.

a Name **three** materials extracted from the Earth and describe what they are used for.

Figure 1 We benefit from products made from the Earth's resources every day but often forget the activities that make them

As well as the consumer benefits, society benefits from mining in other ways. More than 100 000 people in the UK alone work in the mining and quarrying industries. This is a lot when you consider that most materials in the UK are imported.

What are the problems with taking materials from the ground?
Fossil fuels, stone and metal ores are just some of the materials we mine, dig and pump out of the Earth. Environmentalists argue that mining activity is harming the environment. Mines and quarries damage the appearance of the landscape and destroy wildlife habitats. Then there is the noise and dust that affects the nearby area. Also, mining activity increases traffic near the mine. This leads to more air pollution.

b List all the ways mining can harm the environment. Put your ideas into three groups: land, water and air.

Air pollution isn't just caused by the dust and traffic. Processing ores pollutes the air as well. Metals such as iron and lead produce carbon dioxide when they are purified. Lead, in particular, is very poisonous. Sulfur dioxide gas is also produced when metal sulfide ores are heated. This causes acid rain.

Figure 2 Transporting mined materials

In some mines, rain can cause toxic chemicals to wash into rivers and lakes. This causes even more environmental damage.

There can be social drawbacks with mines as well. The material being mined runs out, eventually. This means that mines don't stay open forever. This can cause problems in communities relying on mines for employment.

Managing resources responsibly
The Planning and Compulsory Purchase Act 2004 set out new rules for developing land into mines and quarries. Any development plans now need to be assessed for **sustainability**. Sustainable development is using resources in a way that finds a balance between human and environmental needs.

Sites that will cause the least environmental damage are preferred. Plans are discussed with all the people they will affect before a mine is started. Also, the after-effects of the mine on the community are considered.

Issues with taking materials from the atmosphere
Like any other industrial process, extracting gases from the atmosphere requires a lot of energy. The more gas is needed, the more energy must be used. Burning fossil fuels is still the most common way to generate electricity. Extracting gases from the atmosphere contributes to the problems of burning fossil fuels. Also, gas generation plants take up land, and some of the products, such as ammonia, can be harmful.

Extracting and using gases from the atmosphere can have benefits. Tens of thousands of people are employed directly working in gas plants. Air is readily available, so there is no need to mine, drill or quarry. The gases being used can easily be recycled back into the air.

Phytomining
Another way to reduce the impact of making products is to find useful ways to remove toxic waste from the environment. Some plants are able to absorb large amounts of harmful materials from the soil and store them. This process is called **phytomining**. It can be used to improve the quality of soil around coal and metal mines. For example, a South African plant called *Berkheya coddii* can safely absorb nickel from the ground. The plants can then be harvested and even used as **bio-ores**, reducing the need for mining.

Summary questions
1 What are:
 a the potential social benefits of opening a mine?
 b the social costs of opening a mine?
2 Describe three ways mining can damage the environment.
3 How can we manage mining sustainably?
4 Give two reasons why extracting iron ore damages the environment more than extracting nitrogen.
5 Describe two ways in which phytomining reduces harm to the environment.

Key points
- We benefit from the products, services and jobs provided by taking materials from the Earth.
- The environment can be damaged by mining operations.
- Managing mining sustainably reduces long-term environmental damage.
- Plants can be used to phytomine areas, absorbing potentially toxic minerals from soil.

Further teaching suggestions

Propaganda
Students could be divided into two groups to produce 'propaganda' posters for both sides of the debate. They need to use evocative images and persuasive writing to get their points across.

Mining and employment in the 20th century
Use graphs to show how employment was affected by the closure of mines in the UK in the 1980s. News websites have plenty of information on the social implications of UK pit closures.

Summary answers

1 **a** local employment, products obtained
 b unemployment when mine runs out.
2 Toxic salts washed into soil; sulfur dioxide and carbon dioxide from smelting and ore processing released into air; pollution from increased traffic.
3 Choose sites that minimise environmental damage, consult with stakeholders, consider after effects on a community.
4 Nitrogen requires less energy to extract – mines are not needed. The extraction of nitrogen does not create toxic waste or greenhouse gases (other than those produced generating the energy for a nitrogen plant to work).
5 Phytomining reduces the pollution of soils and can also be used to produce bio-ores.

2.8 Changing materials to make new products

Unit 1, Theme 1 – My wider world

Learning objectives

Students should learn:
- that mass is conserved during a chemical reaction
- that atoms are rearranged during a chemical reaction.

Learning outcomes

Most students should be able to:
- perform simple calculations to find the mass of a product or reactant
- state that atoms are rearranged during a chemical reaction
- with reference to atoms, explain why mass is conserved during chemical reactions.

Some students should also be able to:
- calculate the mass of reactant or product from information given about the other substances in an equation.

AQA Specification link-up: Science B 3.3.1.4

- Explain why mass is conserved in chemical reactions and that during a reaction products with different properties are formed as a result of atoms rearranging. No atoms are lost or gained.

Lesson structure

Starters

Chemical names – review students' knowledge of chemical formulae with a match-up activity. Students connect formulae with their names. To support students, use familiar formulae such as H_2O, O_2, CO_2. To extend students, use less familiar but still systematically named examples and ask students to explain how they named them (i.e. KCl, MgO, Al_2O_3). *(5 minutes)*

Balancing act – produce a sheet depicting a pair of scales and a selection of different masses. In pairs, students must suggest combinations of masses on each side of the scales that would balance. *(10 minutes)*

Main

- Demonstrate the burning of matches in a large sealed conical flask (see practical support). Record the mass of the sealed flask before burning the matches. Ask students whether they think the mass will have increased or decreased after the matches have burned. Invite students to explain their thinking. Weigh the flask again and determine the mass has stayed the same.
- Students can investigate the law of the conservation of mass themselves by reacting calcium carbonate with hydrochloric acid with and without the means to capture the CO_2 produced (see 'Practical Support').
- Students can investigate the law of the conservation of mass by reacting sodium carbonate with calcium chloride. They will see a white precipitate, indicating there has been a chemical change, whilst the mass remains the same (see 'Practical Support').
- Use models or paper cut-outs to illustrate how the chemicals are reacting with each other in the students' practical work.
- If molecular model kits are available, these could be combined with a pair of scales to show the products and reactants literally balancing each other. Refer back to the 'Balancing act' starter to reinforce this idea.
- Demonstrate the thermite reaction to the class (see 'Practical Support').

Plenaries

Tweet – Individually, students write a 'tweet' (140 characters or less) about the Law of the Conservation of Mass. To support students, provide key words. To extend students, invite them to draw a picture to accompany their tweet. *(5 minutes)*

Explaining – In pairs, students devise a way to explain or illustrate the Law of the Conservation of Mass to a younger class. *(10 minutes)*

Answers to in-text questions

a In a chemical reaction, the total mass of reactants is equal to the total mass of products.
b Aluminium is more reactive than iron.
c 130 g zinc

Support

- Pre-prepare cut-out cards of all the atoms involved in this lesson. Given the formulae, students can build the structures they are working with, making it easier to consider the rearrangement of atoms.

Extend

- To extend students, present them with more complex mass calculations, such as using ratios to scale up a reaction.

Summary answers

1 reaction, rearranged, reactants, atoms
2 Reactants: aluminium and iron oxide
   Products: iron and aluminium oxide
3 162 g zinc oxide
4 32 g oxygen

Practical support

Sealed burning of matches

Equipment and materials required: Eye protection, three matches, 500 cm³ conical flask, balloon, Bunsen burner, safety mat, top-pan balance (accurate to 2 d.p.).

Details
Put the matches together into the conical flask, stretch the balloon over the neck and record the mass. Heat the base of the flask under the matches just long enough to ignite at least one. Allow the matches to burn out, drawing students' attention to the changes taking place. Observe that the mass remains the same.

Safety: Wear eye protection.

Conserving mass

Equipment and materials required: Eye protection, three 100 cm³ conical flasks, calcium carbonate powder, 1 mol/dm³ hydrochloric acid, balloon, 1 mol/dm³ calcium chloride solution, 1 mol/dm³ sodium carbonate solution, universal indicator, top pan balance (accurate to 2 d.p.).

Details
Students react 20 cm³ 1 mol/dm³ hydrochloric acid with 1 g calcium carbonate, recording the mass before and after (the 1 g calcium carbonate can be first measured onto a filter paper, which is included in the final measurement). They will observe

$$CaCO_3 + 2HCl \longrightarrow H_2O + CaCl_2 + CO_2$$

and a loss of mass. They then repeat the experiment, putting the calcium carbonate into a balloon which is stretched over the lip of the flask and lifted to empty the calcium carbonate into the acid, having recorded the total mass first. This time the mass will remain the same.

Students measure 10 cm³ calcium chloride solution and 10 cm³ sodium carbonate solution into separate conical flasks and record the total mass. They then add one to the other, recording the total mass again. No mass change should be observed.

Safety: Wear eye protection. CLEAPSS Hazcard 19A Calcium chloride – irritant. CLEAPSS Hazcard 95A Sodium carbonate. CLEAPSS Hazcard 47A Hydrochloric acid – corrosive.

Thermite reaction

Equipment and materials required: Eye protection for teacher and students, safety screen, three filter papers, Bunsen burner, clay triangle, tripod, one litre beaker, sand, mat, bar magnet, thermite mixture (9 g iron(III) oxide, 3 g aluminium powder) igniter mixture, 0.2 g magnesium powder, 2 g barium nitrate, approximately 10 cm magnesium ribbon (CAUTION: FLAMMABLE).

Details
This experiment should be set up using CLEAPSS Hazcard 11 or CLEAPSS guide (the under water method). Teacher must wear eye protection. Students must remain at least 4 m distance from the demonstration, and wear eye protection. Light the top of the magnesium ribbon and step back. Take care not to look directly at the burning magnesium. The reaction is highly exothermic and the resulting iron will fall into the sand. After the reaction has finished, the water can be poured away, the iron retrieved with the magnet and washed before handling.

Safety: Wear eye protection and use a safety screen while demonstrating. CLEAPSS Hazcard 1 Aluminium powder – highly flammable. CLEAPSS Hazcard 11 Barium nitrate – oxidising and harmful. CLEAPSS Hazcard 59A Magnesium powder – highly flammable. Take care with hot equipment.

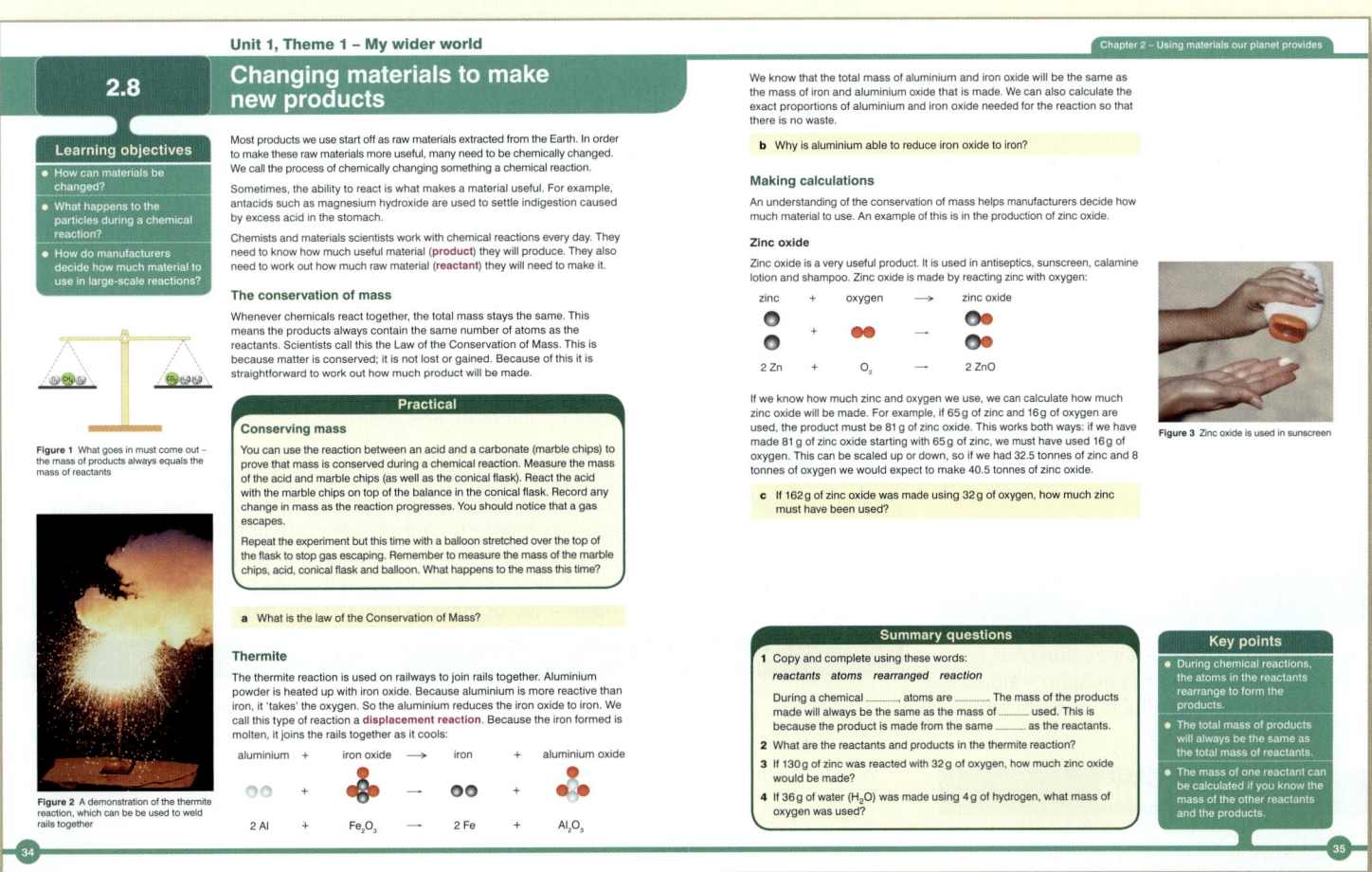

Unit 1, Theme 1 – My wider world

2.9 Using equations

Learning objectives

Students should learn:
- that chemical formulae and equations help chemists understand how much of a material they need
- that balanced chemical equations enable chemists to decide accurately the proportions of reactants they need.

Learning outcomes

Most students should be able to:
- use word and symbol equations to describe chemical reactions.

Some students should also be able to:
- construct balanced equations to describe chemical reactions. [HT only]

AQA Specification link-up: Science B 3.3.1.4
- Know that, when producing new products, chemical reactions can be represented using balanced chemical equations.

Lesson structure

Starters

Molecular bingo – Produce bingo cards with the numbers 1 to 12 on them. Show students slides of molecular formulae; the number of atoms present being the number they must mark off on their bingo card. Ar = 1, O_2 = 2, H_2O = 3 Na_2O_2 = 4, CH_4 = 5. *(5 minutes)*

LCM – Prepare students for the calculations they will use in this lesson by practising finding the lowest common multiple (LCM) of various combinations of numbers. To extend students, groups could work with groups of three or even four members. To support students, stick to finding the LCM of two numbers and arrange the task as multiple choice questions. *(10 minutes)*

Main

- Start by practising interpreting chemical formulae. Use examples such as $Ba(NO_3)_2$ to give students more practice working with subscripts and brackets. A table with the headings 'Name of chemical', 'Formula', 'Total number of atoms', 'Breakdown by element' may be useful to structure this task.
- Students should remember from the previous lesson that the Law of the Conservation of Mass states that all the mass of reactants must appear in the products. Remind them of this and give them sets of equations to decide whether they are balanced or not. For each equation that is not balanced, students should write the reason why (i.e. there are four oxygen atoms in the reactant molecules, but six in the products).
- Provide pairs of students with molecular modelling kits and show them the formula of methane. They must combine the model with oxygen to produce only water and carbon dioxide. Students will quickly see this is impossible in a 1:1 ratio. They can then experiment with different ratios to find a combination that will work, i.e.

$$CH_4 + 2O_2 \longrightarrow CO_2 + 2H_2O$$

- If molecular modelling kits are not available, card cut-outs representing atoms can be used instead, as can Lego™ blocks or similar items.
- If it was not used in the previous lesson, a pair of scales may be employed to demonstrate the balanced equation in a more concrete way.
- In small groups or pairs, students can create an 'equation poster' for a given equation. Using small items such as beads or other craft items, they must show the reaction taking place, clearly explaining the way in which the atoms rearrange and the correct balanced proportions of reactants and products.

Plenaries

Race – Divide the class into teams. Each team competes to balance equations you display on the board. Each team member can only answer one question. *(5 minutes)*

Explain – In pairs, students use their own words to write instructions for balancing equations. Pairs can then share their work and compare their descriptions. To support students, provide sentence fragments for them to use. *(10 minutes)*

Support
- Practise turning a description of a reaction into a word equation – focus on identifying products and reactants from texts.

Extend
- Balancing equations becomes more challenging the more chemicals are involved. Favouring reactions such as acid + carbonate for more able students will extend them further. [HT only]

Chapter 2 – Using materials our planet provides

Further teaching suggestions

Relative atomic and molecular (formula) mass

Although not on the specification, understanding relative atomic and molecular mass will help a great deal as it enables practical examples to be used and understood. For example, knowing the relative formula mass of magnesium oxide and the relative atomic mass of magnesium and relative molecular mass of oxygen would enable a student to use a balanced equation to predict the mass of magnesium oxide formed when the metal is burned.

Unit 1, Theme 1 – My wider world

2.9 Using equations

Learning objectives
- How do manufacturers decide the amount of raw materials they need?
- How can we balance chemical equations? [H]

Choosing the right amount of materials to react with each other is very important. If the wrong amounts are used, then some reactants may be wasted. Waste means lost money to manufacturers and could also cause damage to the environment.

Making fertiliser

If manufacturers understand a chemical reaction well they can choose the right quantities of reactants. The best way to start is with a word or symbol equation.

Ammonium nitrate is used as a fertiliser. It is an important chemical because it helps crops grow. It has the formula NH_4NO_3. The formula tells us it contains nitrogen, hydrogen and oxygen. The formula also tells us there are two nitrogen atoms, four hydrogen atoms and three oxygen atoms.

a How many different elements are present in ammonium nitrate?
b How many atoms are there in ammonium nitrate?

Ammonium nitrate is made by reacting ammonia (NH_3) with nitric acid (HNO_3). We can show this with the equation:

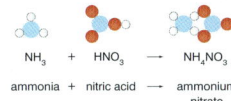

NH_3 + HNO_3 ⟶ NH_4NO_3
ammonia + nitric acid ⟶ ammonium nitrate

The symbol equation helps us see which atoms come from which reactant. We can also count how many atoms there are in the reactants and products. This helps us check whether the equation is '**balanced**'. For this reaction, the equation is already balanced. There are two N, four H and three O atoms on either side of the equation.

Making ammonia

Making ammonia is an example of a reaction that has an equation that needs balancing. To produce ammonia (NH_3), hydrogen (H_2) is reacted with nitrogen (N_2).

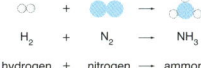

H_2 + N_2 ⟶ NH_3
hydrogen + nitrogen ⟶ ammonia

Can you see that the numbers of atoms that react do not balance the number of atoms that are in the product?

Balancing equations

Count the atoms in the reactants and the products of the equation for the formation of ammonia.

There are two atoms of hydrogen and two atoms of nitrogen producing three atoms of hydrogen and only one atom of nitrogen. This is impossible!

The solution is to increase the number of molecules of hydrogen being used and also increase the number of molecules of ammonia being produced.

There have to be two ammonia molecules made, because two nitrogen atoms are being used. Two ammonia molecules would contain six hydrogen atoms. That means three hydrogen molecules are needed. This is what the balanced equation looks like:

$3 H_2$ + N_2 ⟶ $2 NH_3$

Balancing symbol equations

Writing a number in front of a chemical formula tells chemists how many molecules take part in a reaction. This is crucial for making ammonia. It tells chemists they need three times as much hydrogen as nitrogen for the reaction to work properly. If there is not enough hydrogen, some of the nitrogen will be wasted. This would waste money, making ammonia more expensive.

The next example shows how ammonium sulfate (a fertiliser) is made by adding sulfuric acid to ammonia:

NH_3 + H_2SO_4 ⟶ $(NH_4)_2SO_4$

In the reactants: one nitrogen atom, five hydrogen atoms, one sulfur atom, four oxygen atoms.
In the products: two nitrogen atoms, eight hydrogen atoms, one sulfur atom, four oxygen atoms.
So the reactants need one more nitrogen atom and three more hydrogen atoms.
Therefore, there must be two ammonia molecules, not one.
So the balanced equation is:

$2 NH_3$ + H_2SO_4 ⟶ $(NH_4)_2SO_4$

Symbol	Atoms in reactants	Atoms in products
N	1	2
H	5	8
S	1	1
O	4	4

Summary questions

1 **a** What reactants are used to make ammonia?
  b Name a fertiliser can be made from ammonia.
2 Hydrogen gas (H_2) is produced by reacting methane (CH_4) with steam (H_2O). Carbon dioxide (CO_2) is also produced.
  a Write a word equation to show this.
  b Write a balanced symbol equation to show this. [H]
3 Balance the following equation, showing the reaction between hydrochloric acid and sodium carbonate:
  $HCl + Na_2CO_3 ⟶ H_2O + CO_2 + NaCl$ [H]

Key points

- Chemical equations are used to show how chemicals react and what is produced.
- Balanced chemical equations help chemists calculate how much raw material they need. [H]

Summary answers

1 **a** hydrogen and nitrogen
  b ammonium nitrate/ammonium sulfate
2 **a** methane + water ⟶ hydrogen + carbon dioxide;
  b $CH_4 + 2H_2O ⟶ 4H_2 + CO_2$
3 $2HCl + Na_2CO_3 ⟶ H_2O + CO_2 + 2NaCl$

Answers to in-text questions

a 3
b 9

Unit 1, Theme 1 – My wider world

2.10 The cost of a product

Learning objectives

Students should learn:
- that the cost of making new products is affected by the quantities of material used, the amount of waste and the amount of energy required
- that the cost of raw materials affects the consumer.

Learning outcomes

Most students should be able to:
- describe the costs affecting the manufacture of products
- relate quantities of raw materials to manufacturing costs
- use word equations to describe chemical reactions.

Some students should also be able to:
- evaluate the financial impact of different features of the manufacturing process.

AQA Specification link-up: Science B 3.3.1.4

- Explain why, in order to produce a product economically and safely it is important that the correct amount of material is used.

Within this context, candidates should be able to use scientific data and evidence to discuss, evaluate and suggest implications of the following:
- material costs when making products
- costs of energy consumption when making new products
- the 'value for money' of a range of products.

Lesson structure

Starters

What is it? – To prepare for the main activity, students could name/label pieces of laboratory equipment. To support students, provide names they can match to pictures. To extend students, invite them to write a sentence about what the equipment is used for. *(5 minutes)*

How much? – Find out the cost per unit to manufacture some well known items. Challenge students to estimate the prices (i.e. trainers around £8, iPad™ around £150). Why are they not sold at these prices? Where does the money go? *(10 minutes)*

Main

- Working in groups, students model a factory manufacturing process. Their task is to produce 40 cm^3 of carbon dioxide as cheaply (and safely) as possible, using sodium hydrogencarbonate and citric acid. Allow students either to collect their product over water or in a gas syringe. Each group is given a budget of £100 000. Groups purchase equipment and chemicals from this budget, keeping a record of how much money they spend. The following prices could be used:

Salary per group member	£10 000
Conical flask with delivery tube (their factory)	£15 000
Safety glasses (legal requirement for all employees)	£2000
0.5 g sodium hydrogencarbonate	£10 000
0.5 g citric acid	£14 000
Access to distilled water	£3000
Waste disposal (per trip to sink/bin)	£6000
Gas syringe	£8000
Measuring cylinder	£2000
Water trough (filled)	£1200

- Give group members different roles (safety inspector, accountant, equipment manager, chief chemist, etc.) so they can track different aspects of the group's performance. Allow between 40 minutes to 1 hour for this activity. Afterwards, add up all the costs and divide them into raw materials, equipment and services. Groups can make and compare bar charts of their performance. Which costs were easier to keep down? What extra information would have helped with decisions? Use this to lead onto the importance of knowing how much raw material you need from the start.
- Examine the balanced equation for the reaction between sodium hydrogencarbonate and citric acid. Students will see that more carbonate is needed than acid (an understanding of moles is not required). Would this knowledge have affected their purchases? Explain how.

Plenaries

Adding up – Work through the maths skills section in the Student Book, drawing parallels with the main activity. *(5 minutes)*

Keeping costs down – Students write tips for future groups on keeping costs down for the carbon dioxide activity. To support students, they could write tips for the method, to help the process go more smoothly. To extend students, the absolute minimum price could be calculated. *(10 minutes)*

Support

- Prepare a card sort activity listing factors that might make products more expensive and factors that would reduce the cost, for example 'factory is close to where raw materials come from', 'worker wages are high', 'most processes are run by machines', 'raw materials get wasted', etc.

Extend

- Students may have noticed that the reaction between sodium hydrogen carbonate and citric acid is endothermic. Using ammonia as an example, they could research how conditions are manipulated to produce the most products possible (a detailed understanding of equilibrium reactions is not needed). This is also an opportunity to practise balancing equations.

Chapter 2 – Using materials our planet provides

Further teaching suggestions

Temperature

As an alternative to the open-ended activity given here, groups of students could simply investigate the relationship between temperature and CO_2 production, given a standard way to collect the gas. A water bath could be used, with higher temperatures costing more money to use. Students will see that in this case, temperature does not increase the yield, but it does increase the rate of reaction, enabling them to make money faster.

Unit 1, Theme 1 – My wider world

2.10 The cost of a product

Learning objectives
- What factors affect the cost of producing chemicals?
- How does the cost of raw materials affect us all?
- How can we calculate the cost of a product, and the cost of any waste?

When making new products, manufacturers need to consider the amount of money being spent on raw materials. As we have seen, any waste will cost the manufacturer money. Other factors also affect the cost of making new products. They include paying workers, using energy and building the facilities to make the product safely.

Value for money

Reducing waste and the costs of raw materials affect consumers. For instance, if ammonium nitrate can be made cheaply, it can be sold for less. This means it is cheaper for farmers to fertilise their land. In turn, that means more crops can be grown for less money, keeping the price of food down. In this way, industrial chemistry affects us all as consumers.

As well as affecting the price of fuel, the cost of separating hydrocarbons from crude oil will affect a huge number of products. Everything from plastics to makeup uses hydrocarbons as raw materials. This means that the cost of fractional distillation of crude oil will be passed on to the consumer on a wide range of products.

Factors affecting the cost of a product

There are a few factors to think about when producing materials: the cost of the reactants, the price the products can be sold for, the cost of any waste, and 'overhead' costs. Overhead costs include the energy needed in production, maintaining a safe plant, any taxes, the wages of workers and the cost of transporting the final product.

So, the financial cost of making a product = cost of reactants + cost of waste + overhead costs.

To make a profit, these costs must be less than the price the final product is sold for:

Profit = price the product is sold for − total costs

Figure 1 The cost of producing fertiliser affects the cost of our food in the supermarket

Maths skills

In the following example, a factory is making chemical X from chemicals A and B:
- X can be sold for £1500 per tonne.
- A costs £800 per tonne; B costs £200 per tonne.
- Overheads cost £5000 per day.
- The factory makes 20 tonnes of X every day.
- To make X, the factory uses 15 tonnes of A and 5 tonnes of B every day.

The total daily cost of A would be 15 × £800 = £12000

The total daily cost of B would be 5 × £200 = £1000

With no waste, the total daily cost would be:

£12000 A + £1000 B + £5000 overheads = £18000

20 tonnes of chemical X is worth 20 × £1500 = £30000

So in this example, the factory makes a profit of £30000 − £18000 = £12000 every day.

Activity

Calculating the cost

A chemical plant makes 80 tonnes of ammonium nitrate every day. The ammonium nitrate can be sold for £3000 per tonne.

The equation for the reaction is:

ammonia + nitric acid → ammonium nitrate
NH_3 + HNO_3 → NH_4NO_3

The costs of the product

a If the factory used 17 tonnes of ammonia every day, how much nitric acid would it use every day?

b If ammonia costs £4000 per tonne, how much will the ammonia cost each day?

c If nitric acid costs £1200 per tonne, how much will the nitric acid cost each day?

Running costs

The energy needed, worker wages, maintenance and other costs of running the chemical plant are £16400 per day.

d What is the total cost of running the chemical plant each day?

e How much money would be wasted in a week if 1 tonne too much ammonia was used every day?

Is it making money?

f How much money can be made in one day from selling the ammonium nitrate?

g Is the factory making a profit? If so, how much per day?

Figure 2 This UK chemical plant makes chemicals for use in new products. In order to make the chemicals as cheaply and as safely as possible, chemists must work out the correct amounts of materials to use.

Summary questions

1 Copy and complete using these words:
cost transportation cheaper waste more overheads

The cost of making a product depends on the .... of the reactants, ..., and the cost of any .... materials. The less waste, the .... the product is to make. Overheads include wages for workers, the cost of energy and other costs such as taxes or .... To make a profit, the product must be worth .... than the total cost of making it.

2 List the costs of running a chemical plant.

3 Name **two** products that would be cheaper if the cost of fractional distillation could be reduced.

4 How can the cost of ammonium sulfate affect the price of a bag of crisps?

Key points
- Using the right amount of raw materials prevents wastage and saves money.
- The cost of raw materials affects a wide range of consumer products.
- The cost of industrially made products depends on the costs of the reactants, overheads and waste.

Summary answers

1 cost, overheads, waste, cheaper, transportation, more

2 cost of materials, energy bills, maintenance, taxes, wages, transport, also factors such as machinery or R&D

3 plastics, make-up (any other petrochemical products)

4 If the price of ammonium sulfate increases, the price of a bag of crisps may increase too. That's because it will cost the farmer more to fertilise potatoes which will be passed along the buying chain to the consumer.

Activity answers

a 80 tonnes − 17 tonnes = 63 tonnes of nitric acid

b 17 tonnes × £4000 = £68000

c 63 tonnes × £1200 = £75600

d £68000 + £75600 + £16400 = £160000

e £4000 × 7 days = £28000 (or £20000 if a five day week is assumed)

f 80 tonnes × £3000 = £240000

g Yes, the profit is: £240000 (total cost) − £160000 (ammonium nitrate) = £80000 profit

Unit 1, Theme 1 – My wider world

Summary answers

1

Atom	A charged particle
Compound	Made of more than one type of atom not chemically combined
Element	The building block of all chemicals
Ion	Made of one type of atom
Mixture	Made of two or more atoms stuck together
Molecule	Made of two or more types of atom chemically combined

(Atom → Made of one type of atom; Compound → Made of two or more types of atom chemically combined; Element → The building block of all chemicals; Ion → A charged particle; Mixture → Made of more than one type of atom not chemically combined; Molecule → Made of two or more atoms stuck together)

2 a 6
 b 6
 c 6

3 a An ion has a charge.
 b positive and negative
 c Cu – atom Ca^{2+} – ion H_2 – molecule
    Fe – atom N_2 – molecule Cl^- – ion
    Na^+ – ion H_2O – molecule

4 a Straight from the ground: gold, marble, limestone, sulfur
    Need separation: petrol, salt
 b Because salt is soluble and rock is not.

5 a oxygen
 b carbon dioxide and carbon monoxide
 c because it is more reactive.

6 a ammonia, made from nitrogen and hydrogen, used to make fertiliser
    light bulbs, filled with argon because it is unreactive
 b nitrogen

7 a Local employment, but only for the duration of mining. National economy when materials are sold. Consumers using and living in the products of these raw materials
    Positive impacts: see above
    Negative impacts: environmental damage, possible eventual unemployment
 b pollution from increased transport, toxic salts washing into soil, noise and dust from quarry

8 a calcium carbonate ⟶ calcium oxide + carbon dioxide
 b 52g
 c [particles demonstrating 1 Ca, 1 C and 3 O atoms being rearranged appropriately]

9 a cheaper products due to lower worker wages
 b cheaper products
 c more expensive products due to wastage

Kerboodle resources

- Bump up your grade: What am I? (2.1)
- Interactive activity: Atoms (2.2)
- Practical: Extraction of salt (2.4)
- On your marks: Crude oil
- Examination-style questions
- Answers to examination-style questions
- Test yourself: My wider world

Summary questions

1 Match each word to its definition.
  Atom — A charged particle
  Compound — Made of more than one type of atom not chemically combined
  Element — The building block of all chemicals
  Ion — Made of one type of atom
  Mixture — Made of two or more atoms stuck together
  Molecule — Made of two or more types of atom chemically combined

2 The mass number of carbon is 12. Its atomic number is 6.
 a How many protons does a carbon atom have?
 b How many neutrons does a carbon atom have?
 c How many electrons does a carbon atom have?

3 Particles can be either molecules, ions or atoms.
 a What is the difference between an atom and an ion?
 b What are the two types of ion?
 c Decide whether the following particles are atoms, ions or molecules.
    Cu Ca^{2+} H_2 Fe N_2 Cl^- Na^+ H_2O

4 We can use some materials from the Earth straight from the ground. Others must be separated from other materials first.
 a Put the following materials into two groups, straight from the ground or needing separation:
    salt gold limestone petrol sulfur marble
 b Crude oil can be separated. This is because the different materials in it have different boiling points. Why can rock salt be separated into rock and salt?

5 Metals are obtained from rocks called ores.
 a What element often has to be removed from ores in order to extract the pure metal?
 b What waste gases are produced when iron and lead are produced?
 c Why is carbon used as a reducing agent for lead and iron oxides?

6 a Describe some important products made using gases extracted from the atmosphere.
 b Which gases can be extracted from the atmosphere and used to make other materials?

7 The materials we extract from the Earth have impacts on the environment, people and the economy.
 a Who would be affected by the opening of a new limestone quarry? What are the positive and negative impacts?
 b How would the environment be affected by the opening of a new quarry?

8 When calcium carbonate ($CaCO_3$) is heated, it decomposes into calcium oxide (CaO) and carbon dioxide (CO_2).
 a Write a word equation to show this happening.
 b If 88g of calcium carbonate releases 36g of carbon dioxide when it decomposes, how much calcium oxide is left?
 c Use a diagram to explain how mass is conserved during this reaction.

9 Explain the effect the following factors would have on the cost of making a product:
 a more automatic machinery
 b reducing energy costs
 c using too much of one reactant

AQA Practical suggestions

Practicals	AQA	k	📖
Compare less reactive metals with more reactive metals, e.g. in acid.	✓		
Heat metal oxides with carbon to compare reactivity, e.g. CuO, PbO, Fe_2O_3.	✓		✓
Heat copper carbonate with charcoal to produce copper.		✓	
Displacement reactions, e.g. $CuSO_4$(aq) + Fe.		✓	✓
Ignition tube demonstration of blast furnace – potassium permanganate, mineral wool plug, iron oxide mixed with carbon.		✓	
Demonstrate fractional distillation of crude oil using CLEAPPS mixture (take care to avoid confusion with the continuous process in a fractionating column).		✓	✓
Grow brassica plants in compost with added copper sulfate or spray brassica plants (e.g. cabbage leaves) with copper sulfate solution, ash the plants (fume cupboard), add sulfuric acid to the ash, filter and obtain the metal from the solution by displacement or electrolysis.		✓	

Chapter 2 – Using materials our planet provides

AQA Examination-style questions

The diagrams show the atoms in various substances.

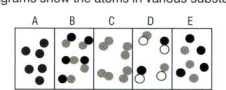

a Explain which diagram represents a mixture of different compounds. (3)
b Explain which diagram represents molecules of one element. (3)

Crude oil is a mixture of hydrocarbons that need to be separated before use.
a Describe the process used to separate crude oil. (4)
b Name **two** liquids separated from crude oil that can be used in a car. (2)

Use chemicals from the box to answer the questions.

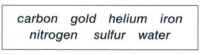

a Name and explain which chemical is a compound. (3)
b Name an element that can be used straight from the ground and explain why it does not need to be processed first. (3)
c Give the chemical symbol for the element that needs to be extracted from its ore. (1)

The table shows the price of some metals that are used to make brass.

Metal	Price in £ per kg
Copper	0.96
Iron	0.14
Tin	2.28
Zinc	0.25

a The brass used for making metal hoses is called low brass. It is 80% copper and 20% zinc. Calculate the cost of making 10kg of low brass.
£1.21 £7.68 £8.18 £12.24 (1)
b Aich's brass is used for boats as it is corrosion resistant. It is made from 60.66% copper, 36.58% zinc, 1.02% tin and the rest is iron. Calculate how much iron is in Aich's brass. (1)

Scientists have studied the structure of the atom so they can understand how reactions occur.
a Draw and label a lithium atom: $^{7}_{3}Li$ (4)
b Describe how a lithium atom is turned into a lithium ion: Li^+. (1)

6 A manufacturing company was given many small samples of lead oxide.
They did an experiment to find how much carbon to use to extract the lead from each sample.
Their results are shown in the graph below.

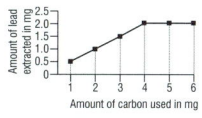

a How much lead was in the samples? (1)
b Suggest **two** things that the company could have done to make sure they collected accurate results. (2)
c Suggest and explain the amount of carbon the company should use. (2)
d *In this question you will be assessed on using good English, organising information clearly and using specialist terms where appropriate.*
Describe the reaction that the company could have used to separate lead from its ore. (6)

7 Air is a mixture of gases that can be separated. First carbon dioxide, water and dust are removed. The air is then compressed, cooled and allowed to expand.
This is repeated until it turns into a mixture of liquids all with different boiling points.

Gas	Boiling point in °C
Argon	−185
Nitrogen	−195
Oxygen	−182

Use information from the table to explain how pure oxygen can be separated from liquid air. (3) [H]

5 a nucleus labelled in the centre
3 protons labelled inside the nucleus
4 neutrons labelled inside the nucleus
3 electrons labelled orbiting outside of the nucleus (4 marks)

b Remove one electron. (1 mark)

6 a 2g (1 mark, amount and unit needed)

b Any **two** from: repeated, kept the temperature the same, kept the size of the pieces the same, (allow any sensible factor kept the same) (2 marks)

c 4g
Because this amount got the most out and any more carbon added did not increase the amount of lead extracted. (2 marks)

d Marks awarded for this answer will be determined by the Quality of Written Communication (QWC) as well as the standard of the scientific response.

There is a full and detailed description of the reaction that the company could have used to separate lead from its ore. The answer is well structured with minimal repetition or irrelevant points. There is an accurate, fluent and clear expression of ideas. It contains only minor errors in the use of technical terms, spelling, punctuation and grammar. (5–6 marks)

There is some description of the reaction that the company could have used to separate lead from its ore with some omissions. The answer shows some attempt at structuring and the ideas are expressed with reasonable fluency and clarity. There are some errors in the use of technical terms, spelling, punctuation and grammar. (3–4 marks)

There is a brief description of the reaction that the company could have used to separate lead from its ore. The answer is largely incomplete and contains some valid points which are not clearly structured. It lacks fluency and/or clarity. It contains errors in the use of technical terms, spelling, punctuation and grammar. (1–2 marks)

No relevant content (0 marks)

Examples of points made in the response:
- Lead is less reactive than carbon
- Carbon can be used as a reducing agent
- Reduction reaction
- $2PbO + C \longrightarrow 2Pb + CO_2$
- $PbO + CO \longrightarrow Pb + CO_2$
- Heat the PbO and C/CO

7 The mixture of liquids enters a fractionating column (1 mark) and allowed to warm up to almost −182°C. (1 mark) Oxygen has a higher boiling point than the other gases and can be removed from the column as a liquid. (1 mark)

AQA Examination-style answers

1 a D
A compound contains two or more different types of atom chemically combined. A mixture contains more than one type of molecule or atom, not chemically combined. (3 marks)

b C
An element has only one type of atom. Molecules are more than one atom chemically combined. (3 marks)

2 a Fractional distillation (1 mark)
Each fraction has a different boiling point (1 mark)
Different fractions condense at different points in the distillation column (1 mark)
The smaller / lighter fractions rise to the top (accept the converse) (1 mark)

b petrol (1 mark)
diesel (1 mark)

3 a Water
It is made up of two different types of atom
OR
It is made up of two different elements (1 mark)
These are hydrogen and oxygen (1 mark)

b gold / sulfur (1 mark)
They are unreactive (1 mark)
So do not form compounds. (1 mark)

c Fe (1 mark)

4 a (0.96 x 80) ÷ 100 = £7.68; (0.25 x 20) ÷ 100 = 50p;
0.5 + 7.68 = £8.18 (1 mark)

b 60.66 + 36.5 + 1.02 = 98.26; 100 − 98.26 = 1.74% (1 mark)

Unit 1, Theme 2

Life on our planet

AQA Specification link-up: Science B 3.3.2

In this theme there are three contexts:

3.3.2.1 Life on our planet

3.3.2.2 Biomass and energy flow through the biosphere

3.3.2.3 The importance of carbon

Life on our planet

There are a wide variety of flora and fauna currently present on the Earth. Students should be made aware that the species we see today have evolved over many years. Species with the most favourable characteristics for the prevailing conditions have survived, whereas others have become extinct.

The distribution and abundance of organisms is dependent on the environmental conditions present. Even in extreme environments such as the Arctic, or Death Valley in California, some species of organism can be found. In preparation for an activity on this topic, students could be asked to produce a 'conditions on the ground' report, in the style of a weather forecast, for the environment of their choice.

How can plants and animals survive in extreme environments?

Activity:

- What makes this environment extreme? Produce a poster illustrating an extreme environment. It must explain why the environment is extreme, and name some examples of organisms that live there. Higher level students should explain the adaptations of these organisms, explaining how they can survive.
- Produce a 3D model of an animal that can live in an extreme environment. The animal can be completely fictitious, but must be fully adapted to its specific environment.

Do we look like our ancestors?

Activity:

- Produce a timeline of the evolution of a species.
- Find out how humans have evolved.

Possible misconceptions

When studying adaptations, some students think that individual organisms have 'developed' characteristics to enable them to survive in an environment. These characteristics can then be passed on to their offspring, hence the species 'evolves'. For example, some students believe that a species needs long legs (e.g. to escape from a predator) and so grows them. Because this offers an advantage to the species, the species therefore retains this feature.

Unit 1, Theme 2

Life on our planet

In Unit 1, Theme 2 you will work in the following contexts, covered in Chapters 3 and 4:

Life on our planet

How can plants and animals survive in extreme environments?

Plants and animals have specific characteristics that allow them to live in a particular environment. These are known as adaptations. For example, to survive in the extreme cold of the Arctic most mammals are covered in layers of fat and blubber, and have very thick fur. Plants and animals that live in desert environments have different characteristics. Here, animals are adapted to transfer energy in order to cool down – for example, by not having much fur.

Do we look like our ancestors?

The further back in time you go, the more different we look from our ancestors. This is true for all species. The wide variety of life present on the Earth has evolved over time and is still changing. Organisms are continually evolving to become better adapted to their environment. This process is known as natural selection.

Biomass and energy flow through the biosphere

How do organisms gain energy?

Plants obtain their energy from the Sun in the form of light. This is transferred into chemical energy during photosynthesis. Animals can only gain energy by eating other organisms. Some of the energy is used to power body reactions, such as movement. Some is used for growth, which increases an organism's biomass.

How do scientists monitor the flow of energy and biomass through the biosphere?

Ecologists study the movement of energy and biomass within food chains. To achieve this, data is collected on the numbers and sizes of organisms within a feeding relationship. Generally, as you move along a food chain the size of organism increases. However, fewer and fewer organisms exist at each trophic level.

The importance of carbon

Why is carbon so important?

Carbon is an essential element in all organisms. In fact, it is the major element within your body. Carbon forms the basis of all organic molecules, such as carbohydrates, fats and proteins. These are the essential building blocks of life.

Carbon is constantly cycled through the environment. For example, it is removed from the atmosphere when plants photosynthesise, and returned when organisms respire.

How are humans affecting the carbon cycle?

Carbon is stored underground, in fossil fuels. It is also present in the Earth's atmosphere, in the form of carbon dioxide which is also dissolved in the oceans.

Over the past 40 years, large areas of forest have been cleared to provide extra space for farming. This means that there are now fewer trees to remove carbon dioxide from the atmosphere. In addition, many countries burn fossil fuels to generate electricity and power transport. These processes release extra carbon dioxide into the atmosphere. Both of these processes are disrupting the natural carbon cycle.

Biomass and energy flow through the biosphere

Ecologists study how organisms interact with other organisms and with their environment. Through their knowledge of species and through observations, they provide an insight into how human activity affects ecosystems. For example, if a new factory was to be built in a rural area, ecologists might consider:

- the effect on individual species in the area
- the effect on the ecosystem as a whole
- the steps that could be taken to reduce the impact of such a development.

Ecologists also carry out research and provide advice about sustainability and how to preserve the habitats of endangered species.

How do organisms gain energy?

Activity:

- Determine how much energy is in a particular food source by burning the food and calculating the temperature rise of a fixed quantity of water.
- Find out what an owl eats by dissecting an owl pellet and studying the remains of the animals in it.

Life on our planet

Bacteria and fungi play an essential role in decomposition. Decomposers are nature's recyclers and form an essential part of the world's ecosystems. The nutrients they release are used by plants for growth. Without decomposers, dead remains would not be broken down. In fact, without them the world would still be covered with the dead remains of dinosaurs!

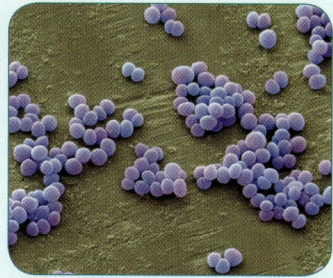

In order to survive, species need to be able to cope with changes to the environment. Through mutations, individuals in a species can develop characteristics that provide survival advantages. For example, some bacteria have developed a resistance to antibiotic drugs. This is an advantage for the bacteria, as they cannot be killed using traditional treatments. These bacteria are therefore more likely to survive, and pass on this advantage to the next generation. Over recent years, increased numbers of resistant bacteria have been detected. This is an example of natural selection going on today.

Agricultural scientists study ways to maximise plant growth. For example, they suggest the ideal conditions in which a crop should be grown. This results in higher crop yields and cropping over longer periods of the year, helping keep supplies high and costs low. Auxins are plant hormones responsible for controlling plant growth. For example, they make plants grow towards the light (phototropism). Farmers use products containing auxins to control weed growth, and to help ripen fruit.

When drawing food chains or webs, many students have difficulty in putting the arrows in the correct direction to demonstrate energy or biomass flow. Often, students add arrows pointing backwards, for example showing that grass eats rabbits! Provide students with lots of practice of drawing their own food chains/webs to ensure that they orientate arrows correctly.

Some students believe that energy is lost when it is transferred between trophic levels. Energy is never 'lost'; however, it is often transferred as energy heating the surroundings, which is no longer able to be passed on to the next organism in the chain.

The importance of carbon

The concepts of global warming and the greenhouse effect are often covered in the media. Provide students with newspaper clippings, internet news articles, or watch short news bulletins to look at evidence for this phenomenon. Students should note the increasing levels of carbon dioxide in the atmosphere. This would form an ideal introduction to the carbon cycle.

Why is carbon so important?
Activity:
- Using chemical formulae, show how respiration and photosynthesis are opposite processes.
- Research how other chemicals are cycled through the environment, such as nitrogen or water.

How are humans affecting the carbon cycle?
Activity:
- Write a balanced magazine article, explaining the pros and cons of deforestation.
- Write a school environmental policy, describing ways the school could save energy. Ask students to explain how this would help to reduce carbon dioxide emissions.

Possible misconceptions

Some students believe that the presence of any carbon dioxide in the Earth's atmosphere is bad. It is important to explain to students that the presence of greenhouse gases (such as carbon dioxide) is essential for life on Earth. Without greenhouse gases, the Earth would be too cold to support life.

How do scientists monitor the flow of energy and biomass through the biosphere?

Activity:
- Produce a 3D model of a food chain using a variety of media, e.g. Lego™, building blocks, plasticine.
- Play the 'web of life', where students represent the animals and plants in a particular food web. Students sit in a circle with the person representing the Sun in the middle. Pass string between students to represent energy transfer between organisms in the food web. Many situations can be modelled in this manner. For example, what would happen if a particular species contracted a disease that killed them all? To represent death, students could drop their piece of string. Anyone who is touching this string would therefore be directly affected.

Possible misconceptions

Students struggle with the correct use of the terms mass and weight. When calculating biomass, students will often refer to 'weighing' an object. The correct terminology, which should be reinforced regularly, is mass – and hence the use of the term biomass. This misconception has mainly arisen from the everyday use of the word weight.

Unit 1, Theme 2 – Life on our planet

3.1 Classification

Learning objectives

Students should learn:
- that organisms are classified into taxonomic groups based on their physical characteristics
- that classification is important for naming and identifying organisms.

Learning outcomes

Most students should be able to:
- describe how organisms are classified
- state the advantages of classifying organisms into groups.

Some students should also be able to:
- explain why classification is used to name and identify organisms.

AQA Specification link-up: Science B 3.3.2.1

- Understand that there is a huge variety of life, which is categorised into kingdoms.
- Understand that animals and plants can be classified according to their physical characteristics.
- Explain why classification is important as an international method of grouping living organisms with similar characteristics to aid naming and identification.

Within this context, candidates should be able to use scientific data and evidence to discuss, evaluate or suggest implications of the following:
- the advantages of classifying the range of species that exist on the planet and the methods used.

Lesson structure

Starters

What do people classify? – Ask students to make a list of common objects that are classified. Students can then use their results to share ideas on why we choose to classify objects. *(5 minutes)*

Learning to classify – Split the class into small groups. Provide each group with a small box full of stationery (pens, pencils, felt tips, coloured paper, etc.). Ask students to sort the contents out in whichever manner they feel is appropriate. To support students, give each group a system to follow – for example, sort by colour, material, use, shape, size. To extend students, groups should discuss the quality of their solution compared to other groups, and whether or not there is an 'ideal' classification system. *(10 minutes)*

Main

- Before introducing students to the classification of living organisms, ask them to think about situations where they use classification. Examples include books in a library, food in a supermarket, or the various ways students are grouped within their own school setting. Discuss why classification is such a useful tool.
- Ask students to sort a range of clothes into groups, depending on their function. For example, a group could include jumpers and coats. Students could then take one of these groups and classify it further to demonstrate the hierarchical nature of taxonomy. For example jumpers could then be sorted according to their colour, and then sorted again into the material they are made from.
- Introduce the broadest categories of living things as kingdoms, limiting these to animals, plants and microbes (fungi and single-celled organisms). Students will not need to know the five kingdoms of animals, plants, fungi, protoctists and prokaryotes.
- Provide students with images of a range of vertebrates. These should include some species that are not immediately obvious – for example, a student could mistakenly place a bat with birds. They then need to use the table in the Student Book to classify them.

Plenaries

Hierachy of taxonomical categories – Provide pairs of students with sticky notes containing the taxonomical categories, kingdom, phylum, class, order, family, genus, species. They need to organise the categories into the correct hierarchy. *(5 minutes)*

A guide to classification – Students should produce a guide for year 8 students into how to classify organisms. It should define any scientific words using age-appropriate language. To support students, provide a writing frame with question prompts. To extend students, ask students to include a key in their guide, to help younger students classify an unknown organism. *(10 minutes)*

Support

- Provide more opportunities to practise classifying objects. For example, a selection of images on cards, a selection of material samples, pieces of Lego™.

Extend

- Ask students to choose a species of animal and explain the reasons why it is classified into its genus, family, order, class, phylum and kingdom. They could then explain their findings to another individual in the class as a peer teaching exercise.

Chapter 3 – Life on our planet

Further teaching suggestions

Names
Ask students to carry out some research into the naming of species – binomial nomenclature. Each organism's Latin name consists of two parts – the first indicates the genus name, the second indicates the species within that genus.

Classification of invertebrates
Provide students with the classification system for some invertebrates. Supply a range of specimens or images for students to classify within this system.

Unit 1, Theme 2 – Life on our planet

3.1 Classification

Learning objectives
- What is classification?
- Why are organisms classified?
- How can we classify organisms

Did you know ... ?
The diagram below shows how humans are classified according to the Linnaean classification system. This is how we get our name *Homo sapiens*. A species name is specific. It refers to a single organism rather than a group.

Kingdom – *Animalia* — Broadest category
Phylum – *Chordata*
Class – *Mammalia*
Order – *Primates*
Family – *Hominoidea*
Genus – *Homo*
Species – *sapiens* — Most specific category – contains only one type of organism

Classification means sorting things into groups based on similar features. Lots of everyday objects are classified. For example, in a supermarket, semi-skimmed milk is classified as a dairy product. If you wanted to find a similar product such as cream, you know it would be located nearby.

a What does 'classify' mean?

Living organisms are classified into **taxonomic groups**. All species within a taxonomic group share similar characteristics. This system of classification was introduced by Carl Linnaeus in the eighteenth century. It is now used by scientists around the world.

There are seven main groups. They are arranged in order, from **kingdom** – the broadest category (organisms share some characteristics), to **species** (organisms' characteristics are almost identical).

Within each kingdom, organisms are further subdivided into smaller and smaller groups.

Ways to classify living things
The main groups of living things include:
- Plants – organisms that make their own food by photosynthesis. For example, flowering plants.
- Animals – organisms that cannot make their own food. For example, insects.
- Microbes – which would include fungi and single-celled organisms.

b What is the main characteristic of the plant kingdom?

Organisms in the animal kingdom can be divided into two groups depending on whether or not they have a backbone:
1 Vertebrates – animals with a backbone.
2 Invertebrates – animals without a backbone.

c What is the difference between a vertebrate and an invertebrate?

The table shows the characteristics of the five vertebrate groups.

Characteristics	Vertebrates				
	Mammals	Birds	Fish	Reptiles	Amphibians
Blood	Warm blooded	Warm blooded	Cold blooded	Cold blooded	Cold blooded
Reproduction	Live young	Lay hard-shelled eggs	Lay eggs in water	Lay soft-shelled eggs	Lay jelly-coated eggs in water
Skin covering	Have hair	Have feathers	Have wet scales	Have dry scales	Have moist skin

Why classify organisms?
Scientists classify organisms for a number of reasons:
1 To name and identify species – it makes it easier to find out which species an organism belongs to if everything is organised.
2 To predict characteristics – if several members in a group have a particular characteristic, another species in the group may have the characteristic.
3 To find evolutionary links – species in the same group probably share characteristics because they have evolved from a common ancestor.

Activity

The Natural History Museum
The Natural History Museum in London, houses the **largest** and most **important natural history collection** in the world. This diverse collection has been gathered over the last 400 years. It contains over **70 million** specimens ranging from microscopic specimens to dinosaur skeletons. The specimens are organised into 'collections' – groups of items that have something in common. These are constantly being reorganised and developed. They try to take into account the latest scientific thinking on the classification and relationships between organisms.

The collections are used by scientists to investigate the natural world. They also provide a point of reference and authority for wider investigations by scientists around the world. By studying the specimens, scientists gain knowledge of animals and plants. They can also discover the processes that have shaped the world and our solar system.

- Find out about the classification work carried out at the Natural History Museum.
- Write a leaflet for fellow students explaining the work scientists carry out there.

Summary questions

1 Copy and complete using these words:
groups organisms animals characteristics plants classification taxonomic
.... means sorting things into ....
.... are sorted into .... groups based on similar physical ....
The main groups or kingdoms include .... and ....
2 How could you tell the difference between an amphibian and a reptile?
3 Why is it important that scientists around the world use the same classification system?

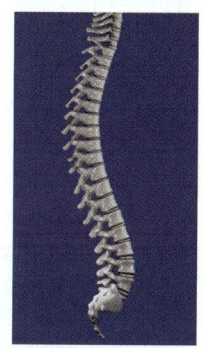

Figure 1 The name vertebrate comes from the vertebrae present in the spinal column

Did you know ... ?
It has been estimated that approximately two-thirds of all living species are insects. Only around 1 per cent of all animal species are larger than a bumble bee!

links
Variation exists between members of the same species but their main characteristics are the same. For more information on variation within species see 6.2 Variation.

Key points
- Classification sorts things that share similar features into groups.
- Organisms are classified into groups to aid naming and identification.
- Organisms are classified into groups based on their physical characteristics.

Summary answers

1 classification, groups, organisms, taxonomic, characteristics, plants (animals), animals (plants).
2 Amphibians have moist skin and lay jelly-like eggs in water, whereas reptiles have dry scales and lay soft shelled eggs.
3 To help scientists worldwide to identify and name organisms.

Answers to in-text questions
a To sort objects into groups.
b organisms that make their own food by photosynthesis
c Vertebrates have a backbone; invertebrates do not have a backbone.

Activity

The Natural History Museum
Organise a trip to the Natural History museum, to observe the collections of materials at first hand. Arrange for a talk to students on how the specimens are classified and displayed. A visit to a local museum or collection near the school could provide a similar learning experience.

Unit 1, Theme 2 – Life on our planet

3.2 Using evolutionary and ecological relationships

Learning objectives

Students should learn:
- that evolutionary trees represent the evolutionary relationships between species, and illustrate how they evolved from a common ancestor
- that predator and prey population size have an oscillating relationship of dependency

Learning outcomes

Most students should be able to:
- describe what is meant by a predator and its prey
- describe simply the relationship between predator and prey populations
- interpret evolutionary trees

Some students should also be able to:
- explain in detail the data shown in evolutionary trees and predator–prey graphs

AQA Specification link-up: Science B 3.3.2.1
- Understand that there is a huge variety of life which is categorised into kingdoms.

Within this context, candidates should be able to use scientific data and evidence to discuss, evaluate or suggest implications of the following:
- the similarities and differences between species to gain an understanding of evolutionary and ecological relationships.

Lesson structure

Starters

Why is classification important? – Ask students to state three reasons why classification is important (revision from previous lesson). *(5 minutes)*

What is the difference between a predator and its prey? – Ask students to define a predator and a prey. They should then state some of the features a predator and a prey organism would normally possess. To support students, provide students with a photo of a tiger and a rabbit. Students should be asked how these animals are adapted to being a predator or a prey organism. To extend students, ask students to illustrate their answers with examples of organisms that possess a particular feature. *(10 minutes)*

Main

- Explain to students that evolutionary relationships exist between organisms. Choose an example, such as the evolution of the horse (also referred to in Spread 3.5 Evolution), to describe how different species of organism have slowly changed over time to develop characteristics that are more suited to their environment. These relationships can be shown on an evolutionary tree.
- Look at a range of evolutionary trees such as human, or elephants (from the Student Book). Explain the key features of the chart – the timeline goes up the side; if a line to an organism stops before the present day the organism has become extinct; the organism that the branches radiate from is the common ancestor. Students should practise reading information from these charts.
- Define what is meant by a predator organism and a prey organism. Discuss some of the key features which make them specialised for their role.
- Explain to students that interactions between species are referred to as ecological relationships. Introduce students to predator–prey graphs, which demonstrate the oscillating relationship between the size of a predator and a prey population. The Student Book illustrates the relationship between the lynx and the snowshoe hare.

Plenaries

True or false –
1. Identification keys show evolutionary relationships. [F]
2. Tigers, lions and foxes are examples of predator organisms. [T]
3. Prey organisms normally have large claws and teeth. [F]
4. You can discover the ancestors of a species by looking at an evolutionary tree. [T]
5. The interactions between predators and prey are an example of an ecological relationship. [T]
(5 minutes)

Evolutionary tree – Provide students with an evolutionary tree, along with a series of questions. These could include students being asked to identify the common ancestor, which species are alive today, the names of species that are now extinct. To support students, ask them to colour-code an evolutionary tree. One colour could be used for the common ancestor, one for those species alive today, and one for those that are extinct. To extend students, ask them to evaluate what an evolutionary tree tells them about a specific species. *(10 minutes)*

Support

Provide students with a number of predator–prey graphs to analyse. Stress the importance in looking at the shape of the graphs to describe the relationship between the species – a peak in population of prey organisms is shortly followed by a peak in population of predator organisms.

Extend

Provide students with information to create their own evolutionary tree. For example, the evolutionary tree of the horse.

46

Further teaching suggestions

Biological control
Students could carry out research into biological control – the commercial exploitation of predator–prey relationships. An example is the relationship between ladybirds and aphids when growing tomatoes. Students could provide information on this technique in the form of an article, which is to be placed in a gardening magazine.

Tree of Life
Provide students with an image of the 'Tree of Life' – an evolutionary tree of all species. Discuss with students the implication of how life began on Earth.

Science in context
By studying the characteristics of organisms, scientists know how species are related to each other. The same classification system is used all over the world so that scientists can share their research. This means that links between different organisms can be seen, even if they live on different continents.

Unit 1, Theme 2 – Life on our planet

3.2 Using evolutionary and ecological relationships

Learning objectives
- What is an evolutionary tree?
- How do predator and prey populations depend on each other?

By studying the characteristics of organisms, scientists know how species are related to each other. The same classification system is used all over the world so that scientists can share their research. This means that links between different organisms can be seen, even if they live on different continents.

Evolutionary relationships

Fossil records have enabled scientists to produce **evolutionary trees**. These are branched diagrams that show how different species have evolved from a common ancestor. Evolutionary trees are produced by looking at similarities and differences in a species' physical characteristics and genetic makeup.

a What does an evolutionary tree show?

The elephant's evolutionary tree shows that all species of elephant have evolved from a palaeomastodon. This organism was alive at the time of the dinosaurs. Over time, the appearance of an elephant has changed. That's because it has adapted to the environment it now lives in. Today, only two species of elephant survive – the Indian elephant and the African elephant.

links
For more information about evolution and fossils go to 3.5 Evolution.

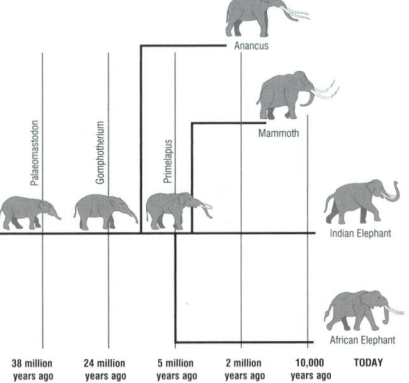

Figure 1 The elephant's evolutionary tree

AQA Examiner's tip
To answer questions about evolutionary trees, you need to look both at the drawings of the organism, and the timeline. Often, the organisms that are alive today are drawn at the top of the diagram. The common ancestor from which all the species develop is found at the bottom. This will be where all the branches meet. If the species is not present in the highest level of the diagram then it is now extinct.

b Around what time did Anancus become extinct?

Figure 2 The hare is the lynx's preferred food source

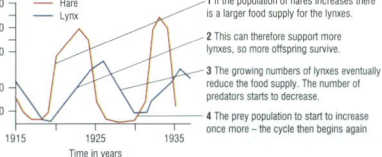

1 If the population of hares increases there is a larger food supply for the lynxes.
2 This can therefore support more lynxes, so more offspring survive.
3 The growing numbers of lynxes eventually reduce the food supply. The number of predators starts to decrease.
4 The prey population to start to increase once more – the cycle then begins again

Figure 3 Changes in the populations of lynx and snowshoe hares, 1915-35

Ecological relationships

Scientists also study how species depend on each other. These interactions are known as ecological relationships. One example is the relationship between a predator and its prey.

Predator species need to be adapted to hunt and kill, to ensure they catch enough food to survive. **Prey** species must be adapted to escape their predators. These features ensure enough organisms survive for the species to continue.

c What is the difference between a predator and its prey?

Summary questions

1 Copy and complete using these words:
  ancestor evolutionary evolved
  Scientists use _____ trees to show how organisms have _____ from a common _____ .

2 Using the elephant's evolutionary tree on the previous page, answer the following questions:
  a What is the elephant's common ancestor?
  b Which two species of elephant are alive today?
  c Which species of elephant was alive around 24 million years ago?
  d What was the last species of elephant to become extinct?

3 Describe the relationship between aphid (prey) and ladybird (predator) populations.

Key points
- An evolutionary tree shows the evolutionary relationship between species. It shows they have evolved from a common ancestor.
- Predator populations depend upon the population of their prey. If the number of prey organisms decreases, so does the number of predator organisms.

Summary answers

1 evolutionary, evolved, ancestor

2 **a** *Palaeomastodon*
  b African and Indian elephants
  c *Gomphotherium*
  d mammoth

3 If the aphid population increases the ladybird population will increase. Eventually there will be many ladybirds eating the aphids, so the aphid population will decrease. The ladybirds will now not have enough food, so the ladybird population will fall. The cycle will then be repeated.

Answers to in-text questions

a the evolutionary relationship between species
b around two million years ago
c Predator eats prey; prey is the food source.

Unit 1, Theme 2 – Life on our planet

3.3 Competition

Learning objectives

Students should learn:
- that organisms live in habitats
- that animals need food, water, mates and a suitable territory
- that plants need sunlight, water and nutrients to survive.

Learning outcomes

Most students should be able to:
- describe an organism's habitat
- state what plants and animals need to survive
- explain why organisms compete for resources in an ecosystem.

Some students should also be able to:
- use scientific data to explain the complex reactions between organisms in an environment.

AQA Specification link-up: Science B 3.3.2.1

- Know that, to survive, organisms require a supply of materials from their surroundings and from other living organisms:
  (a) plants need sunlight, water and nutrients to survive
  (b) animals need food, mates, shelter and a suitable territory.

Within this context, candidates should be able to use scientific data and evidence to discuss, evaluate or suggest implications of the following:
- the factors for which organisms are competing in a given environment.

Lesson structure

Starters

What do animals need to survive? – Ask students to produce a list of factors an animal needs to survive. *(5 minutes)*

What is in a habitat? – Ask the class to choose a generalised habitat, such as an ocean. Each student is then asked to name five organisms that live within the habitat. To support students, they could be supplied with a series of images of organisms which do or do not live in a given habitat. Students are asked to select which organisms live within the habitat, and which ones do not. To extend students, ask them to name five smaller habitats within the main habitat. For example, an ocean may include rock pools, sandy shore, deep ocean, sea floor and lagoon as sub-habitats. *(10 minutes)*

Main

- Throughout this section of teaching, it is worthwhile spending time defining and revisiting a range of vocabulary. Students often find the following terms difficult to use correctly – community, ecosystem, habitat, environment, population, organism and species.
- Students should discuss what it would be like living in one of the habitats discussed in the starter. They should be challenged to think about how conditions could change – for example, with the time of day, or season of the year.
- The basic conditions for the survival of plants and animals should then be discussed. Students could think about how an organism may satisfy these needs from the chosen habitat.
- Working in small groups, students should carry out the line transect activity (see 'Practical support').
- Each group could be asked to report back to the class. Which organisms were present at their section of the line transect? What were the environmental conditions at this point? How do the type of organisms present link to the conditions?
- Conclusions should be made as a class about how environmental conditions affect the variety and number of species which exist in an environment.

Plenaries

True or false –
1. Nocturnal animals are active during the day. [F]
2. Rock pools are an example of a habitat. [T]
3. Animals need sunlight to survive. [F]
4. Oxygen is required by both plants and animals to survive. [T]
5. Competition means fighting for a resource. [T]

(5 minutes)

Ecology definitions – Ask students to write definitions for the key ecological terms – species, organism, population, ecosystem, habitat, environment and community. To support students, provide them with a card sort of terms and their definitions. To extend students, ask them to illustrate their definitions with examples from a forest habitat. *(10 minutes)*

Support

- Pre-prepare the line transect in an area that displays significant environmental variation, for example, a region of hedge, un-cut grass and trampled grass.

Extend

- Ask students to carry out a belt transect of the chosen region, to study a wider range of organisms within the area. Identification charts could be used to assist students in working independently.

Answers to in-text questions

a ocean, desert, pond, forest
b To avoid cold and dark conditions.
c They are materials they need for their survival.
d microorganism – algae, plant – lily/reeds, animal – pond skater/minnow/perch/heron

Chapter 3 – Life on our planet

Further teaching suggestions

Fieldwork
The line transect activity could be extended to form a short piece of fieldwork – for example, by carrying out the line transect at a beach or rocky shoreline. Students could be asked to work in small groups; with each group performing a transect. Groups could then compare and contrast their results back in the classroom.

Presenting data
The line transect activity also allows the opportunity to discuss how data should be presented to an audience. Students could discuss what the most appropriate graph or chart would be to present their data. Time could be spent preparing a short presentation, perhaps using PowerPoint, to display their findings to the rest of the group.

Food web display
The diagram of a pond habitat could be used to revise the concept of a food web. A visual approach, suspending photographs of organisms at different trophic levels from a coat-hanger, can form an excellent class display.

Unit 1, Theme 2 – Life on our planet

3.3 Competition

Learning objectives
- What is a habitat?
- What do animals need to survive?
- What do plants need to survive?

A **habitat** is the place where an organism lives. There are many different habitats including ocean, desert, forest, pond or even a garden.

a Name **three** examples of a habitat.

Each habitat has different **environmental conditions**. These include temperature and amount of rainfall. The environmental conditions in most habitats vary throughout the day and throughout the year:

- **Daily changes** – these include changes in light levels and temperature. Night time is darker, and generally colder, than day time. To cope with this change most animals sleep during the night. However, nocturnal animals such as foxes and owls use this as an advantage and hunt during the night.
- **Seasonal changes** – these include changes in temperature and rainfall. For example, in the winter it is colder and the days are shorter. There is often more rainfall. Some animals, like hedgehogs, cope with this change by hibernating.

b Why do most animals sleep during the night?

An **ecosystem** is the name given to a habitat and all the living organisms that live there. To survive, the plants and animals need a number of different materials from their surroundings. If materials are limited, organisms have to **compete** for these resources. For example, plants compete for access to light. In many cases, only the strongest species will survive.

c Why do plants and animals compete for resources?

Figure 1 Limpets have to cope with regular changes in sea level. When they are underwater, limpets move over rocks eating small pieces of seaweed. When the tide goes out, they cling tightly to the rock. This stops them being washed away or being eaten by predators.

What do plants need to survive?
1 Sunlight ⎫
2 Water ⎬ – Needed for **photosynthesis**, to produce food for growth
3 Nutrients – Needed for healthy growth

What do animals need to survive?
1 Food ⎫ – To grow
2 Water ⎭
3 Mates – To reproduce
4 Suitable territory – For safety and shelter

A pond ecosystem

Figure 2 The animals and plants found in a pond ecosystem

Many kinds of plant and animal life can be found in a pond. The different **populations** of species live together in a **community**. The variety and number of organisms in this community are determined by the amount of oxygen, light and shelter that is available. Some plants and animals can live in the pond, others live on or near the pond.

d Name a microorganism, plant and animal species that can live in a pond habitat.

The community also depends on the way plants and animals live together in the pond. Organisms may compete for resources such as light. If the water lily covers part of the surface of the pond, other plants living below it will die. That's because not enough light can reach the plants below.

Other organisms in a community may provide a food source. For example, the heron may feed on the minnow population. This means that the perch may no longer be able survive in this habitat as its food source has been used up.

Ecosystems are in a constant state of balance. Small changes can have dramatic effects.

Practical
Surveying organisms using a line transect

A line transect is a sampling technique that compares the conditions in a habitat with the species present at that point. A number of samples are taken along a pre-marked path. The range of data collected allows scientists to link environmental factors with the type and number of a species.

Carry out a line transect of a habitat – for example, you could look at an area of your school grounds. Stretch a measuring tape for a distance of 10 metres. At every metre along the tape:

1. List the plants and animals that are present at each metre along the line.
2. Note the environmental conditions at each sampling point. For example, the light level, or how trampled the ground is.
3. How do the environmental conditions affect the distribution of plants and animals along the line?

Summary questions

1 Match the following terms to their definition:
- ecosystem — place where an organism lives
- population — the conditions that surround an organism
- environment — the habitat and living organisms within it
- habitat — the different species living in a habitat
- community — numbers of the same species living in a habitat

2 **a** What do plants need to survive?
 b What do animals need to survive?

3 Most fish that live in a river cannot survive in an ocean habitat. What is the main difference between these two habitats?

4 **a** What are the differences in habitat between the surface and the bottom of a pond?
 b How might these differences affect the types of organisms that live there?

Key points
- A habitat is a place where a living organism lives.
- Plants need sunlight, water and nutrients to survive.
- Animals need food, water, mates and a suitable territory to survive.

Practical support

Surveying organisms using a line transect

Equipment and materials required: Tape measure, quadrats or pooters.

Details
Carry out a line transect, marked out using a tape measure. Samples should be taken at regular intervals using, for example, quadrats or pooters. A suitable area should be chosen – an area including hedgerow, shade, uncut and mown or trampled grass would be ideal. Class results should then be pooled, and conclusions formed about how the conditions of a habitat affect the range and number of species present.

Safety: Follow local guidelines on outdoor activities, wash hands after experiment.

Summary answers

1.
habitat	place where an organism lives
environment	the conditions that surround an organism
ecosystem	the habitat and living organisms within it
community	the different species living in a habitat
population	numbers of the same species living in a habitat

2 **a** Plants require sunlight, water and nutrients.
 b Animals need food, water, mates and a suitable territory

3 ocean – salt water
 river – fresh water

4 **a** For example, more light received at the surface than at the bottom of the pond.
 b For example, the suggestion of larger leaves at the bottom of a pond as light levels decrease.

Unit 1, Theme 2 – Life on our planet

3.4 Adaptations

Learning objectives

Students should learn:
- that organisms have developed special adaptations to enable them to survive in a specific habitat
- that plants and animals that live in the desert are adapted to survive hot, dry conditions
- that plants and animals that live in the arctic are adapted to survive cold, windy conditions.

Learning outcomes

Most students should be able to:
- define adaptation
- describe the key features of a habitat
- identify how organisms are adapted to their habitat, explaining some of their adaptations.

Some students should also be able to:
- explain in detail how an organism's adaptations ensure their survival.

Support
- Provide students with a picture of a made up organism which is very badly adapted to an environment. For example a green, fat, camel who walks on its tiny feet! Students should find it fairly easy to identify what is wrong with this organism. This will then help them to identify the characteristics that make a camel adapted to live successfully in a desert

Extend
- Provide images of organisms that have adaptations for specific environments. Students analyse the images, and suggest the type of environment within which the organism lives.

AQA Specification link-up: Science B 3.3.2.1

- Explain how animals, plants and microbes may be adapted for survival in the conditions where they normally live: (a) Plants adapt to conditions through changes in surface area, water storage tissues and extensive root systems (b) In the case of animals factors should include surface area, insulation, body fat and water storage (c) Microbes (extremophiles) have been found living in the arctic, volcanic vents, very dry environments and severe chemical environments.

Within this context, candidates should be able to use scientific data and evidence to discuss, evaluate or suggest implications of the following:
- the reasons for the distribution of animals or plants in a particular habitat
- how organisms have adapted to the conditions in which they live.

Lesson structure

Starters

What does adaptation mean? – Students write their own definition of the term adaptation. *(5 minutes)*

How are camels and polar bears adapted? – Students need to draw/annotate a picture of a polar bear and a camel, stating how they are adapted to the environment in which they live. To support students, provide them with a list of the adaptations of polar bears and camels. They need to cut and paste them onto the correct animal. To extend students, ask them to explain why each adaptation increases the animal's chance of survival. *(10 minutes)*

Main

- Define with the class what is meant by an adaptation. Show students a photo of a cactus. As a class, identify its adaptations and explain how these enable it to survive. It is worth reminding students at this point that adaptations are not 'chosen' by an organism, but arise through the inheritance of mutated genetic material (this is covered in more detail in Chapter 6).
- Show students images of a range of arctic plants. In groups identify their similarities. For example – small plants, grow close to the ground, small leaves and hairs. Discuss as a class why these features are important.
- Show students a photo of a snowshoe hare in the winter and the summer. Discuss with students why these changes are an advantage to the hares.
- Introduce students to the concept that the size and mass of an organism defines its surface area to volume ratio. This is a very important adaptation for organisms in very cold or very hot environments.
- Students complete the 'Practical' activity to find out how surface area to volume ratio affects the rate of cooling. Students should conclude from this practical that animals living in a very hot environment must have a large surface area to volume ratio to maximise heat loss. Animals living in a very cold environment must have a small surface area to volume ratio, to minimise heat loss, thus keeping the animal warm.
- Ask students to design their own organism, which is specially adapted to survive in an environment of their choice. The organism can be as unusual as they wish, as long as its features display characteristics that would enable it to be perfectly adapted to survive in their chosen habitat.

Plenaries

Sharing adaptations – Ask students to think of an organism that has an adaptation which has not been discussed during the lesson. Students share their ideas with the class, stating one key adaptation, and how it enables the organism to be successful in its environment. *(5 minutes)*

Maintaining body temperature – Provide students with two different sized 'organisms', each represented by a cube of known dimensions. Students have to calculate the surface area to volume ratio of each organism, to work out which would lose heat the most rapidly. To support students, provide them with a sheet that has spaces for completing the relevant figures. Parts of the calculation should be pre-completed. To extend students, ask them to suggest a shape that would have the optimum surface area to volume ratio for an organism that is trying to conserve heat energy. *(10 minutes)*

Chapter 3 – Life on our planet

Practical support

Measuring surface area to volume ratios

Equipment and materials required: Thermometer, boiling water, cold water, measuring cylinder, different sized beakers, stopwatch.

Details

Place 50 cm^3 of water at $70\,°\text{C}$ into beakers with different surface areas. Take the temperature of the water every 30 seconds for 5 minutes.

Is there a relationship between the surface area of the flask and the rate of heat loss? Students should be encouraged to use data logging equipment to monitor heat loss as the water cools (a discussion could be held at this point on the advantages of using data logging over traditional analogue apparatus). This would provide precise continuous data, which should be represented graphically. Students will then gain a clear visual representation of the relationship between surface area and heat loss.

Safety: Take care when using boiling water.

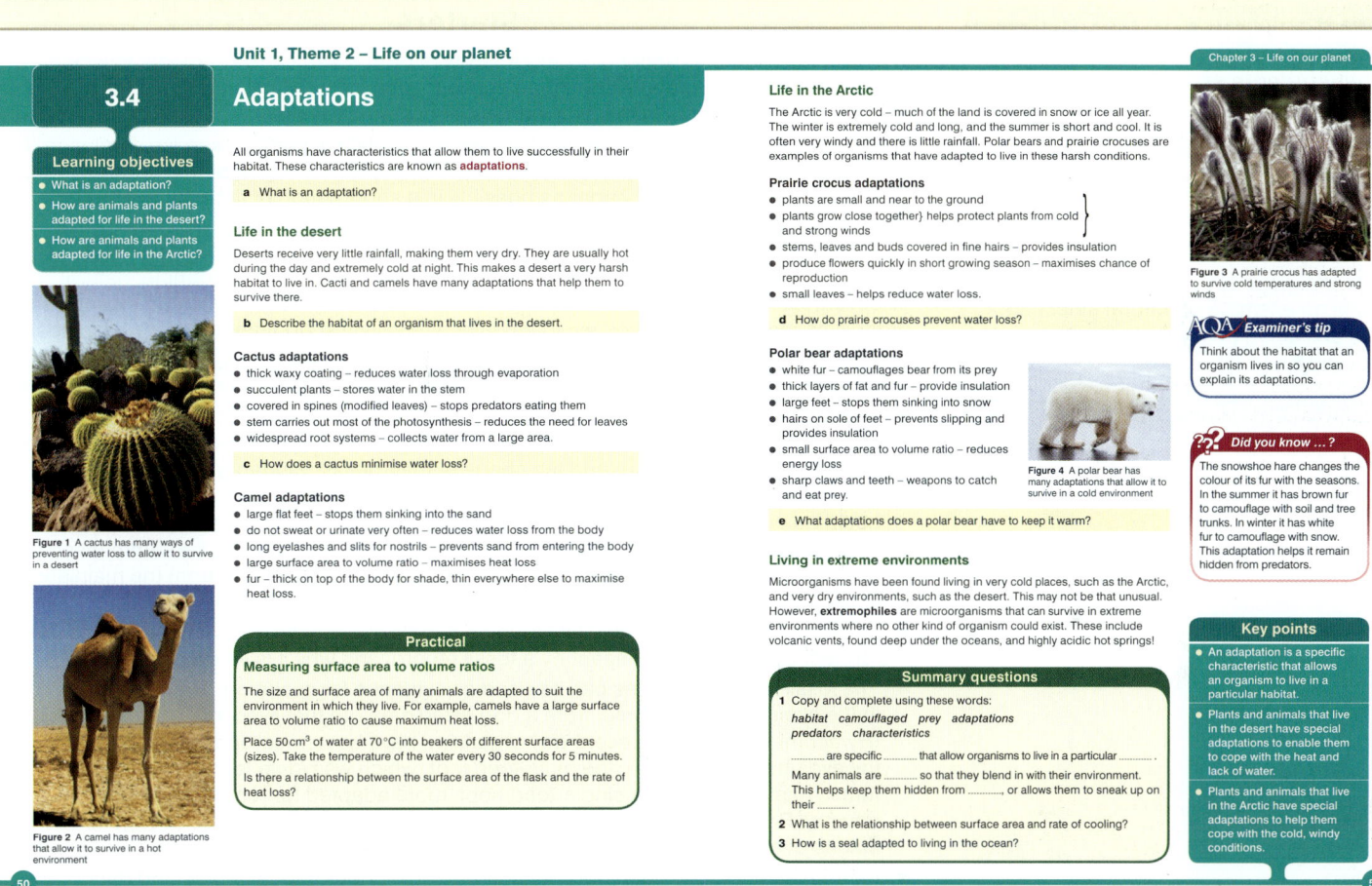

Further teaching suggestions

Extremophiles

Ask students to carry out some research into a group of extremophiles. For example, thermophiles and halophiles. Students could present their findings as a poster.

Cactus

The 'Practical' activity could be extended to show the advantage of a cactus having no leaves. Connect a digital balance to a data logger. Suspend a wet paper towel on the balance and record mass loss as the water evaporates. Repeat the experiment with the same type of wet paper towel but this time folded up so the volume is the same but the surface area is less. It is possible to produce a range of surface area to volume ratios.

Answers to in-text questions

a a specific characteristic, which enables an organism to survive in its habitat

b dry, very hot during the day, and cold during the night

c covered in a waxy layer, spines, few leaves

d small leaves

e thick layers of fat and fur, hairs on soles of feet, small surface area to volume ratio.

Summary answers

1 adaptations, characteristics, habitat, camouflaged, predators, prey

2 The larger the surface area, the greater the heat loss.

3 For example – blubber and hair for insulation, flippers for swimming, streamlined shape.

Unit 1, Theme 2 – Life on our planet

3.5 Evolution

Learning objectives

Students should learn:
- that organisms have evolved from a common ancestor millions of years ago
- that organisms evolve by the process of natural selection
- that fossils provide evidence for evolution.

Learning outcomes

Most students should be able to:
- define evolution
- explain the process of evolution and of natural selection
- describe what a fossil is

Some students should also be able to:
- use an example of natural selection to explain in detail evidence for evolution.

AQA Specification link-up: Science B 3.3.2.1
- Explain how evolution occurs via natural selection.
- Explain how individuals with characteristics most suited to the environment are more likely to survive and breed successfully.
- Know that the genes that have enabled these individuals to survive are then passed on to the next generation.

Lesson structure

Starters

Why will it become extinct? – Supply images of organisms in environments to which they are not suited. Students should be asked to discuss what will happen to the population of organisms in this environment and why. *(5 minutes)*

What is evolution? – Small groups of students complete a 'graffiti' exercise. Evolution is written in large letters in the centre of a piece of paper. Each group then adds as many words or terms to the paper as possible in a short period of time (for example 90 seconds). Each group then passes its paper onto the next group, which then adds to the previous graffiti. Misconceptions and relevant points can then be discussed. To support students, provide them with key words to start them off, for example; natural selection, fossil, genes and Darwin. To extend students, ask them to include examples of organisms that have displayed evolution over a relatively short time period. *(10 minutes)*

Main

- Introduce the theory of evolution by studying Darwin's research on finches. Darwin noticed that the finches on the different Galapagos islands showed wide variation. For example, they differ in size and the shape of their beaks. Students could be asked to predict why they think this is the case, and then carry out research (relevant articles could be provided) to find out the reason why. For example finches have different shaped beaks depending on their food source. The activity should be summarised by discussing Darwin's conclusions that as the islands are so distant from the mainland, finches that had arrived there in the past had changed over time.
- Introduce fossils – if possible let students handle specimens. Explain how fossils are formed. Discuss what is meant by a fossil record, and why it is incomplete.
- The fossil record of horses can be used as evidence for evolution. The main stages in the evolution of the horse, from a dog-sized creature that lived in a rainforest (*Hyracotherium*) to the present-day organism, have been recorded in fossils. During the evolution of the horse, the multi-toed foot, which was ideal for walking in the forest, has evolved into a single-toed hoof, more adapted for running over open country. Provide students with a series of photos/diagrams of horses at various stages of evolution for them to sequence into a timeline.
- Introduce the process of natural selection. Show photographs of pale and dark peppered moths on pale trees and those darkened by pollution. (The images in the Student Book could be used.) Ask students which one has an advantage in each situation. If the country became more and more polluted which species of moth would evolve? How would a reduction in the level of pollution affect the species?
- Ask students to discuss what they think a 'super bug' is and why these are such a problem in hospitals. Explain to students how a resistant bacterial strain may develop as the result of a mutation. Using the information in the paragraph on antibiotic resistance in the Student Book, students could be asked to produce a cartoon strip depicting the stages in the evolution of these new species. A brief sentence of explanation could accompany each picture.
- Discuss what factors could cause extinction. These should include both human activity and natural processes. Ensure that students are aware that the term 'extinct' means that no organisms of that species exist anywhere on the planet, not just in one place.

Plenaries

Definitions – Students produce simple definitions of the terms evolution, natural selection, and extinction. *(5 minutes)*

What is natural selection? – Students produce a flow diagram to explain how evolution occurs through natural selection. To support students, provide them with statements to rearrange into the correct order. To extend students, ask them to illustrate their flow chart with an animal example. *(10 minutes)*

Support
- Provide students with a range of photos of different animals in different habitats. Ask students to describe what characteristics would make the organism most likely to survive and reproduce. This will reinforce the concept of survival of the fittest.

Extend
- Ask students to research how bacteria can develop antibiotic resistance. This is an example of evolution occurring in a relatively short period of time.

Further teaching suggestions

Extinction
Ask students to research an organism that has become extinct. Students should produce a presentation explaining what it was, where it lived and why it became extinct. The dodo would be an excellent example to suggest to students.

Making fossils
Students can make their own 'fossil' using plaster of Paris. Fill a small (petri dish) container with plaster of Paris, then use a leaf to make an imprint.

Did you know …?
You share about 95% of your genes with a gorilla and 50% with a banana! This is because all living things evolved from the same ancestor millions of years ago.

Unit 1, Theme 2 – Life on our planet

3.5 Evolution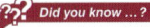

Learning objectives
- What is evolution?
- What is natural selection?
- What is a fossil?

Scientists believe that all living organisms have gradually developed, from a common ancestor over millions of years. This process is called **evolution**.

Darwin's theory of evolution
Charles Darwin was a famous English scientist. In 1859 he published the theory of evolution in his book on the origin of species. He based it on his observations of variation in plants and animals on a voyage around the world. Darwin's ideas caused controversy as they conflicted with religious views on creation.

Charles Darwin's theory of evolution states that all species have evolved from simple life forms. These simple organisms lived in water more than three billion years ago. They were similar to bacteria found today. These organisms evolved to become more complex. Eventually, organisms developed that could live on land and in the air.

Evidence for evolution – fossils
Fossils were formed when animal and plant remains were preserved in mud. Over millions of years the mud turned into rocks. The fossil record provides most of the **evidence** for evolution. Different fossils show that organisms have gradually changed over time to become more adapted to their environment. The best adapted organisms survived to reproduce. This process is known as **natural selection**. However, the fossil record is not complete as not all organisms fossilise well. Many fossils have also been destroyed by the Earth's movements, or lie undiscovered. This means that some evidence for evolution has not yet been discovered.

Figure 1 Charles Darwin

a What is a fossil?

Natural selection
Organisms evolve through the process of natural selection. Natural selection takes years (sometimes millions of years) to occur, as it has to take place over a number of generations. It follows the steps below:

Organisms in a species show a wide range of variation
(caused by genetic differences).
↓
The organisms with the characteristics that are most suited
to the environment are most likely to survive and reproduce.
This is often referred to as 'survival of the fittest'.
↓
Genes from successful organisms are passed
to the offspring in the next generation.

Did you know …?
You share about 95% of your genes with a gorilla and 50% with a banana! This is because all living things evolved from the same ancestor millions of years ago.

AQA Examiner's tip
Make sure you know the meanings of the terms 'natural selection' and 'evolution'.

links
Farmers exploit the process of natural selection when they selectively breed their plants or animals. You can find out more in 13.1 Selective breeding.

This process is then repeated many times. Living organisms are continually evolving to become better adapted to their environment. This means many species are still evolving by natural selection.

b What does 'survival of the fittest' mean?

Antibiotic resistance
Antibiotic-resistant bacteria cause many problems in hospitals. MRSA is an example species.

Bacteria reproduce very rapidly and so can evolve in a relatively short time. When bacteria divide, their DNA can be damaged or altered. This usually results in the bacteria dying. However, the **mutation** (altered DNA) is sometimes an advantage for the bacteria. It can increase their chance of survival. For example, the mutation may cause resistance to an antibiotic. The antibiotic will no longer be able to destroy the bacteria.

c What term is used to describe DNA that has been damaged or altered?

Peppered moths
Originally, most peppered moths in Britain were pale coloured. This made them camouflaged against tree trunks. A mutant form of these moths is dark coloured. These were easily seen by birds and eaten. The pale moths were therefore more likely to survive and reproduce.

In industrialised areas soot started to coat trees, turning the bark black. This meant that the black moths were now better camouflaged, and so more of them survived. After several years, dark peppered moths became more common in towns and cities than pale moths.

Extinction
The fossil record shows that many species have become **extinct** since life began. If a species is poorly adapted to its environment it will not survive and will eventually become extinct. Extinction occurs naturally but is often made more likely through human activity. Factors that can cause a species to become extinct include:
- changes to environmental conditions
- new diseases
- new predators and competitors
- destruction of habitat
- human use of organisms for food or materials.

Summary questions
1. Copy and complete using these words:
   selection fossils evolved adapted evidence
   Organisms have .... from a common ancestor. .... for this is gained from .... These show that organisms have gradually become better .... to their environment, through the process of natural ....
2. Why might a species become extinct?
3. Which type of peppered moth would you be most likely to find near a steel factory? Explain your answer.

AQA Examiner's tip
To answer a question on natural selection successfully, ask yourself the following question. Which features help this organism to survive, and therefore be able to reproduce?

Figure 2 Methicillin-resistant *Staphylococcus aureus* (MRSA) – a hospital superbug

links
For more information about bacteria, the use of antibiotics and antibiotic resistant strains see 10.2 Antibiotics

Key points
- Evolution is the theory that all organisms have gradually evolved from a common ancestor over millions of years.
- Natural selection occurs because only the organisms with the characteristics most suited to their environment survive and reproduce. They pass on these characteristics to their offspring through genes.
- A fossil is formed when a plant or animal's remains are preserved in rock. The fossil record provides evidence for evolution.

Summary answers
1. evolved, evidence, fossils, adapted, selection
2. new diseases, new predators, new competitors or environmental changes
3. Black peppered moth, because the steel factory will produce air pollution, which will darken the trees. Black moths will therefore be more camouflaged.

Answers to in-text questions
a animal or plant remains, preserved in rock
b Organisms that have the characteristics best suited to their environment are more likely to survive and reproduce.
c mutated

Unit 1, Theme 2 – Life on our planet

3.6 Plant growth

Learning objectives

Students should learn:
- that environmental factors affect plant growth
- that plants use auxins to respond to environmental stimuli
- that plant shoots grow towards the light (phototropism) and roots grow towards gravity (gravitropism).

Learning outcomes

Most students should be able to:
- name some environmental factors which affect plant growth
- explain how plants use auxins to respond to environmental stimuli
- describe the difference between phototropism and gravitropism.

Some students should also be able to:
- explain in detail how auxins enable plant shoots to bend towards the light.

Specification link-up: Science B 3.3.2.1

- Explain the effect of the external features, light (phototropism), temperature, day length and gravity (gravitropism) on plant growth.
- Explain the role of auxins in controlling plant growth.

Lesson structure

Starters

What does a plant need to grow? – Ask students to produce a list of factors that plants require to grow effectively. *(5 minutes)*

What environmental factors affect plant growth? – In small groups, ask students to produce a spider diagram showing which environmental factors affect plant growth, and the effect that each of these factors has on the plant. To support students, provide images of different environmental conditions. Students select which conditions they think would promote plant growth and which would reduce growth, or even kill the plant. To extend students, ask them to select from their diagrams what they feel would be the optimum conditions for plant growth. *(10 minutes)*

Main

- Discuss the main environmental factors that affect plant growth – focus on light intensity, temperature and the presence of gravity. Students could set up a series of growth experiments to study the effects of different temperatures (room temperature, incubator set at various temperatures) and light intensity (cupboard, light from all directions, in a box with a window). Basil or cress seeds would be ideal for this purpose. Students should make justified predictions about what they think will happen. They should revisit their experiments at regular intervals – you may wish to have set up some experiments in advance of this lesson.
- Introduce the concept of tropisms. By looking at a plant (ideally grown in a glass beaker so the roots can be seen), explain what is meant by the terms phototropism and gravitropism. Explain how auxins control these responses and why these processes are beneficial to plant growth.
- Set up the experiment described in the practical activity. Students will need to monitor results from this experiment over a period of several days. Alternatively, if set up well in advance of the lesson, students could be asked to take a range of observations of the cress seeds in the two agar plates, and explain the differences in their findings.
- Show students some examples of fruit or vegetables that cannot be grown outdoors in the UK in winter. Peppers and tomatoes are excellent examples. Ask students to discuss the reasons why many crops cannot be grown in the winter, and how they might overcome these reasons if they were a commercial grower. It would be useful to share images of real examples when discussing their solutions – students might be surprised to see the scale of a commercial tomato farm, for example!
- Controlling environmental factors and the use of auxin-containing products could be discussed.

Plenaries

How are auxins used commercially? – Students need to list at least three ways auxins can be used commercially. *(5 minutes)*

How do auxins allow plants to respond to light? – Provide students with a drawing of two plants. One which is exposed to light on all sides and one which is only exposed to light on one side for two weeks. How will their appearance differ? Why is this the case? To support students, provide them with a series of diagrams, showing internal cell sizes. Students arrange these diagrams into 'exposed to light on all sides' and 'exposed to light on one side' categories. To extend students, ask them to draw a simple diagram of the internal structure of the plant to explain why the two plants have grown differently. *(10 minutes)*

Support

- Provide students with a series of diagrams showing how plants respond to the light. They need to sequence them into the correct order to describe how auxins enable plants to respond to a stimulus.

Extend

- Ask students to draw a diagram with a couple of sentences of explanation, to explain how roots grow towards the centre of the Earth.

Further teaching suggestions

Environmental controls
Organise a trip to a commercial greenhouse, to study how they carefully control the environment to maximise rates of photosynthesis and the growing season of their plants.

Use of plant hormones
Ask students to research one use of plant hormones. For example, a weed killer or some rooting powder. Students could produce a leaflet or advert to promote sales of the product – it must include an explanation of how the product works.

Practical

Investigating gravitropism
Equipment and materials required: Two agar plates, cotton wool, 20 cress seeds, a clinostat.

Details
Students should take two agar plates, fill the bottom with cotton wool and dampen it sufficiently. The cotton wool should be secure enough to enable the plate to be turned upside down. Students then add 10 seeds and allow time for germination. Once the seeds have germinated, one of the agar plates should be attached to the clinostat, and one left as a control. After a few days the small cress plants should be growing horizontally, whereas the plants on the control plate should be growing vertically.

Safety: Do not touch mains electrical equipment with wet hands.

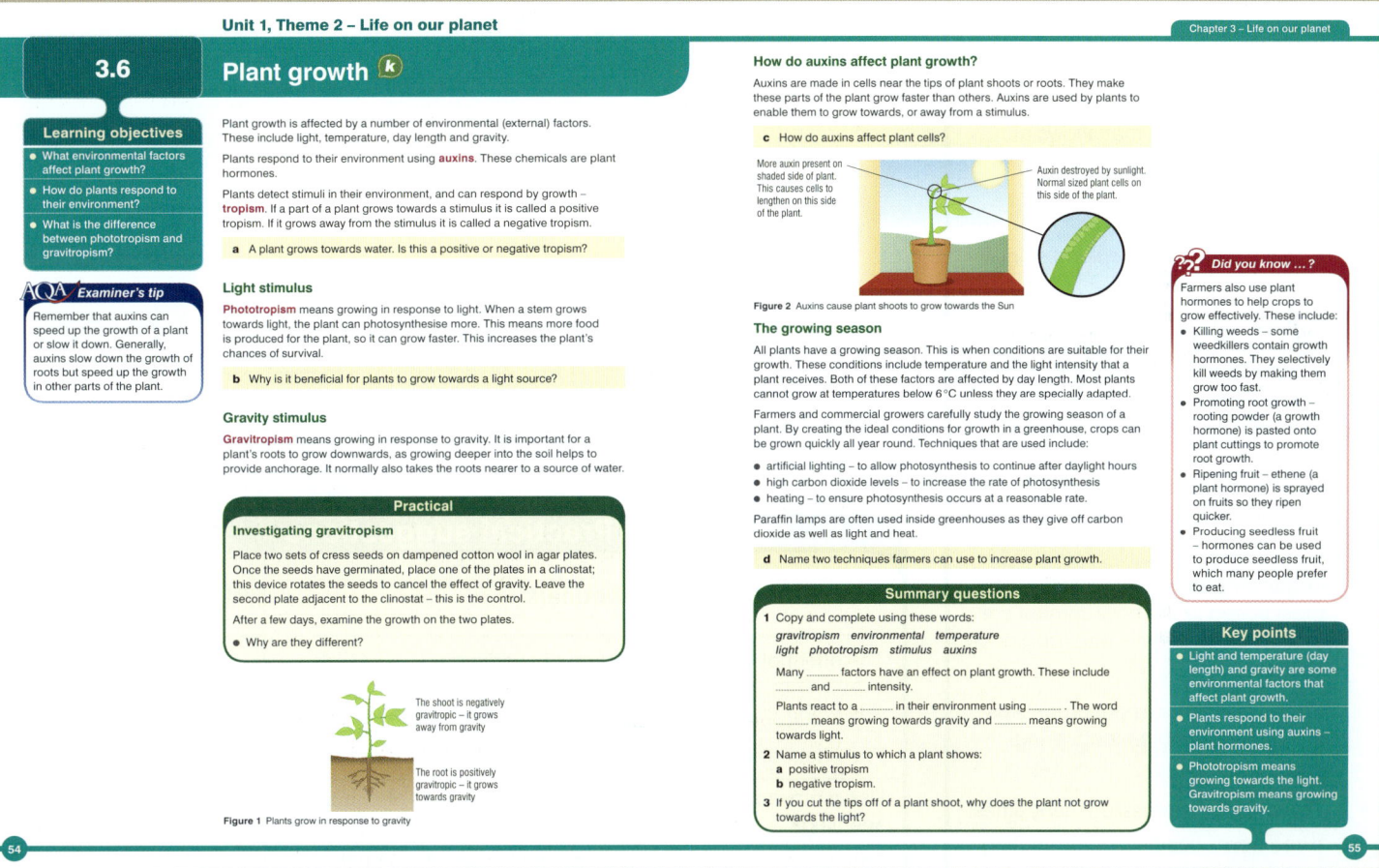

Summary answers

1 environmental, temperature, light, stimulus, auxins, gravitropism, phototropism

2 a Gravity – roots show positive (gravi)tropism
  b Light – roots show negative (photo)tropism.

3 Auxins are produced in the tips. No hormone means no cell elongation, so the plant cannot bend towards the light.

Answers to in-text questions

a positive tropism
b Increases rate of photosynthesis – more food produced – greater rate of plant growth.
c Make cells grow longer.
d heating, artificial lighting, increasing CO_2 levels, use of paraffin lamps

Unit 1, Theme 2 – Life on our planet

Summary answers

1 a sorting organisms into groups based on their physical characteristics

b

Kingdom	Example organisms
Animal	horse, jellyfish
Plant	oak, seaweed
Microbe	yeast, *E. coli*

c for example, presence of a backbone

d **two** from: to name and identify species, to predict characteristics, to find evolutionary links

2 a sunlight, water and nutrients

b **three** from: food, water, mates and suitable territory

3 a a characteristic that allows an organism to live successfully in its habitat

b cactus, for example

c waxy coating, succulent, spines, large root system

d waxy coating – reduces water loss; succulent – large water store; spines – protection from predators; large root system – collect water from a large area.

4 Organisms in a species show variation, caused by genetic differences
↓
Organisms with the characteristics most suited to the environment are most likely to survive and reproduce
↓
Genes from successful organisms are passed on to their offspring
↓
This process is repeated many times

5 **three** from: weedkillers, promoting root growth, ripening fruit, producing seedless fruit

6 a **two** from: light, gravity, water, day length

b auxins (plant hormones)

c tips of shoots

d Phototropism: growing towards the light
Gravitropism: growing towards gravity

e Part of plant facing the light should have normal sized plant cells with a label stating that auxin has been destroyed by sunlight. On the part of the plant facing away from the light the cells should be larger with a label stating more auxin is present, and this causes cells to lengthen on this side of the plant.

Unit 1, Theme 2 – Life on our planet

Summary questions

1 To aid in the research and understanding of organisms, scientists classify organisms.
 a What is meant by the term *classification*?
 b Using the kingdoms (groups) in the table below, classify the following organisms:

 oak   horse   jellyfish   yeast
 seaweed   *E. coli*

Kingdom	Example organisms
Animal	
Plant	
Microbe	

 c Describe **one** way in which organisms within the animal kingdom can be further classified.
 d Describe **two** benefits of scientists classifying organisms.

2 List **three** factors that are needed for the survival of:
 a plants
 b animals.

3 Animals and plants are adapted to live successfully in their environments.
 a What is meant by an adaptation?
 b Name a plant that is adapted to live in the desert.
 c List **four** adaptations for this plant.
 d Explain how each adaptation allows the plant to survive in the desert.

4 Sort the statements below into the correct order, to describe the process of natural selection:

 Genes from successful organisms are passed on to the offspring
 ↓
 This process is repeated many times
 ↓
 Organisms in a species show variation, caused by genetic differences
 ↓
 Organisms with the characteristics most suited to their environment are most likely to survive and reproduce

5 Give **three** examples of how plant hormones can be used to help crops grow effectively.

6 Plant growth is affected by a number of different factors.
 a Name **two** factors that affect plant growth.
 b What chemicals do plants use to respond to their environment?
 c Where are these chemicals produced?
 d What is the difference between phototropism and gravitropism?
 e Draw a labelled diagram to explain how plants grow towards the light.

AQA Practical suggestions

Practicals	AQA	k	📖
The effect of light on the growth of seedlings.	✓	✓	
The effect of gravity on growth in germinating seedlings.	✓	✓	✓
The effect of water on the growth of seedlings.	✓	✓	
Use a movement sensor to measure the growth of plants and seedlings.	✓		
The effect of rooting compounds and weed killers on the growth of plants.	✓		
Size and rate of diffusion – acid penetration of indicator jelly blocks.	✓		
Carry out a European banded snail survey.	✓		
Use of choice chambers, e.g. with woodlice.	✓		
Plant growth, varying the conditions, e.g. degrees of shade, density of sowing, supply of nutrients.	✓		
Investigate the effect of phosphate on oxygen levels in water using jars with algae, water and varying numbers of drops of phosphate, then monitor oxygen using meter.	✓		
Look at variation in leaf length or width, pod length, height. Compare plants growing in different conditions – sun/shade.	✓		✓
Use samples of organisms, identify their features and classify them.	✓		✓
Size and surface area (different-sized flasks and how quickly they cool).	✓		✓

Chapter 3 – Life on our planet

Examination-style questions

Acidophiles are a type of extremophile that lives best in very acidic conditions. The table shows the percentage of species that survive in different pH levels.

pH	Amount of species that survive (%)
1	87
2	75
3	70
4	
5	49

a Predict the percentage survival at pH 4. (1)
b In which pH level do the acidophiles have the best chance of survival? (1)
c Calculate the percentage increase in the amount of species that survive between pH 2 and pH 1. (1)

Choose words from the box to complete the sentences. Words can be used more than once, once or not at all.

| compete | food | short | survive | water |

Living things .... for resources that are in .... supply, such as .... and .... Those plants and animals that .... successfully will .... to breed. (6)

Auxin is a hormone that controls plant growth. Use the diagram of a root to help you answer the questions.

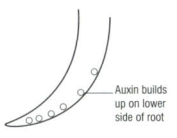
Auxin builds up on lower side of root

a How does auxin cause roots to grow downwards? (3)
b What role do roots play in the plant's survival? (1)

4 Dandelions have many adaptions that allow them to survive.
Match each adaption to how it helps the dandelion to survive.

Adaption	How it helps
Deep roots	Offspring spread far away so not in competition with parent
Grows quickly on bare soil	Chemical methods will not kill it
Leaves spread out over ground	No low lying plants can get sunlight
Resistant to many weedkillers	Collect more water
Seeds are spread by the wind	Can grow anywhere

(4)

5 A penguin and an osprey look very different.

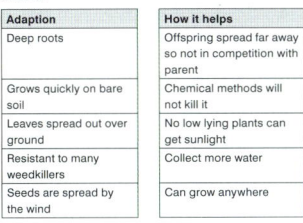

a *In this question you will be assessed on using good English, organising information clearly and using specialist terms where appropriate.*
Give reasons for and against scientists classifying them in the same group. (6)
b Suggest **one** thing that penguins might compete with other animals for. (1)

6 Some seed-eating finches from the mainland colonised the Galapagos Islands thousands of years ago. These had thick beaks to crush seeds. There are many types of finches on the islands now, some with thick beaks, but others have slender beaks to crush insects.
a Why would a slender beak be an evolutionary advantage on the islands? (1)
b Suggest what might have happened to the population of seed-eating finches on the islands since their colonisation. (2)
c Explain the term '**natural selection**'. (2)

4

Adaption	How it helps
Deep roots	Offspring spread far away so not in competition with parent
Grows quickly on bare soil	Chemical methods will not kill it
Leaves spread out over ground	No low lying plants can get sunlight
Resistant to many weedkillers	Collect more water
Seeds are spread by the wind	Can grow anywhere

(All 5 correct = 4 marks, 4 correct = 3 marks, 3 correct = 2 marks, 1 or 2 correct = 1 mark)

5 a Marks awarded for this answer will be determined by the Quality of Written Communication (QWC) as well as the standard of the scientific response.

There are full, balanced and detailed reasons for and against scientists classifying penguins and ospreys in the same group. The answer is well structured with minimal repetition or irrelevant points. There is an accurate, fluent and clear expression of ideas with only minor errors in the use of technical terms, spelling, punctuation and grammar. *(5–6 marks)*

There are some reasons for and against scientists classifying penguins and ospreys in the same group with some omissions. The answer shows some attempt at structuring and the ideas are expressed with reasonable fluency and clarity. There are some errors in the use of technical terms, spelling, punctuation and grammar. *(3–4 marks)*

There are some brief reasons for and against scientists classifying penguins and ospreys in the same group. The answer is largely incomplete and may contain some valid points which are not clearly structured. It lacks fluency and/or clarity. It contains errors in the use of technical terms, spelling, punctuation and grammar. *(1–2 marks)*

No relevant content *(0 marks)*

Examples of points made in the response:
- Both lay eggs
- Both have feathers
- Both have beak
- Both have wings
- Both eat fish
- Penguin swims and osprey flies
- Live / nest in different places
- Penguin has layer of fat under skin
- Penguins' wings are more like flippers.

b food/mates/territory (1 mark)

6 a less competition for food (1 mark)
b Increased at start. (1 mark)
Then remained stable. (1 mark)
c individuals with characteristics most suited to environment
more likely to breed successfully (2 marks)

Examination-style answers

1 a Any number between 50 and 69 (1 mark)
b 1 (1 mark)
c 87 − 75 = 12; 12 ÷ 75 = 0.16 = 16% (1 mark)

2 compete (1 mark)
short (1 mark)
food/water (1 mark)
water/food (1 mark)
compete (1 mark)
survive (1 mark)

3 a slow growth at bottom of root (1 mark)
more growth at top of root (1 mark)
root bends down (1 mark)
b anchorage
OR
absorption of water / minerals (1 mark)

Kerboodle resources

- Theme map: Life on our planet
- How Science Works: What effect does oxygen have on the distribution of fish? (3.3)
- Video: Evolution (3.5)
- Revision podcast: Survival (3.5)
- Practical: Plant growth (3.6)
- Examination-style questions
- Answers to examination-style questions

Unit 1, Theme 2 – Life on our planet

4.1 Biomass and food chains

Learning objectives

Students should learn:
- that biomass is all the living organic matter present in an ecosystem
- that producers make their own food by photosynthesis but consumers eat food to gain energy
- how food chains show the flow of energy between organisms in a community.

Learning outcomes

Most students should be able to:
- define the term biomass
- describe the difference between producers and consumers
- produce a simple food chain.

Some students should also be able to:
- explain how producers and consumers transfer energy through the processes of respiration and photosynthesis
- interpret food webs.

AQA Specification link-up: Science B 3.3.2.2

- Know that energy enters the biosphere as sunlight.
- Know that sunlight is converted to chemical energy and stored in organic compounds (biomass) by producers.
- Know that biomass is broken down to release energy through respiration by consumers.
- Understand that food chains show the flow of matter and energy between all the producers and consumers in a given ecosystem.

Lesson structure

Starters

Make a food chain – Ask students to rearrange the organisms: bird, pondweed, fish, pond snail into a food chain. *(5 minutes)*

Name some producers and consumers – Ask students to produce a list of at least three producers and three consumers. They then need to write a sentence to explain the difference between a producer and a consumer. This is a straightforward revision of previous material, and will allow the opportunity to recap some key concepts before beginning new work. To support students, provide them with names and images of some organisms, for them to sort into producers and consumers. To extend students, ask them to write a description of what a decomposer is and illustrate their answer with an example. If images have been supplied as support material, more able students could be asked to classify organisms further into primary and secondary consumers. *(10 minutes)*

Main

- Begin by discussing the fact that ecology involves the use of a wide range of scientific terms, many of which will be new to students. Suggest that students start to produce an ecology dictionary (perhaps in the back of their exercise book/working folder) to use in this series of lessons. Each time a new term is introduced, they can add it to their dictionary.
- Ask students to discuss what they think the term biomass means – suggest they split the term into two parts, 'bio' and 'mass'. Produce a class definition.
- Revise with students where plants get their energy from, and how this becomes fixed into chemical compounds during photosynthesis. Introduce the term producer. Students should be made familiar with using the word equation for photosynthesis; higher tier students should also be familiar with the balanced chemical equation.
- Discuss how energy gained by a producer becomes transferred to consumers when they eat. Students should discuss the role of this energy – some is used to produce chemical compounds that are used for growth and result in increased biomass, whereas some is used in respiration to release energy required for movement, etc. Students should be familiar with using the word equation for respiration. Higher level students should also be familiar with the balanced chemical equation.
- Introduce the concept of a food chain. That it begins with a producer, then the different levels of consumers – herbivore (primary consumer), then carnivore (secondary and also tertiary consumers). Stress the importance of the direction of the arrows, showing the direction in which the energy and biomass are transferred.
- Students should be given ample opportunities to make their own food chains. For example, provide them with a series of organism cards for them to rearrange (you may also wish to include the Sun in these, to demonstrate ultimately the source of all the energy in the food chain). Students could act as the organisms producing human food chains; or provide students with information on organisms in a particular habitat – using this information they can then produce a series of food chains.
- Discuss the fact that food webs provide a more realistic representation of the flow of energy and biomass within an ecosystem, as most animals eat more than one type of food. Using some of the food chains already produced in the lesson, students should be asked to link them together, to produce a food web to illustrate this concept.
- Provide students with a series of food webs from different habitats for them to decipher which organism eats what, and produce a series of food chains from the web.

Answers to in-text questions

a Producers make their own food; consumers cannot.
b light energy from the Sun
c Herbivores eat plants; carnivores eat meat.

Support

- Supply students with a simple food chain and/or web, featuring organisms that will be familiar to them, but with the arrows linking the organisms removed.
- Students should be asked to complete the food chain/web, using arrows to show which direction energy flows between organisms.

Extend

- Ask students to analyse the feeding relationships within a food chain or web. For example, if one of the primary consumers died, what could happen to the number of secondary consumers which fed on them? Or, what could happen to the number of other primary consumers?

Chapter 4 – Biomass, energy flow and the importance of carbon

Plenaries

How do producers transfer light energy into chemical energy? – Ask students to write the word and chemical equation for photosynthesis. To support students, provide them with the reactants and products of the equation for them to rearrange. To extend students, ask them to balance the chemical equation. *(5 minutes)*

What is wrong with these food chains? – Provide students with a series of incorrect food chains. They then need to spot the mistakes (for example, a carnivore before a herbivore, a food chain that begins with an animal and a food chain with the arrows pointing in the wrong direction) and then correct them. *(10 minutes)*

Science in context

Ecologists can find out what happens to energy and biomass as it passes along the food chain by observing the numbers and sizes of the organisms in food chains. Different sampling techniques are employed to collect this data, as it would usually be impossible to count all of the organisms present within a habitat without them. Common examples include the use of quadrats, transects, pitfall traps, pooters, sweep nets, D-vacs, kick sampling and tree beating.

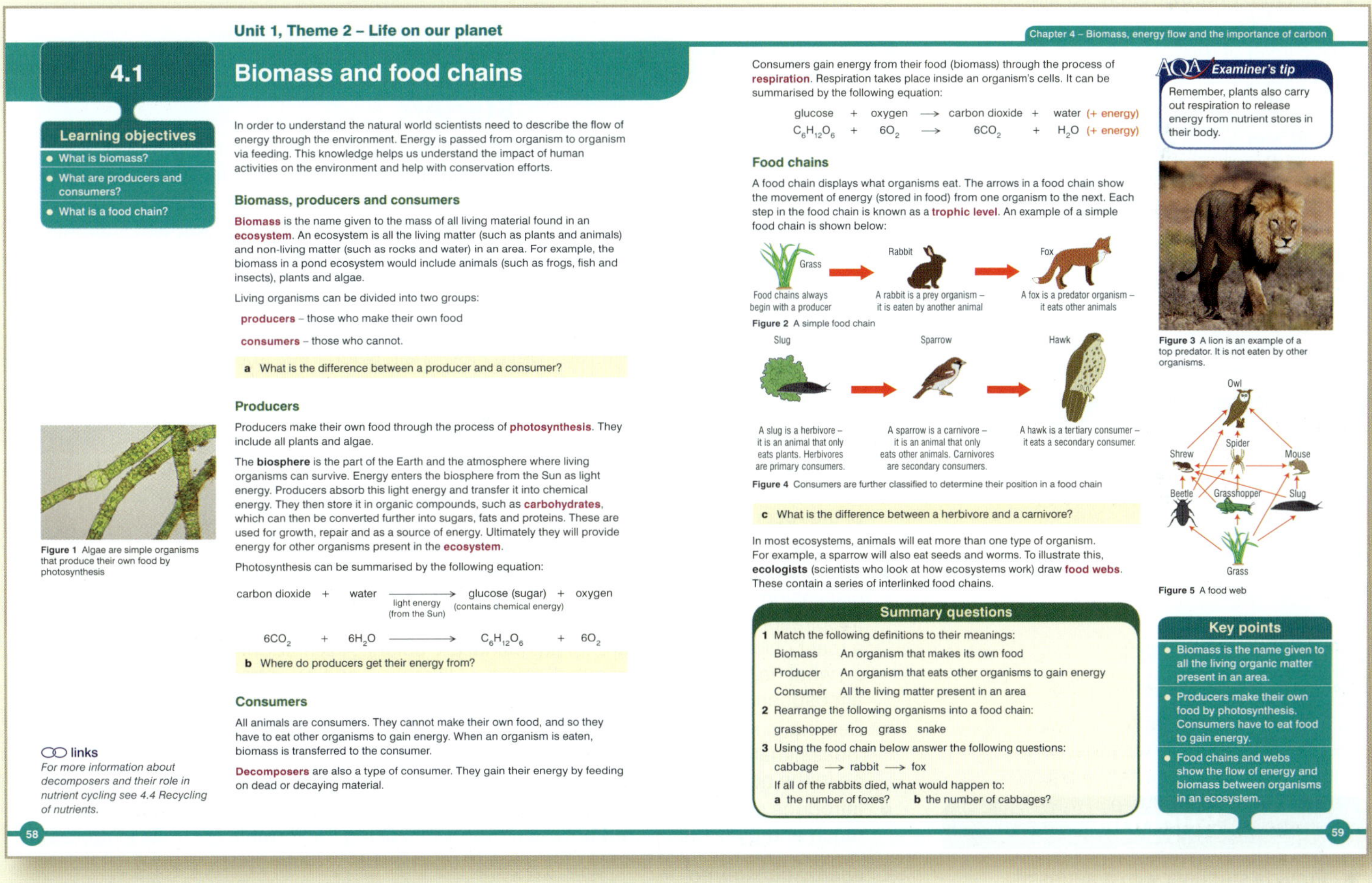

Summary answers

1. Biomass — All the living matter present in an area
   Producer — An organism that makes its own food
   Consumer — An organism that eats other organisms to gain energy
2. grass → grasshopper → frog → snake
3. a Their numbers would decrease, as they would have nothing to eat.
   b Their numbers would increase, as nothing would be eating them.

Unit 1, Theme 2 – Life on our planet

4.1 Biomass and food chains

Learning objectives
- What is biomass?
- What are producers and consumers?
- What is a food chain?

In order to understand the natural world scientists need to describe the flow of energy through the environment. Energy is passed from organism to organism via feeding. This knowledge helps us understand the impact of human activities on the environment and help with conservation efforts.

Biomass, producers and consumers

Biomass is the name given to the mass of all living material found in an **ecosystem**. An ecosystem is all the living matter (such as plants and animals) and non-living matter (such as rocks and water) in an area. For example, the biomass in a pond ecosystem would include animals (such as frogs, fish and insects), plants and algae.

Living organisms can be divided into two groups:
- **producers** – those who make their own food
- **consumers** – those who cannot.

a What is the difference between a producer and a consumer?

Producers

Producers make their own food through the process of **photosynthesis**. They include all plants and algae.

The **biosphere** is the part of the Earth and the atmosphere where living organisms can survive. Energy enters the biosphere from the Sun as light energy. Producers absorb this light energy and transfer it into chemical energy. They then store it in organic compounds, such as **carbohydrates**, which can then be converted further into sugars, fats and proteins. These are used for growth, repair and as a source of energy. Ultimately they will provide energy for other organisms present in the **ecosystem**.

Figure 1 Algae are simple organisms that produce their own food by photosynthesis

Photosynthesis can be summarised by the following equation:

carbon dioxide + water —light energy (from the Sun)→ glucose (sugar) (contains chemical energy) + oxygen

$6CO_2 + 6H_2O \longrightarrow C_6H_{12}O_6 + 6O_2$

b Where do producers get their energy from?

Consumers

All animals are consumers. They cannot make their own food, and so they have to eat other organisms to gain energy. When an organism is eaten, biomass is transferred to the consumer.

Decomposers are also a type of consumer. They gain their energy by feeding on dead or decaying material.

○○ **links**
For more information about decomposers and their role in nutrient cycling see 4.4 Recycling of nutrients.

Chapter 4 – Biomass, energy flow and the importance of carbon

Consumers gain energy from their food (biomass) through the process of **respiration**. Respiration takes place inside an organism's cells. It can be summarised by the following equation:

glucose + oxygen → carbon dioxide + water (+ energy)
$C_6H_{12}O_6 + 6O_2 \longrightarrow 6CO_2 + 6H_2O$ (+ energy)

AQA Examiner's tip
Remember, plants also carry out respiration to release energy from nutrient stores in their body.

Food chains

A food chain displays what organisms eat. The arrows in a food chain show the movement of energy (stored in food) from one organism to the next. Each step in the food chain is known as a **trophic level**. An example of a simple food chain is shown below:

Grass → Rabbit → Fox

Food chains always begin with a producer | A rabbit is a prey organism – it is eaten by another animal | A fox is a predator organism – it eats other animals

Figure 2 A simple food chain

Slug → Sparrow → Hawk

A slug is a herbivore – it is an animal that only eats plants. Herbivores are primary consumers. | A sparrow is a carnivore – it is an animal that only eats other animals. Carnivores are secondary consumers. | A hawk is a tertiary consumer – it eats a secondary consumer.

Figure 4 Consumers are further classified to determine their position in a food chain

c What is the difference between a herbivore and a carnivore?

In most ecosystems, animals will eat more than one type of organism. For example, a sparrow will also eat seeds and worms. To illustrate this, **ecologists** (scientists who look at how ecosystems work) draw **food webs**. These contain a series of interlinked food chains.

Figure 3 A lion is an example of a top predator. It is not eaten by other organisms.

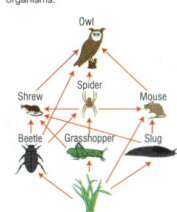

Figure 5 A food web

Summary questions

1. Match the following definitions to their meanings:
   Biomass — An organism that makes its own food
   Producer — An organism that eats other organisms to gain energy
   Consumer — All the living matter present in an area
2. Rearrange the following organisms into a food chain:
   grasshopper frog grass snake
3. Using the food chain below answer the following questions:
   cabbage → rabbit → fox
   If all of the rabbits died, what would happen to:
   a the number of foxes? b the number of cabbages?

Key points

- Biomass is the name given to all the living organic matter present in an area.
- Producers make their own food by photosynthesis. Consumers have to eat food to gain energy.
- Food chains and webs show the flow of energy and biomass between organisms in an ecosystem.

Further teaching suggestions

Investigating habitats
Ask students to look at the organisms that are present in their garden/local park and arrange their findings into a food chain and/or food web. Students should be provided with an identification chart, and individuals/groups should pool findings to gain a more comprehensive list of organisms present in a particular habitat.

Food web mobiles
Food webs can be illustrated clearly by creating a mobile, using coat hangers suspended from the ceiling of your classroom. Organisms at each tropic level are suspended using particular lengths of string so that, for example, all primary consumers hang at the same height. A series of food chains can then be organised simply into a potential food web.

Did you know …?
Respiration is the 'opposite' reaction to photosynthesis.

You can demonstrate this to students using a cycle diagram, which shows the link between photosynthesis and respiration and the flow of energy from the Sun through the organisms.

Unit 1, Theme 2 – Life on our planet

4.2 Energy transfer within food chains

Learning objectives

Students should learn:
- that around 90% of the energy received at each trophic level is ultimately lost in heating the organism and its surroundings
- that producers do not convert all of the light energy they receive into chemical energy
- that consumers lose energy through movement, respiration and waste.

Learning outcomes

Most students should be able to:
- state that approximately 10% of the received energy is transferred from one level of the food chain to the next
- explain why light energy is not all transferred to chemical energy by a producer
- name some ways in which energy is transferred from consumers in a food chain.

Some students should also be able to:
- calculate the percentage of energy transferred between trophic levels in a food chain.

AQA Specification link-up: Science B 3.3.2.2

- Know that energy leaves the biosphere as heat.
- Understand that food chains show the flow of matter and energy between all the producers and consumers in a given ecosystem.
- Know the mass of living material (biomass) and amount of energy at each stage in a food chain is less than it was at the previous stage.
- Be able to calculate the percentage of energy transfer at each stage of a food chain.
- Explain the reasons for the inefficiency of the energy transfer:
  (a) some plant material passes out of the body of a herbivore as faeces without being digested
  (b) energy is transferred to the environment in respiration …

Within this context, candidates should be able to use scientific data and evidence to discuss, evaluate or suggest implications of the following:
- efficiency of energy transfer at different stages of a food chain.

Lesson structure

Starters

Draw a food chain – As revision from the previous lesson, ask students to produce a food chain and label: producer, primary consumer, secondary consumer, herbivore and carnivore. *(5 minutes)*

How can energy enter and be lost from a food chain? – In small groups, ask students to divide a piece of paper into two halves, and list ways energy can enter a food chain and ways energy can be transferred. To support students, provide them with a list of ways energy can be added to, or transferred from a food chain, for them to sort into the correct category. To extend students, ask them to illustrate each idea with an example. For example, energy is transferred from a food chain when trees shed their leaves in the autumn. *(10 minutes)*

Main

- Discuss ways energy can enter the food chain – through photosynthesis and consumption. Looking first at photosynthesis, students should discuss why not all of the received light energy is converted to chemical energy. Students could be provided with a number of leaves to prompt discussion.
- The discussion should then be extended to other parts of a food chain. To stimulate discussion, show students a series of images. For example, a man running (respiration), a 'no fouling' sign (waste), a pile of leaves (loss of material), people eating a roast chicken (not all parts of an animal are eaten). Ask students to hypothesise on what happens to this 'wasted' energy. Ultimately, energy is transferred to the atmosphere via decomposers as energy heating up the surroundings.
- Bearing in mind the energy transfers previously discussed, ask students to 'guesstimate' the proportion of the energy received by an organism that gets passed on to the next trophic level. Many students will be far too optimistic in their guesses!
- Introduce students to energy transfer diagrams. First, explain how to calculate how much energy is transferred between organisms using the formula (all units in Joules):

$$\text{energy transferred} = \text{energy taken in} - \text{energy transferred in waste} - \text{energy transferred through respiration}$$

- Students should be given opportunities to manipulate the equation. For example, to calculate how much energy is transferred through respiration by a particular organism.
- Provide students with data in a number of forms – pictorially, in data tables, and in written questions. Higher ability students should also practise converting how much energy is transferred from a unit of energy into a percentage, using the formula:

$$\% \text{ energy transferred} = \frac{\text{energy transferred}}{\text{total energy taken in}} \times 100$$

Support

- Provide students with a series of energy transfer diagrams, with partially completed worked examples for them to complete.

Extend

- Ask students to carry out research into how light frequency affects photosynthesis. Through data or experiments, students should deduce that plants only use certain wavelengths of light for photosynthesis.

Chapter 4 – Biomass, energy flow and the importance of carbon

Plenaries

Why is not all light energy transferred into chemical energy? – Ask students to list three reasons why producers cannot transfer all the light energy they receive into chemical energy. *(5 minutes)*

Calculate how much energy is transferred between a chicken and a human – Provide students with information to calculate how much energy is transferred to a human when they eat a portion of roast chicken. To support students, provide worked calculations for them to follow and complete. To extend students, ask them to explain what happens to the energy that is transferred from the food chain – how is it transferred back into the environment? *(10 minutes)*

Unit 1, Theme 2 – Life on our planet

4.2 Energy transfer within food chains

Learning objectives
- How much energy is transferred between organisms?
- How is energy wasted during photosynthesis?
- How is energy lost from a food chain?

Energy conversion and transfer
The energy contained in food originally comes from the Sun in the form of light energy. This is then transferred to chemical energy by producers, during photosynthesis.

Energy is transferred when one organism eats another. However, not all of the energy is transferred from one organism to the next – energy is wasted in a number of different ways. Eventually, all of the energy will be transferred from the biosphere, heating the surroundings.

Energy transfer through producers
Not all of the light energy coming from the Sun is transferred into chemical energy by photosynthesis. Only about 1 per cent of the energy a plant receives is transferred. This is because:
- some of the light is reflected from the leaf back into the atmosphere
- some light passes through the leaf
- not all the light is of the correct wavelength for the plant to use
- some energy is used in photosynthesis reactions.

The chemical energy gained is used by the plant for growth. This increases a plant's biomass and provides food for consumers.

AQA Examiner's tip
The percentage of energy transfer will never be more than 100 per cent. If your calculation gives an answer over 100 per cent, you have made a mistake, so check your working.

a What process transfers light energy to chemical energy?

Energy transfer through consumers
When one organism eats another, only around 10 per cent of the energy is transferred. Therefore, at each trophic level, less and less energy is available. This means that food chains do not consist of many stages. Most have only four levels.

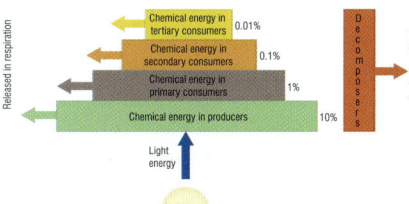

Figure 1 Energy flow through a typical food chain

Energy might not be transferred between organisms because:
- Some parts of a plant or animal might not be eaten. For example, leaves may be lost from trees. These may then be broken down by decomposers.
- Some parts of the plant or animal cannot be digested and these will be lost from the body in faeces.
- Energy released by respiration is used for movement and other body processes. It is eventually transferred to energy heating the surroundings.
- Energy is transferred from the body in urine – waste products.

b Approximately how much energy is transferred from one trophic level to the next?

Maths skills

Energy efficiency in a food chain
Figure 2 shows that the chicken has taken in 110 kJ of energy from its food. Of this, 66 kJ are lost through waste, and 32 kJ are lost through respiration. How much energy would be available to a fox that ate this chicken?
To calculate how much energy is transferred from one organism to the next, you can use the following equation:

energy transferred = energy taken in − energy transferred in waste − energy transferred through respiration
= 110 kJ − 66 kJ − 32 kJ
= 12 kJ

Or, expressed as a percentage:

% energy transferred = $\frac{\text{Energy transferred}}{\text{Total energy taken in}} \times 100 = \frac{12}{110} \times 100$
= 10.9%

c How does a chicken transfer energy that is not passed along a food chain?

Summary questions

1 Copy and complete using these words:
respiration chains energy ten waste

Food _____ show how energy is transferred between organisms. At each trophic level, less _____ is available – only _____ per cent, approximately, is transferred between organisms. Energy is 'wasted' through _____ products and _____.

2 Using the table below, answer the following questions:

Energy in:	Food (eaten by organism)	Biomass (contained in organism)	Faeces and waste (lost from organism)	Respiration (released as energy)
Energy (kJ)	150	20	120	?

a How much energy is used in respiration and given out as heat to the surroundings?
b How efficient is the energy transfer in this organism, expressed as a percentage?

? Did you know ...?
Although less energy is available at each trophic level, energy is never destroyed. Energy may be transferred to a different and possibly less useful form, such as heating up the organism and its surroundings. This energy is no longer available to be passed on to the next organism in the chain. Energy is also transferred to the environment from the food chain through faeces. This energy is not passed on but can be used by other organisms that eat the faeces, such as dung beetles.

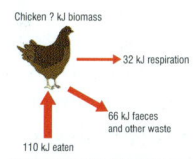

Figure 2 Energy transfer in a chicken

Key points
- Approximately 10 per cent of energy is transferred from one level of the food chain to the next. The remaining energy is transferred, heating organisms and their surroundings eventually.
- Producers do not convert all of the Sun's light energy into chemical energy. Some is reflected and some is not of the correct wavelength.
- Consumers do not pass all of their energy onto the next organism. Some is lost through respiration and waste and some parts of the organism may not be eaten.

Further teaching suggestions

Role of dung beetles in decomposition
Students could research how dung beetles are involved in breaking down waste.

Efficiency of energy transfer
More able students could be asked to look into the efficiency of energy transfer between different trophic levels in a food chain, and explain why such significant differences occur. For example, one study measured the efficiency of transfer between producers and primary consumers at 17%, but only at 4.5% from primary to secondary consumers.

Answers to in-text questions
a photosynthesis
b 10%
c primarily through waste and respiration

Summary answers
1 chains, energy, ten, waste, respiration
2 a 10 kJ
b $\frac{20}{150} \times 100 = 13.3\%$

Unit 1, Theme 2 – Life on our planet

4.3 Pyramids of numbers and biomass

Learning objectives

Students should learn:

- how pyramids of numbers represent the number of organisms at each trophic level of a food chain
- how pyramids of biomass represent the total amount of biomass present at each trophic level of a food chain
- that the size of each bar in a pyramid of biomass is calculated from the average mass of an organism, multiplied by the number of organisms present at that trophic level.

Learning outcomes

Most students should be able to:

- state the differences between a pyramid of numbers and pyramid of biomass
- describe how scientists calculate biomass
- construct a pyramid of biomass, when provided with appropriate data.

Some students should also be able to:

- explain the advantages of pyramids of biomass over pyramids of numbers.

Support

- Provide a partially completed worked example of a pyramid of biomass, using data from the Student Book. Students should complete key pieces of information, and complete bars on the pyramid (a scaled axis should be provided).
- You could set this up as a card sort activity and give them all the labels to place on each bar of the pyramid.

Extend

- Ask students to research and draw pyramids of numbers and biomass for food chains involving parasites.
- Students should be asked to compare and contrast the shapes of the pyramids, and explain why the pyramids of numbers may be misleading.

AQA Specification link-up: Science B 3.3.2.2

- Use data to construct pyramids of biomass.

Within this context, candidates should be able to use scientific data and evidence to discuss, evaluate or suggest implications of the following:

- interpreting and constructing pyramids of biomass.

Lesson structure

Starters

What does this food chain show? – Show students a simple pyramid of numbers for a typical food chain. Ask students to write out the food chain that the diagram is showing. *(5 minutes)*

Construct a model pyramid of numbers – Provide students with a blank pyramid of numbers. Ask them to annotate the pyramid with as many terms as possible. For example, producer and herbivore. To support students, provide them with terms to cut and stick, to annotate their diagrams. To extend students, ask them to add an example of an organism that could be present at each level of the pyramid. Challenge students to produce a pyramid of numbers for more than one habitat. *(10 minutes)*

Main

- Show students a variety of standard pyramids of numbers – ask them to list similar features between the pyramids. For example, all have a producer at their base and the organisms decrease in number but increase in size as you move up the pyramid. Ask students to relate their findings to energy transfer – there is less energy available as you travel up the food chain, therefore fewer organisms can be supported.
- Show students the inverted pyramid, produced when a tree is the producer. The example in the Student Book could be used. Ask students why they think the pyramid is inverted. Ask students to come up with other examples where an inverted pyramid could be produced.
- Introduce the concept of a pyramid of biomass. Discuss with students what is the difference between the data these diagrams show, compared with a pyramid of numbers? Why might a pyramid of biomass give a better indication of the energy flow through a food chain?
- It is worthwhile spending time as a class working through the example in the Student Book, to produce a scale pyramid of biomass. The example uses a food chain present on an oak tree. Less able students would benefit from a visualisation of the food chain – perhaps through projecting images of the organisms involved. Students will find it easier to draw pyramids of biomass on squared or graph paper.
- Provide students with data for another pyramid – easily available via a search from www.google.co.uk – or use question 2 in the Student Book for students to produce a pyramid of biomass independently.

Plenaries

Which food chain? – Display a range of different shaped pyramids of numbers, along with an example food chain for each. Students need to match each food chain to the correct pyramid of numbers. *(5 minutes)*

Construct a pyramid of biomass – Provide data for students to produce their own pyramid of biomass. To support students provide them with a suitable scale. They should then use this scale to calculate how wide the bars should be. To extend students do not provide them with the total biomass at each trophic level; instead, provide data on the number of organisms at each trophic level, and their typical mass. *(10 minutes)*

Chapter 4 – Biomass, energy flow and the importance of carbon

Further teaching suggestions

Calculating biomass
Students could measure the biomass of a plant by taking a sample of its leaves. They should calculate the dry mass of an average leaf, and multiply this value by the number of leaves on the plant. This value could then be compared to the measured dry mass of the leaves. Students should consider how accurate the sampling technique was. Is the inaccuracy inherent in the sampling technique important? What would happen if several hundred plants were sampled?

Inverted pyramids
Students could consider whether there are any conditions under which a pyramid of biomass could be inverted. This could be extended into a research-based task.

Unit 1, Theme 2 – Life on our planet

4.3 Pyramids of numbers and biomass

Learning objectives
- What is a pyramid of numbers?
- What is a pyramid of biomass?
- How can you draw a pyramid of biomass?

Pyramids of numbers
Food chains show the flow of energy and biomass through a community. However, they do not show how many organisms are involved, or how much biomass is being transferred.

A pyramid of numbers is used to show the population at each level in the food chain. The producer in the food chain is placed at the base of the pyramid. The width of each bar in the pyramid represents the number of organisms present.

The diagram usually has the shape of a pyramid. An organism normally eats more than one organism in the trophic level below. For example, a blue tit needs to eat many caterpillars to survive.

a What generally happens to the number of organisms as you move up a food chain?
b What generally happens to the size of the organisms as you move up a food chain?

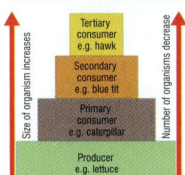

Figure 1 A typical pyramid of numbers. As you move from one trophic level to the next, the size of organisms generally increases. However, there are fewer and fewer organisms at each level.

Inverted pyramids of numbers
Not all pyramids of numbers are pyramid shaped. This is because these diagrams do not take into account the *size* of the organisms present at each trophic level.

For example, a single tree can support a large number of living organisms. Therefore, when a tree is the producer in a food chain it will result in an inverted pyramid of numbers.

c Why is the bar for an oak tree smaller than the one representing caterpillars?

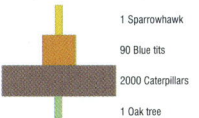

Figure 2 An inverted pyramid of numbers

Pyramids of biomass
Many scientists choose to represent population data in a pyramid of *biomass*. Here the biomass of the organisms at each trophic level is calculated. This takes into account both the *number* and *size* of the organisms present at each trophic level. Pyramids of biomass are *never* inverted. Figure 3 shows the oak tree pyramid of numbers, as a pyramid of biomass.

d How is a pyramid of biomass different from a pyramid of numbers?

Calculating biomass
To collect data to construct a pyramid of biomass, scientists:
- take samples of organisms from each trophic level
- measure the average mass of each of these organisms
- use this data to calculate the total biomass at each trophic level.

Scientists normally calculate the dry mass of an organism, as water content can vary between individuals. This often requires the organisms to be killed and dried in a kiln.

Figure 3 A pyramid of biomass from the pyramid of numbers in Figure 2

Organism	Number present	Mass of one organism (kg)	Total biomass (kg)
Oak tree	1	100	100
Caterpillar	2500	0.004	10
Blue tit	50	0.02	?
Sparrowhawk	1	0.10	0.10

The total biomass contained in a trophic level can be calculated from:

Total biomass = Mass of one organism × Number of organisms

e.g. Biomass of blue tit = Mass of one blue tit × Number present
= 0.02 × 50 = 1 kg

Drawing a pyramid of biomass
Pyramids of biomass are scale diagrams. The width of a bar in the diagram represents the biomass of the living organisms in the trophic level. Using the data from the table above, a sensible scale would be 1 cm = 10 kg. The width of each bar can then be calculated by dividing each biomass by the scale factor (10 kg/cm in this case). For example:

Oak tree: $\frac{100\,kg}{10} = 10\,cm$ Blue tit: $\frac{1\,kg}{10} = 0.1\,cm = 1\,mm$

e Create a scale pyramid of biomass for the food chain:
Oak tree → Caterpillar → Blue tit → Sparrowhawk

AQA Examiner's tip
When drawing pyramids of biomass, graph paper should be used to ensure that the diagram is drawn as accurately as possible to scale.

Figure 4 The sparrowhawk is at the top of a pyramid of numbers or biomass

Summary questions

1 Copy and complete using the following words:
trophic   inverted   biomass   numbers   size

Pyramids of .... represent the number of organisms at each .... level. However, they can sometimes be ....

Pyramids of .... take into account both the number and .... of organisms present at each level.

2 a Complete the table below. Then draw its pyramid of biomass.

Organism	Number present	Mass of one organism (kg)	Total biomass (kg)
Rose bush	1	4	
Aphid	2000	0.0001	
Ladybird	5	0.002	

b What would a pyramid of numbers look like for this food chain?

Key points
- Pyramids of numbers represent the number of organisms present at each trophic level.
- Pyramids of biomass represent the total amount of biomass present at each trophic level.
- To calculate the size of each bar in a pyramid of biomass, multiply the average mass of the organism by the number of organisms present at that trophic level.

Summary answers

1 numbers, trophic, inverted, biomass, size

2 a

Organism	Number present	Mass of one organism (kg)	Total biomass (kg)
Rose bush	1	4	4
Aphid	2000	0.0001	0.2
Ladybird	5	0.002	0.01

b inverted pyramid

Answers to in-text questions

a The number of organisms decreases.
b The size of organisms increases.
c The oak tree is a significantly larger organism than a caterpillar, therefore one tree can support many caterpillars.
d A pyramid of biomass takes into account both the size and number of organisms.
e An accurate scale diagram should be drawn – it will have a classic pyramid shape with bars of width 10 cm, 1 cm, 1 mm, 0.1 mm.

Unit 1, Theme 2 – Life on our planet

4.4 Recycling of nutrients

Learning objectives

Students should learn:
- that decomposers are microorganisms that break down dead organic material
- the optimum conditions for microorganisms to function effectively
- that nutrients are constantly cycled through the ecosystem, so they can be used over and over again.

Learning outcomes

Most students should be able to:
- state the best conditions for decomposition
- describe generally how nutrients are cycled through the ecosystem
- explain how organic waste products from the garden or kitchen can be recycled.

Some students should also be able to:
- explain in detail the optimum conditions for decomposition.

Summary answers

1. decomposers, dead, bacteria, detritivores, small
2. by using enzymes
3. A fridge is cold, dry and lacking in oxygen. The best conditions for decay are warm, moist and oxygen-rich.

Support
- Provide students with a range of nutrient cycles with labels missing. Students should practise completing these diagrams.

Extend
- Ask students to research a specific decomposer or detritivore and find out the exact role they play in decomposing material.
- Students should present their findings in a format they choose to the rest of the class.

AQA Specification link-up: Science B 3.3.2.2

- Explain that some energy passes to decomposers in dead remains.
- Know that microorganisms function better in warm, moist conditions and in a plentiful supply of oxygen.
- Know that when living things die their bodies are broken down by decomposers, so releasing the elements they contain.
- Know that these minerals can be used by plants to grow so that the cycle repeats over again.

Within this context, candidates should be able to use scientific data and evidence to discuss, evaluate or suggest implications of the following:
- the recycling of organic waste products from the garden or kitchen.

Lesson structure

Starters

Which organisms break down dead remains? – Ask students to make a list of as many organisms as they can think of that help to break down dead material. You may wish to provide some projected images as stimuli. *(5 minutes)*

Nutrient cycling graffiti – Provide students in groups with a large piece of paper. Ask them to write as many facts as they know about nutrient cycling on the paper in two minutes. This could include examples of nutrients that are recycled and the organisms and processes involved. They then need to pass the paper onto the next group for them to add to. To support students, provide them with some sub headings to help trigger ideas such as decay, decomposers and food chains. To extend students, once the graffiti exercise has been completed, ask students to try to link together their ideas to form a 'nutrient cycle' – this can then be revisited in the plenary section, and amended or added to as appropriate. *(10 minutes)*

Main

- Without further introduction, set students a short creative writing challenge – to describe a world in which no decomposers existed. Students should be invited to read their descriptions to the class.
- Discuss with students the essential role played by decomposers – not only in removing waste material from our world, but just as importantly in recycling the nutrients back to the Earth. Analogies should be drawn with human attempts at recycling – why do students feel that recycling is an important contribution to a person's place in society? Include within this discussion the complementary role played by detritivores. If time allows, it is well worth students scouring the school site for examples of these – perhaps earthworms or woodlice.
- If one is available on the school site, show students a real compost bin – the school gardener or eco-team should be able to provide advice on their location and are usually more than willing to demonstrate their action to students. If this is not possible, provide students with samples of material from different levels of a compost bin. Students should make observations of the material – many students have difficulty equating waste vegetation with a bag of compost from a garden centre.
- Carry out the practical, as described in the 'Practical support' box opposite. It would be ideal if this activity were set up well in advance (1–2 weeks) of the lesson, so that the results can be discussed during this lesson. Students should be encouraged to explain the results.
- If the nutrient cycling starter and plenary activities are not being used in this lesson, summarise the cycle in the Student Book with students; the key concept for students is that there is a finite supply of nutrients available – only by constant natural recycling processes occurring can life be sustained.

Plenaries

What is the difference between a decomposer and a detritivore? – Ask students to write a definition for a decomposer and a detritivore. To support students, provide them with key terms to include in their definitions. To extend students, ask them to illustrate their definitions with examples of decomposers and detritivores. *(5 minutes)*

Nutrient cycling – Ask students to revisit their nutrient cycles produced in the starter activity. In light of the work covered in this lesson, are students happy that their cycles are complete? Ask groups to produce what they feel is a comprehensive cycle on a large sheet of paper, and share it with the rest of the class. *(10 minutes)*

Chapter 4 – Biomass, energy flow and the importance of carbon

Practical support

What are the optimum conditions for decay?

Equipment and materials required: five petri dishes, materials to decay (bread), anhydrous calcium chloride, sticky tape.

Details
Students should be given the opportunity to investigate the optimum conditions for a material to decay. Each group should set up a series of petri dishes, which are then left for a period of time before being examined for the degree of decay that has occurred. Bread forms a good subject to study, and should be left for between 1 and 2 weeks.

Station 1 – Open to the environment
Station 2 – Open to the environment, and dampened
Station 3 – Sealed, but having been previously open to the environment
Station 4 – Sealed, dry (use anhydrous calcium chloride to absorb water vapour from within the dish)
Station 5 – Sealed, stored in a fridge or freezer.

Following the incubation period, students should study the petri dishes and order the stations in terms of the degree of decay that has occurred.

Safety: Wear eye protection. Once mould starts to grow on the bread, it should be kept in a sealed plastic box, or in sealable plastic bags. Dispose of these correctly by autoclaving the unopened plastic boxes or bags. Spores can cause allergic reactions in some people. CLEAPSS Hazcard 19A Calcium chloride (anhydrous) – irritant.

Answers to in-text questions

a Decomposers break down dead organic material (at microscopic level), releasing nutrients. Detritivores break down material into smaller pieces, speeding up decomposition.
b warm, moist and oxygen-rich
c the food they eat

4.4 Recycling of nutrients

Learning objectives
- What are decomposers?
- What are detritivores?
- What is meant by nutrient cycling?
- How can we use recycled organic waste from gardens and kitchens?

What are decomposers and detritivores?
Imagine what the world would be like if plants and animals didn't decay when they died. Not only would it soon be swamped with waste, but all the nutrients within that waste would never be available for use.

Thankfully, this doesn't happen. When plants and animals die, their bodies are broken down by **decomposers**. This releases the elements they contain back into the environment. These elements can be used again when they are absorbed by plants in the form of soluble mineral salts. For example, nitrogen is obtained from nitrates.

Decomposers are microorganisms – bacteria and fungi (moulds) – which break down dead organic material. They also break down animal waste – faeces and urine.

Detritivores are small animals that help to break down organic material. This speeds up decomposition. They shred the dead material into very small pieces, which makes it easier for decomposers to break down. Examples include:

- earthworms – break down dead leaves
- woodlice – break down wood
- maggots – break down animal material.

a What is the difference between a decomposer and a detritivore?

Figure 1 A dung beetle is an example of a detritivore. It helps to break down organic waste.

How do decomposers release nutrients?
Bacteria and fungi have **enzymes** that break down complex chemicals in organic matter. They produce soluble nutrients, which they can then absorb into their bodies. The nutrients are used for growth and energy. They include amino acids and sugars. If the bacteria and fungi are then eaten by other organisms the nutrients are passed on. In addition, some of the nutrients are released directly into the environment.

What are the optimum conditions for decomposition?
Microorganisms decompose materials most efficiently in conditions that are:

- **Warm** – at high temperatures, the enzymes used by microorganisms are destroyed. At low temperatures, the process of decay is slowed down by reducing the enzymes' rates of reaction.
- **Moist** – if not enough water is available, reactions within the microorganisms will slow down or be prevented altogether.
- **Oxygen-rich** – oxygen is needed for the microorganisms to respire. Anaerobic (oxygen-free) conditions will prevent most forms of decay.

b What are the optimum conditions for decay to take place?

Nutrient cycling within an ecosystem
Plants need nutrients (minerals, elements and organic compounds) for growth. They obtain these from the soil. These are then passed on to animals when the plant is eaten. When plants shed material, such as leaves, and when plants and animals die, decomposers release the nutrients trapped in them. Many of the nutrients find their way back into the soil, where they are absorbed by plants.

This process is known as nutrient cycling. It is summarised in Figure 2.

c Where do animals get their nutrients from?

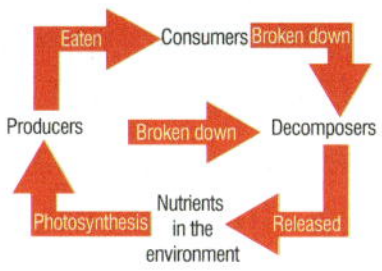

Figure 2 Nutrients are constantly cycled through an ecosystem

Activity

Make your own compost

Compost is nutrient-rich decayed organic material. It provides an excellent growing medium for plants and composting is a very good way of recycling waste. Instead of throwing away kitchen scraps and garden waste, try making your own compost. Follow these four simple steps.

1 Collect kitchen scraps, such as fruit and vegetable peelings, egg shells, tea bags and toilet roll tubes. Do not use the remains of meat or fish or cooked food.
2 Add the kitchen scraps to garden waste (grass clippings, etc.), in approximately equal amounts, to your compost bin. (A normal plastic dustbin with air holes drilled into it can be used.)
3 Wait! It takes between 9 and 12 months to become compost, but during this time you can keep adding to the bin.
4 Once the compost has turned into a crumbly dark material, which resembles thick moist soil, it is ready to use.

Questions
a Why is compost good for your garden?
b Name at least **two** advantages of making your own compost, compared with buying compost from a garden centre.
c Are there any disadvantages of making your own compost?
d Why do some people add worms to their compost bins?
e Why do compost bins often feel warm?

Summary questions

1 Copy and complete using the following words:
small bacteria detritivores decomposers dead

Microorganisms called .... return nutrients to the environment by breaking down .... organic material. They include .... and fungi. Small animals called .... speed up decomposition by shredding the organic material into .... pieces.

2 How do decomposers break down organic material?
3 Why would vegetables keep longer in an airtight container in the fridge, rather than in a fruit bowl on the kitchen table?

Key points

- Decomposers are microorganisms that break down dead organic material.
- Detritivores are small animals that speed up decomposition by shredding organic material into smaller pieces.
- Nutrients are constantly cycled through the ecosystem, so they can be used over and over again.
- We can recycle organic waste from gardens and kitchens.

Activity answers

Make your own compost

a It returns nutrients to the soil, which helps the plants to grow more effectively.
b Cheaper, recycles your own garden waste; no transport costs; uses no chemicals in its production.
c Can attract rats to garden; requires (small amount of) space; takes time to produce (9–12 months approximately).
d Worms shred up the organic material, increasing the surface area for decomposition.
e Microorganisms release energy (by warming their surroundings) when they respire.

Further teaching suggestions

Role of detritivores in decomposition

The action of detritivores could be studied. Students could collect woodlice from the school site and provide a controlled environment and a food source, such as rotten wood. This could be examined at intervals to see the effect of the woodlice on the pieces of wood.

Making your own compost

Students could produce a leaflet to explain to people how they can make their own compost heap in their garden. It should list the benefits of using home-produced compost both to the gardener, and to the wider environment.

Unit 1, Theme 2 – Life on our planet

4.5 The carbon cycle

Learning objectives

Students should learn:

- how carbon is continually cycled throughout the biosphere
- that plants and algae remove carbon from the atmosphere via photosynthesis
- that carbon dioxide is released into the atmosphere when living organisms respire, and when fossil fuels are burned.

Learning outcomes

Most students should be able to:

- state that carbon is removed from the atmosphere via photosynthesis
- describe how carbon is returned to the atmosphere, through respiration and burning
- describe simply how carbon is cycled through the biosphere.

Some students should also be able to:

- explain in detail how carbon is cycled through the biosphere.

AQA Specification link-up: Science B 3.3.2.3

- Know that carbon dioxide is removed from the environment by green plants and algae for photosynthesis.
- Know that the carbon from the carbon dioxide is used to make carbohydrates, fats and proteins, which make up the bodies of plants and algae.
- Know that when green plants and algae are eaten by animals some of the carbon becomes part of the fats and proteins that make up their bodies.
- Understand that when green plants, algae and animals respire some of this carbon becomes carbon dioxide and is released into the atmosphere.
- Understand that when plants, algae and animals die, some animals and microorganisms feed on their remains and release carbon dioxide into the atmosphere when they respire.
- Know that carbon is stored in fossil fuels and is released as carbon dioxide when they are burnt.

Lesson structure

Starters

Photosynthesis and respiration – Ask students to work in pairs. One student writes the word (and, at higher level, symbol) equation for photosynthesis, the other respiration. Students assess each others' equations, and correct any errors. *(5 minutes)*

A very simple carbon cycle – Ask students to arrange the following terms into a simple carbon cycle. Carbon in the atmosphere, photosynthesis, carbon in plants, eaten, carbon in animals, respiration. To support students, provide them with a completed diagram (in a different format) to refer to. To extend students, ask them to add extra information to their cycle – on decomposition, and the burning of fossil fuels. *(10 minutes)*

Main

- Provide students with a large, blank diagram of the carbon cycle. It should have arrows pre-printed, with boxes for labels and diagrams – but the content of these should be missing for students to complete as the lesson progresses.
- Starting at the top of the diagram, ask students to label the form in which carbon is found in the atmosphere – students should label carbon dioxide in the atmosphere. They should then revise the concept of photosynthesis to explain how carbon is removed from the atmosphere. Students should then annotate their diagrams with the appropriate information.
- Discuss and recap work from previous lessons on how carbon passes from one organism to the next through the food chain. Students should then annotate their diagrams with the appropriate information.
- Students should demonstrate to each other that they are breathing out carbon dioxide as a result of respiration, by breathing out through a straw into limewater. Demonstrate to the class the products of combustion using the 'Practical support' activity, details opposite. Both practicals also need a control set up in order to show it is not just the carbon dioxide already in the air that turns the limewater cloudy in the time observed. Students should then annotate their carbon cycle diagrams with the relevant information. It is well worth introducing the idea at this stage that this has only become a significant part of the carbon cycle since the industrial revolution. The next lesson will focus more closely on this aspect of the carbon cycle.
- Revise the process of decomposition from the previous lesson. Students should then annotate their diagrams with the appropriate information.

Plenaries

How can you test for the presence of carbon dioxide? – Ask students to describe the limewater test for carbon dioxide, in a way that would be easily understood by a Year 7 student. They can explain the test using words or pictures. *(5 minutes)*

How does carbon move from one part of the cycle to another? – Provide students with a diagram of the carbon cycle (ideally in a slightly different format to the one used in the lesson), with pictures and arrows but no annotations. Ask students to complete the diagram. To support students, provide them with labels to add to the diagram; they need to organise them into the correct positions. To extend students, ask them to add any relevant words or chemical equations to their diagrams. *(10 minutes)*

Support

- Provide students with different forms of the carbon cycle, each with some information missing. Students should then complete the diagrams.

Extend

- Ask students to produce a cartoon strip, to show how a specific fossil fuel is produced. It should explain how carbon becomes trapped in the fossil fuel.

Practical support

The products of combustion

Equipment and materials required: Glass funnel, glass delivery tubing (arranged as in apparatus diagram), two boiling tubes, boiling tube rack, cobalt chloride paper, limewater, tap operated suction pump, tin lid, range of fuels – small pieces of wood, paper, crushed coal pieces, eye protection.

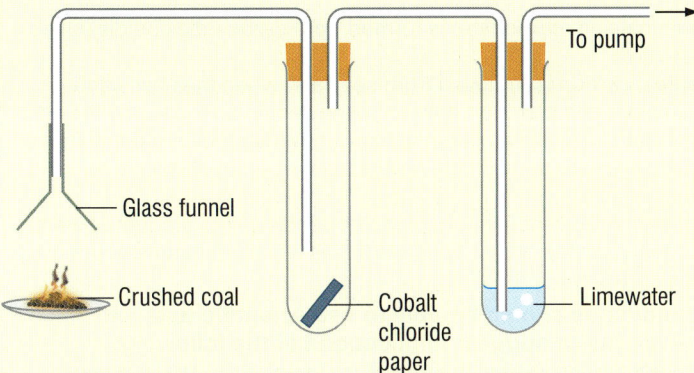

A tap-operated or hand pump provides adequate suction to draw the gases through the apparatus.

Details

A range of materials can be burned to complete the demonstration – good examples include wood, paper and crushed pieces of coal. As the fuel burns, the gases produced are drawn through the delivery tubing. Water vapour in these gases will change the colour of the cobalt chloride paper from blue to pink. Carbon dioxide will change the appearance of the limewater from clear to cloudy. The same products will be produced from any organic-based fuel.

Safety: Wear eye protection. CLEAPSS Hazcard 25 Cobalt chloride – toxic (handle as little as possible). CLEAPSS Hazcard 18 Limewater – irritant.

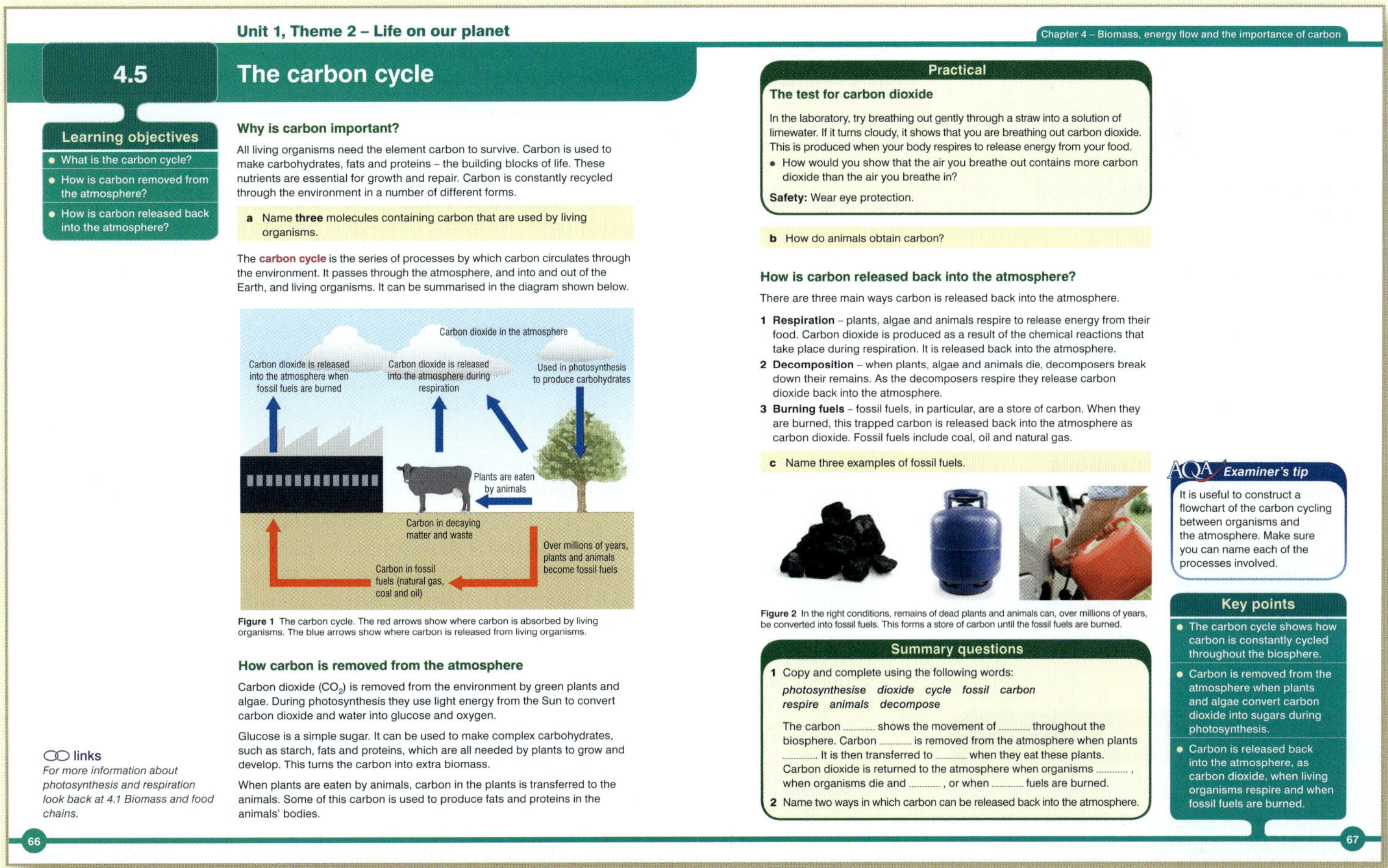

4.5 The carbon cycle

Learning objectives
- What is the carbon cycle?
- How is carbon removed from the atmosphere?
- How is carbon released back into the atmosphere?

Why is carbon important?

All living organisms need the element carbon to survive. Carbon is used to make carbohydrates, fats and proteins – the building blocks of life. These nutrients are essential for growth and repair. Carbon is constantly recycled through the environment in a number of different forms.

a Name **three** molecules containing carbon that are used by living organisms.

The **carbon cycle** is the series of processes by which carbon circulates through the environment. It passes through the atmosphere, and into and out of the Earth, and living organisms. It can be summarised in the diagram shown below.

Figure 1 The carbon cycle. The red arrows show where carbon is absorbed by living organisms. The blue arrows show where carbon is released from living organisms.

How carbon is removed from the atmosphere

Carbon dioxide (CO_2) is removed from the environment by green plants and algae. During photosynthesis they use light energy from the Sun to convert carbon dioxide and water into glucose and oxygen.

Glucose is a simple sugar. It can be used to make complex carbohydrates, such as starch, fats and proteins, which are all needed by plants to grow and develop. This turns the carbon into extra biomass.

When plants are eaten by animals, carbon in the plants is transferred to the animals. Some of this carbon is used to produce fats and proteins in the animals' bodies.

links
For more information about photosynthesis and respiration look back at 4.1 Biomass and food chains.

Practical

The test for carbon dioxide

In the laboratory, try breathing out gently through a straw into a solution of limewater. If it turns cloudy, it shows that you are breathing out carbon dioxide. This is produced when your body respires to release energy from your food.
- How would you show that the air you breathe out contains more carbon dioxide than the air you breathe in?

Safety: Wear eye protection.

b How do animals obtain carbon?

How is carbon released back into the atmosphere?

There are three main ways carbon is released back into the atmosphere.

1. **Respiration** – plants, algae and animals respire to release energy from their food. Carbon dioxide is produced as a result of the chemical reactions that take place during respiration. It is released back into the atmosphere.
2. **Decomposition** – when plants, algae and animals die, decomposers break down their remains. As the decomposers respire they release carbon dioxide back into the atmosphere.
3. **Burning fuels** – fossil fuels, in particular, are a store of carbon. When they are burned, this trapped carbon is released back into the atmosphere as carbon dioxide. Fossil fuels include coal, oil and natural gas.

c Name three examples of fossil fuels.

Figure 2 In the right conditions, remains of dead plants and animals can, over millions of years, be converted into fossil fuels. This forms a store of carbon until the fossil fuels are burned.

Examiner's tip
It is useful to construct a flowchart of the carbon cycling between organisms and the atmosphere. Make sure you can name each of the processes involved.

Key points
- The carbon cycle shows how carbon is constantly cycled throughout the biosphere.
- Carbon is removed from the atmosphere when plants and algae convert carbon dioxide into sugars during photosynthesis.
- Carbon is released back into the atmosphere, as carbon dioxide, when living organisms respire and when fossil fuels are burned.

Summary questions

1. Copy and complete using the following words:
   photosynthesise dioxide cycle fossil carbon respire animals decompose

   The carbon .... shows the movement of .... throughout the biosphere. Carbon .... is removed from the atmosphere when plants .... It is then transferred to .... when they eat these plants. Carbon dioxide is returned to the atmosphere when organisms ...., when organisms die and ...., or when .... fuels are burned.

2. Name two ways in which carbon can be released back into the atmosphere.

Further teaching suggestions

Carbon cycling
Ask students to imagine that they are a molecule of carbon. They should write an essay explaining their journey from the atmosphere, back to the atmosphere via plants, animals and decomposers.

Act out the carbon cycle
Students perform a role-play of the carbon cycle. Provide students with A4 sheets of paper labelled with different stages of the carbon cycle. Students arrange themselves in the correct order to pass along a ball labelled as carbon.

Answers to in-text questions
a. carbohydrates, fats and proteins
b. by eating plants and/or other animals
c. coal, oil and natural gas

Summary answers

1. cycle, carbon, dioxide, photosynthesise, animals, respire, decompose, fossil
2. **two** from: respiration, burning fossil fuels, decomposition

Unit 1, Theme 2 – Life on our planet

4.6 Human influence on the carbon cycle

Learning objectives

Students should learn:
- that carbon may be stored in the bodies of living organisms, fossil fuels and limestone rock
- that limestone is formed under the sea when marine animals and plankton die
- how human activity has increased carbon dioxide levels in the atmosphere through the burning of fossil fuels and deforestation.

Learning outcomes

Most students should be able to:
- name some 'carbon stores' – places where carbon is stored
- explain how limestone originates
- list ways in which human activity has increased carbon dioxide levels in the atmosphere.

Some students should also be able to:
- explain why deforestation contributes to increased atmospheric carbon dioxide levels in two different ways.

AQA Specification link-up: Science B 3.3.2.3

- Know that carbon is stored in fossil fuels, and is released as carbon dioxide when they are burnt.
- Explain how limestone (calcium carbonate) is formed from carbon dioxide dissolved in water:
  (a) over long time scales, carbon is removed from seawater when the shells and bones of marine animals and plankton collect on the sea floor. These shells and bones are made of limestone, which contains carbon. When they are deposited on the sea floor, carbon is stored from the rest of the carbon cycle for some amount of time.
  (b) the amount of limestone deposited in the ocean depends on the amount of warm, tropical, shallow oceans on the planet because this is where limestone-producing organisms such as corals live.

Within this context, candidates should be able to use scientific data and evidence to discuss, evaluate or suggest implications of the following:
- human interference in the natural carbon cycle, e.g. the destruction of rainforests and other forms of vegetation without replanting.

Lesson structure

Starters

Name some stores of carbon – Ask students to name some stores of carbon. This forms a brief revision of content from the last lesson. *(5 minutes)*

How can humans interfere with the carbon cycle? – Ask students in pairs to make a list of the ways humans can increase the levels of carbon dioxide in the atmosphere. To support students – supply them with a series of statements to sort into 'Could affect carbon dioxide levels in the atmosphere' and 'Does not affect carbon dioxide levels in the atmosphere'. Examples of statements could include: burning fossil fuels, driving a car, gardening, cycling, wind farms, heating a house. To extend students, ask them to explain how each activity increases the levels of carbon dioxide in the atmosphere. *(10 minutes)*

Main

- Using their knowledge of the carbon cycle, discuss ways in which carbon is stored and temporarily prevented from being cycled – a tree would make a good example as the basis of this discussion. Introduce students to the concept that carbon is stored in limestone.
- Show students some samples of limestone. Ask them to compare it with some other samples of rock. For example, is it soft? What colour is it? What is it made of (grains or crystals)? Students should also add a few drops of dilute hydrochloric acid to the samples.
- Provide students with information explaining how limestone rock is formed. The Student Book contains the basic information. Ask students to find out more about the formation of sedimentary rock and use this information to explain how limestone is formed. They should display the information in a way that best suits their learning. For example, a bulleted list, a cartoon strip, a podcast or a flow diagram.
- Students should then test shells and the samples of rock, with hydrochloric acid (see 'Practical support' for further details).
- Discuss what is meant by the term 'carbon sink' and how the carbon in limestone is eventually returned to the environment. It is well worth spending time at this point asking students to contemplate the carbon cycle without any human interference – a balanced cycle, which has maintained atmospheric carbon dioxide levels over a long period of time.
- Ask students to list as many uses of fossil fuels as they can – discuss how their use is so important to the energy-reliant developed world. Remind students that fossil fuels are carbon-rich and hence their combustion leads to increased levels of carbon dioxide. Although it is worth mentioning the link to global warming at this stage, this is covered in more detail in Chapter 14.
- Introduce the concept of deforestation. Students should research why this has occurred and what steps could be taken to reduce the environmental impact of removing trees with reference to carbon.
- For higher attaining students, introduce the concept of 'slash and burn' agriculture – a rapid clearing (by burning) of a forest area for use in agriculture. Ask students to explain why this has a double effect on the carbon cycle – and what the more developed nations in the world could be doing to prevent this as an environmental issue.

Answers to in-text questions
a **three** from: plants, animals, fossil fuels (coal, oil, natural gas), limestone
b calcium carbonate
c skeletons and shells
d More fossil fuels have to be burnt, releasing more carbon dioxide.

Support
- Provide a large carbon cycle to be projected or displayed throughout the lesson. Constantly refer back to the relevant parts of the diagram, as processes such as burning fossil fuels are discussed.

Extend
- Ask students to contrast the impact on the carbon cycle of a more developed country with that of a less developed country. Students could debate the responsibilities of each group of nations and produce an action plan to control the levels of carbon dioxide in the atmosphere.

Chapter 4 – Biomass, energy flow and the importance of carbon

Plenaries

How is limestone formed? – Provide students with a scrambled flow diagram of limestone formation. They need to arrange the sentences into the correct order. *(5 minutes)*

How can we reduce carbon dioxide levels? – Ask students in small groups to come up with ten ways we can reduce carbon dioxide levels. Encourage students to think of ways we can increase carbon dioxide removal from the atmosphere. To support students, provide them with the list of activities that contribute to carbon dioxide emissions. Students could suggest how each one could be reduced. To extend students, ask them to explain how this reduces carbon dioxide levels – either by reducing reliance on a carbon dioxide producing process, or by increasing the storage of carbon dioxide in a carbon sink. *(10 minutes)*

Summary answers

1. carbon, bodies, millions, limestone, deforestation, fossil, dioxide
2. Burning trees releases carbon dioxide into the atmosphere; less carbon dioxide is removed from the atmosphere through photosynthesis.
3. Greater demand for electricity and more travel (especially air travel) has resulted in more fossil fuels being burned. This leads to higher levels of carbon dioxide being released.

Unit 1, Theme 2 – Life on our planet

4.6 Human influence on the carbon cycle

Learning objectives
- How is carbon stored?
- How is limestone formed?
- How do humans influence the carbon cycle?

Carbon in the environment
Natural processes that maintain the levels of carbon dioxide in the atmosphere include photosynthesis, respiration and combustion.

Stores of carbon
There are four main ways in which carbon is stored away from the atmosphere:
- **In the oceans** – The oceans hold 50 times as much carbon dioxide as the atmosphere. They also absorb a lot of carbon dioxide created by human activities, such as burning fossil fuels.
- **In the bodies of plants and animals** – This carbon is only stored temporarily and is released by decomposers when the organism dies and its body is broken down. Rain forests are particularly vital carbon stores.
- **In fossil fuels** – Oil, coal and natural gas deposits lock up carbon for millions of years. The carbon is released back into the atmosphere in the form of carbon dioxide when the fuel is burned.
- **In rocks** – Carbon can be stored in rocks, such as limestone and chalk, which are composed mainly of calcium carbonate.

a Name **three** stores of carbon.

Limestone
Limestone is formed mainly from the remains of marine animals, such as molluscs, coral and plankton. They convert the carbon dissolved in sea water into calcium carbonate to make shells or other hard parts. When these organisms die, their shells are deposited on the sea bed. Over long periods of time this sediment builds up and forms limestone.

Figure 1 Limestone (calcium carbonate) acts as a store of carbon. It can take millions of years before this carbon is returned to the atmosphere.

Figure 2 Coral is an example of a limestone-building organism

Carbon may be stored in limestone for millions of years. This is an example of a 'carbon sink'. Eventually, through movements of the Earth, the limestone may become exposed to the air. As a result of chemical weathering or thermal decomposition by volcanic activity, carbon may then be released back into the atmosphere as carbon dioxide.

b What is the chemical name for limestone?

c Which parts of marine animals might form limestone?

Many of the marine organisms that are capable of producing limestone live in warm, shallow, tropical waters. However, it is these areas of the oceans that humans have exploited widely. This has endangered the habitats of these creatures, potentially reducing this carbon store.

Human influence
Human activity has increased the amount of carbon dioxide being released back into the atmosphere. This has mostly occurred in two ways:

- **Burning fossil fuels** – Carbon dioxide is produced when we burn fossil fuels. Due to increases in the human population and a desire for a higher standard of living, more energy is being used. We use lots of electrical appliances. More fossil fuels have been burned to generate electricity. This has increased the amount of carbon dioxide released into the atmosphere.
- **Deforestation** – Large areas of forest have been cleared for farming, to make space for roads and buildings, agriculture and animal grazing, as well as for the timber itself. This means that there are fewer trees to remove carbon dioxide from the atmosphere by photosynthesis. Therefore, more carbon dioxide remains in the atmosphere.

In many cases the felled trees are burned or left to decompose. This further increases the levels of carbon dioxide in the atmosphere.

d Why does an increasing use of energy lead to an increase in the amount of carbon dioxide in the atmosphere?

Figure 3 Large-scale deforestation is still taking place in many areas to create space for farming and supply timber for industry/furniture

Summary questions

1. Copy and complete using the following words:
   fossil   dioxide   limestone   deforestation   bodies   carbon   millions

   .... is stored temporarily in the .... of plants and animals. However, it can be stored for .... of years in .... rocks. .... and the burning of .... fuels is increasing the level of carbon .... in the atmosphere.

2. Explain two different ways that burning forests to clear land for farming affects the atmospheric level of carbon dioxide.

3. Explain why an increased standard of living has led to an increase in the amount of carbon dioxide being released.

Key points
- Carbon may be stored in the bodies of plants and animals, fossil fuels and limestone rock.
- Limestone is formed under the sea when marine animals, such as coral and plankton, die and their shells and skeletons build up on the sea floor. Over millions of years, these deposits turn to rock.
- Human activity can increase the levels of carbon dioxide in the atmosphere through deforestation and the burning of fossil fuels.

Further teaching suggestions

Saving energy
Ask students to produce a poster depicting ways to save energy as this will reduce the amount of fossil fuels being burnt, in turn reducing the levels of carbon dioxide in the atmosphere.

Pros and cons of deforestation
Research and produce a balanced argument into why less developed countries may wish to deforest large areas. The writing should include practical suggestions on how more developed countries could help to preserve forests as essential carbon sinks.

Dissolved carbon dioxide
The solubility of carbon dioxide in water can be demonstrated. Bubble carbon dioxide through deionised water; the water becoming acidic. This can be visualised using universal indicator or, ideally, a pH probe connected to a datalogger. A pH versus time graph can then be produced and projected 'live' to the class.

Practical support

Testing shells for the presence of carbon

Equipment and materials required: Crushed shells (e.g. cockle, oyster), and a range of rock samples, hydrochloric acid (1 mol/dm^3), limewater, two test tubes, test tube racks, bung and delivery tube, eye protection.

Details
Students can test crushed shells (e.g. cockle, oyster), and a range of rock samples, with dilute hydrochloric acid. The resulting gas should be bubbled through limewater to demonstrate that the sample is a store of carbon, which has been released in the form of carbon dioxide.

Safety: Wear eye protection. CLEAPSS Hazcard 47A Hydrochloric acid – corrosive. CLEAPSS Hazcard 18 Limewater – irritant.

Unit 1, Theme 2 – Life on our planet

Summary answers

1. **a** oak tree → slug → robin → owl
   b oak tree
   c Herbivores eat plants, carnivores eat animals.
   d They would be likely to increase.

2. **a** 140 kJ – 85 kJ – 35 kJ = 20 kJ
   b $\frac{20}{140} \times 100 = 14.3\%$
   c respiration, waste products, not all parts of the rabbit are eaten

3. **a** increases
   b decreases
   c Pyramids of numbers display the numbers of each organism in a food chain; pyramids of biomass take into account the number and size of organisms at each level of the food chain.

4. **a** bacteria/fungi
   b Shred organic material into smaller pieces.
   c moist, warm and oxygen-rich
   d i Little oxygen is available for bacteria to respire, limiting the rate of decay.
   ii Cool temperatures limit the rate of decay in two ways – by limiting the rate of bacterial replication, and by slowing the rate of enzyme reactions.

5. **a** A – photosynthesis, B – eaten, C – respiration, D – burning/combustion.
   b Pass it through limewater. Carbon dioxide will turn the limewater cloudy.
   c Burn less fossil fuels; reduce use of cars; reduce energy usage.

6. **a** limestone
   b Carbon dioxide dissolves in sea water. Marine animals including plankton, convert this carbon into calcium carbonate to form skeletons or a hard outer shell. When these organisms die, their shells and skeletons are deposited on the sea bed. Over long periods of time this sediment is compressed, forming limestone.
   c As a result of earth movements, limestone may become exposed. Physical and chemical weathering of the rock will result in the carbon being released.

7. **a**

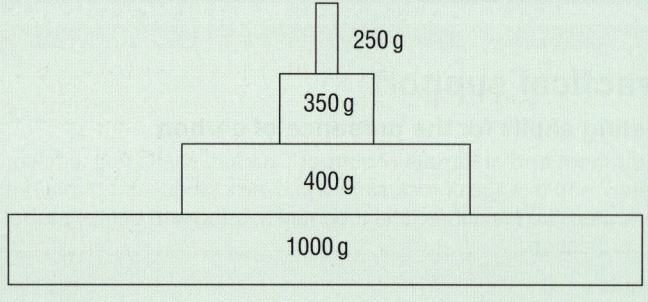

   b $\frac{250}{350} \times 100$
   $= 71\%$
   c more leaves

Unit 1, Theme 2 – Life on our planet

Summary questions

1. These organisms form a food chain.

   a Rearrange the organisms above into a food chain.
   b Which organism is a producer?
   c What is the difference between a herbivore and a carnivore?
   d What could happen to the number of robins if all the owls died?

2. A rabbit has consumed 140 kJ of energy from its food. Of this, 35 kJ are lost through waste and 85 kJ are lost through respiration.
   a How much energy would be available to a fox that ate this rabbit?
   b What percentage of the energy that a rabbit consumes is transferred to the fox?
   c Name three ways energy is 'lost' from the rabbit, and not transferred to the fox.

3. This is a typical pyramid of numbers.

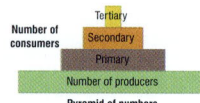

   a In general, what happens to the size of the organisms as you move up a food chain?
   b In the pyramid shown above, what happens to the number of organisms at each level as you move up the food chain?
   c What is the difference between a pyramid of numbers and a pyramid of biomass?

4. When plants and animals die, their bodies are broken down by **decomposers**. This releases the nutrients they contain back into the environment, where they can be used again.
   a Name a type of organism that decomposes materials.
   b How do detritivores help decomposition occur more quickly?
   c What are the optimum conditions for decomposition?
   d Why does food keep fresh for longer in:
   i an airtight container?
   ii a fridge?

5. This is a diagram of the carbon cycle.

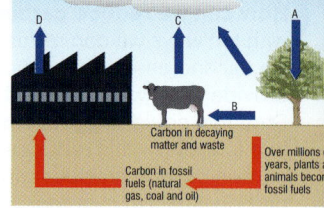

   a Which processes do arrows A to D represent on the carbon cycle?
   b How can you test for the presence of carbon dioxide?
   c What steps can be taken to limit the amount of carbon dioxide that is being added to the atmosphere through human activity?

6. Carbon is temporarily stored in parts of the carbon cycle – sometimes for millions of years.
   a Calcium carbonate is the chemical name of which rock?
   b Describe how calcium carbonate is formed.
   c Explain how carbon is released from calcium carbonate, back into the atmosphere.

7. The pyramid of numbers for a farmland food chain is given below.

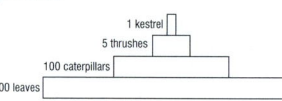

   a Draw and label a pyramid of biomass for the food chain using information from the table.

Organism	Mass in grams
Leaf	5
Caterpillar	4
Thrush	70
Kestrel	250

   b Calculate the percentage of biomass that is passed from the thrushes to the kestrel.
   c Suggest why the biomass of a tree is greater in summer than it is in winter.

Kerboodle resources

- Simulation: Microorganisms and decay (4.4)
- Practical: Composting (4.4)
- Interactive activity: Exploring energy in biomass
- Examination-style questions
- Answers to examination-style questions
- Test yourself: Life on our planet

AQA Practical suggestions

Practicals	AQA	k	📖
Role play – A4 sheets of paper labelled with different stages of the carbon cycle. Students arrange themselves in the correct order to then pass along a ball labelled as carbon.	✓		
Test crushed shells (e.g. cockle, oyster) with dilute hydrochloric acid to show they contain carbonates.	✓		✓

Chapter 4 – Biomass, energy flow and the importance of carbon

AQA Examination-style questions

1 Choose the correct words from the box to complete the sentence.

| food | fox | grass | heat | movement |

In a food chain of: grass ⟶ rabbit ⟶ fox, all the energy that the rabbit gets from the .... is not passed on to the .... as the rabbit transfers some of its energy into .... and .... (4)

2 Match the correct pyramid of numbers to its food chain.

Food chain	
a 1 beech tree ⟶ 50 bark beetles ⟶ 5 blue tits ⟶ 1 sparrowhawk	(1)
b 1 oak tree ⟶ 100 caterpillars ⟶ 10 blue tits	(1)
c 10 lettuce plants ⟶ 2 rabbits ⟶ 100 fleas	(1)
d 300 oak leaves ⟶ 100 caterpillars ⟶ 10 shrews ⟶ 1 owl	(1)

Pyramid of numbers

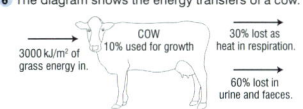

A   B   C   D

3 The element carbon can move through plants, animals, the air and the ground. This movement is called the carbon cycle.
a Describe the role that plants play in absorbing and storing carbon dioxide. (3)
b Name and describe a process by which carbon dioxide is released into the atmosphere. (2)

4 Energy flows through living things in a food chain. The organisms below are part of a woodland food chain.

| fox | hedgehog | leaves | snail |

a Explain which organism is the producer. (2)
b Explain why the fox is called a tertiary consumer. (2)
c Suggest how the biomass of a leaf could be found. (2)

5 Energy enters the food chain as sunlight.
a The Sun provides 100 000 kJ/m² per year to producers.
Of this, plants waste 90 000 kJ/m² per year as heat and 500 kJ/m² per year is inedible.
Calculate the percentage of energy transferred to the next stage of the food chain. (4)
b Explain why not all of the energy in a prey transfers to its predator. (2)

6 The diagram shows the energy transfers of a cow.

[Diagram: COW, 3000 kJ/m² of grass energy in, 10% used for growth, 30% lost as heat in respiration, 60% lost in urine and faeces.]

a Calculate how much of the energy that the cow gets per m² of grass is lost. (2)
b Use the equation to calculate how efficient the cow is at making biomass.

efficiency = (useful energy / total energy) × 100% (2)

c Suggest **one** reason why cows spend most of the time grazing. (1)

7 Part of the carbon cycle is shown in the diagram.
a Name the processes labelled A, B and C.

[Diagram: Carbon in the atmosphere → (C) → Carbon in an animal → (B) → Carbon in a plant → (A) → Carbon in the atmosphere] (3)

b *In this question you will be assessed on using good English, organising information clearly and using specialist terms where appropriate.*

Explain how carbon is locked away in the formation of limestone. (6)

AQA Examination-style answers

1 grass (1 mark)
fox (1 mark)
movement and heat (in either order) (2 marks if 2 correct, 1 mark if 1 correct)

2 a B
b C
c D
d A (3 marks if 3 or 4 correct, 2 marks if 2 correct, 1 mark if 1 correct)

3 a Plants take in carbon dioxide from the air (1 mark)
They use it during photosynthesis (1 mark)
To make fats / proteins / carbohydrates (1 mark)
b Animals and plants release it
In a process called respiration
OR
It is released from fossil fuels (1 mark)
In a process called combustion / burning (1 mark)

4 a leaves because they make their own food. (2 marks)
b It is a called a consumer because it needs to eat to get energy. (1 mark)
It is called tertiary because it is the third consumer in the food chain. (1 mark)
c Dry the leaf. (1 mark)
Weigh it. (1 mark)

5 a 100 000 − 90 000 − 500 (1 mark)
= 9500 (1 mark)
9500 ÷ 100 000 × 100 (1 mark)
= 9.5 % (1 mark)
b Some of the prey is inedible. (1 mark)
The prey transfers some of its energy to the environment during respiration. (1 mark)

6 a 60 + 30 = 90% OR 100 − 10 = 90%
90% of 3000) = 2700 kJ (both answer and unit required for the mark) (2 marks)
b $\frac{300}{3000} \times 100 = 10\%$ (2 marks)
c Only use 10% of energy from the grass for growing OR waste most of the energy they gain from the grass. (1 mark)

7 a A – Photosynthesis (1 mark)
B – Nutrition/feeding/eating (1 mark)
C – Respiration (1 mark)
b Marks awarded for this answer will be determined by the Quality of Written Communication (QWC) as well as the standard of the scientific response.

There is a full and detailed explanation of how carbon is locked away in the formation of limestone. The answer is well structured with minimal repetition or irrelevant points. There is an accurate, fluent and clear expression of ideas with only minor errors in the use of technical terms, spelling, punctuation and grammar. (5–6 marks)

There is some explanation of how carbon is locked away in the formation of limestone with some omissions. The answer shows some attempt at structuring and the ideas are expressed with reasonable fluency and clarity. There are some errors in the use of technical terms, spelling, punctuation and grammar. (3–4 marks)

There is an attempt at a brief explanation of how carbon is locked away in the formation of limestone. The answer is largely incomplete and may contain some valid points which are not clearly structured. It lacks fluency and/or clarity. It contains errors in the use of technical terms, spelling, punctuation and grammar. (1–2 marks)

No relevant content (0 marks)

Examples of points made in the response:
- Carbon dioxide is soluble in water
- Carbon dioxide is used to form calcium carbonate (limestone)
- Shells and bones of marine animals and plankton are made of limestone
- These shells and bones are deposited on the sea floor when the organism dies
- The amount of limestone deposited in the ocean depends on the amount of warm, tropical, shallow oceans because this is where limestone producing organisms live.

Unit 1 – My world

AQA Examination-style answers

1 a Any **two** sensible suggestions, e.g. *(1 mark for the adaptation and 1 for the reason)*
- long eyelashes; to stop sand going into eyes
- does not sweat; to avoid losing too much water
- large feet; so they do not sink into sand *(4 marks)*

b in water *(1 mark)*
to help mobility *(1 mark)*

2 a increases surface area *(1 mark)*
quicker *(1 mark)*
OR
more salt can be dissolved *(1 mark)*

b Add water *(1 mark)*
to dissolve the soluble salt from the rock salt *(1 mark)*
filter *(1 mark)*
to separate the insoluble rock from the salty water *(1 mark)*
then evaporate / heat to remove the water *(1 mark)*

3 a Nucleus in the centre labelled *(1 mark)*
Nucleus contains 7 protons and 7 neutrons *(1 mark)*
There are 2 electrons in the first shell (energy level) and 5 electrons in the second shell around the outside of the nucleus. *(1 mark)*

b Electrons are removed. *(1 mark)*

4 a 2Fe *(1 mark)*
$3CO_2$ *(1 mark)*

b Carbon monoxide is used as a reducing agent. *(1 mark)*
It reacts with the oxygen from the iron oxide. *(1 mark)*

5 a Air is a mixture. *(1 mark)*
It is made up of many different elements and compounds *(1 mark)*
that are not chemically combined. *(1 mark)*

b Oxygen is an element. *(1 mark)*
It is made up of just one type of atom. *(1 mark)*

6 a Marks awarded for this answer will be determined by the Quality of Written Communication (QWC) as well as the standard of the scientific response.

There is a clear and detailed description of the structure of the Earth. The answer is well structured with minimal repetition or irrelevant points. There is an accurate, fluent and clear expression of ideas with only minor errors in the use of technical terms, spelling, punctuation and grammar. *(5–6 marks)*

There is some description of the structure of the Earth with some omissions. The answer shows some attempt at structuring and the ideas are expressed with reasonable fluency and clarity. There are some errors in the use of technical terms, spelling, punctuation and grammar *(3–4 marks)*

There is a brief description of the structure of the Earth. The answer is largely incomplete and may contain some valid points which are not clearly structured. It lacks fluency and/or clarity. It contains errors in the use of technical terms, spelling, punctuation and grammar. *(1–2 marks)*

No relevant content *(0 marks)*

Examples of points made in the response:
- The outer part of the Earth's surface is called the crust
- It is split into a number of large pieces called plates
- Just below that there is the mantle
- Mantle is a liquid
- In the centre there is the core
- The core is made from molten rock.

AQA Examination-style questions

1 Animals and plants become adapted to their environment.

  a Suggest and explain **two** adaptions of animals to help them live in the desert. (4)
  b Suggest and explain an environment where webbed feet would be an advantage. (2)

2 Rock salt has to be separated before it is used.
  a Explain why it might be an advantage to crush the rock salt first. (2)
  b Explain the process that could be carried out in the lab to separate salt from rock salt. (6)

3 Scientists need to know the structure of the atom so they can understand how reactions happen.
  a Nitrogen has a mass number of 14 and an atomic number of 7. Draw a labelled diagram of a nitrogen atom. (3)
  b Explain how an atom can become a positive ion. (1)

4 Iron can be separated from its ore using a reduction reaction.
  a Complete the symbol equation for the reduction of iron oxide.
    $Fe_2O_3 + 3CO \rightarrow \ldots Fe + \ldots CO_2$ [H] (2)
  b Use the word equation to explain how iron is separated from its ore. (2)

5 Materials can be classified as element, compound or mixture.
  a Explain how air could be classified. (3)
  b Explain how oxygen molecules could be classified. (2)

6 Earth scientists have mapped the structure of the Earth.
  a *In this question you will be assessed on using good English, organising information clearly and using specialist terms where appropriate.*
    Describe the structure of the Earth. (6)
  b Explain what happens inside the Earth to make a volcano (4)

7 Choose the correct word from each box to complete the sentences.
  a Astronomers believe that the universe is ....

| expanding | shrinking | warming |

  b The evidence for this is that light coming from distant stars is ....

| absorbed | squashed | stretched |

  c This change in the light is called .... shift.

| blue | green | red |

(3)

---

  b Convection currents *(1 mark)*
inside the mantle *(1 mark)*
move the plates. *(1 mark)*
When they move apart volcanoes occur *(1 mark)*

7 a expanding *(1 mark)*
  b stretched *(1 mark)*
  c red *(1 mark)*

8 a It has increased. *(1 mark)*
Oxygen has been added. *(1 mark)*

  b Any **two** sensible answers, e.g.:
- repeat experiment and find an average
- use a more precise pan balance reading to 2 decimal places *(2 marks)*

9 a 300 million years ago *(1 mark)*
Carbon dioxide levels start to increase. *(1 mark)*
Animals breathe it out. *(1 mark)*

  b i Keeps temperatures on planet warm and stable *(1 mark)*
so as to support life. *(1 mark)*
    ii increasing *(1 mark)*
produced during the combustion of fossil fuels *(1 mark)*
OR
deforestation

End of Unit 1 questions

Magnesium burned in air produces magnesium oxide.

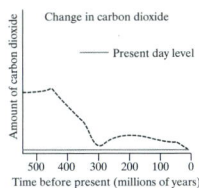

a Explain what has happened to the mass of the magnesium. (2)
b How could the accuracy of the experiment be improved? (2)

The level of carbon dioxide has changed a lot over the history of this planet.

Change in carbon dioxide

(graph: Amount of carbon dioxide vs Time before present (millions of years) 500 400 300 200 100 0, with Present day level indicated)

a Using the graph suggest the year that animals first started to appear and explain your choice. (3)
b i Carbon dioxide is a greenhouse gas. Describe the need for greenhouse gases on a planet. (2)
ii Explain what is happening to the carbon dioxide levels now. (2)

Bump up your grades

Make sure your C/D borderline candidates can describe the evidence for the Big Bang in terms of change of wavelength and red-shift. They need to know all the chemical formulae for the gases in the early atmosphere. They should be able to explain how greenhouse gases cause global warming and draw a labelled diagram of the atom. Calculation of the energy flow through a food chain is expected but in the higher tier paper, the equation may not be given so students would have to come up with it themselves.

Grade A candidates need to be able to write balanced equations for all the processes in this unit. Using balanced symbol equations in an explanation is perfectly acceptable as an answer to an exam question as long as they are correct.

Grade F candidates should be able to describe how animals are adapted to a particular environment, and write a conclusion based on a graph or a table.

Unit 2, Theme 1
My family

Specification link-up: Science B 3.4.1

In this theme there are three contexts:

3.4.1.1 Control of body systems

3.4.1.2 Chemistry in action in the body

3.4.1.3 Human inheritance and genetic disorders

Control of body systems

Doctors use their knowledge of physiology, biochemistry and drugs to diagnose and treat disease. Before they can diagnose a problem they need to understand how the body functions normally. They need to be aware of how humans respond to external changes, through the use of nerves. They also need to understand how a constant internal environment is maintained through the process of homeostasis.

How do we react to our environment?

Activity:

- Ask students to write down as many situations as possible that the body would respond to with a reflex response. What is the result of the reflex response? Why was this response necessary?
- Working in pairs, ask students to study their forearms using a magnifying lens. Ask them to discuss with their partner its appearance. Then ask one of the pair to wrap their forearm in an icepack or bag of ice for two minutes and the other to place a hot water bottle or heated wheat pack on their arm. Students should then re-examine their arms – how has the appearance changed? Why has this occurred?

Can loud sounds damage your hearing?

Activity:

- Ask students to produce a list of all the jobs and situations in which continual exposure to loud sounds could lead to hearing damage. For each occupation or situation, students should state how the damage could be prevented. They should try to think of a number of different prevention or control measures.
- Students should carry out a short piece of research work into sound intensities on the decibel scale. Using a large piece of poster paper, ask students to produce a display that demonstrates the loudness of everyday sounds on this scale. Students could also include a range of relevant quantities on the scale – the threshold of human hearing, the maximum permissible noise levels at home and at work and the pain threshold, for example. More able students could be asked to describe what an increase of 10 dB implies – an increase in sound intensity by a factor of 10.
- Blow a dog whistle, in conjunction with a microphone and oscilloscope, so that students can 'see' a sound wave, but not hear it. Use this stimulus to discuss animals that can detect a wider, or different, range of frequencies from humans. How does the frequency range of, for example, a dolphin compare with those of a human?

Possible misconceptions

The representation of sound as a wave, usually by using an oscilloscope connected to a microphone, can reinforce the misconception that sound waves are of the same type as light waves (transverse). It is worth keeping a demonstration slinky available throughout the lesson to reinforce the nature of longitudinal waves. A little more fiddly, but highly effective, technique is to attach a data logging position sensor, on a long arm, to a part of the slinky. As you demonstrate longitudinal waves using the slinky, the position sensor produces the familiar sine waveform on a computer screen for students to observe. The similarities and differences can then be discussed.

Chemistry in action in the body

Many chemical reactions are constantly occurring inside the body, for example, during indigestion. It is essential that the stomach remains acidic as the hydrochloric acid present helps enzymes to break down protein. It also kills off microorganisms that may be present in food. However, at times this acid can also result in indigestion.

Doctors can prescribe antacids to relieve indigestion. Pharmacologists test the effectiveness of antacids, to determine how efficiently they neutralise excess stomach acid before they are sold to the consumer.

My family

Possible misconceptions

Many students think that alkalis are safer than acids. This is often reinforced at KS3 by the explanation that alkalis are 'often used in cleaning products' – students equate this with soaps. Using some carefully selected student safety sheets, students could list the harmful effects of acids and alkalis, before discussing as a class what they have found out.

Some students also think that antacid tablets make your stomach contents neutral. Although they have a neutralising affect, they only reduce the acidity of the stomach. It is essential that the stomach remains acidic for enzymes to work effectively in digestion.

Human inheritance and genetic disorders

Individuals in a family share some similarities with their relatives and some differences. These occur as a result of genetic and environmental causes. Genetic variation results from the combination of genes present within the nucleus of a cell. Geneticists study the structure and function of genes. This means they can explain differences between family members. Sometimes, people inherit a faulty gene, which results in a genetic disorder. Geneticists working on the human genome project are researching treatments and ultimately, cures for genetic disorders.

What are genes?

Activity:

- Provide students with an outline of a cell. Ask them to annotate the features of the cell. Where possible they should name the feature and describe its function. Students could also draw in any other structures they are aware of and explain their purpose e.g. chromosomes and mitochondria.
- Provide students with anagrams of the key genetic terms – DNA, nucleus, chromosome and gene. When students have solved the anagrams they should put them in order of size, starting with the smallest.

Can you change the way you look?

Activity:

- Provide students with an outline of a teenager. Ask students to alter the image to make the person appear as different as possible to the original image. Then, using a red pen they should label any characteristics that would be determined by a person's genes (genetic variation). Characteristics that a person can change themselves (environmental variation) can be highlighted in green. Characteristics affected by both can be labelled in blue. Students' images could be collated and displayed on the board, to emphasise the variation that exists as a result of each of the factors. This would also form a fun interactive whiteboard activity.
- Ask students to work in small groups. Using display materials and a volunteer from within each group, challenge students to highlight characteristics that can be affected by environmental factors. The volunteer students model the highlighted features to the class. For example, using red poster paper, a head of red hair could be created, and mounted onto the 'model'.

Possible misconceptions

Many students think that all cells contain a nucleus, however this is untrue. Red blood cells do not contain a nucleus. This increases their oxygen-carrying capacity.

My family

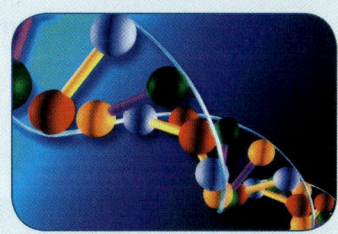

If there is a family history of a genetically inherited disorder, couples often seek advice from a genetic counsellor. Genetic counsellors calculate the risk of a couple having a child with a disorder, such as cystic fibrosis or haemophilia. These are incurable diseases which can have a large effect on the children and their family. Genetic councillors also offer advice on how to manage the conditions. Due to medical advances in the field of genetics, scientists are now able to screen embryos for the presence of genetic disorders.

The hormones insulin and glucagon control the amount of sugar that we have in our blood. Some people suffer from diabetes. They are unable to keep their blood sugar at the correct level. Some types of diabetes can be managed through a careful diet and exercise, but some diabetics have to manage their condition through regular injections of insulin.

As you get older, your ears become less sensitive to sound. This applies to both the loudness and the pitch of the sound. The typical human hearing range is between 20Hz and 20000Hz. As your hearing starts to deteriorate, quiet sounds and higher pitched sounds become more difficult to distinguish. With an ageing population, science and engineering have come together to develop a range of hearing aids to help overcome these problems.

How can you stop indigestion?

Activity:

- Split students into groups of three. Ask one person in each group to write a definition of the terms acid, alkali and neutralisation. After a minute, students should pass their definitions on to the next member of the group so they can try to improve it. Repeat for the third member of the group. After three minutes, discuss as a class the definitions produced, to arrive at an 'agreed' series of definitions.
- Students could carry out a short piece of research into stomach acid. Which type of acid is stomach acid? What are the roles of stomach acid?

Why are hazard warning labels used on cleaning products?

Activity:

- Provide students with a pack of cards containing a list of standard hazard warning symbols and their definitions. Shuffle the packs and ask students to match the correct warning symbol to its definition.
- Using student safety sheets, students should produce a standard procedure for safe working with acids and with bases. These instructions should include safety equipment (such as wearing goggles at all times), advice on dealing with spills, and emergency procedures.

Unit 2, Theme 1 – My family

5.1 Nerves

Learning objectives

Students should learn:
- that nerves transmit electrical impulses around the body, so that the body can react to changes in the external environment
- that during a controlled nervous reaction, information about the environment is sent to the brain to be processed, and an appropriate response triggered
- how the body reacts to danger using reflex actions.

Learning outcomes

Most students should be able to:
- give examples of receptor cells
- name the three types of neuron
- describe the differences between, and steps involved in, controlled nervous reactions and reflex actions.

Some students should also be able to:
- explain why reflex actions occur more quickly than controlled reactions.

Support
- Provide students with a blank table, to record the results of the practical activity 'Reaction times'.

Extend
- Students could research a disease that affects the nervous system, e.g. motor neuron disease.
- The symptoms can be related to, and explained by, the flow diagram of a nervous response.

AQA Specification link-up: Science B 3.4.1.1
- Give examples of receptor cells that detect stimuli (light, sound, smell, taste, touch, heat).
- Describe how information from receptors passes along cells (neurons) in nerves to the brain. The brain coordinates our response.
- Know that some responses to stimuli are automatic and rapid and are called reflex actions.
- Describe how reflex actions involve three neurons called sensory, relay and motor neurons.

Lesson structure

Starters

Factors in your environment – Ask students to produce a list of the factors that change in their environment. An example of temperature could be given to start them off. *(5 minutes)*

Nervous system graffiti – Split students into small groups. The words 'nervous system' are written in large letters in the centre of the paper. Students add as many related words or terms to the paper as possible in a short period of time (for example 90 seconds). Each group then passes its paper onto the next group, adding to the previous responses. *(10 minutes)*

Main

- Make a very loud noise, such as slamming books on to the desk near to someone. Discuss with students how they responded (jump, blink, change of skin colour, heart beats faster) and why. Which part of their body detected these changes in their environment? Introduce the terms sense organs and receptor cells.
- Get students to measure their 'reaction time' in pairs. One student drops a metre rule, the other catches it between two fingers. The distance travelled is measured. Repeat this activity ten times and record the results. Students then analyse their results and relate this to athletes practising race starts.
- If time permits, repeat the experiment, but this time with the bottom of the metre rule resting gently against the catcher's finger. Repeat ten times as before. They should be able to catch the rule faster because they are relying on two senses – sight and touch. The experiment could be repeated again saying 'go' each time the rule is dropped. Students predict what will happen to their reaction time. (It should decrease as they are using another sense.)
- Discuss the flow chart of a nervous reaction in the Student Book, and relate this to the practical that students just carried out. The flow chart could be visualised by role play. Students are given cards to represent each step of the flow diagram. One student acts as the electrical impulses and follows the path, another student could time their movement.
- Discuss reflex actions. Repeat the role play but with the brain removed. When a student acts as an electrical impulse, it takes them a shorter time to follow the path, demonstrating that reflexes occur much quicker than nervous reactions.

Plenaries

Reflex reaction? – Ask students to sort a list of situations that the body responds to into controlled and reflex reactions. *(5 minutes)*

Nerve definitions – Ask students in pairs to write definitions of the following terms – sense organ, receptor cell, effector, sensory neuron, motor neuron and relay neuron. They should take three terms each, then swap their definitions and see if their partner can improve them. *(10 minutes)*

Chapter 5 – Body systems

Further teaching suggestions

Researching features and functions

Show students illustrations of a sensory, motor and relay neuron. They could then research how their differing appearance is related to their function.

Fight or flight!

Students could research what is meant by the term 'fight or flight' response. This links together the actions of the endocrine and nervous systems. Students need to write an explanation for what is meant by a fight or flight response. They could illustrate their explanation by a few situations in which this could occur.

Extension activity

As an extension idea you could provide students with the conversions below and ask them to prove the statement in the 'Did you know' box in the Student Book.

Average neurone length = 10 micrometres (μm)
1000 μm = 1 millimetre (mm)
1000 mm = 1 metre (m)
1000 m = 1 kilometre (km)

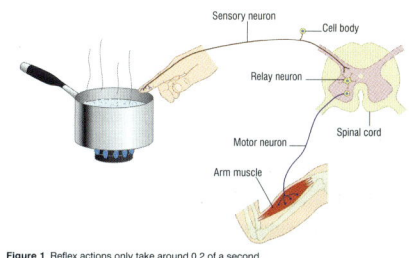

Answers to in-text questions

a as electrical impulses
b light, sound, chemical and heat
c central nervous system
d relay neuron
e reflex action
f Touching a hot iron / cooker / fire. Stepping on a pin, glass, etc.

Activity

Improving reaction times

An activity to improve reflexes by catching a metre rule is detailed in the main lesson plan. As a longer term alternative, students could be asked repeatedly to attempt 'brain training' exercises, to see if their brain can learn a response to a situation.

Summary answers

1.

Receptor cells	Detect changes in your environment.
Effectors	Cause a response.
Neurons	Carry electrical impulses.
Brain	Decides on a response.

2 a reflex
  b controlled
  c controlled
  d reflex

3 Example for touching a pin:
pressure receptors in the skin ⟶ sensory neuron ⟶ spinal cord ⟶ motor neuron ⟶ muscles in arm pull hand away.

Unit 2, Theme 1 – My family

5.2 Hearing

Learning objectives

Students should learn:
- that sound waves travel as longitudinal waves to our ears
- that the human hearing range is from 20–20 000 Hz
- the implications that loud sounds can have on the environment, society and our health.

Learning outcomes

Most students should be able to:
- explain how a sound waves travels
- state the range of human hearing
- give examples and describe the main problems caused by loud sounds
- explain some ways to reduce the damage caused by loud sounds.

Some students should also be able to:
- explain why a sound waves needs particles to travel

Support
- Include as many practical demonstrations as possible.
- You can demonstrate that sound is caused by vibrating objects by demonstrating musical instruments such as a plucked guitar string, tuning fork, struck cymbal.
- Remind students that sound can only travel if there are particles present – in solids, liquids or gases.
- Use a decibel meter to compare the loudness of sounds in quiet and loud places. Also measure the sound coming from an MP3 player at different settings.

Extend
- Explain the decibel scale in more detail. The reason it uses a logarithmic scale is that the range of sounds from quietest to loudest is enormous. The logarithmic scale means that 0 decibels on the scale = 1 unit of loudness; 10 decibels = 10 units of loudness, 20 decibels = 100 units of loudness, 30 decibels = 1000 units of loudness, etc.
- Keep in mind that students will not have met logarithms – you can tell them that these scales are called logarithmic scales and make links with other uses of these scales.

AQA Specification link-up: Science B 3.4.1.1
- Explain how longitudinal waves travel from vibrating objects to our ears for us to hear sounds.
- Know that the human hearing range is 20–20 000 Hz.

Within this context, candidates should be able to use scientific data and evidence to discuss, evaluate or suggest implications of the following:
- the environmental, social and health implications of loud sounds (e.g. from MP3 players, night clubs).

Lesson structure

Starters

What can you hear? – Ask students if they think we can hear sounds in different situations, e.g. can we hear sounds from our Sun or when we are underwater or if we are on the Moon's surface or underground or in a room with triple glazing? (We cannot hear sounds on the Moon or from the Sun. In the other cases there is a medium for the sound to travel through. The triple glazing helps to reduce sound, but vibrations still come through the window frames. Astronauts communicated using radio waves.) *(5 minutes)*

Waves on a slinky – Demonstrate waves using a slinky spring about 1–2 m long. Students hold each end so the spring is extended. A person at one end moves their hand back and forwards in the same direction as the spring. The longitudinal wave travels along the spring, with rarefactions and compressions. Moving their hand sideways causes a transverse wave along the spring – this is how electromagnetic waves travel. To extend this starter for more able students, explain that wave speed = wavelength × frequency. The wavelength is the distance between successive rarefactions or between successive compressions. The frequency is the number of compressions passing a point per second (and is also how many times the person moves their hand forwards per second). Measure these quantities for different frequencies of wave – the speed should stay constant, as the wavelength will shorten at higher frequencies. This is a good introduction to wave speed – which comes up in 9.5 How fast do waves travel? – but early exposure to the concept may be useful. It also helps students put the idea that sound travels at different speeds in different materials into context. *(10 minutes)*

Main

- Demonstrate the range of audible sounds using a loudspeaker and a signal generator set on a sinusoidal wave. Increase the frequency from below 20 Hz to discover when students can hear the sound. Move through the frequencies reasonably quickly until you reach the higher frequencies. Students should indicate when they stop hearing the sounds. Occasionally turn the dial with the amplitude (volume control) at zero to discover who really is hearing the highest pitches. This can lead onto a discussion about damaged hearing – it is quite possible to find students whose hearing cuts off at about 15 kHz.
- After this activity, it may become clear that some students have some noise-related hearing damage. This is an opportunity to discuss sensitively sensible listening habits with all students. This is a good way into the topic of noise-related hearing damage. Students research the causes of noise-related hearing damage and precautions against this. In small groups they can produce a short PowerPoint presentation or a leaflet explaining the causes, effects and prevention measures to reduce damage from MP3 players, night clubs and concerts, or work-related noise. This is a big problem and there are many excellent sites online. Search for 'hearing loss young people'.
- Demonstrate a decibel meter. Either take sound measurements in advance in different situations (e.g. quiet classroom, dining hall, cheering at a match, on a car journey) or demonstrate this in the classroom. Use this information to log how much noise students are exposed to over the course of a day.

Plenaries

What I know about hearing loss – Students present three points that they discovered during their research on noise-related hearing loss. *(5 minutes)*

My daily noise exposure – More able students log their typical day, indicating durations and levels of noise they are exposed to. They suggest ways they can cut down their level of exposure. Draw up a typical school day on the board to use to support students. As a group, suggest activities and indicate the typical sound levels they are exposed to. Help them spot times when they should be careful to protect their hearing. *(10 minutes)*

Chapter 5 – Body systems

Further teaching suggestions

'How the ear works' animation

Show an animation that demonstrates how the ear works. You will find suitable animations on the Bupa website.

Hearing damage

Play a news clip about noise-related hearing damage – this is very topical and searching for news clips about hearing loss will give you a choice of clips that can be used at any suitable point in the lesson.

How often do you use your earphones?

Students should prepare a log of their usage of earphones during a typical day. Identify times their hearing may be at risk, for example, many people turn up the volume when on public transport.

Science in context

There are many concerns about the damage young people do to their hearing with MP3 players and at concerts. DJs and well known artists are now reporting irreversible hearing loss as a result of earlier exposure to excessive noise.

Students may have experienced temporary tinnitus or pain in their ears after concerts – it is important that they realise that there are solutions (e.g. personal earplugs, not standing by the loudspeakers, taking regular breaks from the excessive sound) and that they can reduce the extent of the damage by changing their habits as soon as possible. It is also important that they realise that the next very loud concert they attend could be the one to tip the balance to permanent damage.

Unit 2, Theme 1 – My family

5.2 Hearing

Learning objectives
- How do sound waves travel?
- What is the range of human hearing (in hertz)?
- What is the impact of loud sounds on the environment, society and our health?

How do you hear a drum playing?

The sound energy travels as a wave from the drum through the air to your ear. When the drum is hit, it vibrates. These vibrations make the air particles (molecules) next to the drum vibrate. The vibrations pass energy on to neighbouring particles, which also vibrate.

- As the vibration passes, some particles are squashed together. This is called a **compression**.
- In other places particles become spread out. This is called a **rarefaction**.

a Sketch a diagram showing how a sound wave travels through air.

Sound waves are one example of **longitudinal waves**. In this type of wave the particles vibrate in the same direction that the energy is travelling in.

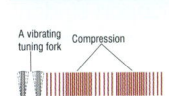

Figure 1 The sound wave is a series of compressions and rarefactions

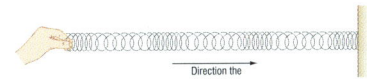

Figure 2 This shows a longitudinal wave travelling along a slinky spring

Vibrating objects cause sound waves. The vibrations can only travel when there are particles that can vibrate, passing the energy on to their neighbours. This means sound travels fast through solids, slowly through gases and not at all in space.

We measure how quickly something vibrates (called its **frequency**) in **hertz**. A vibration of one cycle per second is one hertz. We cannot hear sound waves if objects vibrate at less than 20 hertz. The frequency of the sound is too low for us to hear. The highest pitched sound humans can hear is 20 000 hertz (also called 20 **kilohertz**).

Sounds above 20 000 hertz are called **ultrasound**. As people get older, they cannot hear these very high pitched sounds.

b Write down the frequency of a sound that is too low pitched for us to hear.

Noise levels are measured in decibels. As the decibel reading goes up by 10, the noise level is ten times louder.

The table on the next page shows that a loud rock concert (120 decibels) is 100 times louder than a night club bar (100 decibels). Hearing loss can be caused by any sound above 85 decibels. The damage depends on:

- the loudness of the sounds,
- how long you are exposed to the sounds, and
- how often you hear them.

The table is only a rough guide as sounds are quieter if you stand further away from the source.

AQA Examiner's tip

Make sure that you know the difference between longitudinal waves (sound) and transverse waves (light).

?? Did you know …?

Some shopping centres have installed 'Mosquitoes'. These are speakers that play very high-pitched noises to persuade groups of young people to move on without disturbing older people who cannot hear such high-frequency sound waves. What frequency sound do you think a mosquito device should produce?

Activity	Sound level in decibels	Maximum exposure time to avoid hearing damage
Whispering	20	No harm
Speaking	50–60	No harm
Loud radio	65–70	No harm
Busy city traffic	78–85	8 hours
Power drill	95	4 hours
Night club bar	100	2 hours
Road drill	105	1 hour
Loud rock concert, MP3 player at maximum volume	120	1 minute
Pneumatic drill	125	No safe limit
Jet engine	140	No safe limit

c What affects the amount of hearing damage a person experiences?

The table shows that an MP3 player can be as damaging to your hearing as a loud rock concert. On a bus or train journey, many people listen at high volume to drown out the background noise. The longer you are exposed to the noise, the more damage is likely to occur to your hearing. Specialists are worried that many young people listen to their MP3 players for several hours a day.

The damage to hearing is reversible to begin with. However, after prolonged exposure to loud sounds, people start to suffer from tinnitus (a permanent ringing in the ears). Sometimes they become unable to hear conversations in busy rooms as they cannot hear certain frequencies.

You can prevent this damage by using headphones that fit inside the ear. These can block out the background noise so you can turn your MP3 player down to a safe level.

d Why do hearing specialists worry about young people using MP3 players?

By law, employers must not expose their employees to sounds above 87 decibels on a daily basis, or to a peak noise level of 140 decibels. In noisy areas, they must train their staff and provide ear protection to reduce the risk of damage to hearing. They can also sound-proof equipment.

People disturb their neighbours if they cause too much noise. This can include loud music, parties and barking dogs. The noise becomes officially a nuisance at certain decibel levels. Many neighbours manage to sort problems out before calling in the Council. But some cases do reach the courts and people are ordered to stop the noise nuisance.

?? Did you know …?

Astronauts in space communicate by radio as sound waves cannot travel in a vacuum. Victims of earthquakes attract attention more effectively by tapping on pipes rather than shouting.

Figure 3 Headphones like these can protect your hearing by blocking out background noise so the volume on your MP3 player can be set at a lower level

Summary questions

1 Complete these sentences by choosing the right word:
   a Sound waves travel as **longitudinal/electromagnetic** waves.
   b With longitudinal waves, particles vibrate **parallel/at right angles** to the way the energy travels.
2 Explain why sounds cannot travel through space.
3 Research and prepare a poster warning young people of the dangers of loud noise. Your poster can be about one particular aspect, such as concerts, or more general.

Key points
- Sound travels as longitudinal waves through gases, liquids and solids.
- Humans can hear sounds between 20 hertz and 20 000 hertz.
- Very loud sounds can damage hearing or disturb other people.

Summary answers

1 **a** longitudinal
   b parallel
2 Sound waves need particles to transfer sound energy from one particle to the next.
3 The poster should highlight the safe upper limit of sound (85 decibels), and explain levels above this can only be heard safely for shorter times. Suggestions such as reducing the volume of sound, wearing ear protection or reducing the time they are exposed to the sound should be included.

Answers to in-text questions

a The diagram should include regularly spaced regions of compressions and rarefactions.
b any number between 1 and 19
c the level of noise in decibels, the duration of exposure, the frequency of exposure.
d The sound is going directly into the young person's ear; it can be very loud, and exposure can be for several hours daily.

Unit 2, Theme 1 – My family

5.3 Hormones

Learning objectives

Students should learn:

- that hormones are secreted by glands and are chemical messengers that travel around the body in the blood
- that hormones control body processes that need constant adjustment, maintaining a constant internal environment (homeostasis)
- that this is achieved by hormones and nerves working together.

Learning outcomes

Most students should be able to:

- name some hormones and the glands they come from, and describe simply how they act
- define the term homeostasis
- give examples of body systems that are controlled in homeostasis.

Some students should also be able to:

- explain why homeostasis is an example of a negative feedback control system. [HT only]

AQA Specification link-up: Science B 3.4.1.1

- Know that the body needs to maintain a constant internal environment and that this is called homeostasis.
- Know that chemical substances called hormones control many processes within the body. Hormones are secreted by glands and are transported to their target organs in the bloodstream.
- Explain the principle of negative feedback in maintaining a constant internal environment. **[HT only]**

Lesson structure

Starters

Hormone definition – To assess the extent of students' KS3 knowledge, ask students to rearrange these phrases to produce a definition for 'hormone':

- they are secreted by glands
- into the blood,
- where they travel to the target organ
- they cause an effect in this organ
- hormones are chemical messengers

[Hormones are chemical messengers. They are secreted by glands into the blood, where they travel to the target organ. They cause an effect in this organ.] *(5 minutes)*

Nervous responses – Ask students to describe the steps involved in a nervous response, which involves the brain. For example, what are the steps involved in kicking a football? This starter provides a link back to work covered previously on nerves. Support students by providing them with the steps required, in a jumbled order for students to rearrange. Extend students by asking them to list the specific parts of the body involved in each stage of the response – for example, what are the receptor cells involved? *(10 minutes)*

Main

- Students discuss their ideas about hormones – what they do, where they are produced, how they travel around the body and their role in the body. This will enable the teacher to evaluate students' prior knowledge. Key words that need to be introduced include – gland, target organ and a definition of the term 'hormone' (if not completed in the starter activity, above).
- If the 'matching glands to hormones' starter has not been completed, students should be provided with an image of the body, with the major glands highlighted. Students should then label the main hormones that a gland produces, and the role of those hormones in the body.
- Introduce the concept of homeostasis. Ask students in pairs to produce a list of factors in the body, which they think may need to be controlled. Images could be provided to stimulate discussion, if required.
- Explain that homeostasis is an example of a feedback control system. The concept can be modelled using the example of temperature control in a central heating system – a thermostat (representing the brain) decides whether to switch the heating on or off; a temperature sensor acts as the receptor; the boiler acts as the effector. Students could produce a flow diagram to show how this system works. Higher Tier students should be made aware of the concept of negative feedback in this control system.
- Ask students to try to maintain the temperature of a beaker of water at 37 °C. Students should produce a flow diagram to show how their feedback system works. They should appreciate that tiny modifications are needed regularly, to maintain a constant temperature. Further details are provided in 'Practical support' opposite.

Plenaries

Differences between nerves and hormones – Ask students in pairs to produce a table summarising the main differences between nerves and hormones. *(5 minutes)*

Flow diagram – Ask students to rearrange the words into the correct order – response, brain, stimulus, gland, target cell, blood, hormone – to produce a generalised flow diagram for a hormonal response. [stimulus ⟶ brain ⟶ gland ⟶ hormone ⟶ blood ⟶ target cell ⟶ response]. Support students by projecting an ordered sequence of images, each of which represents one of the steps of the flow diagram. Extend students by asking them to produce a flow diagram similar to example above, for a specific hormone. *(10 minutes)*

Support

- Provide students with a card sort activity, to match the hormone to the gland or organ which produces it, and its role in the body.

Extend

- Ask students to research a different example of negative feedback. For example the role of ADH in controlling water urine levels. Students should present their findings in a flow diagram.

Chapter 5 – Body systems

Further teaching suggestions

Role-play
Students could role-play a feedback system. For example the concept of a central heating system, introduced in the lesson. This could be varied to demonstrate how internal temperature or blood glucose levels are maintained.

Understanding a simple negative feedback control system
A more visual example of a simple negative feedback control system would be to use a tea urn in the lesson. Students should be asked how the urn maintains the water at the required temperature, without wasting energy and heating the water unnecessarily. [HT only]

Practical support

Maintaining a constant body temperature

Equipment and materials required: 0.1 °C resolution temperature probe (or, failing this, a thermometer – ideally a medical thermometer with high resolution), beaker of water, tripod and heating safety equipment, Bunsen burner.

Details
Ask students to work in pairs. Challenge each pair of students to maintain the temperature of the water at exactly 37.0 °C – normal body temperature. Which pair can manage the smallest deviation from this temperature? Are there any improvements students could make to their system – for example, the introduction of a cooling system if the temperature of the water rises too high?

Ask students to construct a flow diagram of their system, and correlate this to systems within the body which regulate temperature.

Safety: Wear eye protection. Take care not to let the temperature rise too much or a medical thermometer will break.

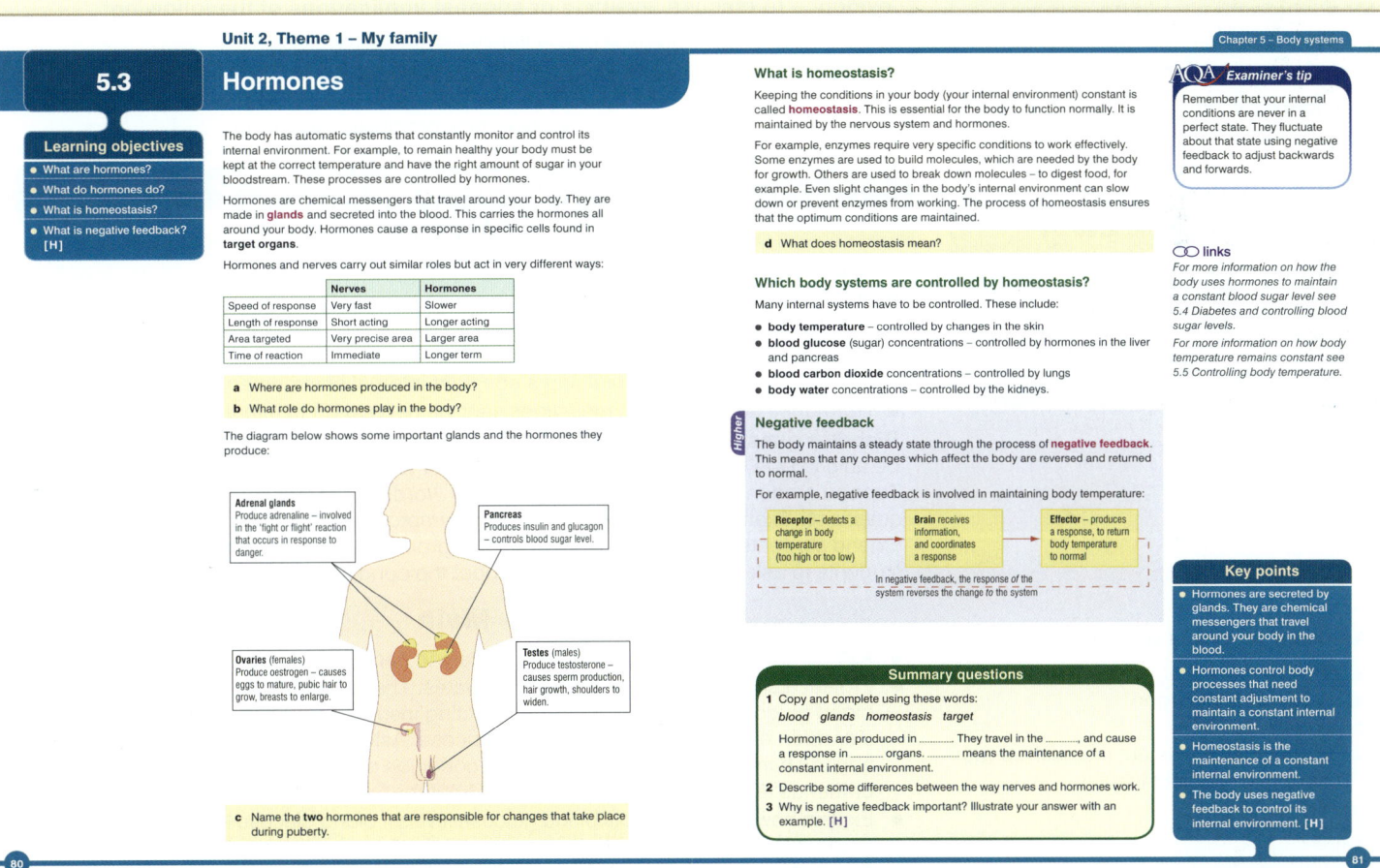

Answers to in-text questions
a in glands
b They help to keep conditions in the body constant (large role in homeostasis).
c oestrogen and testosterone
d maintaining a constant internal environment

Summary answers

1 glands, blood, target, homeostasis, negative

2 Nerves are very fast acting, providing an immediate response in a very precise area. Hormones take longer to cause a response, which then acts over a longer term and a larger area.

3 Negative feedback is important, as it restores body systems to normal by reversing the change in the body. For example, if the body is cooled too much by sweating, the body will stop sweating (and promote shivering if required) to return the body temperature to normal.

81

Unit 2, Theme 1 – My family

5.4 Diabetes and controlling blood sugar levels

Learning objectives

Students should learn:
- that insulin reduces blood glucose levels
- that glucagon increases blood glucose levels [HT only]
- that diabetics cannot control their blood glucose levels
- that diabetes is controlled through diet and exercise, or by using insulin.

Learning outcomes

Most students should be able to:
- describe simply how insulin controls blood glucose levels
- describe the difference between Type 1 and Type 2 diabetes
- explain how Type 1 and Type 2 diabetes can be controlled.

Some students should also be able to:
- explain in detail how insulin and glucagon control blood glucose levels. [HT only]

Specification link-up: Science B 3.4.1.1

- Explain how the hormone insulin controls the blood glucose levels. High blood glucose levels are a symptom of diabetes. They should be aware that some forms of diabetes (Type 2 diabetes) may be controlled by a change in lifestyle (diet and exercise). Type 1 diabetes is controlled by insulin dosage and is sometimes termed insulin-dependent diabetes. Candidates should be able to describe how blood glucose levels are monitored and controlled by cells in the pancreas:
  a if the blood glucose concentration is too high, the pancreas releases the hormone insulin into the blood, which causes the liver to remove glucose from the blood and store it as insoluble glycogen
  b if the blood glucose concentration is too low, the pancreas releases glucagon, which causes the liver to convert glycogen back to glucose and release it into the blood. [HT only]

Within this context, candidates should be able to use scientific data and evidence to discuss, evaluate or suggest implications of the following:
- the social, economic, and health implications of diabetes
- the personal and social choices in lifestyle in terms of a balance of risk and benefit to health
- what happens when normal physiological processes go wrong.

Lesson structure

Starters

True or false –
- Hormones are made in glands. [T]
- Plasma is an example of a hormone. [F]
- Hormones travel around the body in the blood. [T]
- Homeostasis means maintaining a constant internal environment. [T]
- Hormones act faster than messages sent via nerves. [F] *(5 minutes)*

Diabetes questionnaire – Ask students to construct a short questionnaire they would be able to give to a sufferer of diabetes, in order to find out more about their condition. Support students by providing a series of key words to prompt students in devising their questions. Examples could include causes, symptoms, health problems, treatment. Extend students by providing a sheet of background information on diabetes, to enable students to ask higher order questions about the condition. *(10 minutes)*

Main

- Ask students, in pairs, to revise and explain to each other the process of homeostasis.
- Project a series of images related to diabetes. These could include, for example, blood glucose testing kits, a person injecting insulin, a range of sugar-rich foods and a cartoon of starch being digested to glucose. Ask students to discuss these images and to share their understanding of the condition of diabetes. (Be aware of students or friends and family with diabetes, and treat the issue with sensitivity.)
- Explain how insulin controls blood glucose levels if blood glucose levels are too high. Students could be provided with the steps in the flow chart shown in the Student Book in a mixed up order. They then need to cut and paste the relevant steps into the correct order, to describe how high glucose levels are controlled by the body.
- Explain how glucagon controls blood glucose levels, if blood glucose levels are too low. Students could then add these steps onto their flow chart. Students often muddle up the terms glycogen and glucagon. It is essential that they understand which one is a glucose store (glycogen) and which is the hormone (glucagon). These are two words where correct spelling is essential in an examination. **[HT only]**
- Explain the main differences between Type 1 and Type 2 diabetes (these issues may have arisen as a result of the student discussion above). Ask students to explain how diet and exercise could be used to control diabetes, and under what circumstances this course of treatment alone would be insufficient to control the condition.
- Ask students to act as medical analysts, checking 'urine samples' for abnormalities. Three samples should be given – one normal, one with excess sugar and one with excess protein. Students will be testing the samples for signs of diabetes. Further details are provided in 'Practical support', opposite. Provide students with an appropriate advice sheet, to be given to a patient who displays the relevant positive result. Students should summarise this advice into their results table for the practical activity.
- Ask students to produce an information leaflet, aimed at adults, which could be available in a doctor's surgery. Leaflets should provide clear and concise information on all aspects of diabetes, including symptoms, causes and how to control the condition.

Answers to in-text questions
a energy
b insulin and glucagon
c diabetes
d So that they can inject themselves with the correct dose of insulin.

Support
- Produce a tick-box type results table for the 'urine samples' for students to fill in.

Extend
- Research in detail the different types of diabetes and the treatments that are available. Explain why different treatments are needed for the different types of diabetes. A useful website is the Diabetes UK website www.diabetes.org.uk

Chapter 5 – Body systems

Plenaries

Type 1 or Type 2 diabetes? – Provide students with each of the following statements. Students need to relate each one to Type 1 or Type 2 diabetes, or both.
- Inability to produce insulin. (Type 1)
- Poor quality insulin produced. (Type 2)
- Can be controlled only by diet and exercise. (Type 2)
- Regular insulin injections required. (Type 1)
- Sufferers need to be aware of the sugar content of their foods. (Both) *(5 minutes)*

Definition sentences – Ask students to write a sentence for each of the following key words, explaining how it is involved in the process of controlling blood sugar levels: pancreas, liver, insulin, glycogen, glucagon, glucose. Support students by providing sentences with the key term missing in each case. Students need to select which word correctly completes each sentence. Extend students by asking them to explain within their sentences why glucose cannot be stored in the body, but glycogen can. *(10 minutes)*

Activity

Implications of obesity

Divide the class into groups of three, to research the implications of obesity for the UK. Split each group, so that one student studies the social implications, one the economic implications, and one the health implications of obesity. Results for each group should be summarised, and the class results collated to produce a factsheet describing the issues surrounding the UK's obesity problem. Further support is provided on a worksheet.

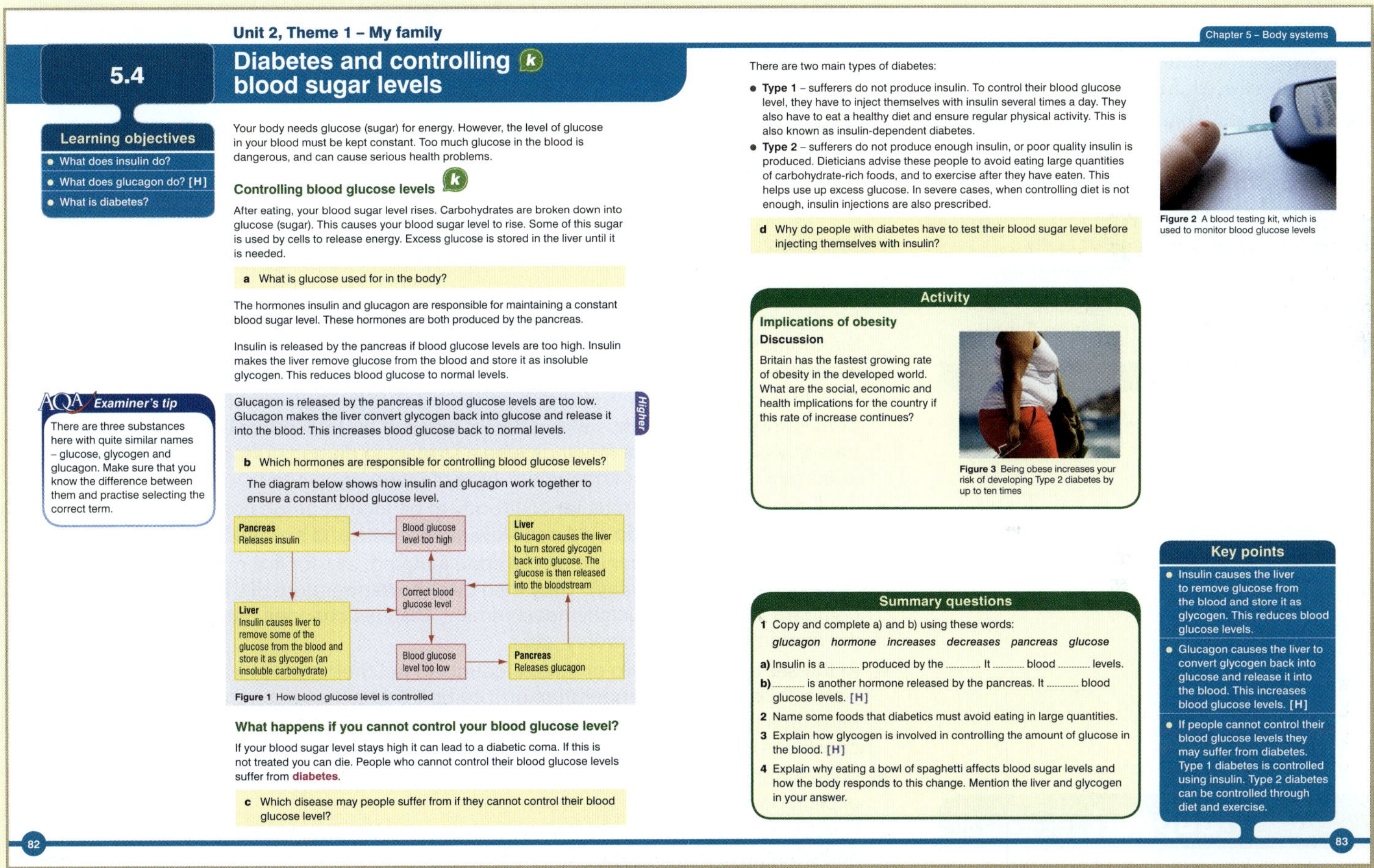

Unit 2, Theme 1 – My family

5.4 Diabetes and controlling blood sugar levels

Learning objectives
- What does insulin do?
- What does glucagon do? [H]
- What is diabetes?

Your body needs glucose (sugar) for energy. However, the level of glucose in your blood must be kept constant. Too much glucose in the blood is dangerous, and can cause serious health problems.

Controlling blood glucose levels

After eating, your blood sugar level rises. Carbohydrates are broken down into glucose (sugar). This causes your blood sugar level to rise. Some of this sugar is used by cells to release energy. Excess glucose is stored in the liver until it is needed.

a What is glucose used for in the body?

The hormones insulin and glucagon are responsible for maintaining a constant blood sugar level. These hormones are both produced by the pancreas.

Insulin is released by the pancreas if blood glucose levels are too high. Insulin makes the liver remove glucose from the blood and store it as insoluble glycogen. This reduces blood glucose to normal levels.

AQA Examiner's tip
There are three substances here with quite similar names – glucose, glycogen and glucagon. Make sure that you know the difference between them and practise selecting the correct term.

Glucagon is released by the pancreas if blood glucose levels are too low. Glucagon makes the liver convert glycogen back into glucose and release it into the blood. This increases blood glucose back to normal levels. *[Higher]*

b Which hormones are responsible for controlling blood glucose levels?

The diagram below shows how insulin and glucagon work together to ensure a constant blood glucose level.

Pancreas Releases insulin → **Blood glucose level too high** → **Liver** Glucagon causes the liver to turn stored glycogen back into glucose. The glucose is then released into the bloodstream → **Correct blood glucose level** → **Liver** Insulin causes liver to remove some of the glucose from the blood and store it as glycogen (an insoluble carbohydrate) → **Blood glucose level too low** → **Pancreas** Releases glucagon

Figure 1 How blood glucose level is controlled

What happens if you cannot control your blood glucose level?

If your blood sugar level stays high it can lead to a diabetic coma. If this is not treated you can die. People who cannot control their blood glucose levels suffer from **diabetes**.

c Which disease may people suffer from if they cannot control their blood glucose level?

There are two main types of diabetes:
- **Type 1** – sufferers do not produce insulin. To control their blood glucose level, they have to inject themselves with insulin several times a day. They also have to eat a healthy diet and ensure regular physical activity. This is also known as insulin-dependent diabetes.
- **Type 2** – sufferers do not produce enough insulin, or poor quality insulin is produced. Dieticians advise these people to avoid eating large quantities of carbohydrate-rich foods, and to exercise after they have eaten. This helps use up excess glucose. In severe cases, when controlling diet is not enough, insulin injections are also prescribed.

d Why do people with diabetes have to test their blood sugar level before injecting themselves with insulin?

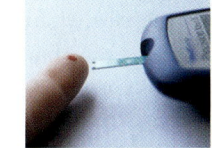
Figure 2 A blood testing kit, which is used to monitor blood glucose levels

Activity

Implications of obesity
Discussion
Britain has the fastest growing rate of obesity in the developed world. What are the social, economic and health implications for the country if this rate of increase continues?

Figure 3 Being obese increases your risk of developing Type 2 diabetes by up to ten times

Summary questions

1 Copy and complete a) and b) using these words:
glucagon hormone increases decreases pancreas glucose
   a) Insulin is a .... produced by the .... It .... blood .... levels.
   b) .... is another hormone released by the pancreas. It .... blood glucose levels. [H]
2 Name some foods that diabetics must avoid eating in large quantities.
3 Explain how glycogen is involved in controlling the amount of glucose in the blood. [H]
4 Explain why eating a bowl of spaghetti affects blood sugar levels and how the body responds to this change. Mention the liver and glycogen in your answer.

Key points
- Insulin causes the liver to remove glucose from the blood and store it as glycogen. This reduces blood glucose levels.
- Glucagon causes the liver to convert glycogen back into glucose and release it into the blood. This increases blood glucose levels. [H]
- If people cannot control their blood glucose levels they may suffer from diabetes. Type 1 diabetes is controlled using insulin. Type 2 diabetes can be controlled through diet and exercise.

Practical support

Testing urine

Equipment and materials required: 'Clinistix' (sugar and protein testing) sticks, water, sugar, yellow food colouring, protein powder (available from health food shops).

Details

Students should dip 'Clinistix' (sugar and protein testing) sticks into the 'urine' samples. The 'urine' samples should be made using water and yellow food colouring. Add nothing for normal urine and add sugar for diabetic urine (as an extension idea you could also and add protein powder for kidney failure). The stick will change colour according to the sample it was added to. This needs to be read in conjunction with the label on the testing sticks box, to determine if anything abnormal is present in the urine.

Safety: Dispose of used sticks into a beaker of disinfectant for realism!

Summary answers

1 **a** hormone, pancreas, decreases, glucose
   b glucagon, increases
2 pasta, bread, cakes, sweets, etc.
3 If blood glucose levels are too high, glucose is converted into glycogen. This removes glucose from the blood. If blood glucose levels are too low, glycogen is converted into glucose. This increases the levels of glucose in the blood.
4 Carbohydrates in spaghetti are broken down into glucose. This raises blood glucose levels. Some is used for respiration, but excess is detected by cells in the pancreas. They secrete insulin, which tells cells in the liver to store glucose as glucagon, reducing blood glucose levels to normal.

Unit 2, Theme 1 – My family

5.5 Controlling body temperature

Learning objectives

Students should learn:

- that the brain regulates body temperature by changing skin conditions
- that sweating cools the body as the water evaporates
- how blood flow, close the surface of the skin, can be varied to regulate body temperature.

Learning outcomes

Students should be able to:

- describe the changes that take place in your skin when you are too hot
- describe the changes that take place in your skin when you are too cold.

Some students should also be able to:

- explain in detail how your body maintains a constant temperature
- explain in detail what happens if your body temperature drops significantly.

Support

- Provide students with a pre-prepared worksheet for them to record their practical observations related to exercising. Key words should be given.

Extend

- Ask students to carry out research into hypothermia.
- They could then prepare a booklet warning mountaineers of the dangers of exposure to extreme cold. This should include warnings of the obvious symptoms to look out for, and measures they should take to ensure they are safe in hazardous conditions.

AQA Specification link-up: Science B 3.4.1.1

- Explain how the body maintains a constant temperature, using the thermoregulatory centre in the brain:
  a by increasing or decreasing the amount of sweating, which cools the body by evaporation
  b by dilating the blood vessels supplying the skin capillaries, increasing the blood flow to, and consequently the amount of heat lost from, the skin
  c by constricting blood vessels supplying the skin capillaries, decreasing the blood flow and the amount of heat lost.

Lesson structure

Starters

Structure of the skin – Provide students with a cross-section through the skin – ideally a cartoon cross-section next to a real image. They then need to identify the key features e.g. sweat gland, hair, capillary – on both diagrams. *(5 minutes)*

How the body keeps cool – Ask students to draw a cartoon face of a person who has just run a race. They label as many mechanisms as possible that the body uses to keep itself cool. For each mechanism they should try to state how it helps to cool the body down. *(10 minutes)*

Main

- Students are asked to measure their body temperature using thermometers. Ideally use forehead thermometers (as this is easier than disinfecting standard thermometers) but it is useful to demonstrate the wide variety of thermometers available.
- Collate a range of measurements from the class to illustrate that there is not a 'fixed' body temperature. Discuss why different types of thermometers used on the same person will give different readings (use a student to demonstrate this) and why it is important to let a doctor know which type of thermometer you are using.
- Discuss why it is dangerous for your body to be too hot or too cold. (Fever can cause fits, dehydration, etc., too cold can slow movements and eventually result in a coma).
- Students exercise by running on the spot, jumping, etc. for five minutes (check that students are fit and healthy). They then take the temperature of their skin again, and compare this with their initial measurement. Discuss changes that have taken place in the students' skin (red, sweaty, etc.).
- At the same time, ask two or three students to put their arms in a bucket of icy water. Take their skin temperature and compare what has happened to their skin. Introduce the concept of dilation and constriction of blood vessels supplying the skin capillaries. Use diagrams in the Student Book to help.
- Wipe ethanol onto the back of the students' hands. Why does it feel cold? Use this to explain how sweat cools you down.
- Discuss what would happen if they had stayed in icy water for even longer – shivering and eventually hypothermia.

Plenaries

Which is which? – Provide students with schematic diagrams of hot and cold skin. Students have to decide which diagram is which and explain to a partner their choice of diagram. *(5 minutes)*

Word definitions – Ask students to write definitions for the three key terms – vasodilation, vasoconstriction and sweat. They should swap their definitions with a partner to see if they can improve them further. To support students, allow access to the internet once the definitions have been swapped, but give each group a short time limit. *(10 minutes)*

Chapter 5 – Body systems

Further teaching suggestions

Adaptations for extreme environments
Look at the adaptations animals have developed to live in extreme environments, e.g. polar bears or camels. Why do dogs and other animals grow thicker fur in the winter? Why do birds fluff up their feathers? Relate these to the changes which take place in human skin.

Manufacturing clothes for arctic explorers
Students could carry out a short investigation into choosing the ideal material to manufacture clothes for an arctic explorer. This can be carried out simply by wrapping the fabric around a test tube of hot water, and measuring the drop in temperature. Data logging can be used to record the data produced. The data should be displayed graphically and groups can then compare results to choose the 'ideal' insulating fabric.

Unit 2, Theme 1 – My family

5.5 Controlling body temperature

Learning objectives
- How does your body control its temperature?
- Why do we sweat?
- Why does the amount of blood near the surface of our skin change?

Your body works best at 37°C. Whatever conditions are like outside, your body will try to maintain this temperature.

The thermoregulatory centre in your brain is responsible for controlling body temperature. Your brain monitors the temperature of your blood, and receptors in your skin receive information about the external temperature. The thermoregulatory centre processes this information and sends nerve impulses to the skin and muscles to tell the body how to respond.

a What is the normal body temperature?

Sunbathing and physical activity can cause the body to overheat. Exposure to cold weather can cool the body down. Just a couple of degrees difference in your body's temperature can stop the body from working efficiently. This happens especially to the brain. For example, if you are too hot it can cause fits and dehydration.

Your body is designed to protect itself from changes in temperature in several ways, mainly using the skin.

What happens when you get too hot?
- Hairs on your skin lie flat.
- Sweat glands produce sweat.
- Blood vessels supplying capillaries near the surface of your skin widen (dilate). This increases blood flow through the capillaries, increasing heat loss.

Figure 1 Skin's appearance when a person is too hot

What happens when you get too cold?
- Hairs on your skin stand on end.
- This traps a layer of air close to the skin, preventing heat loss.
- Sweat glands do not produce sweat.
- Blood vessels supplying capillaries near the surface of your skin narrow (constrict). This reduces blood flow through the capillaries, reducing heat loss.
- Shivering (rapid muscle contractions). This requires extra energy, so your cells respire more. This produces extra heat.

Figure 2 Skin's appearance when a person is too cold

b Why do the hairs on your arm stand up when you are cold?

c Why does shivering help you warm up?

AQA Examiner's tip
Just remember what colour your face goes after exercise – bright red. This shows that the capillaries close to the surface of the skin have more blood than normal as the body is trying to cool down. When you are cold your skin looks white as most of your blood is redirected away from the surface of your skin. This is so that you transfer less energy to your surroundings.

Why do you go red when you are hot?
When you are hot the capillaries in your skin widen (vasodilation). This allows more blood to flow close to the surface of the skin and makes you look red. So more energy is transferred from the blood by radiation, cooling you down.

Why do we sweat?
Sweat is mainly water, but it also contains salt and urea (a waste material). When the water in sweat evaporates it absorbs energy from your body. As energy is lost from your body, your temperature falls so you feel cooler.

The more you sweat, the more you cool down but you also lose more water and salt. These substances must be replaced by drinking and eating, otherwise you will dehydrate.

d What is sweat made of?

What happens if your temperature drops too much?
As your body cools it starts to function at a slower rate. If your temperature drops by 2°C your brain will be affected. Body movements will slow and your speech will begin to slur.

If your body temperature continues to drop, you will go into a coma. Eventually you will die. This condition, when body temperature drops to below 35°C, is called hypothermia. It is a major problem for explorers in extreme weather conditions. It can also affect the elderly if they have poor heating.

e How would you know if someone is suffering from hypothermia?

This flow diagram summarises how the thermoregulatory centre in the brain controls body temperature.

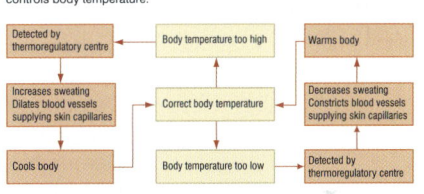

Figure 3 Explorers need to make sure they don't get too cold by wearing protective clothing

Summary questions

1 Copy and complete the table with the following words:
   lie flat   make sweat   narrow   stand up   do not make sweat   widen

	Body too hot	Body too cold
Hairs		
Blood vessels		
Sweat glands		

2 Explain what the words dilation and constriction of blood vessels mean.

3 If you tasted your skin after you exercised, why would it taste salty?

4 Suggest how a polar bear is adapted so that it doesn't get hypothermia.

Key points
- The thermoregulatory centre in your brain monitors body temperature. If it is too high or low, it causes changes in the skin that return body temperature to normal.
- The water in sweat evaporates from the surface of the skin. This cools your body.
- When blood flows near the surface of the skin, energy is transferred by radiation to the surroundings. If you are hot, blood flow increases near the surface of your skin increasing the rate of energy transfer. This cools you down.

Summary answers

1

	Body too hot	Body too cold
Hairs	lie flat	stand up
Blood vessels	widen	narrow
Sweat glands	make sweat	do not make sweat

2 dilation – blood vessels widen.
   constriction – blood vessels narrow.

3 During exercise you sweat. When the water evaporates from your skin, the salt is left behind.

4 It has thick fur, a layer of blubber under its skin.

Answers to in-text questions

a 37°C

b To trap a layer of air close to the skin – this prevents energy transfer to the surroundings.

c Shivering requires energy. This energy is produced by extra respiration, which warms muscles up.

d water, salt and urea

e They would be cold, their movements would slow down and their speech would be slurred.

Unit 2, Theme 1 – My family

5.6 Body chemistry

Learning objectives

Students should learn:
- that many chemical reactions take place in the human body
- that acids and bases can harm our bodies
- that hazard symbols are used to tell us how a chemical could harm us.

Learning outcomes

Students should be able to:
- describe how many chemical reactions are essential in the body
- explain that acids and bases can be harmful
- recognise the hazard symbols for harmful, irritant and corrosive and suggest means to reduce risk
- explain how risks from acids and bases can be reduced by the use of hazard symbols and protective equipment.

AQA Specification link-up: Science B 3.4.1.2

- Know that the body functions properly due to a series of complex chemical reactions.
- Name some hazards of acids and bases and some control measures that can be put in place to minimise risks from them.

Lesson structure

Starters

Naming molecules – Students can match up names of biological chemicals to symbols or diagrams, e.g. glucose, starch, water, carbon dioxide, urea, an enzyme. *(5 minutes)*

Reviewing biological molecules – Students can discuss and brainstorm what they already know about the processes that go on in the human body. To support students, focus on describing the seven life processes. To extend students, ask them to suggest the types of molecule/structures involved in each of the life processes to produce a mind map. [For example, reproduction – hormones, genes; movement – muscle tissue, glucose, oxygen.] *(10 minutes)*

Main

- Run a straightforward class practical to examine the effects of acids on different materials. Students then share their observations, e.g. no change, temperature change, change of colour, fizzing, dissolving. Materials to use can include a penny, chalk, baking powder, plastic and wood.
- Show students domestic cleaning packaging, displaying hazard symbols. Discuss what the symbols are warning us about.
- Demonstrate the dehydration of sugar by concentrated sulfuric acid to introduce the dangers of acids and bases.
- Demonstrate the denaturing of egg albumen by hydrochloric acid to show the effect of strong acids on proteins.
- Demonstrate the action of hydrochloric acid on eggshell to show the effects of acids on teeth.
- Use the above demonstrations to introduce the concept of risk assessment. Explain the difference between a hazard and a risk, then provide students with a number of scenarios where they must assess the potential hazards and risks, suggesting ways to minimise the risks, concentrating on acids and bases.

Plenaries

Key word recap – Support students by asking them to write definitions for key words (e.g. reaction, digestion, hazard, risk). Extend students by asking them to write revision questions for the key words, [e.g. 'What type of chemical reaction happens between glucose and oxygen?'] *(5 minutes)*

Return to the mind map – If it was used, the mind map produced during the starter activity can be revisited, amended and added to. *(10 minutes)*

Support

- Review the seven life processes with students first to remind them of the central role of digestion.
- Students could either keep a food diary for a day or two, or visual prompts could be used to get students thinking and talking about tastes.
- This can be used as a basis for describing various foods as either sour or bitter, leading on to describing acids and alkalis in the same terms.

Extend

- Students can research the wider roles of acids in the body. Research findings can be used to make presentations. Possible subjects could be DNA, hyaluronic acid, citric acid, vitamin C and acidosis.

Practical support

The effects of acids

Equipment and materials required: Per group: watch glass, 1 mol/dm^3 HCl, dropping pipette, copper coins, sodium bicarbonate, chalk, 1 mol/dm^3 sodium hydroxide, polystyrene, eggshell, wire wool, boiling tubes and rack, thermometer, eye protection.

Concentrated sulfuric acid and sugar: eye protection, fume cupboard, beaker (100 cm^3), 50 g sucrose, 20 cm^3 concentrated sulfuric acid.

Hydrochloric acid and eggs: eye protection, 500 cm^3 1 mol/dm^3 hydrochloric acid, 1000 cm^3 beaker, 250 cm^3 beaker, 2 eggs, stirring rod.

Details

Students make simple observations about the chemical effects of acids on everyday items and materials. They should add drops of acid to each material on a watch glass, recording their observations. Temperature change can be assessed by performing the reaction in 2 cm depth acid in a boiling tube.

Concentrated sulfuric acid and sugar

Wear eye protection throughout this demonstration. In a fume cupboard, add about 50 g sucrose to a 100 cm^3 beaker. Clamp the beaker. Carefully add the concentrated sulfuric acid to the sugar. The sugar will turn yellow, then brown. After a minute it will blacken, and a spongy mass of carbon will rise up the beaker, releasing steam and sulfur dioxide. Explain to students that the acid is removing the hydrogen and oxygen from the carbon in the sugar as water, pointing out that their body tissues contain a great deal of molecules containing carbon, hydrogen and oxygen.

Hydrochloric acid and eggs

Wear eye protection throughout. Demonstrate the effect of acid on the shell of the egg by carefully submerging an egg in hydrochloric acid in the 1000 cm^3 beaker. Explain to the students that the shell is made of a similar substance to our teeth. Carbon dioxide will be evolved from the shell as it reacts with the acid. It takes around an hour for the shell to completely react with the acid, after this the protein inside will slowly start to denature. An egg can be prepared in this way before the lesson and students could be invited to suggest how the shell could have been removed before starting the demo.

Demonstrate the effect of acid on the egg albumin by separating an egg and adding about 50 cm^3 of the hydrochloric acid in the 250 cm^3 beaker. The protein will denature, becoming white and insoluble. Again, this can be likened to the effect of acid on proteins in the body.

Safety: Wear eye protection. Nitrile gloves must be worn when using concentrated sulfuric acid. CLEAPSS Hazcard 47A Hydrochloric acid; CLEAPSS Hazcard 91 Sodium hydroxide; CLEAPSS Hazcard 98A Sulfuric acid – corrosive.

Take care when disposing of the resulting lump of carbon. Soak in water and wrap it in plastic bags before putting in the waste bin.

Unit 2, Theme 1 – My family

5.6 Body chemistry

Learning objectives
- How do chemical reactions help our bodies function properly?
- How can acids and bases harm our bodies?
- How can we reduce the risk of harm from acids and bases?

Useful chemical reactions

Just think about the processes taking place in living organisms:
- Movement
- Respiration
- Sensing
- Growth
- Reproduction
- Excreting waste
- Nutrition

Each of these processes is controlled by highly specialised chemical reactions. Scientists who study these reactions are called biochemists.

Our bodies are continually breaking chemicals apart and putting them back together in different ways. A good example of this is nutrition. Starches, proteins and fats are all complex molecules we use to feed our bodies every day. They are broken into smaller molecules by enzymes in the body's digestive system. These small molecules are then reassembled to make new materials for the body. For example, proteins are broken down into amino acids. The amino acids are then used to make new, different proteins for the body.

Acids and bases play an important role in these chemical changes. Hydrochloric acid in the stomach helps digestive enzymes to work. Other cells in the body produce soluble bases called alkalis. Some stomach cells produce alkaline mucus (slime) which protects the stomach lining from the hydrochloric acid and enzymes. Cells in the liver produce bile, which contains the weak alkali sodium bicarbonate. This neutralises the stomach acid when it leaves the stomach.

a Name **two** organs of the body that produce an alkali.

Hazards of acids and bases

Acids and alkalis can also have harmful effects on the body. Strong acids are very reactive. They react with other chemicals very easily. If those other chemicals are part of your body, it can be very painful! Acids in the home vary from weak acids, such as vinegar, to stronger acids, such as acids found in toilet cleaners.

As well as burning your skin, acids can damage you on the inside. By cleaning your teeth, you help prevent bacteria growing in your mouth. Bacteria produce acids in their waste. These acids attack your teeth. Stomach ulcers are another example. They are caused by excess stomach acid damaging your stomach lining.

Figure 1 Tooth decay is caused by acids

b What produces the acid that causes tooth decay?

Bases can also do a lot of damage to our bodies. If you accidentally get a base on your skin, it will feel slippery. This is because it is turning the oils in your skin into soap-like substances. Stronger bases such as sodium hydroxide (called **caustic soda** in the home) can burn straight through your skin into the flesh below. This is shown in Figure 2.

c Why might a base feel slippery if you touch it?

Staying safe

Any chemical that can harm you should have a label to warn you. These warnings are called hazard symbols. The symbol usually gives an idea of the kind of harm it can cause. The most common hazard symbols you would find on acids and bases are those meaning irritant, harmful or corrosive.

| Irritant – can cause a rash or itching | Harmful – general damage to living organisms | Corrosive – will burn through materials |

As well as reading the label carefully, you should protect yourself with safety equipment. Goggles should always be worn when working with strong acids or bases. Gloves and masks can be worn when working with harmful chemicals at home. Protective clothing like a lab coat or apron will also give some protection.

d What word describes a chemical that causes a rash or itching?

Figure 2 Bases can be just as dangerous as acids

Did you know …?
Biologists estimate there are around 75 000 enzymes in the human body. Each of them controls at least one chemical reaction.

Summary questions

1. Copy and complete using these words:

   soap burns ulcers reactions bases gloves protect tooth goggles enzymes

   Body cells carry out many different chemical _____. These are controlled by _____. Acids and _____ can damage your body. Acids can cause skin _____, _____ decay and _____. Bases can turn oils in your skin into _____. Hazard symbols tell us if a chemical is dangerous. We can wear _____, _____ or a mask to _____ ourselves.

2. Name **two** chemicals broken down by your body.

3. Explain why brushing your teeth helps prevent tooth decay.

Key points
- Chemical reactions control everything that happens in your body.
- Acids and bases are involved in many of your body's chemical reactions.
- Some acids and bases can damage skin, teeth and internal organs.

Further teaching suggestions

Designing a hazard symbol!

Students could build on the idea of hazard symbols by designing their own. Provide some stimulus material about nanotechnology, including the hazards and risks of this emergent science. Ask students to design an appropriate hazard symbol.

Answers to in-text questions

a liver, stomach
b bacteria
c Oils in your skin are turning into soap-like substances.
d irritant

Summary answers

1. reactions, enzymes, bases, burns, tooth, ulcers, soap, gloves (goggles), goggles (gloves), protect
2. protein, starch, fats
3. Gets rid of bacteria, reducing acid production; neutralises existing acid.

Unit 2, Theme 1 – My family

5.7 Acids and bases

Learning objectives

Students should learn:

- why the stomach produces hydrochloric acid
- that acids produce hydrogen ions when dissolved in water
- that bases such as antacids, can neutralise acids.

Learning outcomes

Most students should be able to:

- state that the acid in the stomach is hydrochloric acid
- describe why stomach acid is useful, and that it can also cause indigestion
- state that antacids can be used to neutralise excess stomach acid.

Some students should also be able to:

- explain that acids are chemicals that release hydrogen ions in solution.

AQA Specification link-up: Science B 3.4.1.2

- Understand that the stomach works most effectively in acid conditions by helping break down food.
- Describe how … hydrogen ions (H^+) make solutions acidic.
- Understand that the stomach works most effectively in acid conditions by helping to break down food.

Lesson structure

Starters

Naming equipment – Students match up names of the apparatus used in today's lesson with diagrams or photos. *(5 minutes)*

Acid or base? – Assess students' knowledge of acids and bases by presenting them with a range of substances and asking them to sort them into groups. To support students, use pictures of household materials such as vinegar and soap; to extend students, ask them to match up the name of the acid or base with the substance that contains it. *(10 minutes)*

Main

- Practical 1 – Students can use universal indicator to identify the pH of various solutions. Use this to reacquaint students with the idea of acids and bases.
- Practical 2 – Students perform a simple neutralisation reaction between hydrochloric acid and a solid base. Link the process to the action of antacids in the stomach. As with practical 1, electronic pH meters may be used to check accuracy.
- Show students a video clip of gastric endoscopy footage. Use diagrams of the stomach to show where acid is formed. Explain why acid is formed in the stomach. Refer back to the endoscopy and ask why so much mucus is produced by the stomach. Discuss what could happen if the mucus was not there.
- Model the concept of acids and bases dissociating in water by drawing their structures as ball-and-stick or space-filling diagrams onto plain paper or card. Then the molecules can be cut apart, demonstrating dissociation before being rearranged as salt and water. Starting with HCl and NaOH will make this concept more manageable. You can relate this to bile (a weakly alkaline liquid) neutralising hydrochloric acid in the small intestine during digestion of food.

Plenaries

Students can summarise their learning about the stomach by completing a table of the benefits and potential problems of stomach acid. *(5 minutes)*

Students can complete word equations of acid + hydroxide. To extend students, use symbol equations. *(10 minutes)*

Support

- To make the lesson more memorable for lower ability students, simulate a stomach by using a 'Ziploc' bag, a small amount of fizzy drink to represent stomach acid, and 2–3 slices of bread. Put the bread in the bag and add about $50\,cm^3$ of fizzy drink. Squeeze as much air out of the bag as possible then seal it. Massage the bag for about a minute until the bread dissolves. Discuss with the students what is happening in the bag, and how the liquid is helping the bread dissolve. Draw parallels with the action of the stomach.

Extend

- More able students can perform research to compare the action of antacids with other stomach medicines, such as ranitidine (Zantac).

Further teaching suggestions

Peptic ulcers

Peptic ulcers contribute to around 5000 deaths in the UK every year. Students can research the factors that make ulcers more likely to occur, and suggest ways to prevent them without the use of antacids.

Chapter 5 – Body systems

Practical support

Measuring pH – Practical 1
Equipment and materials required: Eye protection, univeral indicator liquid or paper, spotting tiles, dropping bottles of (0.1 mol/dm^3) solutions of hydrochloric acid, sulfuric acid, ammonium hydroxide, sodium hydroxide, sodium hydrogencarbonate solution, ethanoic acid.

Details
Test a drop of each solution with universal indicator, using a colour chart to estimate the pH. If available, electronic pH probes could be used to validate results.

Safety: Wear eye protection. CLEAPSS Hazcard 32 Universal indicator solution – highly flammable and harmful. CLEAPSS Hazcard 47A Hydrochloric acid; CLEAPSS Hazcard 91 Sodium hydroxide; CLEAPSS Hazcard 98A Sulfuric acid; CLEAPSS Hazcard 6 Ammonium hydroxide – corrosive. CLEAPSS Hazcard 38A Ethanoic acid – irritant.

Measuring pH – Practical 2
Equipment and materials required: Eye protection, 200 cm^3 beaker, 50 cm^3 (1 mol/dm^3) HCl, universal indicator, approx 10 g Mg(OH)$_2$ in a small beaker or measuring boat, glass rod, spatula, electronic balance, measuring cylinder.

Details
Accurately measure 50 cm^3 acid into the large beaker and add a few drops of universal indicator. Record the mass of the magnesium hydroxide and the pH of the acid. Stirring, add the magnesium hydroxide half a spatula at a time. Stop when the mixture reaches pH 7. Measure the mass of magnesium hydroxide left and calculate the mass needed to neutralise the acid. Alternatively, the pH could be recorded after each addition of antacid and a chart of spatulas versus pH could be constructed. This would provide an opportunity to evaluate the use of the spatula in making measurements.

Safety: Wear eye protection. CLEAPSS Hazcard 32 Universal indicator solution – highly flammable and harmful. CLEAPSS Hazcard 47A Hydrochloric acid – corrosive.

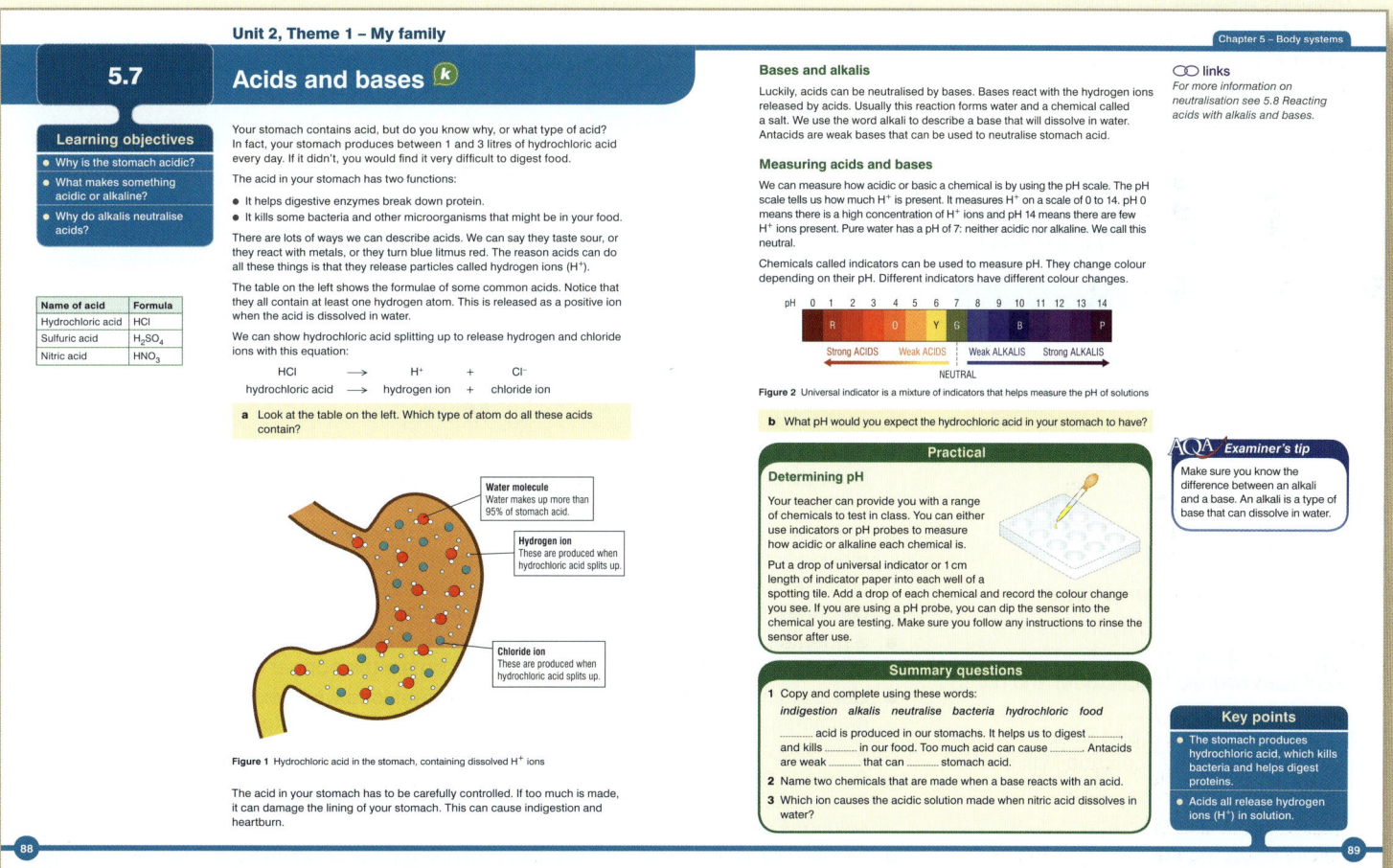

Summary answers

1. hydrochloric, food, bacteria, indigestion, alkalis, neutralise
2. salt + water
3. H$^+$ ion (dissolved in water)

Answers to in-text questions

a. hydrogen
b. 1–3

Unit 2, Theme 1 – My family

5.8 Reacting acids with alkalis and bases

AQA Specification link-up: Science B 3.4.1.2

- Know that acids are neutralised by reaction with oxides, hydroxides and carbonates to form salts and other products.
- Know the patterns in reactions of soluble hydroxides and carbonates with acids.
- Describe how a neutralisation reaction involves an acid and base reacting to form a salt and water.
  - that hydrogen ions (H^+) make solutions acidic
  - that hydroxide ions (OH^-) make solutions alkaline
  - that this reaction can be represented by the equation:
    $H^+(aq) + OH^-(aq) \longrightarrow H_2O(l)$ **[HT only]**
- Explain how an antacid neutralises excess stomach acid to help to treat heartburn and nausea.

Within this context, candidates should be able to use scientific data to discuss, evaluate or suggest implications of:
- the effectiveness of a range of antacid products.

Learning objectives

Students should learn:
- that there are patterns of reactivity between acids plus carbonates, oxides and hydroxides
- that hydroxide ions make solutions alkaline
- that hydrogen ions and hydroxide ions react to produce water.

Learning outcomes

Most students should be able to:
- describe a neutralisation reaction between an acid and an alkali as producing salt + water
- state that hydroxide ions make solutions alkaline.

Some students should also be able to:
- explain the symbol equation:
  $H^+(aq) + OH^-(aq) \longrightarrow H_2O(l)$. **[HT only]**

Lesson structure

Starters

What's in it? – Distribute empty packets of indigestion remedies. Students read the ingredients and construct a table of brand versus active ingredient. Extend students by getting them to discuss what the other ingredients do, such as citric acid in the case of Alka-Seltzer. *(5 minutes)*

Particles – Use a card match activity to review the differences between ions, atoms and molecules. Students can sort examples of each into groups and write definitions in their own words. *(10 minutes)*

Main

- Discuss with the class what the causes and symptoms of indigestion might be. Collectively come up with a list of ways to avoid indigestion and other stomach problems.
- Students can react hydrochloric acid with carbonates and observe that carbon dioxide is produced. This could be performed quantitatively, with the gas either being collected under water or in a gas syringe. Students could investigate whether all carbonates produce the same volume of carbon dioxide per gram.
- Students could explore metal oxide + acid $\longrightarrow$ salt + water by neutralising sulfuric acid with copper oxide and evaporating to obtain the copper sulfate.
- Use particle diagrams and cut-outs to represent ions when explaining neutralisation reactions. Alternatively, use interactive whiteboard facilities to make a drag-and-drop exercise so students can move ions around and arrange them to show the products of neutralisation. Ensure students know that excess hydrogen ions (H^+) are responsible for acidity in solutions and excess hydroxide ions (OH^-) cause alkalinity in solutions.
- Higher Tier students should practise balancing neutralisation equations. Review the 'counting atoms' approach to balancing equations. They also need to know the ionic equation for neutralisation, i.e. $H^+(aq) + OH^-(aq) \longrightarrow H_2O(l)$ **[HT only]**
- Higher Tier students should practise using state symbols in equations. Re-writing all the equations covered in the lesson and adding state symbols should be fairly straightforward as they have seen all the of the reactions first hand. **[HT only]**
- Students can compare the effectiveness of different antacids by titrating against hydrochloric acid.

Plenaries

Equation card sort – Students rearrange examples of acid + carbonate and acid + hydroxide reactions. *(5 minutes)*

A new product – In groups of three, students invent a new antacid. They need to design a logo and an information panel to put on the box. The information panel must explain what is in the antacid and how it works. *(10 minutes)*

Answers to in-text questions

a potassium ions and hydroxide ions
b salt and water
c calcium hydroxide (or oxide) and nitric acid
d carbonates – Tums and Rennie
   hydroxides – Gaviscon and Milk of Magnesia

Support

- Build on the 'how to avoid indigestion' activity by making an informational poster, which explains the points raised in the class discussion.

Extend

- Students can research contributory factors to ulcers and gastritis, such as alcohol, ibuprofen or aspirin.

Practical support

Do all carbonates produce the same amount of carbon dioxide?

Equipment and materials required: Magnesium carbonate, sodium bicarbonate, calcium carbonate, hydrochloric acid (0.5 mol/dm^3), conical flask with bung and delivery tube, spatula, electronic balance.

For testing the gas: test tube, limewater.

For collecting the gas: 100 cm^3 measuring cylinder inverted in a water trough, or gas syringe.

Details
Add 20 cm^3 acid to the conical flask. Add approx 2 g carbonate and replace bung. Bubble through limewater to verify the gas is carbon dioxide. Use preferred method to collect the carbon dioxide and compare the CO$_2$ produced by each carbonate.

Safety: Wear eye protection. CLEAPSS Hazcard 18 Limewater – irritant. CLEAPSS Hazcard 47A Hydrochloric acid – corrosive.

Making copper sulfate

Equipment and materials required: Bunsen burner, safety mat, tripod, wire gauze, 200 cm^3 beaker, clamp stand, filter paper, funnel, 2 mol/dm^3 sulfuric acid, copper(I) oxide, evaporating basin, glass rod.

Details
Set up the apparatus to heat the beaker, add 50 cm^3 acid to the beaker, then an excess of copper oxide (about three heaped spatulas). Carefully warm whilst stirring for about five minutes. Do not allow the mixture to boil. Allow to cool, then filter the resulting black suspension into an evaporating basin and leave the blue solution to evaporate at room temperature.

Safety: Wear eye protection. CLEAPSS Hazcard 98A Sulfuric acid – corrosive.

Comparing the effectiveness of antacids

Equipment and materials required: Eye protection, (0.1 mol/dm^3) hydrochloric acid, burette, 25 cm^3 volumetric pipette, 100 cm^3 conical flask, phenolphthalein indicator, clamp and stand, pipette filler, white tile, funnel, pestle and mortar, selection of antacids.

Details
Fill the burette with hydrochloric acid, crush one dose of antacid, dissolve it in 25 cm^3 distilled water in the conical flask and add a few drops of phenolphthalein. Slowly titrate the acid into the antacid solution; trial runs will be necessary to get an idea of where the end-point is. Compare the amount of acid one dose of each antacid could neutralise.

Safety: Wear eye protection. CLEAPSS Hazcard 32 Phenolphthalein; CLEAPSS Hazcard 47A Hydrochloric acid – corrosive.

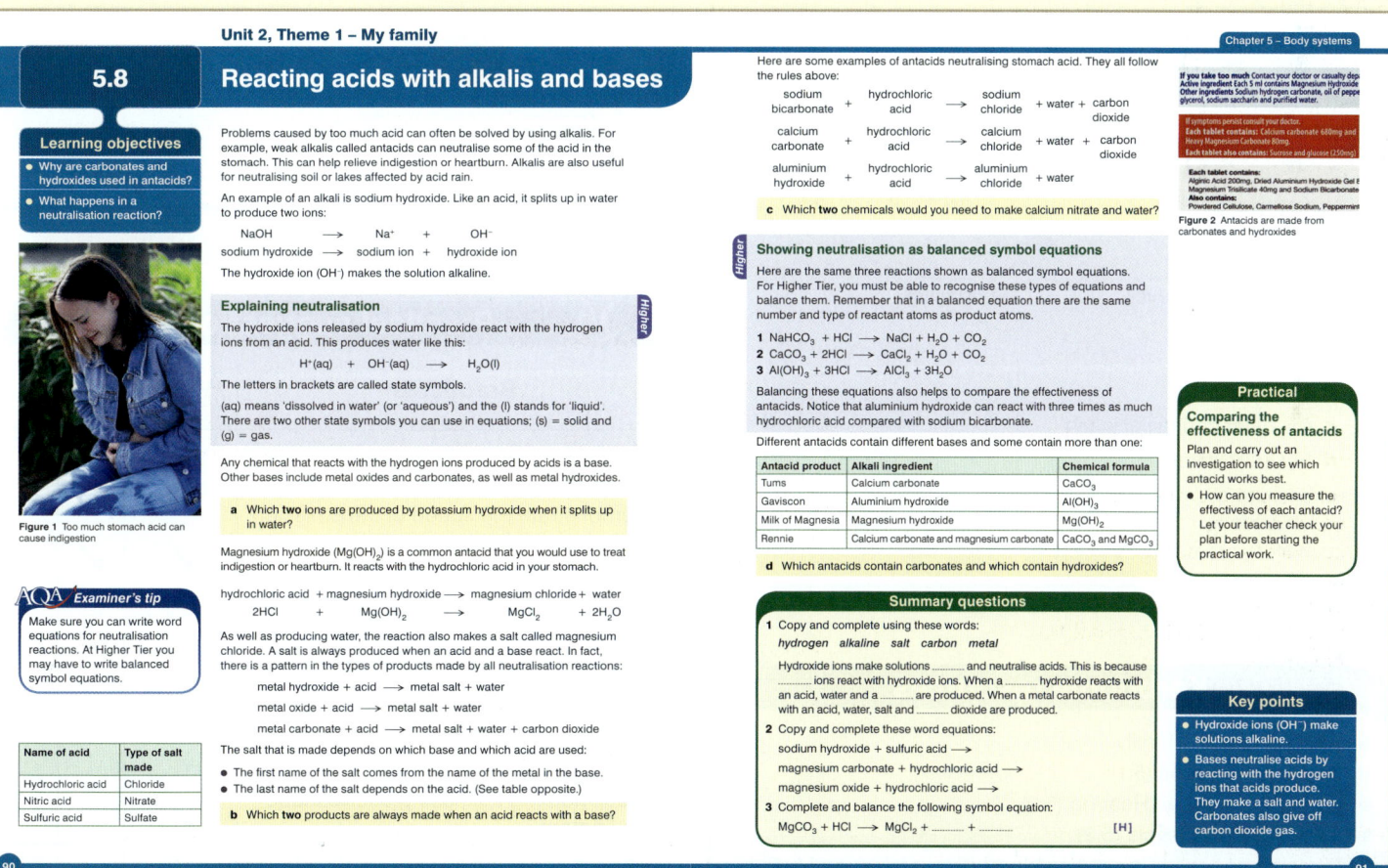

Further teaching suggestions

Neutralisation
The topic of neutralisation is also taught to younger classes. Groups of students could work together to produce activities and support materials to be used lower down the school.

Summary answers

1 alkaline, hydrogen, metal, salt, carbon

2 sodium hydroxide + sulfuric acid $\longrightarrow$ sodium sulfate + water

magnesium carbonate + hydrochloric acid $\longrightarrow$ magnesium chloride + water + carbon dioxide

magnesium oxide + hydrochloric acid $\longrightarrow$ magnesium chloride + water

3 $MgCO_3 + 2HCl \longrightarrow MgCl_2 + CO_2 + H_2O$

Unit 2, Theme 1 – My family

Summary answers

1. **a** temperature
   b muscle
   c electrical impulse
   d reflex

2. **a** Starch is broken down to glucose, and absorbed into the blood.
   b insulin
   c pancreas
   d Insulin causes liver to convert glucose to glycogen. This reduces the blood sugar level.

3. Sweat is being released.
   Vasodilation of blood vessels.
   Hairs are lying flat.

4. Longitudinal waves; must travel through a medium.

5. Harmful — Can cause general damage to people
   Irritant — Can cause a rash
   Corrosive — Can burn through skin and other materials

6. Because the acid releases H^+ ions and the alkali releases OH^- ions. $H^+ + OH^- \longrightarrow H_2O$.
   For example:
   $HCl + NaOH \longrightarrow H^+ + Cl^- + Na^+ + OH^- \longrightarrow NaCl + H_2O$

7.

Venn diagram:
- Nerves: Causes short term changes in the body, Quick
- Overlap: Responds to stimuli
- Hormones: Chemical message, Travels in the blood

8. **a** facial wash
   b blue
   c pH balanced facial wash
   d H_2O *OR* water
   e sodium chloride
        water

Kerboodle resources

- Theme map: My family
- Practical: Reflexes (5.1)
- Interactive activity: Hearing (5.2)
- Support: Noise levels (5.2)
- Revision podcast: Controlling blood glucose (5.3 and 5.4)
- Support: Sugar rush (5.4)
- Practical: Blood glucose (5.4)
- Practical: The effect of acid (5.7)
- Practical: Testing pH (5.7)
- How Science Works: Effectiveness of a range of antacid products (5.8)
- Practical: Neutralisation reaction (5.8)
- Examination-style questions
- Answers to examination-style questions

Summary questions

1. A chef accidentally puts his hand on a hotplate.
   a Which receptor cell detects this?
   b Which effector is stimulated in this reaction?
   c How does an impulse travel along a nerve?
   d Is this an example of a controlled or reflex response?

2. The night before an athlete runs a marathon, she eats a large bowl of pasta. This causes her blood sugar level to rise.
   a Why does this occur?
   b Which hormone regulates the level of sugar in the blood?
   c Where is this hormone made?
   d Explain how this hormone returns the blood sugar level to normal?

3. This picture shows the skin of a builder who has been working hard. What three things in the picture show you that he is hot?

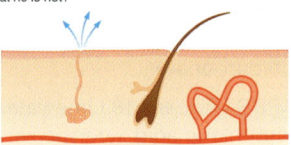

4. State two features of a sound wave.

5. Match these symbols to their meanings

   a Can cause general damage to people
   b Can burn through skin and other materials
   c Can cause a rash

6. Explain why water is one of the products when an acid reacts with an alkali. Use word and balanced symbol equations to help explain your answer.

7. Decide if each of these comments is about hormones, nerves or both. Write your answers in a Venn diagram.
   - Causes short-term changes in the body
   - Chemical message
   - Quick
   - Responds to stimuli
   - Travel in the blood

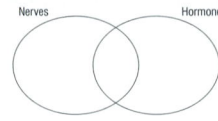

8. The graph shows the pH of various soaps.

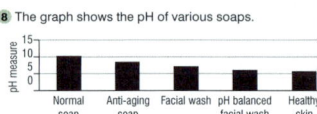

   a Which type of soap is nearest to neutral?
   b What colour would normal soap turn universal indicator?
   c A wasp sting is alkali. Which soap would be best to neutralise the sting?
   d Complete the symbol equation by filling in the missing compound.
       $H^+(aq) + OH^-(aq) \longrightarrow$ ....
   e Complete this word equation to show a neutralisation reaction.
       hydrochloric acid + sodium hydroxide $\longrightarrow$ .... + ...

Practical suggestions

Practicals	AQA	k	📖
Measure reaction times using metre rules, stop clocks or ICT.	✓	✓	
Demonstrate the speed of transmission by nerves: candidates stand in a semi-circle, holding hands and squeezing with eyes closed.	✓	✓	
Use blindfolds and open paper clips to test pressure points and skin sensitivity.	✓		
Demonstrate the knee jerk reaction.	✓		
Test reflexes: elbow, knee, foot, pupils.	✓		
Class hearing test, using oscilloscope equipment.	✓		
The effect of acid on various objects left for a few days.	✓	✓	
Neutralisation titration.	✓	✓	
Reaction of carbonates with acids	✓		

Chapter 5 – Body systems

Examination-style questions

1 Label the neurons in this reflex arc. (3)

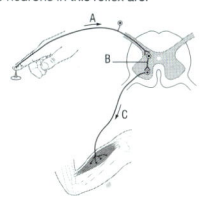

2 Put these sentences in the correct order to show how your body regulates blood sugar level after eating.
A The glucose level in the blood rises.
B The glucose level in the blood then falls.
C The liver turns glucose into glycogen.
D The liver turns glycogen to glucose.
E The pancreas releases the hormone insulin.
F The pancreas releases the hormone glucagon. [H] (5)

3 A nurse did an experiment to find the best antacid tablet to recommend. The nurse put 10 cm³ of acid in a beaker then added antacid powder one gram at a time. The table shows the pH level after 3 mg of antacid tablet was added.

Antacid powder	pH after 3 mg of powder
Alkalon	2
Burpeze	5
Indegon	4
Tummyset	4
Windipops	1

a Which powder was the most effective? (1)
b Predict the pH after 3 mg of Indegon had the nurse used 20 cm³ of acid. (1)
c Choose two things the nurse should have kept the same to make it a fair test.

Same beaker	Same amount of acid	Same number of stirs	Same time left in beaker	Same type of acid

(2)

4 An example of a neutralisation reaction between an acid and a metal carbonate is given:
HCl + MgCO₃ ⟶
a Write a balanced symbol equation for the reaction given. [H] (3)
b Name the two reactants. [H] (1)

5 A recent study has found chocolate to be a benefit to people who are depressed.
a Name the organ that releases the hormones that control blood sugar level. (1)
b Explain how the body reacts when there is too much sugar in the blood. (3)
c In this question you will be assessed on using good English, organising information clearly and using specialist terms where appropriate.
Describe the advantages and disadvantages of a person with diabetes eating chocolate. (6)

6 Body temperature needs to stay constant to within a few degrees otherwise we die.
a What is normal body temperature? (1)
b Choose words from the box to complete the sentences.

| air | body | conducting | hair |
| insulating | skin | | |

When we get cold, our .... stands on end. This traps a layer of .... close to the surface of the .... Air is .... so it does not let heat escape from the .... (5)
c Explain what part the blood vessels close to the surface of the skin have in cooling the body when it is too cold. (3)

7 Our ears convert sound energy into nerve impulses.
The diagram shows a sound wave travelling from one person to another.

a What type of wave is a sound wave? (1)
b Which part of the diagram shows a rarefaction? (1)
c Explain how the sound wave travels from the girl to the man's ear. (3)
d Choose the range of a human's hearing.

Letter	Range
A	Between 0 and 20 Hz
B	Between 20 and 200 Hz
C	Between 20 and 20 000 Hz
D	Between 200 and 20 000 Hz

(1)

Examination-style answers

1 A – sensory
   B – relay
   C – motor (3 marks)

2 A E C B F D (All correct = 5 marks, 5 correct = 4 marks, 4 correct = 3 marks, 3 correct = 2 marks, 1 or 2 correct = 1 mark)

3 a Burpeze (1 mark)
   b 2 (1 mark)
   c same amount of acid
     same type of acid (2 marks)

4 a 2HCl + MgCO₃ ⟶ MgCl₂ + CO₂ + H₂O (3 marks)
   b hydrochloric acid (1 mark)
     magnesium carbonate (1 mark)

5 a pancreas (1 mark)
   b Insulin is released.
     Glucose is converted into glycogen by the liver. (3 marks)
   c Marks awarded for this answer will be determined by the Quality of Written Communication (QWC) as well as the standard of the scientific response.

There is a clear, balanced and detailed description of the advantages and disadvantages of a person with diabetes eating chocolate. The answer is well structured with minimal repetition or irrelevant points. There is an accurate, fluent and clear expression of ideas with only minor errors in the use of technical terms, spelling, punctuation and grammar. (5–6 marks)

There is some description of the advantages and disadvantages of a person with diabetes eating chocolate but lacking in detail and balance with some omissions. The answer shows some attempt at structuring and the ideas are expressed with reasonable fluency and clarity. There are some errors in the use of technical terms, spelling, punctuation and grammar. (3–4 marks)

There is a brief description of an advantage and/or a disadvantage of a person with diabetes eating chocolate. The answer is largely incomplete and may contain some valid points which are not clearly structured. It lacks fluency and/or clarity. It contains errors in the use of technical terms, spelling, punctuation and grammar. (1–2 marks)

No relevant content (0 marks)

Examples of points made in the response:
- Feel better
- Can cope better
- May make them ill as not enough insulin in the blood
- Will need to watch what they eat afterwards
- Bad for teeth
- Weight gain.

6 a 37 °C (both temperature and unit are needed for the mark) (1 mark)
   b hair
     air
     skin
     insulating
     body (5 marks)
   c blood vessels narrow
     less blood allowed to flow to the surface
     less heat lost by radiation (3 marks)

7 a longitudinal (1 mark)
   b where the air particles are more spread out (1 mark)
   c Vibrating voice box
     Causes air to move back and forth
     Air particles move perpendicular to direction of sound wave (3 marks)
   d C (1 mark)

Unit 2, Theme 1 – My family

6.1 Animal cells

Learning objectives

Students should learn:
- what makes up an animal cell
- that chromosomes are found in the nucleus and that they are strands of DNA containing sections known as genes
- how parents pass on characteristics to their offspring through genes and that during fertilisation paternal and maternal genes combine.

Learning outcomes

Most students should be able to:
- state the four features of animal cells
- define the terms gene, chromosome and inheritance
- describe how genetic material is arranged in a cell.

Some students should also be able to:
- explain, using a diagram, how genetic material is passed from parents to their offspring.

AQA Specification link-up: Science B 3.4.1.3
- Know that simple animal cells have a nucleus, cytoplasm and cell membrane.
- Know that the nucleus of a cell contains chromosomes:
  (a) chromosomes carry genes which control the characteristics of the body ...

Lesson structure

Starters

What do animal cells contain? – Ask students to draw a simple animal cell and label as many structures as possible (revision from previous work). To support students, provide them with a simple diagram of an animal cell for them to label. To extend students, ask them to write a sentence describing the role of each feature. *(10 minutes)*

Matching pairs – Put two lists on the board or give students lists of cell structures and their functions in a muddled order. Set students the task of matching the cell structure to its function. *(5 minutes)*

Main

- Discuss the concept of cells with students to ascertain their prior knowledge. How big do they think a cell is? How can cells be seen? What features do cells contain? Students can be asked if they remember any differences between plant and animal cells. The detailed structure of a plant cell is not required.
- Demonstrate how to make skin cell slides. Students then make their own slides and observe them under a microscope.
- Students should draw what they can see down the microscope (under high magnification), labelling as many features as possible.
- A small bag containing 46 pieces of fusilli pasta could be used to represent the nucleus of a human cell. The students could then each pick out one piece of pasta as this represents the chromosomes found in a human body cell. Fusilli pasta illustrates the appearance of a chromosome. The ridges illustrate the ladder of DNA and the students could colour in horizontal sections to represent the genes. This activity should be used alongside conventional diagrams showing the appearance of DNA. (A detailed understanding of the structure of DNA is not required, only that it has regions that code for proteins and therefore characteristics (genes), and in between there are regions of non-coding DNA.)
- Discuss how characteristics are passed on from parent to offspring. Genetic material in the egg and sperm combine during fertilisation. The diagram in the Student Book could be used to illustrate this process.

Plenary

Which is biggest? – Give students a list of the parts of a cell involved in inheritance. They need to arrange them in order of size to show what fits into what, e.g. gene ⟶ chromosome ⟶ nucleus ⟶ animal cell. *(5 minutes)*

Student definitions – Ask students to write their own definitions of nucleus, gene, chromosome and inheritance. To support students, provide students with key terms or phrases they should include when writing their definition. To extend students, ask students to include an annotated diagram of a cell showing a nucleus, gene and chromosome to illustrate their definitions. *(10 minutes)*

Support
- Skin cell slides can be pre-prepared and stained, or video microscope images from scientific suppliers can be projected onto a whiteboard, to give students exemplar slides to inform their own work.

Extend
- Students could prepare large-scale diagrams of their cell observations to feed back to the group. Stage micrometers can be used for observations, and added to the diagrams, to give students an idea of the scale of the different structures within the cell.

Further teaching suggestions

Specialised cells
Show students photos of a range of animal cells, e.g. nerve, blood, sperm, egg. Students need to identify the nucleus, cytoplasm and cell membrane in each example.

Model cells
Provide students with materials to make a 3-D model of an animal cell.

Extracting DNA
Students could extract DNA from a kiwi fruit. DNA can be extracted from kiwis by mashing the fruit and mixing it with washing up liquid and salt. Leave for 20 minutes, then filter the mixture. Slowly add ice-cold methylated (highly flammable/harmful) spirit to the filtrate. DNA appears where the two liquids mix and can be spooled onto a glass rod.

Chapter 6 – Human inheritance and genetic disorders

Practical support
Making and observing animal-cell slides

Equipment and materials required: Slides and cover slips, sticky tape, methylene blue, microscope.

Details
Students make animal-cell slides by sticking a small piece of sticky tape to the back of their hands, pulling it off and placing it upside down on a slide. Then they should add one drop of methylene blue, put on a cover slip and tap with the end of a pencil to ensure there are no air bubbles. Finally, they observe the slide under the microscope, initially under low magnification. Then they increase the magnification to observe structures in the cell. Ask 'What cell features can you identify?'

Safety: Take care not to break the slide on the microscope when using the high power objective. CLEAPSS Hazcard 32 Methylene blue – harmful.

Answers to in-text questions
a millions
b electron microscope
c nucleus
d a strand of DNA
e a small section of DNA that codes for a particular characteristic

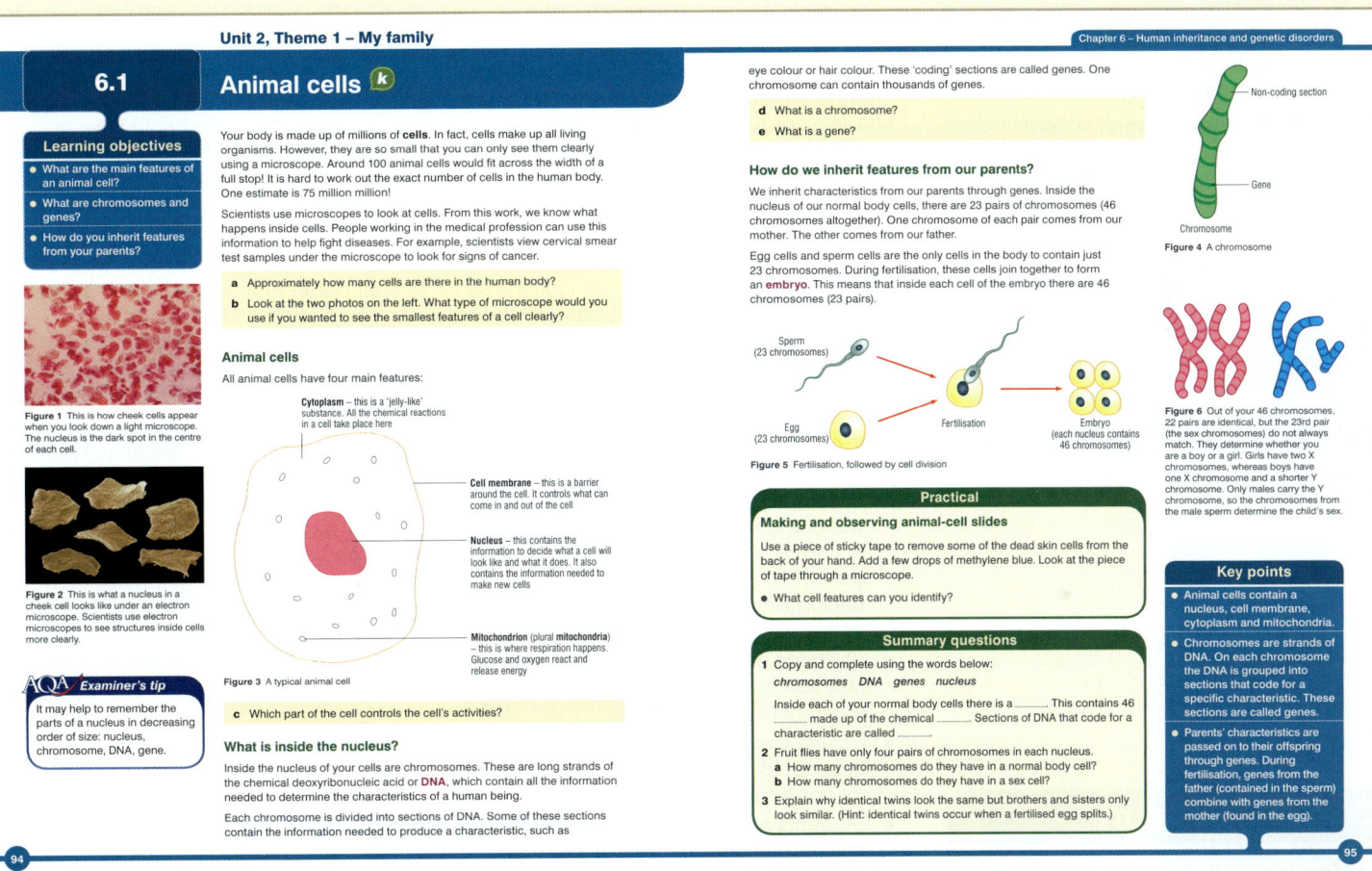

Did you know ...?

Chromosomes can be photographed through a light microscope. These are called photomicrographs. Photomicrographs of the chromosomes of an organism can be arranged in a standard order for study. These are called karyograms.

Students could be given a karyogram to illustrate what chromosomes actually look like and emphasise the fact that there are two copies of every gene (one from each parent).

Students could be asked to match the pairs of chromosomes and then determine the sex of the individual by studying the 23rd pair – the sex chromosomes.

Summary answers

1 nucleus, chromosomes, DNA, genes
2 a eight chromosomes
   b four chromosomes
3 The genetic material found in identical twins is exactly the same. This is because the fertilised egg splits into two – this is after the genetic material from the sperm and egg has already combined. Brothers and sisters inherit genetic material from both their mother and their father, but they inherit slightly different combinations of characteristics.

Unit 2, Theme 1 – My family

6.2 Variation

Learning objectives

Students should learn:
- that different characteristics are present within a species; this is called variation
- that characteristics are a result of environmental variation, genetic variation or both.

Learning outcomes

Most students should be able to:
- name some examples of environmental variation
- name some examples of genetic variation
- describe why some characteristics occur as a result of both genetic and environmental variation.

Some students should also be able to:
- analyse the extent to which characteristics are influenced by the environment.

Support
- Provide students with a set of cards that contain images of environmental factors that can change a person's appearance, e.g. hair dye, the Sun, food and tattoos. In pairs, they should discuss how each factor would change their own characteristics.

Extend
- Ask students to produce a survey to collect data from Year 10 students in their school to help them determine which human characteristics are inherited, which are affected only by environmental factors and which could be affected by both. This will introduce them to the concept that to collect statistically reliable data, sample size is important.

AQA Specification link-up: Science B 3.4.1.3
- Know that differences in the characteristics of individuals (variation) may be due to genetic causes, environmental causes or a combination of both.

Lesson structure

Starters

Animal differences – Show students two or three pictures of a species of animal, e.g. cat or dog. They should produce a list of all the differences they can see between the animals. *(5 minutes)*

Fish variation – Ask students to draw and colour a picture of a fish. They then compare their drawings in small groups. How do their fish vary? They need to write down as many types of variation as possible and think about how this variation is produced. To support students, provide them with a table to complete their results. The table could have the first few possible sources of variation labelled. For example, colour of fish, size of tail fin and shape of tail fin. To extend students, ask them to label on their drawings whether they think each characteristic is a result of inheritance or the environment. *(10 minutes)*

Main

- Student discussion – do you look the same as your parents? If not, why not? During this discussion, try to be aware of student sensitivities. Make a class list on the board of characteristics that show variation. Introduce the differences between environmental (things that they can alter) and genetic variation (things that they cannot). Select students to take it in turns to highlight a factor on the list that shows environmental variation, then a factor that shows genetic variation.

- Show students several pairs of a plant species that display variation. Geranium plants are ideal for this purpose. Ensure differences exist between each pair – for example, ones that produce different coloured flowers, or one with variegated leaves, one without, one that has been grown in nutrient-rich soil and one that has been grown in poor soil and hardly watered. What differences can be seen between the pairs of plants? Students can produce a list describing what variation is present. They then decide whether the difference is a result of genetics or the environment. Is a plant or an animal more affected by the environment?

- Students discuss human characteristics that show variation within their class. The characteristics must be those that they can collect data on and then analyse the differences, e.g. shoe size.

- Collect class data on a range of characteristics and use it to produce graphs, e.g. data could be collected on height and eye colour. (Students should plot line graphs for continuous data and bar charts for discontinuous data.)

- Students analyse their data (e.g. calculate average height, range of variation, differences between sexes) and make conclusions on the factors that affect the chosen characteristics.

Plenaries

What causes the variation? – Give students a list of characteristics and ask them to sort them into those that are caused by the environment, those that are caused by genes and those that are affected by both. *(5 minutes)*

Who is most affected by environmental variation? – Ask students to work in pairs, then fours, then as a class, to discuss which group is most affected by environmental variation – plants or animals [plants]. To extend students, ask them to write a paragraph explaining why the group they have chosen (plants) is most greatly affected by the environment. To support students, provide them with 'extreme' versions of plants and animals to help them visualise the concept. The images of animals should show the least variation. *(10 minutes)*

Chapter 6 – Human inheritance and genetic disorders

Further teaching suggestions

Studying cress growth
Grow some cress seeds. Alter various environmental conditions, such as the amount of water and the direction or amount of light, to study the effects of environmental factors on the appearance of the cress.

Activity

Studying variation
Introduce the concept of continuous and discontinuous variation and the link between genetic and environmental variation. Ask students to collect more data on human characteristics, e.g. hair length, shoe size, ability to roll their tongue. Ask them to determine whether each characteristic shows continuous or discontinuous variation. (Students should plot line graphs for continuous data and bar charts for discontinuous data.)

Unit 2, Theme 1 – My family

6.2 Variation

Learning objectives
- What is variation?
- What can cause variation?

It is easy to tell the difference between a monkey and a fish. This is because they have lots of different **characteristics** (features). However, it is more difficult to tell the difference between two frogs. This is because, within a species, lots of characteristics are shared.

Every person in the world is different – even identical twins are different in some ways. Differences within a species are called **variation**. People vary in many ways including height, build, hair colour and intelligence.

There are two factors that cause variation:
- the characteristics you inherit from your parents – genetic variation
- the environment in which you live – environmental variation.

a What is variation?

How do people vary?

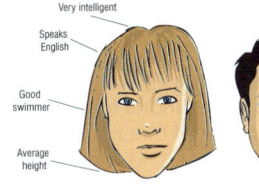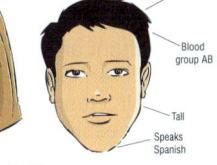

Figure 1 People can vary in a large number of ways

These two children vary in a number of ways. Some of this variation is due to characteristics they have inherited from their parents. However, most is due to factors in their environment. These include where they live and what they learn from their parents, teachers and friends.

b State **three** characteristics that are caused by genetic variation.
c Give **three** characteristics that are influenced by environmental variation.

Hair colour
This characteristic could be classified as an **inherited** feature. People generally have similar colour hair to one of their parents. However, look at the girl in Figure 2. Do you think her parents also have blue hair? Probably not! This is an example of environmental variation. The person has chosen to dye and style her hair in this way.

Figure 2 Is this person's hair colour inherited?

Height
Height is another characteristic that is mostly determined by your genes. If your parents are tall, you are also likely to be tall. However, if you are very poor with very little to eat, your growth is likely to be stunted as a result of poor diet.

Many characteristics are affected by both environmental and genetic variation, such as height and weight.

d Name **two** other characteristics that are influenced by both environmental and genetic variation.

Characteristics that are not influenced by the environment
Here are four examples of characteristics that are not influenced by the environment in humans:
- natural eye colour
- natural hair colour
- blood group
- genetic disorders – such as cystic fibrosis and haemophilia.

Figure 3 These men vary greatly in height

Activity

Studying variation
Carry out a survey of some characteristics of the members of your class to see how they vary.

Is the variation in each characteristic a result of environmental factors, genetic factors or a combination of both?

Summary questions

1 Copy and complete the table using the words below:
weight intelligence blood group skin colour eye colour hair length scar

Type of variation	Characteristic
Genetic variation	
Environmental variation	
Both types of variation	

2 **a** Are plants more or less likely than animals to be influenced by environmental variation?
b Name some factors in the environment that could affect plant growth.

3 Why are identical twins the best people to study if you want to find out how the environment influences characteristics?

Key points
- Variation is the name given to the differences that exist between organisms of the same species.
- Variation occurs as a result of genetic factors, environmental factors or a combination of both.

Summary answers

1

Type of variation	Characteristic
Genetic variation	blood group, eye colour
Environmental variation	scar
Both types of variation	weight, intelligence, skin colour, hair length

2 **a** Plants are more likely to be influenced by environmental variation.
b amount of water/nutrients/sunlight/space available

3 Identical twins have identical genetic material therefore any differences in their appearance will be the result of environmental variation.

Answers to in-text questions
a the differences found within a species
b eye colour, hair colour (natural) and blood group
c scarring, unnatural hair colour, language spoken
d weight, skin colour (sun tan), intelligence, sporting ability, etc.

Unit 2, Theme 1 – My family

6.3 Dominant and recessive alleles

Learning objectives

Students should learn:
- that there are different forms of each gene, called alleles
- that there are dominant and recessive alleles
- how to carry out a simple genetic cross (monohybrid inheritance) using a Punnett square.

Learning outcomes

Most students should be able to:
- name some examples of alleles
- describe the difference between a dominant and recessive allele
- carry out a simple genetic cross using a Punnett square.

Some students should also be able to:
- use information to assign genes with a letter and then use them to carry out genetic crosses
- calculate the probability of inheriting a characteristic, given the parental genotype.

AQA Specification link-up: Science B 3.4.1.3
- Know that genes have different forms called alleles, which produce different characteristics.
- Describe the mechanism of monohybrid inheritance where the dominant and recessive alleles are given.

Lesson structure

Starters

Student ideas – In pairs, students try to come up with a definition for the words dominant and recessive. *(5 minutes)*

Why do we have two copies of each gene? – Ask students to draw a diagram to explain why you have two copies of each gene – fertilisation. Only one set of chromosomes needs to be shown. To support students, provide them with a series of diagrams to arrange into the correct order. To extend students, ask them to annotate their diagrams to show how many chromosomes will be present at each stage (23 chromosomes in the egg and sperm, 46 chromosomes in the fertilised egg). *(10 minutes)*

Main

- Students discuss different variations of characteristics that they possess, e.g. hair and eye colour, and compare these with their own parents. This can be used to introduce the concept of alleles. During this discussion try to be aware of student sensitivities.
- Demonstrate the concept of dominant and recessive alleles, using coloured inks, show what happens if you mix different combinations together (analogy to mixing up alleles during fertilisation). Add half a test tube of black ink (representing a dominant gene) to red ink (representing a recessive gene). The ink will turn black showing that black is dominant. Repeat the demonstration, adding black ink to black ink and red ink to red ink to illustrate all the combinations of alleles that can occur.
- Using a mirror, students should analyse their own features to identify some that are dominant and some that are recessive. They can use the Student Book to identify the features.
- Introduce genetic crosses. As a class, work through a cross, step by step, between a dominant and a recessive gene to determine the characteristics of the offspring (the example given in the Student Book could be used). Students often find it easier to understand if they use a different colour to represent the genes that come from the father from those that come from the mother. Time should be taken to explain how to construct a Punnett square. In its simplest form, this is a square divided into four sections. Often students find it easier if the dividing lines protrude through the top and left hand side. For consistency, the paternal alleles should be written above the grid and the maternal alleles along the left edge. Students then read the square as a grid to complete the Punnett square. They fill in the empty boxes using the allele directly above the box and the one directly to the side.
- Students are then asked to carry out their own genetic crosses. Provide a 'solutions sheet' so that students can peer assess each other's working.

Plenaries

Dominant or recessive – Show students pictures of a series of characteristics. Students record whether they are dominant or recessive characteristics. *(5 minutes)*

Genetic cross – Provide students with a genetic cross diagram with some obvious mistakes. Students highlight and correct the mistakes. To support students, highlight the mistakes in a different colour so students can clearly see where they are. Students can be asked to work out why the cross is wrong. To extend students, ask them to write a help sheet for fellow students learning how to perform genetic crosses using Punnett squares. They should include a section on common mistakes. *(10 minutes)*

Support

- Provide students with a template to solve genetic-cross questions. The worksheet should include prompts for mother/father characteristic, mother/father alleles, Punnett square, offspring alleles, possible offspring characteristics and ratio of possible characteristics.

Extend

- Provide students with information about characteristics in plants and animals. From this they need to assign letters to the alleles that code for these characteristics. They should use the standard notation that a capital letter represents a dominant allele and lower case the recessive allele. Students should then carry out crosses to predict what the offspring will look like.

Chapter 6 – Human inheritance and genetic disorders

Further teaching suggestions

What colour fur will a mouse have?
Ask students to carry out a genetic cross on mice. Tell students the genes that are present in mice that code for fur colour (B = brown fur, b = white fur). The students need to perform the cross, then decide how many mice in a litter of ten have brown fur and how many have white fur. The litter contains five mice with the genes Bb, two with the genes bb, and three with the genes BB (eight will be brown, two will be white).

Mendel's peas
Students study Mendel's work on peas. Ask students to draw genetic crosses to determine the characteristics of pea plants. The allele for tall plants (long stems) is dominant to the allele for short plants (short stems). Students need to assign letters to the genes (guidance could be given for lower ability students) and draw a cross to show what happens if a tall plant and short plant are bred. Students then choose two of these offspring and show what happens when they breed.

Act out a genetic cross
Carry out a role play of a genetic cross – students can represent the alleles to show how they can be arranged into different combinations during fertilisation.

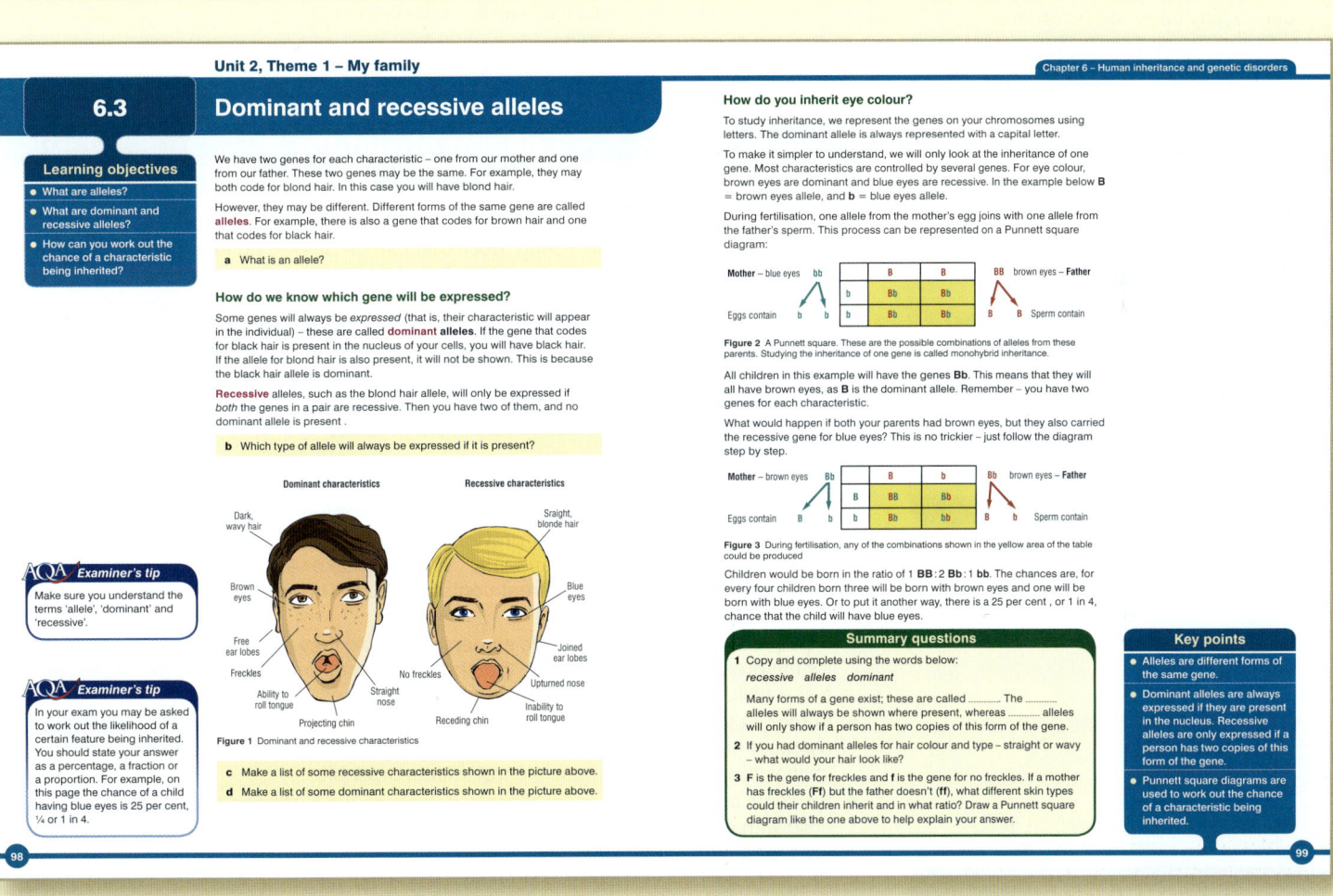

Summary answers

1 alleles, dominant, recessive

2 dark and wavy

3 Mother – freckles   Father – no freckles
    Ff   ff
Eggs contain F f   Sperm contain f f
During fertilisation:

	F	f
f	Ff	ff
f	Ff	ff

Children would be born in the ratio of 1 Ff : 1 ff, i.e. there is a 50% chance of the child having freckles.

Answers to in-text questions

a one form of a gene

b dominant

c straight hair, blond(e) hair, blue eyes, receding chin, joined ear lobe, etc.

d wavy hair, dark hair, brown eyes, projecting chin, free ear lobes, etc.

Unit 2, Theme 1 – My family

6.4 Genetically inherited disorders (1)

Learning objectives

Students should learn:
- that genetically inherited disorders can be passed on from parents to their children
- that cystic fibrosis and sickle-cell anaemia are examples of genetically inherited disorders
- that genetic screening can test for the presence of genes responsible for genetically inherited disorders.

Learning outcomes

Students should be able to:
- describe what a genetically inherited disorder is
- draw genetic crosses to determine how likely it is for a child to be born with cystic fibrosis or sickle-cell anaemia
- list some arguments for and against genetic screening.

Some students should also be able to:
- evaluate the arguments for and against genetic screening.

AQA Specification link-up: Science B 3.4.1.3

- Know that genes have different forms called alleles, which produce different characteristics.
- Describe the mechanism of monohybrid inheritance where the dominant and recessive alleles are given.
- Know that cystic fibrosis, sickle-cell anaemia, haemophilia and polydactyly are genetically inherited disorders.

Within this context, candidates should be able to use scientific data and evidence to discuss, evaluate or suggest implications of the following:
- the use of current research in the treatment of genetic disorders
- the likelihood of a genetically inherited disorder occurring
- the use of genetic screening.

Lesson structure

Starters

Genetically inherited disorders? – Provide students with a list of diseases. Students sort the diseases into a table, listing those that can be passed on from parents to their children in genes, and those that cannot. *(5 minutes)*

Template for genetic crosses – Ask students to produce a blank template that they could complete for any genetic cross. To support students, provide them with the subheadings required to produce a template in the wrong order, for them to rearrange. They should include mother and father characteristics, mother and father alleles, Punnett square, offspring alleles, offspring characteristics. To extend students, ask them to add an explanatory statement for each section. *(10 minutes)*

Main

- Treat your discussions of genetically inherited disorders with sensitivity.
- Show students clips/pictures or provide information about people suffering from cystic fibrosis and sickle-cell anaemia. Students make very brief notes of the symptoms of the disease and treatments available. This information could be summarised in a table.
- Introduce the concept of genetic screening. Would they want to test their unborn baby for the presence of a disorder so that they could prepare themselves for dealing with a sick child (or even terminate the pregnancy)?
- Introduce the role of a genetic counsellor – explain that they work out the likelihood of a couple having a child with an inherited disorder and give advice about bringing up a child with a disease and possible treatments available.
- Students act as genetic counsellors to determine the likelihood of a couple having a child with cystic fibrosis by drawing out genetic crosses. They can repeat the exercise for a range of potential parents.
- Students discuss via role-play how they would feel as a potential parent if they were told that there was a 25% chance of their child suffering from, for example, cystic fibrosis. What would their actions be?

Plenaries

True or false –
- Cystic fibrosis is caused by a dominant allele. [F]
- People with cystic fibrosis produce too much mucus. [T]
- Carriers of a genetic disorder do not suffer from the disease. [T]
- Genetic screening can detect the presence of 'faulty' alleles. [T]
- People with sickle-cell anaemia bleed excessively. [F]

(5 minutes)

Genetic cross – Ask students to work out the likelihood of a healthy (non-carrier) female and male cystic fibrosis sufferer having a healthy child. They should draw out a genetic cross to work out the answer. To support students, provide them with a template to complete their cross. To extend students, ask them to explain why the offspring (all carriers) should find out if their partners have the recessive allele before they have children. *(10 minutes)*

Support

- Provide students with a card-sort activity to match the ratios, percentages and fractions of children inheriting a genetic disorder. For example 1 in 4 is the same as 25% or a quarter of children. They could collate their results in a table with the headings ratio, percentage and fraction.

Extend

- Ask students to carry out a range of genetic crosses to determine whether an individual will suffer from sickle-cell anaemia.

Chapter 6 – Human inheritance and genetic disorders

Further teaching suggestions

Down's syndrome research
Women who are over the age of 35 (and others who are considered at risk) are routinely offered screening for Down's syndrome via an amniocentesis test. Students research what this genetic disorder is and whether or not they would choose (for themselves or their partner) to have an amniocentesis test carried out.

Genetic counselling and screening
Ask students to produce a list of the issues surrounding genetic counselling and screening. What are the advantages of counselling and screening? What are the disadvantages? These could relate to the couple receiving the counselling, the unborn child and society as a whole.

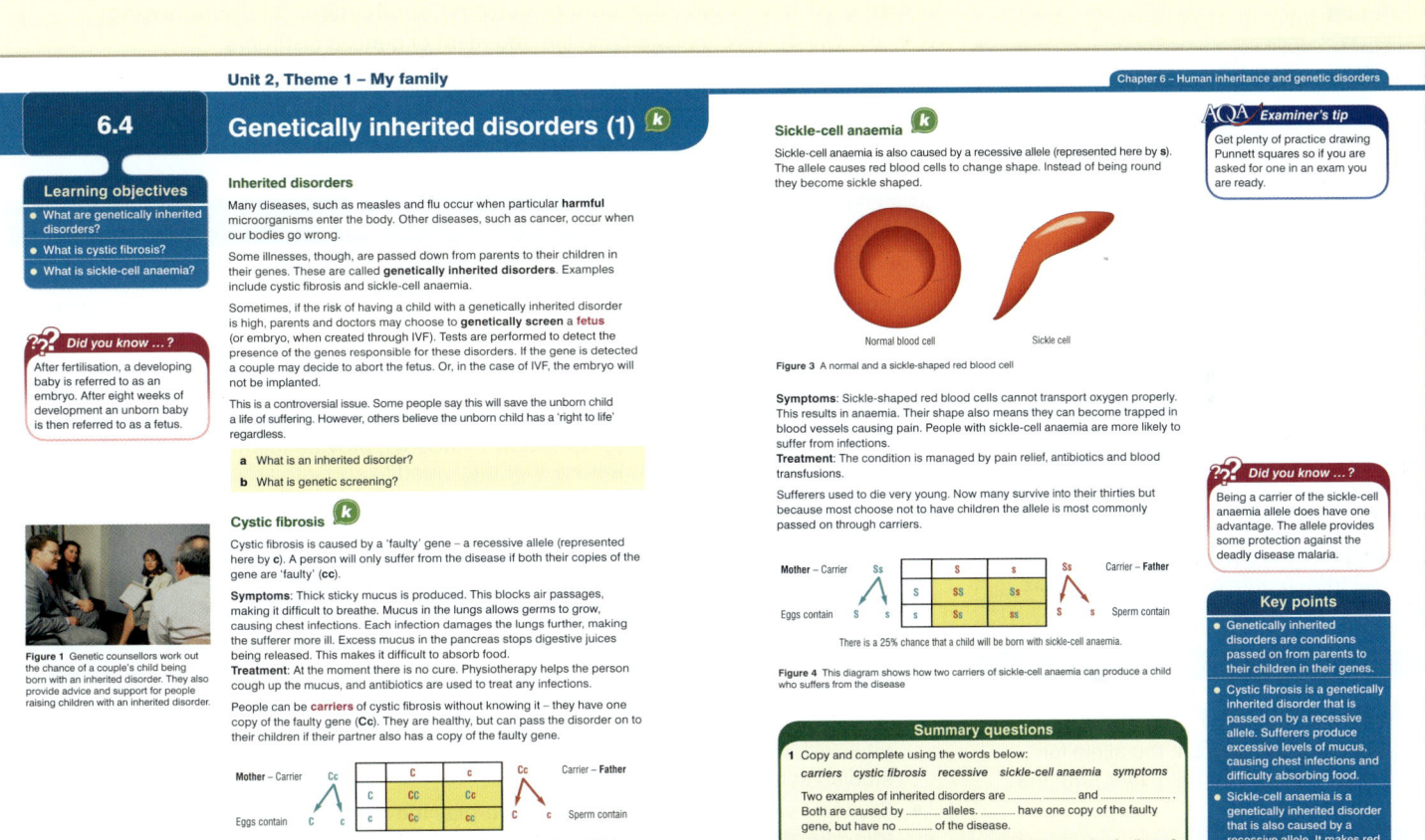

Summary answers

1. cystic fibrosis, sickle-cell anaemia, recessive, carriers, symptoms

2. They would not suffer from the disease, therefore they would have no symptoms.

3. Arguments for: chance to terminate pregnancy and prevent child from suffering, remove worry if they have not got it, start treatments immediately.
   Arguments against: if termination is decided upon the child is not born, sufferers viewed as less worthy citizens, genetic test could damage unborn fetus, potential false negative results.
   Students produce balanced arguments for their views on genetic screening.

Answers to in-text questions

a an illness that is passed on from parents to their children in their genes

b testing a fetus or embryo for the presence of an allele that causes a disorder

c people who have one 'faulty' copy of a gene, but do not suffer from the disorder

Unit 2, Theme 1 – My family

6.5 Genetically inherited disorders (2)

Learning objectives

Students should learn:
- that polydactyly and haemophilia are examples of genetically inherited disorders
- that gene therapy and stem-cell transplants may be able to cure genetically inherited disorders in the future.

Learning outcomes

Most students should be able to:
- name some symptoms and possible treatments for polydactyly and haemophilia
- draw genetic crosses to determine how likely it is for a child to be born with polydactyly
- state two possible developments to treat genetic disorders in the future.

Some students should also be able to:
- draw genetic crosses to determine how likely it is that a child will be born with haemophilia
- explain how gene therapy and stem-cell transplants may be able to cure genetically inherited disorders.

Answers to in-text questions

a Because if you have the allele for the disorder you will suffer from the disease.

b the X chromosome

c A fertilised egg containing two recessive alleles for haemophilia (i.e. one on each X chromosome for a female) will not grow into a baby.

Support

- Provide students with counters to calculate genetic crosses. Different colours or shapes can be used to represent the alleles. Students then move the counters on the genetic cross template to visualise how the alleles combine during fertilisation and the offspring that can be produced.

Extend

- Ask students to carry out genetic crosses to work out the likelihood of a child being born with haemophilia. The notations XH should be used for healthy clotting gene, and Xh for the faulty clotting gene

AQA Specification link-up: Science B 3.4.1.3

- Know that genes have different forms called alleles, which produce different characteristics.
- Describe the mechanism of monohybrid inheritance where the dominant and recessive alleles are given.
- Know that cystic fibrosis, sickle-cell anaemia, haemophilia and polydactyly are genetically inherited disorders.

Within this context, candidates should be able to use scientific data and evidence to discuss, evaluate or suggest implications of the following:
- the use of current research in the treatment of genetic disorders
- the likelihood of a genetically inherited disorder occurring
- the use of genetic screening.

Lesson structure

Starters

Anagrams – Give students anagrams of some common inherited disorders, e.g. cystic fibrosis, haemophilia, sickle-cell anaemia. *(5 minutes)*

Genetic screening – Ask students, in small groups, to add arguments for and against genetic screening onto large sheets of paper. After three minutes, they pass the sheets onto the next group. Students should try to add at least one more argument for and against. Discuss class findings. To support students, provide them with a series of statements about genetic screening that they need to sort into arguments for and against. To extend students, ask them to write a paragraph stating whether or not they would have their unborn offspring genetically screened if there was a family history of a disorder. Their argument must be justified with scientific reasoning. *(10 minutes)*

Main

- Treat your discussions of genetically inherited disorders with sensitivity.
- Show students clips/pictures, or provide information about people with polydactyly and haemophilia. They need to make *very brief* notes of the symptoms of each disease and the treatments available. This information could be summarised in a table. What are the similarities and differences between these disorders and the ones studied in the previous lesson?
- Students work in groups of three to role-play the work of a genetic counsellor (their role was introduced in the previous lesson). Students take it in turns to play the father, mother and counsellor. They should choose their genetic background from a pack of cards provided, e.g. mother – carrier of the disorder, father – normal genes for the condition. The counsellor should complete the cross for any of the four disorders studied (haemophilia should only be used with higher ability students). One parent should ask questions about symptoms, and the other should ask questions about treatments.
- Ask some groups of students to re-enact their role-play for the class. Did they miss out any vital information?
- Explain to students that most genetically inherited disorders have no cure. Although treatment improves quality of life and length, many sufferers die young. Introduce the concepts of gene therapy and stem-cell transplant. (See 'Activity' box in Student Book.)

Plenaries

Polydactyly or haemophilia – Ask students to sort the following statements into facts about polydactyly or haemophilia: caused by a dominant allele/caused by a recessive allele; affects only males/affects males and females; stops blood clotting properly/causes children to be born with extra digits. *(5 minutes)*

Genetically inherited disorders definitions – Ask students to write their own definitions of: carrier, genetically inherited disorder and genetic screening. To support students, provide them with key terms or phrases they should include when writing their definition. To extend students, ask them to write a definition of gene therapy and stem cell transplant. *(10 minutes)*

Chapter 6 – Human inheritance and genetic disorders

Further teaching suggestions

Haemophilia

Mother – Carrier $X_H X_h$ $X_H Y$ Normal – Father

Eggs contain X_H X_h X_H Y Sperm contain

During fertilisation:

	X_H	Y
X_H	$X_H X_H$	$X_H Y$
X_h	$X_H X_h$	$X_h Y$

This diagram shows how females can pass the haemophilia disorder on to their children.
H = healthy clotting gene (dominant)
h = 'faulty' clotting gene (recessive)

Children would be born in the ratio: $1 X_H X_H$: $1 X_H X_h$: $1 X_H Y$: $1 X_h Y$
1 normal female : 1 carrier female : 1 normal male : 1 haemophiliac male

---

Unit 2, Theme 1 – My family

6.5 Genetically inherited disorders (2)

Learning objectives
- What is polydactyly?
- What is haemophilia?
- How is it hoped that current research will provide cures for genetically inherited disorders?

Inherited disorders
Genetically inherited disorders are conditions passed on from parents to their children in their genes. They include polydactyly and haemophilia.

Polydactyly
Some babies are born with extra digits on their hands or feet. These may be fully formed fingers or toes, or no more than fleshy stumps. These extra digits can be removed by surgery but normally there is no medical benefit in removing them.

The most common form of polydactyly is caused by a dominant allele (represented by **P**). A person only needs one copy of the allele to suffer from this disease. People cannot be carriers of this disease – if you have the allele you will have the disease.

a Why is it not possible to be a carrier of polydactyly?

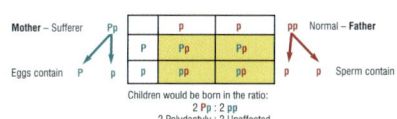

Mother – Sufferer Pp   pp Normal – Father
Eggs contain P p   p p Sperm contain

Children would be born in the ratio:
2 Pp : 2 pp
2 Polydactyly : 2 Unaffected

Figure 1 A genetic cross between a person suffering from polydactyly and an unaffected healthy individual. There is a 50 per cent chance that their children will suffer from the disorder.

Haemophilia
Haemophilia is a genetic disorder that only affects males. It is caused by a recessive allele carried on the X chromosome (one of the sex chromosomes).

Females have two X chromosomes. They are carriers of the disease if they have one dominant and one recessive allele. However, they never suffer from the disease, as fertilised eggs containing two recessive alleles will not develop into a baby.

Men have only one X chromosome, so if the 'faulty' allele is present on their single X chromosome, they will suffer from haemophilia.

Symptoms: Blood does not clot properly. That is because sufferers cannot produce Factor 8, a chemical that aids clotting, in their blood. So, even small cuts can be dangerous, because they keep on bleeding. Sufferers can lose so much blood that they die. Even small knocks can cause internal bleeding, resulting in very large bruises. Internal bleeding can be severe, causing much pain and damage. If sufferers are not treated they are likely to die before they can have children. Men who have the disorder usually have genetic screening to make sure that the allele is not passed on.
Treatment: Regular injections of Factor 8.

AQA Examiner's tip
When you complete a Punnett square, remember to circle or indicate somehow which is the genotype of the affected offspring.

Did you know …?
The genetic information present in an organism is called its genome. In 2001 the Human Genome Project sequenced the order of the human genome. Scientists are now attempting to identify all the genes in the human genome. It is hoped that the information gained from the human genome project will provide prevention and cures for many diseases.

b Which chromosome carries the haemophilia-causing allele?
c Why do females not suffer from haemophilia?

Current research into genetic disorders
Most genetic disorders have no cure. It is hoped that current research may hold the key:
- **Gene therapy**: It may be possible to eventually replace 'faulty' alleles with normal healthy alleles in the tissues where the disease causes damage. For example, to insert healthy alleles into the lung cells of cystic fibrosis sufferers. Alternatively, it may become possible to replace the faulty allele in an embryo.
- **Stem cell** transplants: Embryos contain stem cells. These can grow into any type of cell in the body. Stem cells can be removed from human embryos, or from the umbilical cord after a baby has been born. It may be possible to transplant healthy stem cells into a person who suffers from a genetic disorder.

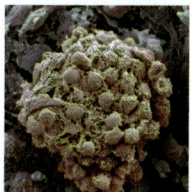

Figure 2 Stem cells

Activity
Stem cell research
Some parents who have a child with an incurable genetic disorder are now choosing to have another child to provide healthy stem cells. The stem cells are taken from the umbilical cord as soon as the sibling is born. These are used to cure their older brother or sister. To ensure the baby does not have the same condition, the embryo would have been genetically screened.

- Research a family who have undergone this treatment. Write a magazine article to explain how and why the family underwent these procedures.

Summary questions
1 Copy and complete using the words below:
   dominant   haemophilia   gene   polydactyly   chromosome
   .... is a genetically inherited disorder caused by a .... allele.
   .... only affects males. It is carried on a sex .... It is hoped that .... therapy may be able to cure these diseases in the future.
2 a Why is Factor 8 important for the body?
  b Can males be carriers of haemophilia? Explain your answer.
3 Draw a Punnett square to determine the likelihood of a child suffering from Huntington's disease, if both parents have the disease – mother (Hh) and father (Hh).

Key points
- Polydactyly is a genetically inherited disorder, caused by a dominant allele. It results in the production of extra digits.
- Haemophilia is a genetically inherited disorder that prevents blood from clotting properly. It is caused by a recessive allele carried on a sex chromosome.
- Current research into gene therapy and stem cell transplants may be able to cure genetically inherited disorders in the future.

Activity

Stem cell research
Suitable families to research include that of Adam Nash, whose older sister Molly suffered from Fanconi anaemia. Students could be provided with a structure to follow in their article – diagnosis of older child with genetic condition, desire to have another child to provide stem cells, how embryo was selected and implanted, birth of sibling, extraction of stem cells, how these were used to cure older sibling, situation now.

Gene therapy research
Research and write a brief report on the use of gene therapy as a potential cure for cystic fibrosis.

Summary answers

1 polydactyly, dominant, haemophilia, chromosome, gene
2 a It is needed for blood to clot.
  b No, they only have one copy of the gene, 'faulty' or normal. Therefore they are either sufferers or healthy.
3

	P	p
P	PP	Pp
p	Pp	pp

1 PP : 2 Pp : 1 pp – 75% chance of suffering from polydactyly

Unit 2, Theme 1 – My family

Summary answers

1. **a** A – nucleus, B – cytoplasm, C – cell membrane
   b cytoplasm
   c nucleus

2. D A C E B

3.
Characteristic	Affected by genes?	Affected by the environment?	Affected by both?
Eye colour	✓		
Weight			✓
Blood group	✓		
Skin colour			✓
Height			✓
Natural hair colour	✓		
Scar		✓	
Intelligence			✓

4. **a** a section of DNA that codes for a characteristic
   b a form of a gene
   c If a person carries two copies of the recessive allele.

5. **a** cc
   b Cc
   c They show no symptoms of the disorder.

6. **a**

	P	p
p	Pp	pp
p	Pp	pp

   b 1 : 1 (50%)

7. genes, alleles, inherited, hair, blond, dominant, recessive

Kerboodle resources

- Practical: Viewing cells (6.1)
- WebQuest: Cystic fibrosis (6.4)
- Support: Genetic Counselling (6.4)
- Support: The helpful gene (6.4)
- Animation: Genetically inherited diseases (6.4 and 6.5)
- On your marks: Sickle cell anaemia and inheritance
- Examination-style questions
- Answers to examination-style questions
- Test yourself: My family

Unit 2, Theme 1 – My family

Summary questions

1. A student is observing an animal cell through a microscope.

   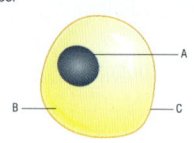

   a Name parts A, B and C.
   b Where in the cell do chemical reactions take place?
   c Which part of the cell contains DNA?

2. Put these in order of size starting with the smallest.
   A: DNA
   B: Cell
   C: Chromosome
   D: Gene
   E: Nucleus

3. Copy and complete the following table. The first one has been completed for you.

Characteristic	Affected by genes?	Affected by the environment?	Affected by both?
Eye colour	✓		
Weight			
Blood group			
Skin colour			
Height			
Natural hair colour			
Intelligence			

4. Many of the characteristics of human beings are affected by their genes.
   a What is a gene?
   b What is an allele?
   c Under which circumstances could a characteristic caused by a recessive allele be expressed?

5. Cystic fibrosis is a disease caused by a recessive allele. People only suffer from the disorder if they carry both copies of the recessive allele. This allele is given the symbol 'c'. The healthy allele is given the symbol 'C'.
   a Which alleles would a sufferer have?
   b Which alleles would a carrier have?
   c Many carriers do not know that they have the recessive allele. Why is this?

6. Polydactyly is an example of an inherited disorder. It is a dominant disorder (P).
   a A man (Pp) and a woman (pp) wish to have a child. Use a Punnett square to show the possible genetic make-up of their offspring.
   b As a ratio, what is the likelihood of their offspring suffering from polydactyly?

7. People can change the colour of their hair.
   a Choose words from the box to complete the sentence about how hair colour is determined.

   > alleles   blond   dominant   genes
   > hair   inherited   recessive

   We have two .... for each of our characteristics. Different forms of the same gene are called .... For example, you might have .... the gene for black .... from one parent and the gene for .... hair from the other. Some genes show up and are called .... and the gene that does not is called ....

Chapter 6 – Human inheritance and genetic disorders

Examination-style questions

The gene for ginger hair is recessive. The diagram shows the inheritance of the ginger gene in a few generations of a family.

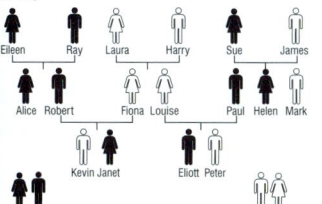

a Which part of the cell carries genetic information? (1)
b Give the genotype of these people (choose from RR, Rr or rr):
  i Sue   ii Louise   iii Kevin (3)
Peter met and had children with someone with ginger hair.
c i Fill in a Punnett square to show the possibility of the children inheriting ginger hair. (2)

	R	r
r		
r		

ii What is the probability of the children inheriting ginger hair? (1)

2 Choose which of these characteristics are as a result of environmental variation, genetic variation or both.

Variation	Environmental	Genetic	Both
Foot size			
Freckles			
Hand span			
Scars			

(4)

3 Gardeners win prizes for making new varieties of plant. A gardener crossed a red flowering sweet pea with a white flowering sweet pea. He collected the seeds and grew them. He noticed that all the seeds grew into red flowering sweet peas.
  a What colour of sweet pea flower was dominant? (1)
  b Some of these flowers self-fertilised. Calculate the ratio of red and white flowered sweet pea plants he would grow if the seeds were planted. (3)

4 Cystic fibrosis is caused by a faulty gene. The disease affects the lungs by causing an excess of thick, sticky mucus to be made. This mucus traps germs so sufferers often get infections and are constantly on antibiotics. Scientists have been finding ways of correcting the faulty genes rather than just treating the symptoms. The best way to do it is with gene therapy – replacing faulty genes with normal ones. Trials on mouse and human lung tissue were effective so pilot studies have been started using people who suffer from the disease.
  a i How would a person first get the disease cystic fibrosis? (1)
    ii Why would some people be against the gene therapy trials completed so far? (1)
  b Each patient in the pilot study had a small dose of the gene therapy product in the nose every 5 minutes. It took 1.5 hours to give the full dose of 20ml of the product. Calculate how much gene therapy product was in each dose. (3)
  c In this question you will be assessed on using good English, organising information clearly and using specialist terms where appropriate.
    Another study was to find the most economical volume of the gene therapy product to use. This would hopefully cure the patient but not cost too much. Plan an investigation that the scientists could carry out to find the smallest volume of therapy product that cures the patients. (6)

Examination-style answers

1 a Nucleus (1 mark)
  b i rr
    ii Rr
    iii Rr (3 marks)
  c i

	R	r
r	Rr	rr
r	Rr	rr

(2 marks)
  ii 50% / ½ / 1 in 2 (1 mark)

2

Variation	Environmental	Genetic	Both
Foot size		✓	
Freckles			✓
Hand span		✓	
Scars	✓		

(4 marks)

3 a red (1 mark)
  b Both parents genotype correct (Rr and Rr)
    All combinations of offspring correct
    ¾ OR 3:1 or 3 in 4 (3 marks)

4 a i inherited (1 mark)
    ii against testing on animals (1 mark)
  b 1.5 hours = 90 minutes
    90 ÷ 5
    = 18
    20 ÷ 18
    = 1.11 ml of gene therapy product in each dose. (3 marks)

6 c Marks awarded for this answer will be determined by the Quality of Written Communication (QWC) as well as the standard of the scientific response.

There is a full, considered and detailed plan of an investigation to find the most economical volume of the gene therapy product to use. The answer is well structured with minimal repetition or irrelevant points. There is an accurate, fluent and clear expression of ideas with only minor errors in the use of technical terms, spelling, punctuation and grammar. (5–6 marks)

There is some attempt at providing a plan of an investigation to find the most economical volume of the gene therapy product to use, with some omissions. The answer shows some attempt at structuring and the ideas are expressed with reasonable fluency and clarity. There are some errors in the use of technical terms, spelling, punctuation and grammar. (3–4 marks)

There is a brief attempt at providing a plan of an investigation to find the most economical volume of the gene therapy product to use. The answer is largely incomplete and may contain some valid points which are not clearly structured. It lacks fluency and/or clarity. It contains errors in the use of technical terms, spelling, punctuation and grammar. (1–2 marks)

No relevant content (0 marks)

Examples of points made in the response:
- Give different doses of the gene therapy to people suffering from cystic fibrosis
- Have more than one person on each dose to spot any anomalous results
- Try to find people of different age ranges and fitness levels to take part in trial so can see effects on different people
- Monitor them over a long period
- Watch for any side effects
- The dosage of any patient that is cured is noted
- The lowest dose that cures the most people can be marketed
- Any dose higher than this is not necessary
- Any lower dose and some patients in the future might not be cured.

Unit 2, Theme 2

My home

AQA Specification link-up: Science B 3.4.2

House design involves many types of science. Construction workers use their knowledge and understanding of materials to build homes that are strong and secure. Energy consultants design energy-saving features to keep us warm in winter. Chemists, using their knowledge of chemical energy, have helped to develop a range of fuels for cooking, heating and transport and have used their knowledge of materials to create a wide variety of products found in the home. Energy workers ensure that the electricity generated by our power stations is distributed all over the country to our homes.

In this theme there are three contexts:
- 3.4.2.1 Materials used to construct our homes
- 3.4.2.2 Fuels used for cooking, heating and transport
- 3.4.2.3 Generation and distribution of electricity

Materials used to construct our homes

Activity:
- Students can visit the websites of construction material companies to find out about the range of jobs involved. Some have very informative case studies about different routes into this kind of work too.
- Another interesting case study is the fairly recent resurgence of cob construction in the UK. Students could find out about the parts of the country where these buildings are becoming more popular. They can find out the factors that make cob buildings attractive from a financial and ecological point of view, contrasting them with conventional methods.
- Giving students access to samples of a wide variety of metals can help reinforce ideas about their general properties and variability between different elements.

Possible misconceptions

Limestone, limewater and quicklime have nothing to do with citrus fruit. It is really surprising how many students think this.

Concrete, mortar and cement are not interchangeable terms. It is worth making sure students can readily tell these three materials apart.

Polymer and plastic are not interchangeable terms. It can help to explain that starch and cellulose are also polymers, or to look at sentences where 'plastic' is an adjective.

Unit 2, Theme 2
My home

In Unit 2, Theme 2 you will work in the following contexts, covered in Chapters 7 and 8:

Materials used to construct our homes
Where do materials like concrete come from?

A long time ago, layer upon layer of dead sea creatures built up at the bottom of the sea. Their shells slowly changed into rock. Millions of years later, humans came along and started mining it out of the ground. We call this rock limestone and it is a vital part of the construction industry. Without limestone, we wouldn't have concrete, or mortar to hold bricks together. Limestone is also used in making glass.

What other materials are used to build our homes?

Metals are also important for making your home work properly. They are used to support concrete, make window frames, wiring, pipes and parts of roofs. By understanding the properties of metals, architects and builders can make the right decisions about which ones to use.

Other materials used to make our homes include polymers, ceramics and composites. As with metals, these materials are chosen carefully for their physical and chemical properties.

Can we build homes without damaging the environment?

Most of the homes you see in the UK are made from bricks and mortar but not all of them. Earth materials such as straw or timber can also be used to build homes. In fact, around a third of the world's population lives in homes made from earth materials. Earth materials are becoming an increasingly attractive choice for building. This is because they have less negative impact on the environment than bricks and concrete. More recently, sustainable building materials such as 'hempcrete' have been developed. These materials are growing in popularity because homes built with them cost less to heat.

Fuels used for cooking, heating and transport
Why is oil so important?

We rely on crude oil for our society to function. Some of the products of crude oil are used as fuels to drive cars and aeroplanes. These include petrol, diesel and kerosene. Some of them are used as fuels in the home, like methane and propane. These materials are called hydrocarbons – they contain only hydrogen and carbon.

What happens when we burn fuels?

When we burn hydrocarbons in plenty of air, we get two products – carbon dioxide and water. This is because the hydrogen and carbon split up and react with oxygen in the air. The carbon dioxide produced by burning fossil fuels is a growing problem. The UK alone produces millions of tonnes of carbon dioxide every day. This and other greenhouse gases are increasing the likelihood of climate change. Also, the fossil fuels are non-renewable and they will run out. This will start happening in your lifetime.

Generation and distribution of electricity
How does crude oil help provide electricity to run my computer?

Hydrocarbons release lots of energy when they react with oxygen. This energy can be used to make electricity in power stations. Generally, this involves turning water into steam, which makes a turbine go round. The turbine is connected to a generator, which uses huge magnets to transfer the kinetic energy into electrical energy.

What alternatives to crude oil are there?

Burning hydrocarbons to make electricity has two problems: the gases it produces and the fact that we're running out of fossil fuels. By using radioactive materials to turn water into steam we can reduce our use of hydrocarbons. This doesn't produce any carbon dioxide at all. However, the deadly radioactive waste will last for tens of thousands of years.

So how else can we make the electricity our society depends on? Wind, sun, tides, rivers and the Earth's crust can all be used to generate electricity. As they will not run out, we say they are renewable. However, each has its own advantages and disadvantages.

How does electricity get to our homes?

Once electricity has been generated, it has to reach the consumer. High currents make wires hot and waste energy, so the current is reduced (stepped down) by devices called transformers. Thousands of miles of wire carry the electricity around the country. However, there are some concerns about whether this is safe.

Fuels for cooking, heating and transport

Activity:
- Students can research what happens on an oil rig and the kinds of careers associated with this kind of work. This can be tied in with a case study of the Deepwater Horizon disaster, possibly comparing it with other major oil spills. This kind of topic lends itself perfectly to examining the social, economic and environmental implications of science.
- Drilling for oil could also be debated in a wider ethical sense, drawing attention to its role in global politics and war.
- Air quality monitoring stations can be used as a resource for teaching about traffic pollution. A number of them publish their data on the internet regularly.

Possible misconceptions

Crude oil is not used for cooking! A number of students leap to the assumption that running out of crude oil means no more fried food.

Nitrogen is also oxidised in car engines. Students are often surprised that oxygen and nitrogen react together in car engines.

My home

As 44 per cent of the UK's CO_2 emissions come from households, we need to make homes more energy efficient. Energy use and sustainability are factors considered whenever new buildings are built. By choosing materials carefully, architects and construction companies can reduce their environmental impact.

Millions of pounds are spent on skiing and snowboarding holidays every year. New boards and skis are being invented all the time using space-age materials. The winter sports industry relies on new composite materials constantly being developed. It makes the sport more fun and ensures people will keep buying new equipment.

I live underneath electrical lines between pylons. Recently I've been reading in news stories that living near pylons could be dangerous. Apparently living here there is more risk of getting serious illnesses like leukaemia. Is this being properly researched? Are the government doing everything they can to make sure my family is safe?

Shell's 'Efficiency Improver' fuel was introduced in 2010 to help petrol last longer. It contains special additives that lubricate and clean engines. This means the engine wastes less of the petrol's energy overcoming friction. The less energy wasted, the further a car can travel on a tank of petrol.

Generation and distribution of electricity

Activity:
- Making model power stations can be a useful way to consider what goes on in each area. This can be especially meaningful if the models are based on existing power stations in the UK.
- Motor kits are readily available from scientific education suppliers. By building a motor and using it to generate a weak current, students can start thinking about ways they could get their generator to turn even faster.
- There are varied attitudes to wind farms across the UK, which means the topic is well suited to debating. Students can role-play different stakeholders, or examine the usefulness of offshore wind farms.

Possible misconceptions

Students often have the idea that renewable energy sources are nothing but good, without considering the environmental impact of flooding areas of land to produce hydroelectric plants or tidal barrages.

Unit 2, Theme 2 – My home

7.1 Limestone as a building material

Learning objectives

Students should learn:
- that limestone is a rock made mainly of calcium carbonate ($CaCO_3$)
- that limestone is quarried from the ground using explosives
- how limestone is a useful raw material and is used in the building industry.

Learning outcomes

Most students should be able to:
- describe how limestone comes from quarries in the ground
- list some uses of limestone.

Some students should also be able to:
- relate the products of limestone to specific purposes in construction.

AQA Specification link-up: Science B 3.4.2.1

- Know that limestone is obtained from the ground by quarrying.
- Give some uses of limestone in the building industry.

Within this context, candidates should be able to use scientific data and evidence to discuss, evaluate or suggest implications of the following:
- the use of quarrying to obtain raw materials for building.

Lesson structure

Starters

Anagrams – Give the students key word anagrams for: limestone, quarry, sedimentary, building, concrete, construction. To support students, provide definitions for them to match up with the key words. To extend students, ask them for their own definitions of the key words. *(5 minutes)*

Sequence activity – Review how sedimentary rock is formed by showing an animation or series of pictures. Students then complete a sentence-sequencing activity describing the process. *(10 minutes)*

Main

- Students can brainstorm the topic of building materials in small groups and then feed back to the rest of the class. After exhausting the obvious ideas such as brick, wood, concrete and steel, encourage students to consider the necessary properties of building materials for particular uses.
- Show students a presentation about the various uses of limestone. Start with quarrying and show some video clips of quarry blasts (readily available on the internet). Go on to show some examples of buildings made from limestone. Use photographs to illustrate limestone's role in glass production, paper production and treatment of areas affected by acid rain.
- Introduce the chemical names and formulae of limestone (mainly calcium carbonate, $CaCO_3$), quicklime (calcium oxide, CaO), slaked lime (calcium hydroxide, $Ca(OH)_2$) and limewater (solution of calcium hydroxide). Reinforce this with a 'domino' activity, where groups or pairs match the names and formulae together. Construct a suitable worksheet to review this knowledge, or alternatively, students could complete summary tables.
- Students can label the limestone-related parts of a picture of a house. Then groups of students could research one particular aspect of limestone's use and present their findings to the rest of the class. Groups could also write questions to accompany their presentation.

Plenaries

Name the material – Ask students to write 'limestone', 'quicklime', 'slaked lime' and 'lime water' on separate pieces of paper. The teacher describes one of the materials (either by chemical formula, chemical name, uses or some other fact) and students hold up the name of the material they think matches. *(5 minutes)*

If there was no limestone – Students amend a diagram of a house, removing all the limestone-related materials. To support students, provide clues, e.g. 'What binds the bricks to each other?' To extend students, they can suggest alternatives that could be used. *(10 minutes)*

Support

- Students can match pictures showing the uses of limestone to corresponding statements about them.
- Give students a list of materials made from limestone. Groups can survey the room or building for these materials.

Extend

- Students can research the formation of limestone in depth. How is it linked to changes in the early atmosphere? Why can we find fossils in it? Where is it quarried?

Further teaching suggestions

Investigating different types of limestone

Students could perform a practical comparing the carbonate content of different types of limestone (oolitic limestone, fossiliferous limestone and chalk). They should use their findings to suggest which would be attacked the least by acid rain.

Practical support

Are all limestones the same?

Equipment and materials required: Chips of oolitic limestone, fossiliferous limestone and chalk, electronic balance (accurate to 2 d.p.), beakers, hydrochloric acid (0.1 mol/dm^3), eye protection.

Details

Ask students to measure approximately 50 cm^3 acid into a beaker on the balance and record the mass. Add 2–3 limestone chips and record the mass. Calculate the mass of the chips. When the students have finished carrying out the reaction, record the new mass and calculate the loss of mass. Express the mass of CO_2 released as a percentage of the original mass of the limestone chips. Repeat for other types of limestone.

Ask: 'Did each limestone have the same carbonate content? Could differences be related to the way the limestone originally formed?'

Safety: Wear eye protection. CLEAPSS Hazcard 47A Hydrochloric acid – corrosive.

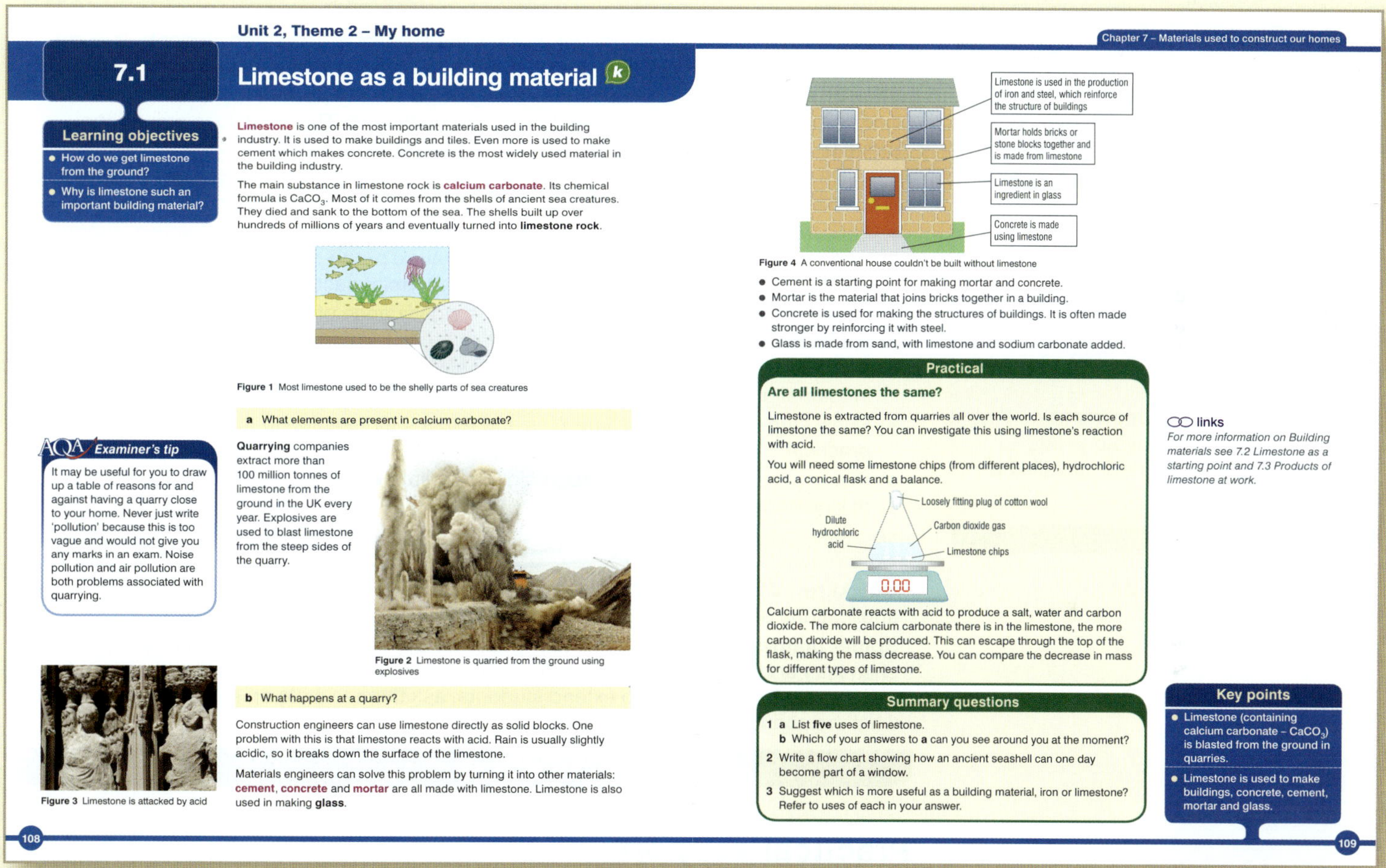

Chapter 7 – Materials used to construct our homes

Unit 2, Theme 2 – My home

7.1 Limestone as a building material

Learning objectives
- How do we get limestone from the ground?
- Why is limestone such an important building material?

Limestone is one of the most important materials used in the building industry. It is used to make buildings and tiles. Even more is used to make cement which makes concrete. Concrete is the most widely used material in the building industry.

The main substance in limestone rock is **calcium carbonate**. Its chemical formula is $CaCO_3$. Most of it comes from the shells of ancient sea creatures. They died and sank to the bottom of the sea. The shells built up over hundreds of millions of years and eventually turned into **limestone rock**.

Figure 1 Most limestone used to be the shelly parts of sea creatures

a What elements are present in calcium carbonate?

Quarrying companies extract more than 100 million tonnes of limestone from the ground in the UK every year. Explosives are used to blast limestone from the steep sides of the quarry.

AQA Examiner's tip

It may be useful for you to draw up a table of reasons for and against having a quarry close to your home. Never just write 'pollution' because this is too vague and would not give you any marks in an exam. Noise pollution and air pollution are both problems associated with quarrying.

Figure 2 Limestone is quarried from the ground using explosives

b What happens at a quarry?

Construction engineers can use limestone directly as solid blocks. One problem with this is that limestone reacts with acid. Rain is usually slightly acidic, so it breaks down the surface of the limestone.

Materials engineers can solve this problem by turning it into other materials: **cement**, **concrete** and **mortar** are all made with limestone. Limestone is also used in making **glass**.

Figure 3 Limestone is attacked by acid

- Limestone is used in the production of iron and steel, which reinforce the structure of buildings
- Mortar holds bricks or stone blocks together and is made from limestone
- Limestone is an ingredient in glass
- Concrete is made using limestone

Figure 4 A conventional house couldn't be built without limestone

- Cement is a starting point for making mortar and concrete.
- Mortar is the material that joins bricks together in a building.
- Concrete is used for making the structures of buildings. It is often made stronger by reinforcing it with steel.
- Glass is made from sand, with limestone and sodium carbonate added.

Practical

Are all limestones the same?

Limestone is extracted from quarries all over the world. Is each source of limestone the same? You can investigate this using limestone's reaction with acid.

You will need some limestone chips (from different places), hydrochloric acid, a conical flask and a balance.

Calcium carbonate reacts with acid to produce a salt, water and carbon dioxide. The more calcium carbonate there is in the limestone, the more carbon dioxide will be produced. This can escape through the top of the flask, making the mass decrease. You can compare the decrease in mass for different types of limestone.

links

For more information on Building materials see 7.2 Limestone as a starting point and 7.3 Products of limestone at work.

Summary questions

1. **a** List five uses of limestone.
   b Which of your answers to **a** can you see around you at the moment?
2. Write a flow chart showing how an ancient seashell can one day become part of a window.
3. Suggest which is more useful as a building material, iron or limestone? Refer to uses of each in your answer.

Key points

- Limestone (containing calcium carbonate – $CaCO_3$) is blasted from the ground in quarries.
- Limestone is used to make buildings, concrete, cement, mortar and glass.

Did you know …?

Use Nobel's discovery of dynamite to start a discussion about the ethical neutrality of science.

Answers to in-text questions

a calcium, carbon and oxygen
b Limestone is extracted from the ground with explosives.

Summary answers

1. concrete walls, mortar between bricks, glass in windows, solid limestone building blocks, cement to make mortar and concrete

2. seashell ⟶ limestone ⟶ extracted from quarry ⟶ used to make glass ⟶ window

3. Limestone is arguably more useful, as it is used to produce lots of other materials, such as concrete or mortar. Iron is also useful as a building material because it can be used to support other structures. Also, limestone is important in the production of iron – it would be hard to produce iron without limestone.

Unit 2, Theme 2 – My home

7.2 Limestone as a starting point

Learning objectives

Students should learn:
- that quicklime is calcium oxide (CaO) and slaked lime is calcium hydroxide ($Ca(OH)_2$)
- that quicklime is obtained from the thermal decomposition of limestone
- that slaked lime is obtained by adding water to quicklime.

Learning outcomes

Most students should be able to:
- state the chemical names and formulae of quicklime and slaked lime
- describe how quicklime is made by heating limestone, including a word equation
- describe how slaked lime is made by adding water to quicklime, including a word equation.

Some students should also be able to:
- write symbol equations to show the conversion of limestone into quicklime and quicklime into slaked lime.

AQA Specification link-up: Science B 3.4.2.1
- Describe the conversion of limestone into quicklime and quicklime into slaked lime, and know the chemical formulae for these materials.
- Outline the manufacturing processes for the production of quicklime …

Lesson structure

Starters

Mix and match – Provide students with a card match activity of common and chemical names of limestone, quicklime, slaked lime and limewater. To extend students, include formulae. To support students, use picture cards showing the material in use. *(5 minutes)*

Limestone facts – Organise students into groups to write down a fact each about limestone and pass it to the student next to them to read. This continues around the group until everyone has their original fact back. *(10 minutes)*

Main

- Perform a practical activity converting limestone to quicklime (see 'Practical support'), then to slaked lime and limewater. A temperature probe can be used during the slaking step to demonstrate that it is an exothermic reaction.
- For each step of the practical, explain the significance of the product – describe its applications and importance to society.
- Students can write an account of the practical, relating each step to its equivalent industrial process.
- Show students images of lime kilns and relate their function to the first part of the practical.
- Discuss the production of carbon dioxide by lime kilns. Is it justified by the importance of limestone to society?

Plenaries

Quickfire – Challenge students to race to write down ten facts about the chemistry and applications of limestone. *(5 minutes)*

Spider diagrams – Ask students to produce spider diagrams or concept maps on the chemistry and applications of limestone. To support students, provide the words and ask them to join them up. To extend students, ask them to write on each connecting line what the connection is. *(10 mintutes)*

Support
- Students could be provided with information on how quicklime and slaked lime are produced (and their uses) and use this information to produce posters.

Extend
- Provided with the molecular mass of limestone and carbon dioxide, students can be guided to work out the carbon dioxide emissions produced by a 500 tonne/day lime kiln. With a little research on particular lime kiln web sites, students could also calculate annual CO_2 emissions.

Practical support

Changing limestone practical

Equipment and materials required: Pieces of limestone, Bunsen burner, heatproof mat, wire gauze, tripod, tongs, watch glass, water, pipette, filter paper, funnel, conical flask, delivery tube or straw, eye protection.

Details
Place the limestone on the wire gauze and heat strongly for several minutes. The limestone should be made to glow white during this step, as it decomposes into calcium oxide. Allow the product to cool for a few minutes before using the tongs to place it on the watch glass.

Ask: 'Are there any changes in appearance?'

Slowly add a few drops of water and observe the reaction as the calcium oxide is turned into calcium hydroxide. A temperature probe could be used at this stage. Transfer the calcium hydroxide to the beaker of water to make limewater, and filter into the conical flask to remove any unreacted limestone and undissolved calcium hydroxide. Carefully blow through the delivery tube into the limewater and observe the change. Point out that the precipitate is calcium carbonate again.

Safety: Wear eye protection. Be careful with hot apparatus. Do not allow calcium oxide produced to come in contact with skin.

Neutralising acid soil

Equipment and materials required: Calcium hydroxide, distilled water, soil samples (filled boiling tubes, each spiked with 5 cm³ sulfuric acid), universal indicator paper, rubber bungs, spatulas, eye protection.

Details
Add approximately 10 cm³ distilled water to the soil sample. Put the bung on and invert the tube a few times. Test the pH with universal indicator paper and record it. Add one small spatula of calcium hydroxide at a time and repeat the last step until the pH becomes neutral. (This practical could be adapted to compare calcium hydroxide with calcium carbonate, but the mixing stage would need to be done with a stirring rod, to avoid carbon dioxide build-up.) It would be worth pointing out that calcium carbonate is better, if slower in practise, because it makes the soil neutral not alkaline.

Safety: Wear eye protection. CLEAPSS Hazcard 32 Universal indicator solution – highly flammable and harmful. CLEAPSS Hazcard 98A Sulfuric acid; CLEAPSS Hazcard 18 Calcium hydroxide – corrosive.

Further teaching suggestions

Neutralising soil samples spiked with acid
Conduct a practical to find out how much calcium hydroxide is needed to neutralise soil samples spiked with acid. Compare this to how much calcium carbonate is needed.

The etymology of the words endothermic and exothermic
Discuss the etymology (the meaning and history of a word) of the words endothermic and exothermic. Show some more examples of such reactions (such as 2 mol/dm³ HCl + 2 mol/dm³ NaOH or sodium hydrogencarbonate + ethanoic acid).

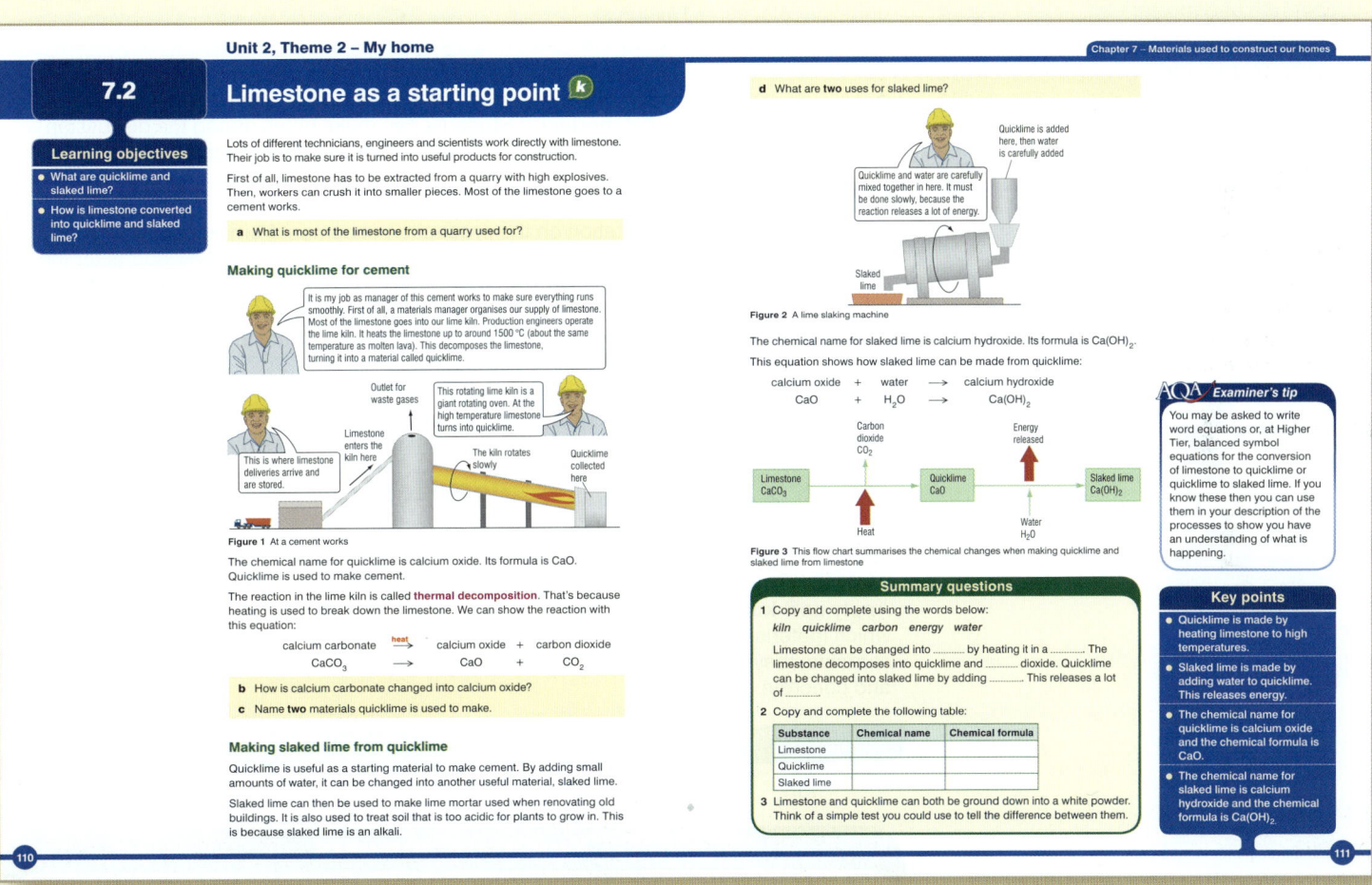

Summary answers

1 quicklime, kiln, carbon, water, energy

2
Substance	Chemical name	Chemical formula
Limestone	Calcium carbonate	$CaCO_3$
Quicklime	Calcium oxide	CaO
Slaked lime	Calcium hydroxide	$Ca(OH)_2$

3 Add water to the powder. If it heats up, it is (or was) quicklime. Add dilute acid, if it fizzes it is limestone.

Answers to in-text questions

a Most of the limestone goes to a cement works.
b heating in a kiln
c cement, slaked lime
d Slaked lime can be used for neutralising acid soil, and making lime mortar.

Unit 2, Theme 2 – My home

7.3 Products of limestone at work

Learning objectives

Students should learn:
- how limestone can be used to produce cement and glass
- the composition and use of concrete and mortar.

Learning outcomes

Most students should be able to:
- state that glass and cement are produced using limestone
- state that cement is used to make concrete and mortar
- compare the composition and uses of concrete and mortar
- describe how glass and cement are produced.

AQA Specification link-up: Science B 3.4.2.1
- Outline the manufacturing processes for the production of ... cement and glass.
- Describe the composition and use of mortar and concrete.

Lesson structure

Starters

Tensile strength – Obtain data on different building materials' strength. Students can practise their graphing skills by representing the data as a bar chart. *(5 minutes)*

Materials in your school – Let the students survey the school for structures built using concrete or glass. To extend students, they could compare the prevelance of concrete with brick-and-mortar. To support students, provide them with a proforma and suggest particular places. *(10 minutes)*

Main

- Discuss cement with the class. Is it different from concrete? How? What does it look like? Do they know what mortar is? Students can make notes during the discussion and use them to produce a visual key to tell the difference between the three materials, e.g. 'Is it a dry powder?', 'Does is contain sand?', 'Does it contain gravel or crushed rock (aggregate)?'
- Show students a presentation on the production of cement from limestone and clay. Groups can then use a written account of the process to produce flow charts for the production of mortar and concrete from their raw materials, starting with limestone.
- Investigate the relative strengths of mortar and concrete as a class practical (see 'Practical support'). This could also be adapted to investigate the best proportions of sand, water, cement and gravel to use in each.
- Students can research glass production on the internet. What chemicals are used to produce the different colours of glass bottles? How much glass does the UK recycle each year? Why is glass recycling important if there is so much sand in the world?

Plenaries

Key word taboo game – Students must describe each of the materials mentioned in this lesson without using particular words (e.g. describing mortar without saying bricks or cement). *(5 minutes)*

Apply new knowledge – Students can use the results of their investigation to advise which material (and in what proportions) to use in different situations. To extend students, they can also include ideas about the differences in the composition of mortar and concrete. *(10 minutes)*

Support
- Follow the instructions to make mortar or concrete, leaving out the investigation into different amounts of sand. Students can write up the practical as a storyboard.

Extend
- Students can research other mortar mixes and test their strength.

Summary answers

1. limestone, clay, gypsum, water, hydroxide
2. Mortar is made of cement, sand and water, but concrete also includes small stones or pebbles.
3.

Material	Ingredients	Uses
Concrete	cement, sand, water, stones	structures
Mortar	cement, sand, water	holding bricks together
Glass	limestone, sand, sodium carbonate	structural panels, windows, screens

112

Chapter 7 – Materials used to construct our homes

Practical support

Which mix of mortar is strongest?

Equipment and materials required: Water, cement, sand, plastic spoons, disposable cups, card, sticky tape, two clamp stands, string, slotted masses with hangers, measuring cylinder, access to balance, eye protection.

Details

Make a 1 cm × 10 cm × 1 cm stick mould by drawing a net template onto the card, cutting it out and fixing together with sticky tape. Put 10 cm^3 water into the cup and add 20 g cement. Mix thoroughly. Add either 40 g, 60 g or 80 g sand to the mixture and continue to mix. Fill the mould with mortar mix and leave to harden (overnight or preferably over a weekend). Carefully remove the mortar stick from the mould and support each end with a clamp stand (tightening the clamps may crush the stick). Loosely tie a string loop around the centre of the mortar stick. Add slotted masses to the string loop until the stick breaks.

Safety: Wear eye protection. Protect the floor or bench top and feet.

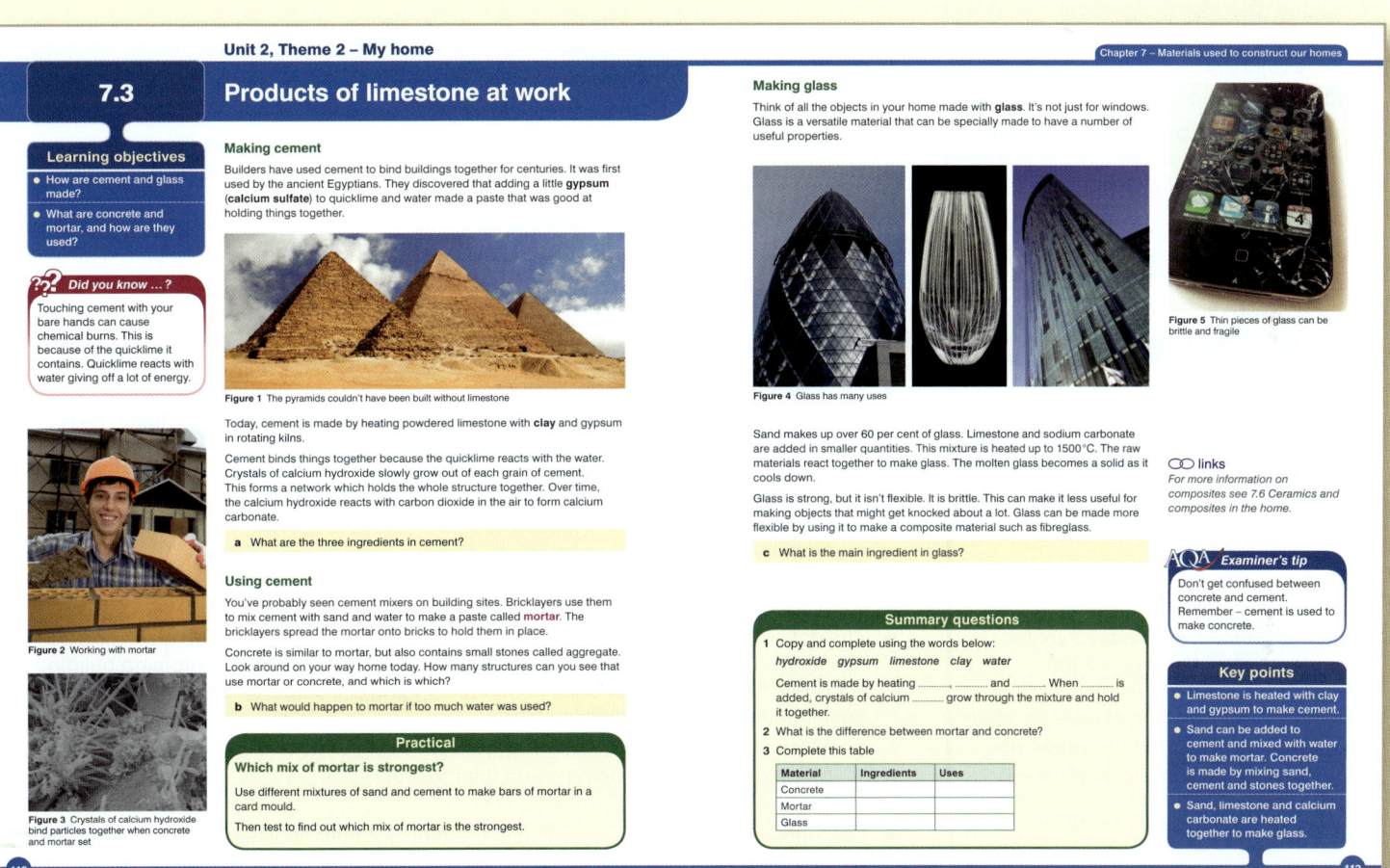

Further teaching suggestions

Making concrete

Adapt the mortar investigation to include a concrete mix (using 40 g sand and 20 g aquarium gravel).

Answers to in-text questions

a limestone, gypsum, clay
b The mortar would not set.
c sand

Unit 2, Theme 2 – My home

7.4 Metals for construction

Learning objectives

Students should learn:
- that metals have characteristic properties that make them useful
- that metals are used extensively in the structures of homes.

Learning outcomes

Most students should be able to:
- list and describe the general properties of metals
- name some uses for metals in the home
- relate a metal's properties to its uses in the home.

Some students should also be able to:
- use data about metals' properties to make decisions about how to use them.

AQA Specification link-up: Science B 3.4.2.1

- Know the characteristic properties of metals (good heat and electrical conductors, malleability, ductility, resistance to corrosion, strength and hardness)
- Relate uses of metals in the building industry to the properties of these metals.

Within this context, candidates should be able to use scientific data and evidence to discuss, evaluate or suggest implications of the following:
- the physical properties of materials
- the most suitable material for a particular use
- changes in the properties of materials resulting from a change of structure.

Lesson structure

Starters

Spot the metal – Give students a list of metals and ask them to locate these on the periodic table. Add steel or solder to the list to generate discussion about non-elemental metals. *(5 minutes)*

Metal structures in school – Organise students to survey the school site for objects or structures made of metals. To extend students, ask them to identify particular metals or try to explain why a metal was chosen for each use. To support students, provide a list of areas to survey and a proforma for collecting data. *(10 minutes)*

Main

- Start by reviewing the characteristic properties of metals. This can be done as a discussion or as a card match activity. For each property, students suggest how it is exploited by metallurgists and construction engineers. Have samples of metals at hand to pass around the class.
- Give students data sheets on some metals (lead, copper, steel, iron, magnesium, tungsten and gold) with information such as relative reactivity, tensile strength, malleability, cost, thermal conductivity and electrical conductivity.
- This information can be used to produce bar charts comparing the different metals, either by hand or using ICT.
- The data exercise could lead into a design brief activity. Students are given briefs for the following items: car doors, water pipes, wiring, roof flashing, reinforcing buildings, light-bulb filaments, ornamental structures. In groups, students decide which metal is best for each job. Questions to ask:
  – Why is it better than the other metals?
  – Which properties are being made the most of?
  – Could any other materials be used?
- Students can then write supporting statements for their choices, justifying their decisions.
- Pick up on steel being an important metal but not an element. Use this to introduce the idea of alloys. Give the class more information sheets about alloys. Individually or in groups, students can complete summary sheets with the headings 'Name of alloy', 'What is in it?', 'Properties of alloy' and 'Uses of alloy'. Discuss how alloying disrupts the regular metallic structure so that layers of atoms find it more difficult to slide past each other, making the metal stronger.

Plenaries

Who am I? – Give students facts about each metal until they can work out which one it is. *(5 minutes)*

Concept map – Ask students to produce a spider diagram of 'Uses of metals in construction'. To extend students, ask them to justify each connection. To support students, provide some words and ask them to suggest the links. *(10 minutes)*

Further teaching suggestions

Lead-free solder

The European Union Restriction of Hazardous Substances issued a directive banning lead-based solder from electronic systems in 2006. Discuss the various reasons why this decision was taken. Why was lead used in the first place? Find out if lead-free solders are as reliable as lead-based ones. Students can also research what flux is. How does it help soldering?

Support

- Use writing frames for students to make statements such as 'We use .... for .... because …' when discussing the properties of metals used for construction.

Extend

- For each of the properties and uses a student has written about metals, they can add a sentence explaining how the structure of the metal enables it to have that property (e.g. for tungsten to have such a high melting point, its atoms must be very strongly bonded to each other).

Practical support

Measuring density

Equipment and materials required: Displacement cans, blocks of various materials that will fit into the cans, appropriately sized measuring cylinders for the displacement cans, access to water, electronic balance.

Details

Set up equipment as in the adjacent diagram. Fill the can to the point that water spills from the spout. Use the balance to measure the mass of the object. Carefully add the object to the displacement can and ensure it is fully submerged, whilst collecting the displaced water in the measuring cylinder. Calculate the density of the object.

Density (g/cm^3) = $\dfrac{\text{mass (g)}}{\text{volume (cm}^3\text{)}}$.

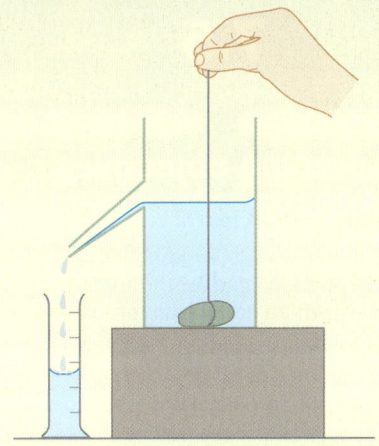

Unit 2, Theme 2 – My home

7.4 Metals for construction

Learning objectives
- What are the properties of metals?
- How are metals used in construction?

AQA Examiner's tip
You need to be able to use data to explain why certain metals are chosen for certain jobs.

Properties of metals

Most of the **elements** in the periodic table are metals. Scientists who work with metals are called **metallurgists**. Their role is to decide which metals are suitable for different jobs. They also combine metals with other metals (and some non-metals) to change their properties. These mixtures of metals, such as steel, are called alloys.

Figure 1 Working with metals

The properties of metals make them ideal for use in the building industry. This table lists some of the properties most metals share:

Property	Meaning
Hard	Metals are difficult to scratch
Strong	Metals don't break easily
Malleable	Metals can be beaten into different shapes
Ductile	Metals can be pulled out into wires
High melting point	Most metals don't melt easily
Good conductor of electricity	Electricity can travel through them easily
Good conductor of energy	Energy can travel through them easily
High density	They are heavy for their size because their atoms are packed closely together

a Lightning rods are designed to channel lightning strikes harmlessly around a building. Which property makes copper a good choice for lightning rods?

Metals in construction

Builders use metals alongside other materials. Look at the reinforced concrete being made in the photo opposite. Concrete is very strong in terms of compression. This means it can withstand large forces squashing it. However, pulling or twisting concrete can cause it to crack. Steel adds tensile strength to the structure, which resists pulling forces.

Figure 2 This worker is making reinforced concrete

b If metal is so useful, why aren't buildings made entirely of metal?

c Why might it be useful to introduce a little flexibility into a building?

Here are some properties of metals found in a home. Concrete and wood have been added so you can compare them to metals.

Material	Strength	Density	Malleability	Ductility	Electrical conductivity
Aluminium	●●●●●	●●	●●●●	●●●	●●●●
Steel	●●●●●	●●●●	●●●●	●●●●	●●●●
Copper	●●●●	●●●●	●●●●●	●●●●●	●●●●●
Lead	●●●	●●●●●	●●●●●	●●●	●●●●
Concrete	●●●	●●			
Wood	●●	●			

Key:
Very high	●●●●●
Moderate	●●●
None	

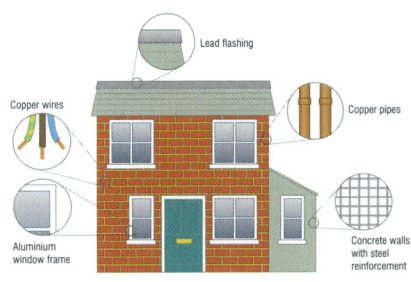

Figure 3 Metals are used extensively in homes

Summary questions

1. Copper is malleable, ductile, a good conductor of electricity and does not react with water.
   a Which **two** of these properties make it good for making pipes?
   b Which **two** of these properties make it good for making wires?
2. Use the information in the table above to explain the different uses of metals in Figure 3.
3. Copper has the lowest chemical reactivity of the metals named on this page. Suggest why this is important for one of its uses in the home.

Practical

Calculating density

Density is what makes the difference between a material being lightweight or very heavy. You can calculate the density of an object if you know its mass and its volume. You can measure mass easily using a balance. To measure volume, you need a displacement can. If you fill a displacement can up to the spout, water will spill out when you put another object in. You can collect the water in a measuring cylinder to find out the volume displaced by the object.

Once you have measured the volume, you can calculate an object's density using this formula:

density = $\dfrac{\text{mass (g)}}{\text{volume (cm}^3\text{)}}$

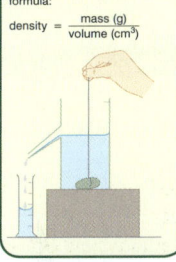

Key points

- Metals all share similar properties. They are malleable, ductile, strong, hard, have high melting points and are good conductors.
- The properties of metals make them useful for many construction tasks.

Summary answers

1 a Malleable, does not react with water.
b Good conductor of electricity, ductile.

2 Aluminium window frames: light and strong, malleable to make it the right shape.
Steel reinforcement: very strong and malleable to make it the right shape.
Copper pipes: see 1a above.
Copper wires: see 1b above.
Lead flashing: dense so it can hold things down, malleable so it can be hammered into shape.

3 Copper's low reactivity is useful for copper water pipes. The copper won't react with the water it is carrying.

Did you know …?

Students could research why lead is no longer used for making water pipes.

Answers to in-text questions

a Copper is a very good electrical conductor.
b Metals are much more expensive than concrete or bricks. They are much denser, so only thin panels could be used. Also, metals a very good thermal conductors – it would be hard for the house to retain energy.
c A little flexibility in a structure means it is less brittle.

Unit 2, Theme 2 – My home

7.5 Polymers in the home

Learning objectives

Students should learn:
- that polymers are large molecules made of many smaller monomers, most of which come from crude oil
- that polymers are useful because they can be easily moulded, are waterproof and excellent insulators.

Learning outcomes

Students should be able to:
- describe the general properties and structure of polymers
- name some polymers (and the monomers that make them) and suggest uses for them.

Some students should also be able to:
- explain how polymers are made up of many monomers, using polyethene and ethene as an example
- evaluate the use of polymers for specific tasks based upon their properties.

AQA Specification link-up: Science B 3.4.2.1
- Know that most polymers are manufactured using chemicals obtained from crude oil
- Describe how polymers are produced when many small molecules (monomers) join together to form very large molecules (polymerisation)
- Know that polymers are flexible, poor conductors of heat and electricity, resistant to corrosion, waterproof and that most of them have low melting points. These properties relate to their use in the home.

Lesson structure

Starters

Anagrams – Give students key word anagrams for polymers, metals, fibreglass, monomers, ceramics, composites. *(5 minutes)*

Polymers around the home – In pairs, students make a list of all the items in their home that are made from polymers. To support students, include pictures on the cards. To extend students, they can suggest alternative materials for these items, which polymers may have replaced. *(10 minutes)*

Main

- Hand out samples of polymers – squares cut from plastic bags or packaging, plastic toys, cups, nylon rope. Alternatively, photographs might be useful. Use these as stimulus material for a discussion about the properties of polymers. What do we already know about polymers? What are they used for? What other examples of polymers can we think of? Why are they useful? What is a polymer? Why might we want to make one?
- Without going into great structural detail, explain how polymerisation occurs. If students link paperclips together it can help them visualise this process, as can video and animations. This could be further visualised by making spaghetti before the lesson and allowing the strands to 'set' together in the fridge.
- Small groups can be given stimulus material and more information on particular polymers, their uses and properties. Each expert group could then feed back to the rest of the class.
- Show the class pictures of conservatories – how much of these structures were made from polymers? Why was a polymer chosen for the job? Most conservatories are made from uPVC – students can research this material and compare it to traditional materials such as wood.
- What about polymers such as polythene, polystyrene and Teflon? What are their properties and how are they used?
- Demonstrate polymerisation to the class with the 'nylon rope trick' (see 'Practical support').

Plenaries

'What have I learned?' race – Set up two flipcharts and divide the class in half. Students in each team take turns to write one fact they have learned on the flipchart (in the style of a relay race). The team with the most (valid) facts after five minutes wins. *(5 minutes)*

Material ideas – Give students a list of the materials discussed in the lesson. Working in small groups, students write one sentence about each material on a piece of paper. The paper is then passed around to the next student, who adds another sentence about the material. To support students, ideas could be limited to one key word each. *(10 minutes)*

Support
- Students are given different cards under the headings 'Type of material', 'Examples of material' and 'Properties of material'. Provide lots of stimulus material such as samples of materials, photos, etc. to help students explore the materials' appearance and properties.

Extend
- Use molecular modelling kits (or paperclips) to show the polymerisation of ethene. For extension work, students can then use the structures of some other monomers to draw the structures of their polymers.

Demonstration support

Nylon rope trick

Equipment and materials required: 10 cm³ aqueous solution of 0.5 mol/dm³ 1,6-diaminohexane (Solution A), 10 cm³ 0.2 mol/dm³ decanedioyl (sebacoyl) chloride in cyclohexane (Solution B), small beaker, glass rod, nitrile gloves, eye protection.

Details

Wearing gloves and eye protection, add A to small beaker. Insert glass rod into the centre of the beaker and slowly pour solution B down the side of the rod. Nylon will polymerise at the interface. Slowly twist the rod to snag the nylon and carefully wind it onto the rod. The nylon can be handled after careful rinsing in water and gentle drying with a tissue.

Safety: Wear eye protection and nitrile gloves. Follow CLEAPSS Hazcard 45 for procedure for making solutions A and B and for disposal. Use in a well-ventilated room with no naked flames. CLEAPSS Hazcard 3B 1,6-Diaminohexane; CLEAPSS Hazcard 41 Decanedioyl (sebacoyl) chloride – corrosive; CLEAPSS Hazcard 45B Cyclohexane – highly flammable and harmful.

Further teaching suggestions

Testing the strength of carrier bags

All materials need to be exhaustively tested before use. Discuss how materials can be tested for strength in compression and in tension. Students can devise a test to compare the tensile strength of different supermarket carrier bags.

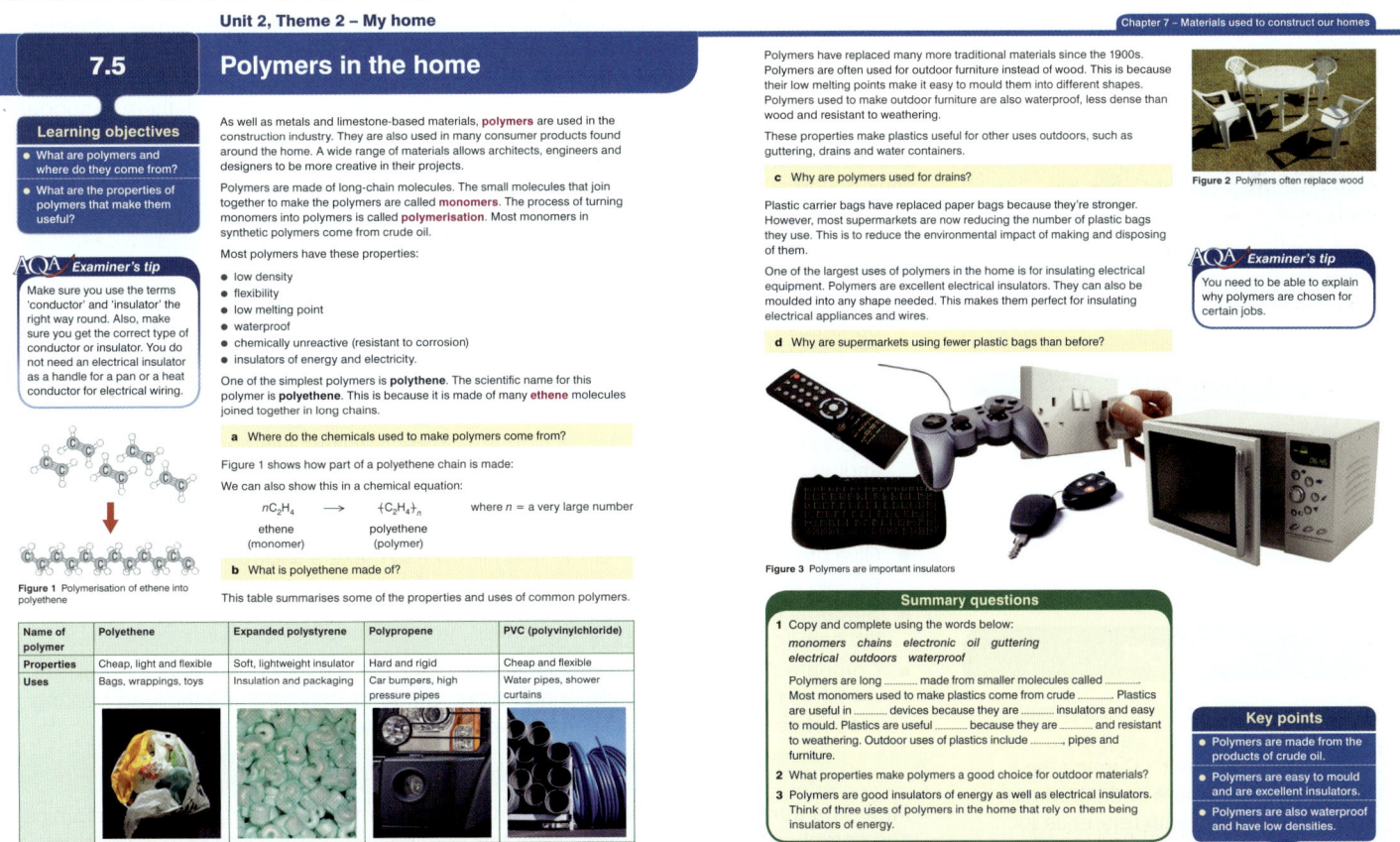

Summary answers

1 chains, monomers, oil, electronic, electrical, outdoors, waterproof, guttering
2 waterproof, resist weathering
3 pan handles, fridge linings, plastic cups, etc.

Answers to in-text questions

a The chemicals used to make polymers come from crude oil.
b Polyethene is made of many ethene molecules joined together in chains.
c Polymers are used for drains because they do not react with water, are tough and easy to mould into shape.
d Using fewer plastic bags reduces the amount of crude oil needed to produce them, so less pollution/environmental damage.

Unit 2, Theme 2 – My home

7.6 Ceramics and composites in the home

Learning objectives

Students should learn:

- that ceramics are very hard, heat resistant brittle materials
- that ceramics are used for plates, tiles, pylon insulation and furnace linings
- that composites are combinations of materials, designed to exploit the materials' best properties
- that examples of composites include glass reinforced plastic, reinforced concrete and reinforced glass.

Learning outcomes

Most students should be able to:

- describe the characteristic properties and some uses of ceramics
- name some composites and their uses.

Some students should also be able to:

- explain how the structures of glass reinforced plastic and reinforced concrete contribute to their properties.

AQA Specification link-up: Science B 3.4.2.1

- Relate the characteristic properties of ceramics (e.g. brittle, high melting point) to their uses in construction.
- Be able to recognise and describe a composite material (e.g. MDF, fibreglass, reinforced concrete).
- Describe the properties of a composite as a combination of the properties of its components.

Lesson structure

Starters

Comparing materials – Students compare the properties of a ceramic and plastic mug, listing the advantages and disadvantages of each. *(5 minutes)*

Examining materials – Hand out samples of a ceramic (small items of crockery perhaps) and a composite (MDF is a good example). Alternatively, photographs might be useful. Use these as stimulus material for a discussion about their properties. What do we already know about these materials? What are they used for? What other examples of ceramics can we think of? What is special about the MDF? Why is it useful? What is a composite? Why might we want to make one? To support students, ask them to suggest similar materials to the ones they have been shown, explaining their choices. To extend students, choose set characteristics to compare and produce a prompt-sheet to guide their observations. *(10 minutes)*

Main

- Use diagrams to show the structure of ceramics. Highlight the strong forces holding the atoms together and relate this to the high melting points of ceramics. Why are ceramics good electrical insulators? Silica is a ceramic, but glass is not – why is this? [The molecular arrangement within glass is amorphous, whilst ceramic is crystalline.]
- Discuss the drawbacks of some traditional materials. How can we combine them with other materials to make the most of their most useful properties?
- Show students images/video of extreme sports activities (watersports are good for finding composite materials) and discuss the demands placed on items such as surfboards, speedboat hulls and snowboards. What are the key properties a designer will be looking for when choosing materials for a jet-ski hull?
- Give groups a list of materials to choose from to make a snowboard – get them to explain the advantages and disadvantages of each.
- Students can make their own composite materials and test their strength compared to the component materials.
- Compare the properties of ice and cotton wool. Ice is hard but shatters; cotton wool has flexibility but can bear no weight. Students can compare the load-bearing properties of ice with the ice/cotton wool composite.

Plenaries

Materials summary – Students can complete a summary table for polymers, ceramics and composites, outlining some properties of each and listing some of their uses (for composites, specific examples will have to be used). To support students, these could be pre-prepared and sorted into the right groups. To extend students, information could be added about alternative materials to the ones listed. *(5 minutes)*

Sorting materials – Give students the properties (but not names) of various polymer, ceramic and composite materials on cards. They must sort them into the correct groups. *(10 minutes)*

Support

- Show students a mug and a plastic toy – what words can we use to describe these items? Remind the students of polymers and introduce the word ceramic with reference to the items being handled. Ask the students how the materials could be made better (e.g. we could make the mug less able to smash and the plastic toy harder to pull apart). Use this angle to lead into the idea of composites and the ice/cotton wool experiment.

Extend

- Students can try to explain the cotton wool/ice composite in terms of structure, by drawing diagrams showing how the cotton wool fibres make it harder for cracks to spread through the ice.

?? Did you know …?

In 2003, the NASA shuttle Columbia was destroyed upon re-entry because its ceramic thermal protection system had been damaged by some debris during take-off. Students could read a news article about the tragedy and then research the types of ceramics used in the shuttle.

Chapter 7 – Materials used to construct our homes

Further teaching suggestions

Extending the composite investigation

The composite practical can be extended to investigate a range of different materials. How would cotton strands or wire compare to cotton wool? This could also be approached quantitatively – how much cotton wool is needed for the strongest sticks?

Answers to in-text questions

a resistant to chemicals, high melting point
b brittleness
c Bricks are stronger than wood and last longer.
d strength of glass and flexibility of plastic

Practical support

Composite ice sticks

Equipment and materials required: Water, cotton wool, ice stick moulds (readily available from large homeware shops), two clamp stands, slotted masses.

Details

Fill the moulds with water and add varying amounts of cotton wool to them. Freeze overnight. Support with two clamp stands and hang masses from the centre.

Safety: Protect the bench, floor and feet from falling masses. Wear eye protection. Try to confine ice chippings to prevent slipping on floor.

Unit 2, Theme 2 – My home

7.6 Ceramics and composites in the home

Learning objectives
- What are ceramics and composites, and what are their properties?
- How do we use ceramics and composites around the home?

Ceramics and composites are materials you may not have heard much about before. However, they have been used in homes for hundreds of years.

Ceramics

Ceramics are materials like clay or china. This is probably the oldest branch of materials science. People have been making pottery for thousands of years.

Builders use ceramics in the bathroom or kitchen. Tiles, sinks, toilets, plates and mugs are all ceramics.

The properties of ceramics are:
- electrical and thermal insulators
- strong
- very hard
- resistant to most chemicals
- very high melting point
- **brittle** (they can shatter if struck sharply).

Did you know …? Ceramic tiles are also used to heat-proof the outside of space shuttles.

a Which of the properties above make ceramics useful for making plates?
b Which of the properties above makes ceramics less useful for making plates?

Why use ceramics?

The most common ceramic objects you will see in buildings are bricks. Bricks are made of clay. This is an ideal material because of its high melting point and its strength.

c What are the advantages of building houses with bricks instead of wood?

You will also see a lot of ceramics in the bathroom. Ceramics are ideal materials for sinks, tiles and toilet bowls. This is because they are hard-wearing, easy to clean and resistant to chemicals in cleaning products.

Composites

Composites are one of the newest and most exciting developments in materials science. A composite is made when two or more different materials are put together to complement each others' properties. This is usually when a material has a really useful property but can't be used by itself.

They are interesting materials because they combine the best properties of different materials.

Glass reinforced plastic (often called **fibreglass**) is a composite material. It is made by painting a resin coating onto sheets of glass fibres. This results in a strong, yet flexible material. Fibreglass is used to make waterslides because it is stronger than plastic and just as light.

d What properties of glass and polymers are important in fibreglass?

Figure 1 Ceramics in action

Figure 2 Tilers use the brittleness of ceramics to crack tiles

Why use composites?

Because composites are made from a wide variety of materials, they have a wide range of properties. Their properties depend on the materials used, and also on how much of each one is used. The possibilities are endless.

When concrete is reinforced with steel it becomes a composite. Steel bars make buildings stronger because they add flexibility. They also slow down the growth of cracks. This is because the steel forms a barrier the crack has to move around.

Thin metal wires can be placed in glass so it doesn't shatter if broken. This composite is called reinforced glass. Another way to reinforce glass is to add thin layers of plastic to its surface. This also holds the glass together if it gets broken. The plastic stops the glass breaking into sharp, dangerous shards.

Another composite found in the home is MDF (medium density fibreboard). MDF is often used to make 'flat-pack' furniture. It is useful because it is light, strong and easy to cut into shapes. It is made by binding individual fibres of wood together strongly with wax and resin. Because it does not have a 'grain' like wood, it does not chip or split easily.

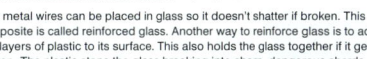

Figure 3 Water slides are made of a composite – glass reinforced plastic (fibreglass)

Figure 4 Steel bars make it hard to break reinforced concrete

Figure 5 Reinforced glass is safer than regular glass

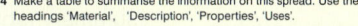
Figure 6 MDF is a composite used to make furniture

Summary questions

1 Which objects in your kitchen are made from ceramics?
2 Which properties of ceramics make them useful for bathroom tiling?
3 Ceramics are used to line kilns and furnaces. Why are they a good choice for this?
4 Make a table to summarise the information on this spread. Use the headings 'Material', 'Description', 'Properties', 'Uses'.

AQA Examiner's tip
In the exam you will need to think about the properties of materials that are needed for the situation given in the question. This will help you to make your choices, descriptions or explanations relevant and gain more marks.

Key points
- Ceramics are hard but brittle. They are good insulators. They are resistant to chemicals and have a high melting point.
- Composites are mixtures of more than one material. They often combine the best properties of each material in the mix.

Summary answers

1 crockery, possibly ovenware, tiles, possibly sink, mugs
2 resistant to chemical attack – easy to clean, brittle – can be snapped to the right size, waterproof, hard
3 very high melting point

4

Material	Description	Properties	Uses
Ceramics	Strong solid materials such as china, clay	Hard, brittle, high melting point, insulators	Insulating very hot things, crockery, tiles, toilets
Glass reinforced plastic	Glass strands set into resin	Hard, strong, light	Water slides, hulls of boats
Reinforced concrete	Steel bars set into concrete	Very strong and resistant to cracking	Structures of buildings
Reinforced glass	Wire mesh set into glass or plastic sheet	Holds together if broken – does not shatter into shards	Door windows, car windows
MDF	Wood fibres set into wax and resin	Light, strong, does not chip	Furniture

Unit 2, Theme 2 – My home

7.7 Building sustainable homes

Learning objectives

Students should learn:
- that timber, cob and straw bale construction are traditional sustainable building methods
- that hempcrete and clay honeycomb blocks are more modern sustainable building materials
- that some building materials are carbon negative because they are plant based.

Learning outcomes

Most students should be able to:
- name some traditional and modern building methods that are sustainable.
- compare the carbon footprints of different construction methods.

Some students should also be able to:
- explain what causes the carbon footprints of particular building materials
- use data to evaluate the impact of particular building materials.

AQA Specification link-up: Science B 3.4.2.1

Within this context, candidates should be able to use scientific data and evidence to discuss, evaluate or suggest implications of:
- the developments in modern (sustainable) building materials, and their advantages and disadvantages when compared with more traditional materials, including straw bale, wood frame, cob construction
- the most suitable material for a particular use.

Lesson structure

Starters

Carbon footprint – Students can start thinking about carbon footprints by visiting one of the many carbon footprint calculators on the internet. Their task is to calculate the carbon footprint of their journey to school. *(5 minutes)*

Energy in the home – Students list all the items and activities in the home that require energy input. To extend students, ask them also to consider the amount and types of energy they think were needed to produce the items. To support students, review the different types of energy first and ask them to think of examples of each type of energy at use in the home (e.g. chemical – gas cooker, electrical – TV, light – light bulbs). *(10 minutes)*

Main

- Compare the timber and brick houses from the Student Book spread. Students can produce a bar chart showing both data series in order to compare them. Which elements of construction show the biggest difference? Why are wooden doors carbon negative?
- Show students pictures and videoclips of cob construction, using it as stimulus for a discussion about alternative designs for houses.
- Split the class into groups – each group must research a different building method or material, choosing from the ones in the Student Book or finding their own. They must find out enough information to write a proposal for building a home using that method or material.
- Students can model passive solar heating with the investigation detailed in the 'Practical support' section.

Plenaries

Match-up – Students must look at pictures of houses and decide by which method they have been built. *(5 minutes)*

Summarise – Students can use the information from the Student Book to produce a summary table with the headings 'Material, Description, Advantages, Disadvantages'. To support students, pre-print the information so the students just have to rearrange it. To extend students, ask them to evaluate each method, giving it a star rating that takes into account all of its features. *(10 minutes)*

Support

- To support students, concentrate on being able to name the different materials and methods of construction. Provide lots of photographs and basic information about sustainable building materials. Students can construct leaflets summarising sustainable building.

Extend

- There are lots of sustainable building projects that have been well documented on the internet. Students can perform case studies of different builds.

Further teaching suggestions

Flow charts of material production

If the corresponding Student Book chapter is being taught as part of a sequence, students will be able to label almost all of the materials used to build a home and produce simple flow charts of how the materials are made. Given information about cob, straw bale, hempcrete and clay honeycomb construction, they can perform the same activity for sustainable materials too. Students can then compare the length and complexity of each flow chart as a way to estimate energy usage and environmental impact. This would also provide a good review of the whole chapter.

Practical support

Passive solar heating

Equipment and materials required: Mounted light bulb or radiant heater, shoeboxes, scissors, glue, acetate sheets, white card, black card, foil, fabric, thermometer or temperature probe with data-logger.

Details

The aim of this investigation is to show that choice of materials and design can affect the amount of energy absorbed by a home.

The thermometer is inserted through a hole in the lid of the shoebox, so that it is fairly central. Different groups experiment with different materials lining the interior of the box and different sized acetate windows. The light source remains the same distance from the showbox so as to be a reasonably fair test. The air temperature inside the box can either be read every minute and plotted on a graph, or just recorded as a final result after ten minutes.

Safety: Take care when using the hot light bulb or radiant heater.

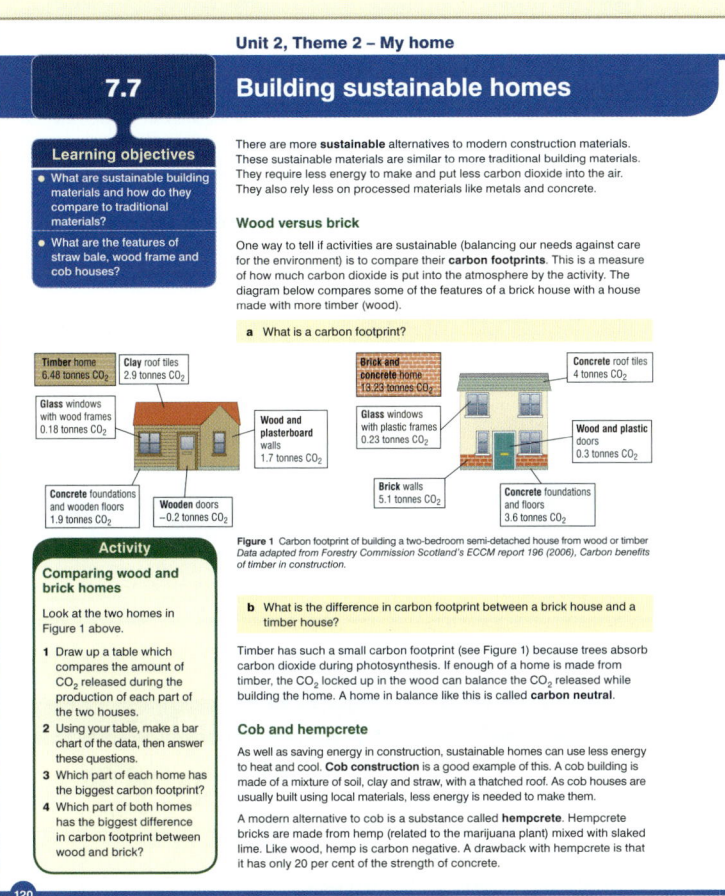

Figure 2 A modern cob home

Figure 3 A straw bale house

Answers to in-text questions

a A carbon footprint is a measure of the amount of carbon dioxide an activity puts into the atmosphere.

b A timber building puts 6.75 tonnes of CO_2 less into the atmosphere than a brick building.

c Straw bale houses cost less money to heat because straw is such a good insulator.

Summary answers

1. footprint, absorb, clay, hempcrete, lower, insulated, honeycomb, longer

2. Their materials and design ensure they absorb the Sun's energy, releasing it slowly when the temperature drops again.

3. **a** Brick lasts a lot longer, but has a much higher carbon footprint for heating. Brick houses also have a bigger carbon footprint in construction. Straw has a lower environmental impact overall.

   b Cob and timber homes have similar carbon footprints for heating, but cob homes last longer and have a smaller carbon footprint in construction. Overall, cob homes are better for the environment.

   c Clay honeycomb lasts a lot longer than hempcrete, but hempcrete is massively carbon negative to build with. However, the heating costs for hempcrete are greater than clay honeycomb, and these would add up over time.

Unit 2, Theme 2 – My home

Summary answers

1. It can be turned into cement, to make concrete or mortar. It is an ingredient in glass. It is used for purifying iron in the blast furnace.

2. Word equation:
   calcium oxide + water ⟶ calcium hydroxide
   Symbol equation:
   $CaO + H_2O \longrightarrow Ca(OH)_2$

3. a cement
   b glass
   c concrete
   d mortar

4. a Flexibility – can be bent; malleability – can be beaten into different shapes
   b i Steel rods are embedded in the concrete to make it stronger.
     ii Steel is too dense.
   c Lighter

5. a Polyethene – carrier bags, wrappers, toys
   b PVC – shower curtains, water pipes
   c Polystyrene – insulation and packaging
   d Polypropene – car bumpers, high pressure pipes

6. plates – ceramic
   kettle – metal
   washing up bottle – polymer

7. a high melting point
   b i

Alpha construction	Limestone, gypsum, clay
Beta Bathrooms	China
Gamma electrical	Copper
Delta sports	Fibreglass
Epsilon glass	Sodium carbonate, limestone

   ii Alpha – baked together in a furnace to make cement
      Epsilon – heated together with sand to make glass
   iii Beta – china resists chemical erosion, is hard and durable
       Gamma – copper is a very good electrical conductor
       Delta – fibreglass is lightweight and hard

8. a Straw bale and clay honeycomb are similar; they contain air spaces what help insulate buildings.
   Cob and hempcrete are similar; they use local resources and are good for passive heating.
   b i Trees lock up carbon dioxide from the atmosphere in photosynthesis.
     ii Straw bale, hempcrete

9. **Advantages:** cheap, renewable, easy to get hold of, good insulator
   Disadvantages: flammable, needs to be thick, not good in strong high winds
   Not relevant: Used for thousands of years, made from bundles of straw tied together with twine or wire

Summary questions

1. Describe three ways limestone is important in building houses.

2. Copy and fill in the spaces in these equations showing the conversion of quicklime into slaked lime:
   Word equation:
   calcium oxide + ... ⟶ calcium hydroxide
   Symbol equation:
   $CaO +$ .... ⟶ ....

3. Which building material is made from:
   a gypsum, clay and limestone
   b sodium carbonate, limestone and sand
   c cement, sand, stones and water
   d cement, sand and water

4. a What is the difference between flexibility and malleability?
   b Steel is an alloy made from iron.
     i How is steel used to reinforce concrete buildings?
     ii Why is steel not used to make aeroplanes?
   c Why is aluminium a better choice for window frames than iron?

5. Suggest a use for the following polymers. Name properties of the polymer that make it a good choice for each use:
   a polyethene
   b PVC
   c polystyrene
   d polypropene

6. The picture shows a kitchen. Suggest the best material to make each of the labelled items out of. Choose from the materials in the box.

   ceramic   composite   metal   polymer

   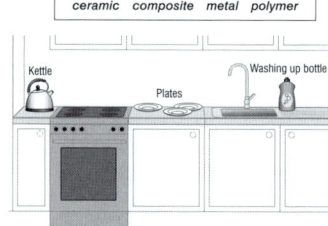
   Kettle   Plates   Washing up bottle

7. a The insides of furnaces are often lined with a ceramic material. Which property of ceramics makes them a good choice for this job?

   b You work for a large corporation that supplies materials to other companies. These are the companies you supply materials to, and what they do with them:

Company	What they produce
Alpha construction	Cement to sell to builders
Beta Bathrooms	Ornamental sinks
Gamma electrical	Electrical wires
Delta sports	Lightweight surfboards
Epsilon glass	Glass for windows

   These are the materials you have in your warehouse
   limestone copper clay gypsum fibreglass
   china quicklime sodium carbonate
   i Which material(s) would you send to each company? (Hint: some companies will need more than one material.)
   ii For Alpha Construction and Epsilon Glass, explain how they use the materials you are sending them
   iii For Beta Bathrooms, Gamma Electrical and Delta Sports, explain why your choices of materials are be[st]

8. a Straw bale and cob construction are traditional ways to build homes. Describe their properties and compare them to hempcrete and clay honeycomb blocks.
   b A carbon footprint is a measure of how much carbon dioxide an activity produces.
     i Why does a timber house have a smaller carbon footprint than a brick house?
     ii What other building materials have a small carbon footprint?

9. A builder made notes about constructing a cob building from a website and now needs to sort them out. Read the notes and sort them in a table as shown.

Advantage	Disadvantage	Not relevant

   Straw is cheap, renewable and flammable.
   Straw walls need to be thick but it is easy to get hold of straw.
   Not strong in high winds, straw has been used as a building material for thousands of years.
   Straw bales are made from bundles of straw tied together with twine or wire and are a good insulator.

Practical suggestions

Practicals	AQA	k	📖
Model the limestone cycle: decomposition of $CaCO_3$ to give CaO, reaction with water to give $Ca(OH)_2$, add more water and filter to give limewater and use limewater to test for CO_2.	✓	✓	✓
Thermal decomposition of $CaCO_3$ to show limelight.	✓	✓	✓
Make concrete blocks in moulds, varying content, and carry out strength tests.		✓	✓
Test physical properties of metals – e.g. density/thermal and electrical conductivity.		✓	✓
Make models of polymer chains from plastic and cocktail sticks.		✓	
Investigate the properties of ceramics.		✓	

Chapter 7 – Materials used to construct our homes

Examination-style questions

1 Metals have many uses.
 a Explain why copper can be used for electrical wiring. Give two reasons. (2)
 b Explain why aluminium would be good for making aeroplanes. Give two reasons. (2)

2 Fill in the table to show the chemical formulae for limestone and its products.

Material	Chemical name	Chemical formula
Limestone		
Quicklime		
Slaked lime		

(3)

3 Use words from the box to describe the composition of some materials.

> sodium carbonate   cement   clay
> gypsum   limestone   sand   water

 a Glass is made from .... + ... + ... (1)
 b Mortar is made from .... + ... + ... (1)
 c Cement is made from .... + ... + ... (1)

4 Plastic has many uses and replaces many traditional materials.
Suggest the traditional material for each of the modern uses of plastic and the advantage of using plastic in each of the situations.
 a Milk bottles (2)
 b Shopping bags (2)
 c Toys (2)

5 Choose the correct words from the box to describe the uses of ceramics.

> are brittle   are resistant to chemicals
> are insulators   are malleable
> have a high melting point

 a Ceramics can be used for bathroom tiles because they .... (1)
 b Ceramics are used as furnace lining because they .... (1)
 c Ceramics can be used on power transmission lines because they .... (1)

6 A shop selling hot drinks got a lot of complaints because the drinks it was selling got cold too quickly. The shop owners decided to do an experiment to find which material kept a drink hot the longest. The table shows their results.

Cup material	Start temperature in °C	Temperature after 10 minutes in °C	Total drop in temperature in °C
China	75	63	
China with lid	75		5
Plastic	75		21
Plastic with lid	75	58	
Polystyrene	75		18
Polystyrene with lid	75	65	

 a Fill in the missing numbers in the table. (2)
 b Make and explain a conclusion based on their results. (3)
 c In this question you will be assessed on using good English, organising information clearly and using specialist terms where appropriate.
   Give a comparison of the advantages and disadvantages of making the cup out of china rather than polystyrene. (6)

7 Glass-reinforced plastic (also known as fibreglass) can also be used in building houses. Fibreglass is made of a polymer reinforced by fine fibres made of glass.
It is strong and lightweight, it has a weather-resistant finish with a variety of surface textures and an unlimited colour range is available.
It can be used for roofing, door surrounds, canopies and chimneys.
 a Suggest why it is important to make sure no air is trapped between the glass fibres when the fibreglass is being made. (1)
 b Why might fibreglass be cheaper to install than other alternatives? (1)
 c Why might someone choose a wooden door surround? (1)

Examination-style answers

1 a good (electrical) conductor (1 mark)
  ductile (1 mark)
 b will not rust (1 mark)
  lightweight (1 mark)

2

Material	Chemical name	Chemical formula
Limestone	Calcium carbonate	$CaCO_3$
Quicklime	Calcium oxide	CaO
Slaked lime	Calcium hydroxide	$Ca(OH)_2$

(1 mark for each correct row filled in.)

3 a sand, limestone, sodium carbonate (1 mark)
 b cement, sand, water (1 mark)
 c limestone, clay, gypsum (1 mark)

4 a glass (1 mark)
   shatterproof (1 mark)
 b paper (1 mark)
   not waterproof (1 mark)
 c wood / metal (1 mark)
   splinter / sharp (1 mark)

5 a Are resistant to chemicals. (1 mark)
 b Have a high melting point. (1 mark)
 c Are insulators. (1 mark)

6 a Three numbers correct in 3rd column. Three numbers correct in 4th column.

China	75	63	12
China with lid	75	70	5
Plastic	75	54	21
Plastic with lid	75	58	17
Polystyrene	75	57	18
Polystyrene with lid	75	65	10

(2 marks)

 b China with lid is the best. The lid prevents heat rising out of the cup. China is a good insulator. (3 marks)
 c Marks awarded for this answer will be determined by the Quality of Written Communication (QWC) as well as the standard of the scientific response.

There is a clear, balanced and detailed comparison of the advantages and disadvantages of making the cup out of china rather than polystyrene. The answer is well structured with minimal repetition or irrelevant points. There is an accurate, fluent and clear expression of ideas with only minor errors in the use of technical terms, spelling, punctuation and grammar. (5–6 marks)

There is some comparison of the advantages and disadvantages of making the cup out of china rather than polystyrene with some omissions. The answer shows some attempt at structuring and the ideas are expressed with reasonable fluency and clarity. There are some errors in the use of technical terms, spelling, punctuation and grammar. (3–4 marks)

There is a brief comparison of an advantage and/or a disadvantage of making the cup out of china rather than polystyrene. The answer is largely incomplete and may contain some valid points which are not clearly structured. It lacks fluency and/or clarity. It contains errors in the use of technical terms, spelling, punctuation and grammar. (1–2 marks)

Examples of points made in the response:
- Polystyrene is made from oil
- Which is non-renewable
- China is heavy / plastic is lightweight
- China is more expensive
- China with lid is the best insulator.

7 a not as strong (1 mark)
 b lightweight (1 mark)
 c Any sensible aesthetic reason, e.g. because they feel it looks better, because it matches the other materials in the building. (1 mark)

Kerboodle resources

- Theme map: My home
- Support: The conversion of limestone (7.2)
- Practical: The limestone cycle (7.2)
- Video: Limestone – building understanding, understanding building (7.3)
- Revision podcast: Metal production (7.4)
- Practical: Properties of materials (7.4 – 7.6)
- Interactive activity: Rocks and building materials
- On your marks: Planning an experiment and using data
- Examination-style questions
- Answers to examination-style questions

Unit 2, Theme 2 – My home

8.1 Everyday fuels

Learning objectives

Students should learn:
- that households use a range of fuels for cooking, heating and transport
- that these fuels are hydrocarbons
- that hydrocarbons contain only carbon and hydrogen.

Learning outcomes

Most students should be able to:
- name a range of fuels and their uses
- state that a hydrocarbon contains only carbon and hydrogen.

Some students should also be able to:
- justify reasons for using particular fuels for particular purposes.

AQA Specification link-up: Science B 3.4.2.2
- Name suitable fuels for cooking and heating our homes and for providing transport.
- Know that hydrocarbons contain carbon and hydrogen only.

Lesson structure

Starters

Match-up activity – Students match up the types of fuel with examples of their uses. *(5 minutes)*

Demonstration – demonstrate a steam engine model to students, focusing on writing out the energy transfers taking place. To support students, just discuss the useful changes. To extend students, consider all of the changes. Invite students to suggest where energy is wasted and how the engine could be made more efficient. *(10 minutes)*

Main

- Students can use a cutaway picture of a home to identify types of energy at work.
- This can be used as an opener to discussing where the energy comes from.
- The information from the table in the Student Book can be used as a basis for a match-up activity, with students explaining why particular fuels are used for particular purposes.
- Using information from the Student Book, students list the costs and benefits of the fuels mentioned in the spread.
- Provided with more information, students can debate the reasons we mostly use petrol in cars, despite diesel being more efficient and less polluting.
- Review how fossil fuels are formed by sequencing a series of statements.

Plenaries

Energy taboo – In threes, students must describe different fuels without using particular key words, e.g. describe diesel without saying petrol or cars. One student describes, the second listens for banned words and the third tries to guess the fuel. Score a point for each fuel correctly identified. To support students, reduce the number of taboo words. To extend students, they can write the taboo cards themselves. *(5 minutes)*

Market research – Put students in the position of an energy company selling fuels to households. What would they need to find out about the household in order to decide which fuels to sell? *(10 minutes)*

Support
- Ask why we store things. Keep this focused on students' own experiences, storing tins, multipacks, etc. Students can list the reasons why stored energy is useful, and then apply these reasons to different fuels.

Extend
- Students could research and compare the costs of different fuels, such as natural gas versus heating oil.

Chapter 8 – Fuels and electricity

> ## Further teaching suggestions
>
> ### Graph skills
>
> Students can visit www.statistics.gov.uk or www.decc.gov.uk to research the UK's fuel use. There is a wealth of statistics on energy usage – students can compare energy consumption by fuel over the past 30 years. This can be displayed graphically and presented to the class.

Unit 2, Theme 2 – My home

8.1 Everyday fuels

Learning objectives
- What fuels do we use and how do we use them?
- What are hydrocarbons and where do they come from?

Our society relies on fuels in order to function. Homes in the UK use the equivalent of 45 million tonnes of **crude oil** every year. The total fuel use in the UK is around 150 million tonnes per year.

Without fossil fuels, we could not live the way we do today. This is because they are such a good source of energy. Burning fossil fuels releases lots of energy.

a Why do we burn fossil fuels?

What fuels do we use?

The fuels you will come across every day include **natural gas** and **petrol**. However, there are many more in use in the UK. Here are a few:

Fuel	Uses	Why do we use it?
Natural gas (methane)	Domestic boilers, cookers	Can be compressed and stored, doesn't wear out pipes like liquids do
Petrol	Cars, taxis	Petrol engines are quieter and faster than diesel
Diesel	Cars, vans, buses, trucks	Cleaner and much more efficient than petrol
Heating oil	Domestic boilers	Can be stored in tanks where there isn't a permanent gas supply
Kerosene (paraffin)	Jet planes	Lower freezing point than diesel and burns more efficiently at high pressure than petrol
Coal	Heating homes, power stations	Easy to store, contains a lot of energy, doesn't easily catch fire
Propane	Gas barbeques, caravans	Easy to compress and store in tanks

b Sort the fuels in the table into solids, liquids and gases.

Figure 1 Natural gas provides energy for cooking

Fuels used at home

Most cookers in the home are powered either by gas or electricity. An advantage of using a gas cooker is that it is more responsive than an electric cooker. As gas cookers take less time to heat up, they are more efficient than electric ones.

Gas is also useful in heating our homes. This gas is either stored in a tank or supplied by mains pipes. The gas is burned in a boiler, which, in turn, heats water. The hot water is then either stored or used right away.

Not all homes are heated by gas, though. Remote places might not have access to a mains supply of gas. These homes often use heating oil as a fuel. This is because it is easy to store, usually in tanks that are buried underground.

Coal was once the main fuel used to heat homes. It was relatively cheap and easy to store, and contained a lot of energy. However, coal is dirty and awkward to use and homes needed to top it up regularly. Natural gas is cleaner and can be piped straight into our homes.

c Why are gas cookers more efficient than electric cookers?

AQA Examiner's tip

If you do not put the word 'only' on the end of your description of a hydrocarbon (a compound containing hydrogen and carbon **only**), then you will not get the mark. Sugar contains hydrogen and carbon but it is not a hydrocarbon because it contains oxygen too.

Fuels used for transport

Petrol and **diesel** are both used as fuels for transport. They are excellent sources of energy. They are denser than natural gas, so they are a more concentrated energy source. They are liquids, so they can flow, making them more useful than solid fuels because they can travel down pipes.

Inside a petrol or diesel engine, the fuel is mixed with air, and then ignited by a spark. This controlled explosion happens many times a second, transferring the energy to make the wheels go round.

Kerosene is used to fuel jet aircraft, but aircraft used to run on ordinary petrol. Over the years, jet engines have become more and more powerful. This means their fuel has to be more efficient. Kerosene burns more efficiently than petrol at high pressures.

Figure 2 Inside a jet engine

What are these fuels made from?

Most of the fuels we use for heating and transport are distilled from crude oil. This is a naturally occurring substance formed from the remains of plankton that died millions of years ago. The crude oil is now trapped in rock formations, usually deep underground.

These **reservoirs** of crude oil have to be drilled into and the oil pumped out of the porous rock formation. Only certain places on Earth have a good supply of crude oil. Billions of pounds are spent finding and exploiting new sources of crude oil. Crude oil is a mixture of **hydrocarbons**. These are chemicals containing carbon and hydrogen only.

d What are hydrocarbons?

Figure 3 Crude oil is the starting point for many fuels

∞ links

For information on fractional distillation of crude oil look back at 2.4 Making products with materials from the Earth.

Summary questions

1 Copy and complete using the words below:
   diesel   cooking   burned   kerosene   energy   carbon   heating   hydrocarbons

   When fuels are ...., they release a lot of .... We use petrol, ... and .... for transportation. Natural gas and .... oil are used for heating and .... in the home. Fuels from crude oil contain ...., which are chemicals containing hydrogen and .... only.

2 **a** Why do we use natural gas for domestic cookers?
   b Why do we use kerosene as jet fuel?
   c When is there an advantage in using heating oil instead of gas for domestic boilers?

3 Where did crude oil originally come from?

Key points
- Fuels release a lot of energy when they burn in air.
- Petrol, diesel and kerosene are used for transport.
- Natural gas, heating oil, coal and propane are used for cooking and heating our homes and buildings.
- Hydrocarbons contain hydrogen and carbon only.

> ## Summary answers
>
> 1 burned, energy, diesel (kerosene), kerosene (diesel), heating, cooking, hydrocarbons, carbon
>
> 2 **a** It can be compressed and stored, it does not wear out pipes like liquids do.
>
>   **b** Lower freezing point than diesel and can be compressed more. This is useful when going to high altitudes and reduces the size of the fuel tank.
>
>   **c** As a supply of fuel to remote homes if there is no mains gas supply.
>
> 3 the remains of sea organisms

Answers to in-text questions

a Because they release energy (used for heating, transport and cooking).

b Solids: coal
Liquids: petrol, diesel, heating oil, kerosene
Gases: methane, propane

c instant heat, easier to control

d chemicals containing carbon and hydrogen only

Unit 2, Theme 2 – My home

8.2 Burning fuels

Learning objectives

Students should learn:
- that carbon dioxide and water vapour are formed when fuels are burned completely
- that we can represent these changes as word equations
- that we can show complete combustion as a balanced symbol equation. [HT only]

Learning outcomes

Most students should be able to:
- state that combustion means burning a fuel in oxygen
- state that hydrocarbon + oxygen $\longrightarrow$ carbon dioxide + water.

Some students should also be able to:
- Explain that combustion involves oxygen reacting with carbon to make carbon dioxide, and with hydrogen to make water
- Write a balanced symbol equation for combustion reactions. [HT only]

AQA Specification link-up: Science B 3.4.2.2.
- Write word and symbol equations for the combustion of hydrocarbons.
- Write balanced symbol equations for the complete combustion of hydrocarbons. [HT only]

Lesson structure

Starters

Burning bubbles demonstration – Perform the demonstration and encourage students to think about it. What did they see? What is the name of the gas in the bubbles? What gas from the air did it react with? What types of energy were given off? *(5 minutes)*

Burning candles – Demonstrate the classic experiment showing water rising up inside a jar when a lit candle is placed in it. Ask students to suggest why the water was pushed up into the jar. (The answer is *not* that it is replacing all the oxygen – remind them that a similar amount of carbon dioxide was also produced. In fact, the water rises because the gases in the jar contract as they cool after the candle goes out. The candle goes out because it is 'drowned' in hot CO_2 filling the jar from the top down). Do not give the class the answer now; return to the demonstration at the end of the lesson. *(10 minutes)*

Main

- Observe a candle. Why is it burning? What is the source of fuel? How could we tell it is not the wick? Discuss with students how the molten wax itself is not burning, but its vapour, which evaporates from the wick. You can demonstrate this by extinguishing the candle – the plume of smoke that appears is paraffin that has evaporated and then re-condensed. If you touch a lit match to this plume, the flame will chase back down and relight the candle.
- Use molecular model kits or circles cut from different colour cards to represent different atoms in fuels and oxygen. Start with methane – students arrange a carbon and four hydrogens together, then take it apart and add as much oxygen as they need to turn the carbon into carbon dioxide and the hydrogen into water. To extend students further, they can try with more complicated hydrocarbons. Students can then draw out their reactions as ball-and-stick diagrams.
- To show that the changes from the last activity are really happening, set up the demonstration in the practical support. Students can describe their observations, decide what they mean, and then explain them scientifically. This could be arranged as a table.

Plenaries

Working with equations – Students can rearrange word equations to show combustion. To extend students, they can work with symbol equations. *(5 minutes)*

Combusting other materials demonstration – Demonstrate the combustion of steel wool and iron filings, then magnesium, drawing on students' observations of oxides to show that new materials have been made. To support students, focus on the physical observations of changes taking place. To extend students, get them to name the products of the reactions. *(10 minutes)*

Answers to in-text questions

a burning a fuel in oxygen
b Limewater would go cloudy, blue cobalt chloride paper would go pink.

Support
- Students can practise getting the word equation right by rearranging cards.

Extend
- Students can research incomplete combustion in more depth. What are the effects of carbon monoxide? How is it formed? How does boiler design try to prevent carbon monoxide poisoning?

Further teaching suggestions

Phlogiston, oxygen and combustion

Students can research the work of Antoine Lavoisier, explain what phlogiston theory was and how it was disproved.

Summary answers

1 burn, oxygen, hydrocarbon, carbon, water
2 a butane + oxygen $\longrightarrow$ water + carbon dioxide
  b pentane + oxygen $\longrightarrow$ water + carbon dioxide
3 The water is a vapour.
4 $2C_2H_6 + 7O_2 \longrightarrow 4CO_2 + 6H_2O$

Chapter 8 – Fuels and electricity

Practical support

Combustion reaction

Equipment and materials required: Funnel, two boiling tubes linked by delivery tubes, two clamp stands, candle, limewater, cobalt chloride paper, suction pump.

Details
Set up as shown in the Student Book. The pump will draw water vapour and carbon dioxide from the flame through the two boiling tubes. Draw students' attention to the cobalt chloride paper changing from blue to pink and the limewater becoming cloudy.

Safety: Handle the cobalt chloride as little as possible and wash hands after use. CLEAPSS Hazcard 18 Limewater – irritant. CLEAPSS Hazcard 25 Cobalt chloride – toxic.

Burning bubbles

Equipment and materials required: Delivery tube, clip valve for delivery tube, bung with pipe to connect to delivery tube, 500 ml drinks bottle with bottom cut off, clamp and stand, 30 cm rule, metre rule, splints, Bunsen burner, water, washing up liquid, glycerine.

Details
Put the bung into the neck of the bottle firmly, attaching the delivery tube to connect the bottle to a gas tap. Mount the bottle upside down in the clamp stand so gas will enter through the neck at the bottom. Fasten the clip valve to the tube to prevent backflow into the gas tap. Make up the bubble mixture with $75\,cm^3$ water, $25\,cm^3$ washing up liquid and $12\,cm^3$ glycerine. Pour the mixture into the upside down bottle. Bubble gas through the mixture. A column of methane bubbles will rise from the top of the bottle. This can be chopped off using the small ruler and ignited in mid-air as it floats to the ceiling. Braver teachers can wet their hands thoroughly and pick up some bubbles to be lit by a student (wearing goggles and using the metre ruler). Keep arms fully extended and away from students and the apparatus if attempting this.

Safety: Follow CLEAPSS supplementary risk assessment SRA 03. Wear eye protection throughout. Students should be one metre away from the apparatus. Ignite bubbles at arm's length, using a lit splint on the end of a metre rule. Light the splint from a Bunsen burner set up far from the 500ml bottle. CLEAPSS Hazcard 45A Methane – extremely flammable.

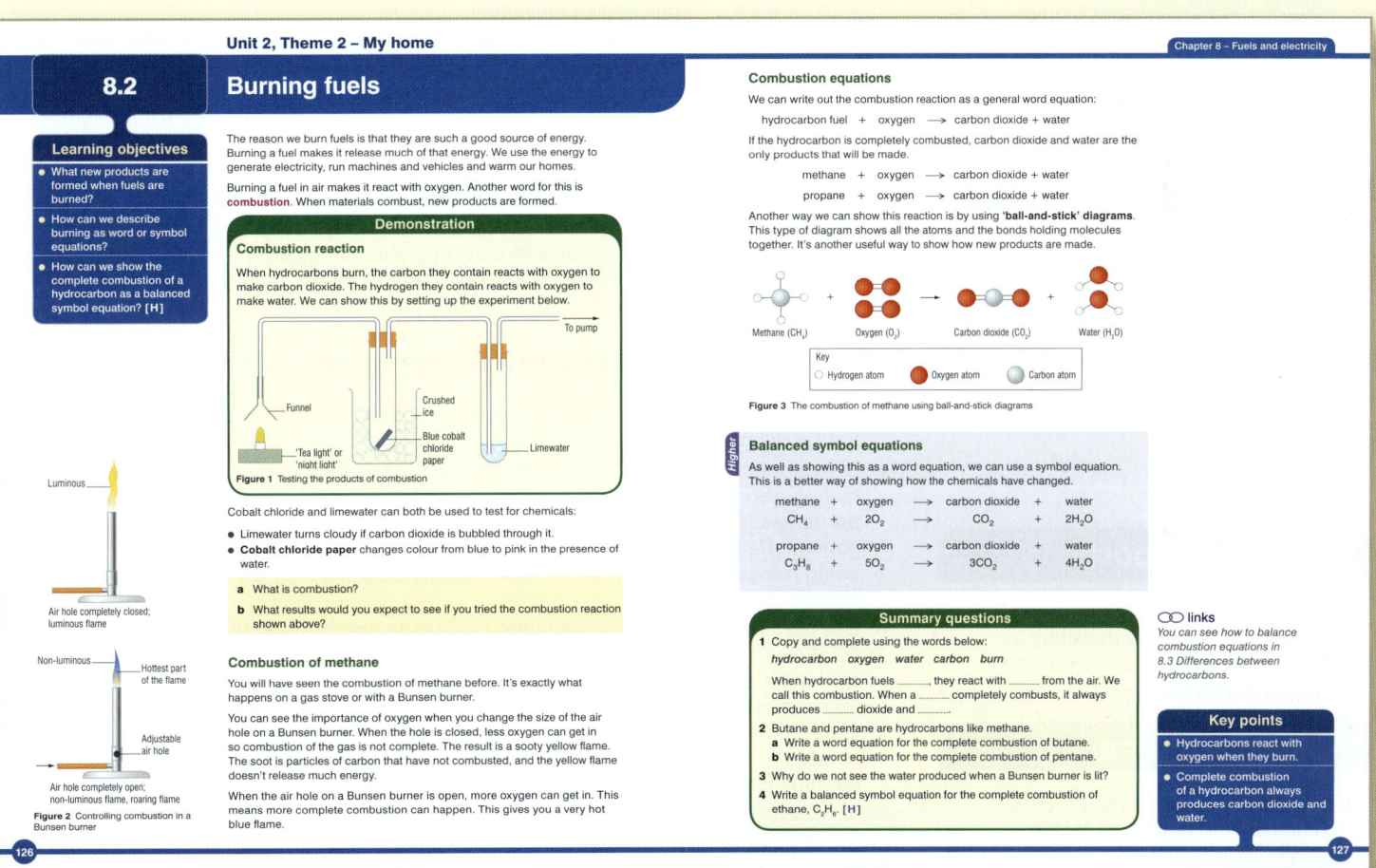

Burning candles

Equipment and materials required: Tea-light candle, large beaker, evaporating basin, glass trough.

Details
Fill the trough 3–4 cm deep with water. A few drops of food colouring can be added to the water to make the demonstration more visible. Put the candle in the evaporating basin and float on the water. Cover with the large beaker. After a few seconds (depending on the size of the beaker), the candle will go out and water will rise up the beaker from the trough.

Combusting other materials

Equipment and materials required: Bunsen burner, safety mat, tongs, steel wool, iron filings, magnesium ribbon, eye protection.

Details
Hold the steel wool over a blue flame with tongs and observe it burning. Observe how it changes in appearance after burning. Sprinkle iron filings over the flame, again observing them combust. Compare burned filings with unburned filings. Use tongs to ignite a 2 cm piece of magnesium ribbon (caution – extreme heat, do not look directly at the flame). Observe the white deposit of magnesium oxide.

Safety: Wear eye protection. Tilt Bunsen to prevent filings falling down the Bunsen tube. CLEAPSS Hazcard 59A Magnesium ribbon – flammable. CLEAPSS Hazcard 55A Iron filings.

Unit 2, Theme 2 – My home

8.3 Differences between hydrocarbons

Learning objectives

Students should learn:
- that hydrocarbons are chains of carbon atoms surrounded by hydrogen atoms
- that there are patterns in the combustion of hydrocarbons.

Learning outcomes

Most students should be able to:
- recognise the pattern in chemical formulae based on C_nH_{2n+2}
- explain patterns in the properties and behaviour of hydrocarbons.

Some students should also be able to:
- write balanced equations for the complete combustion of hydrocarbons. [HT only]

AQA Specification link-up: Science B 3.4.2.2

- Write balanced symbol equations for the complete combustion of hydrocarbons. [HT only]
- Explain patterns in the combustion of hydrocarbon fuels.

Within this context, candidates should be able to use scientific data and evidence to discuss, evaluate or suggest implications of the following:
- the energy content of different fuels

Lesson structure

Starters

Number sequences – Review the skills needed to work with C_nH_{2n+2} by presenting students with a number sequence for them to express in general terms. For example $x_n = 4n$ for the four times table. *(5 minutes)*

Match-up activity – Give students a range of symbol equations, some familiar and some unfamiliar. Challenge them to describe the reactions in words. To support students, provide them with the word equations to match up with the symbol equations. To extend students, provide them with word equations and challenge them to produce the symbol equations. *(10 minutes)*

Main

- Separating crude oil – to gain an appreciation of the structure and variety of the hydrocarbons in crude oil, students can separate molecules of 'model oil' into piles of useful molecules. (See 'Practical support'.) Less messy 'molecules' can be used, such as strips of sugar paper, but it reduces the impact.
- Use the piles of separated hydrocarbons to explore how properties vary according to chain length. Students can pour the different fractions to demonstrate patterns in viscosity.
- Use the model to explain attraction between chains making longer chain molecules harder to boil.
- Give students data sheets on the boiling points and heat of combustion for various hydrocarbons. They can use the data to plot line graphs of chain length against properties. Get students to suggest reasons for the patterns they see in energy content. Relate this pattern to energy being taken in when bonds are broken but more being released when new bonds are made as CO_2 and H_2O are formed. Longer chain equals more bonds.
- Use model kits or multicoloured card cut-outs to help support teaching how to balance equations. To reinforce the steps suggested in the Student Book, first ask the students to disassemble the structure of the hydrocarbon being combusted. They can then rebuild it as carbon dioxide and water using as much oxygen as they need, first starting with the carbon. Students can investigate the number of oxygen molecules needed to combust each hydrocarbon and research the hydrocarbon's heat of combustion to find patterns.

Plenaries

Rank – Give students data on the energy content of a range of similar fuels, such as diesel, kerosene and petrol. Given this information, students can make predictions about the relative size of chains in these fuels. *(5 minutes)*

Identify – Show students ball-and-stick representations of various hydrocarbons. Ask students to give their formulae and name the structures they recognise from the Student Book. To support students, give them photographs of different hydrocarbons (or mock-ups of successively darker liquids). Ask them to put the hydrocarbons into order of chain length, then ask them to identify the most volatile, easiest to burn, etc. *(10 minutes)*

Support

- Return to the demonstrations on the previous spread and use them as a way to reinforce that combustion makes new products.

Extend

- Higher tier students should work on balancing equations for the combustion of different alkanes. Learning them is not enough – they need to learn the principle behind balancing equations.
- Introduce the idea of systematic nomenclature. i.e. meth-, eth-, pro-, but-. Show the class diagrams or models of different alkanes and challenge them to work out the pattern in the formula. Once this has been established, practise predicting the formula of larger alkanes.
- To broaden students' understanding, introduce alcohols as hydrocarbons with –OH on the end. Given methanol as a starting point, students can name the first six (primary) alcohols.

Further teaching suggestions

Other oxidation reactions

Combustion is an oxidation reaction. Compare combustion with other oxidation reactions, such as rusting or respiration. What are the similarities and differences?

Practical support

Model oil

Equipment and materials required: Spaghetti (half a packet per group), newspaper, cooking oil, black food colouring.

Details

Prepare the model oil in advance of the lesson. Snap the spaghetti into different lengths, keeping to four or five different sizes. Boil the spaghetti as normal, dying it with the black food colouring while it cooks. Drain, then add a generous amount of cooking oil to prevent the spaghetti sticking together.

Groups are each given a quantity of model oil (oily black spaghetti strands of varying lengths) and compete to sort the strands into piles according to length. The activity provides a kinaesthetic introduction to the components of crude oil.

Protect surfaces with newspaper when working with the model oil, or do the activities in a plastic tray. As well as using this as an opportunity to compare different chain lengths, you can also introduce the idea of fractional distillation and cracking. Explaining that the smaller chains are more useful than the long ones, ask students how they could make their sample of oil more useful (i.e. chopping up the long pieces).

Unit 2, Theme 2 – My home

8.3 Differences between hydrocarbons

Learning objectives
- What is the structure of a hydrocarbon?
- How can we explain the patterns we see in the combustion of hydrocarbons?
- How can we balance combustion equations? [H]

Hydrocarbons are a very big family of chemicals. Luckily, they are also a very straightforward family. There are strict rules about the names they are given. Also, there are easy to remember patterns in their appearance and behaviour.

As you know, hydrocarbons are compounds made of hydrogen and carbon only. The carbon atoms are joined together in chains. The hydrogen atoms are joined to the outside of the chains, like this:

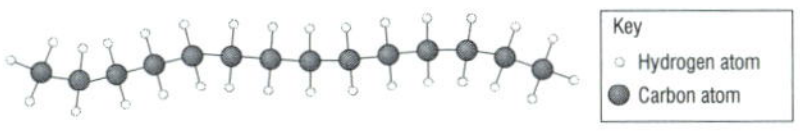

Figure 1 This hydrocarbon would have the formula $C_{14}H_{30}$

Chain length

Crude oil is a mixture of lots of carbon chains of different lengths.

a Draw the structure of a hydrocarbon with the formula C_5H_{12}.

Here are some of the shortest chain hydrocarbons. They are examples of **alkanes**:

Name	Number of carbons in chain	Formula	Structure	Energy released by burning
Methane	1	CH_4		802 kJ/mol
Ethane	2	C_2H_6		1437 kJ/mol
Propane	3	C_3H_8	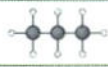	2044 kJ/mol
Butane	4	C_4H_{10}	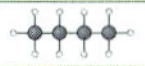	2659 kJ/mol

Look at the link between the number of carbon atoms and the number of hydrogen atoms. However many carbon atoms there are, there will be twice that plus two hydrogen atoms. This is true for all alkanes.

We can write this as a general formula: C_nH_{2n+2}

So for a hydrocarbon with 60 carbon atoms, there would be 122 hydrogen atoms (because n = 60, so (2 × 60) + 2 = 122).

There are patterns in how chains of different lengths appear and behave. This comes in handy for separating them when crude oil is processed.

Short chains	Long chains
Low boiling point	High boiling point
Catches fire easily	Doesn't catch fire easily
Thin and runny	Thick and viscous

b Which is easier to ignite, C_9H_{20} or $C_{20}H_{42}$?

There are also patterns in the combustion of hydrocarbons. The longer the chain, the more oxygen is needed. This means that larger hydrocarbons need a better air supply when they are burning. For example, a methane molecule needs two oxygen molecules during combustion, but a propane molecule needs five.

CH_4 + $2O_2$ → CO_2 + $2H_2O$
methane

C_3H_8 + $5O_2$ → $3CO_2$ + $4H_2O$
propane

Balancing combustion equations [Higher]

You need to be able to work out how much oxygen is needed, using the formula of the hydrocarbon. You also need to work out how much carbon dioxide and water will be produced. Here are some tips:

- There must be the same number of carbon, oxygen and hydrogen atoms on each side of the arrow: i.e. in the propane example, there are 3 × C atoms, 8 × H atoms and 10 × O atoms on each side.
- Balance the carbons first: i.e. C_3H_8 must make $3CO_2$ (3 × C atoms on each side).
- Balance the hydrogens next: i.e. C_3H_8 must make $4H_2O$ (8 × H atoms on each side).
- Calculate how much O_2 is now needed to balance the O in the products: i.e. there are 10 × O atoms in $3CO_2$ plus $4H_2O$, so $5O_2$ are needed to balance the equation.
- If you end up with a 0.5 O_2, just double everything so you have only whole numbers

Let's try butane ... C_4H_{10} + O_2 → CO_2 + H_2O

Balance the carbon: C_4H_{10} + O_2 → $4CO_2$ + H_2O

Now balance the hydrogen: C_4H_{10} + O_2 → $4CO_2$ + $5H_2O$

Calculate the oxygen: C_4H_{10} + $6.5O_2$ → $4CO_2$ + $5H_2O$

Double to remove the 0.5: $2C_4H_{10}$ + $13O_2$ → $8CO_2$ + $10H_2O$

Summary questions

1 What is the formula of an alkane with six carbon atoms?

2 Match up these formulae to the number of oxygen molecules they need to burn completely:
$C_{13}H_{28}$ $14O_2$
C_7H_{16} $20O_2$
C_9H_{20} $11O_2$

3 Complete and balance this equation for the complete combustion of pentane. [H]

$C_5H_{12} + O_2 \rightarrow$ .... + ...

4 **a** Using the information on p128, draw a line graph of the number of carbon atoms in a hydrocarbon and the amount of energy obtained from burning it.
b Use your graph to predict the amount of energy obtained from burning pentane (five carbons).

AQA Examiner's tip
Look at the amount of carbon dioxide released and the amount of oxygen needed for the combustion of different alkanes. Make sure that you will be able to describe the patterns in the exam.

links
For information on balancing equations look back at 2.9 Using equations.

Key points
- Alkane hydrocarbons have the general formula C_nH_{2n+2}
- Longer chains are darker, thicker and harder to boil or burn.
- The longer the chain, the more oxygen it needs when it combusts.

Summary answers

1 C_6H_{14}

2 $C_{13}H_{28}$ $20O_2$
C_7H_{16} $11O_2$
C_9H_{20} $14O_2$

3 $C_5H_{12} + 8O_2 \rightarrow 5CO_2 + 6H_2O$

4 **a**

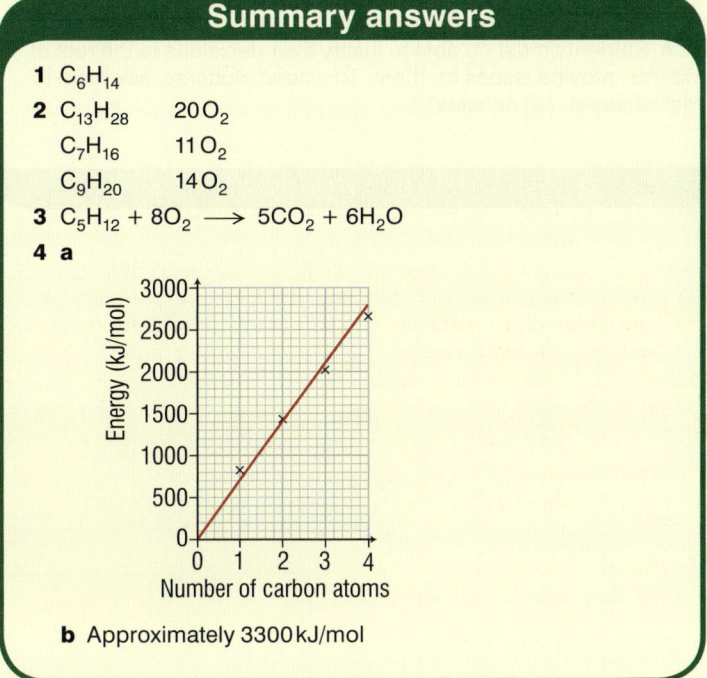

b Approximately 3300 kJ/mol

Answers to in-text questions

a

b C_9H_{20} is more flammable.

Unit 2, Theme 2 – My home

8.4 Problems with fossil fuels

Learning objectives

Students should learn:
- that fossil fuels are a finite resource; they are non-renewable
- that burning fossil fuels contributes to acid rain and global warming.

Learning outcomes

Most students should be able to:
- explain why fossil fuels are non-renewable
- identify problems associated with burning fossil fuels
- suggest the social and economic implications of the continued use of fossil fuels.

Some students should also be able to:
- explain how burning fossil fuels can contribute to acid rain and global warming.

AQA Specification link-up: Science B 3.4.2.2

- Explain some of the problems of burning fossil fuels (pollution, carbon dioxide production and global warming) and that resources of fossil fuels are finite.

Within this context, candidates should be able to use scientific data and evidence to discuss, evaluate or suggest implications of the following:
- the social, economic and environmental impacts of the uses of fossil fuels obtained from crude oil for cooking, heating and transport.

Lesson structure

Starters

Recognising gases – Write the chemical names for the gases in this spread on the board. Students need to give them their proper names. *(5 minutes)*

Assess prior knowledge – Start with the heading 'Problems with fossil fuels'. Students work in small groups to decide what to list underneath. This could be used to guide the flow of the lesson. Groups then feed back to the rest of the class. To support students, provide a list of issues, only some of which relate to fossil fuel use. *(10 minutes)*

Main

- This topic is rich in implications and provides a good opportunity for students to develop skills in identifying economic, social and environmental issues in a given context. Start by showing photographs of various scenes (traffic jam, floods, oil rig, oil disaster, acid rain destruction, production line, plastic toys – there are many possibilities). For each picture, students need to say how the picture affects them, the economy and the environment. Begin by considering a straightforward scene such as a picture of a bus or car. Work through the issues as a whole class. The aim here is to build up the speed with which students identify the issues. Different groups can then be given different pictures, which they later present to the class with their ideas.
- Get students to list items that rely on crude oil. Discuss what might happen when we run out of oil – what can we replace by other materials and what will we have to do without? Are there more sustainable materials to replace plastics with?
- Data analysis – students can visit www.statistics.gov.uk or www.decc.gov.uk and find a breakdown of the activities that produce greenhouse gases in the UK. Using other sources on the internet, students can compare this breakdown with other countries. Which countries have the highest emissions overall? Why might this be the case?

Plenaries

Slogans – Students can think of slogans for use in a public awareness campaign. Their topic can be either using less electricity or using fewer cars. *(5 minutes)*

Prioritising – Students can rank issues surrounding fossil fuel use (such as acid rain, oil slicks, soot) by importance. Students must be able to justify their decisions to the rest of the class. To support students, provide issues for them. To extend students, ask them to come up with their own list of issues. *(10 minutes)*

Support
- Revisit the names and symbols of the chemicals involved in this topic.

Extend
- Students can research global temperature fluctuations. What evidence is there that global warming is a myth?

Further teaching suggestions

Are cows worse than cars?

Methane is a much more potent greenhouse gas than carbon dioxide. Students can research the sources of methane around the world. This could lead to a debate about the environmental impact of switching to a vegetarian diet.

Answers to in-text questions

a Greenhouse gases absorb energy, stopping it being radiated out into space. This raises the temperature of the atmosphere. Changes in temperature can result in changes to weather patterns.

b nitrogen oxides and sulfur oxides

Unit 2, Theme 2 – My home

Chapter 8 – Fuels and electricity

8.4 Problems with fossil fuels

Learning objectives
- Why are fossil fuels non-renewable?
- What are the problems caused by burning fossil fuels?

Fossil fuels take millions of years to form. This means we can't replace the ones we are using now. We say they are **non-renewable**. This also means that there is only a limited amount of fossil fuel in the Earth's crust. Fossil fuel reserves will not last forever.

If we continue to use fossil fuels at our current rate and no new reserves are found, it is estimated that we will run out of oil and natural gas in the next 50 years.

Oil is not just a fuel

An extra problem is that crude oil is not just used for fuel. It is also used for making many important chemicals, including plastics and medicines. As we run out of crude oil, we will need to find alternative ways to make these. Think of the plastic products you use every day – crude oil has a huge impact on your life.

As well as being important for making things, oil creates work for thousands of people in the UK. The petrochemical industry generates billions of pounds in the UK every year.

links
For information about using crude oil to make plastics look back at 7.5 Polymers in the home.

Environmental effects of fossil fuels

Another problem with burning fuels is the pollution caused by the gases that are made. In Chapter 1 you saw that carbon dioxide traps energy inside our atmosphere. This is because it absorbs the energy the Earth radiates out into space.

On the one hand this is vital because it keeps the planet warm. On the other hand, too much greenhouse gas warms the Earth too much. This can cause climate change, resulting in more extreme weather systems all around the world.

a Why can greenhouse gases cause climate change and how does it affect us?

This table shows how much greenhouse gas the UK has produced between 1990 and 2009:

Year	1990	1995	2000	2005	2006	2007	2008	2009
Equivalent amount of CO_2 (millions of tonnes)	774	714	674	655	650	641	628	575

The **Kyoto Protocol** is an agreement between 37 countries (including the UK) to reduce greenhouse gas emissions. The UK has agreed to decrease its emissions to 92 per cent of 1990 levels by 2012.

Fossil fuels don't just produce carbon dioxide when they are burned. Other waste products from fuels can also be harmful to us and our environment.

- **Sulfur dioxide and nitrogen oxides** – produced by cars and power stations. Cause acid rain.
- **Soot** – tiny particles of carbon and unburnt hydrocarbons (also called particulates) made during incomplete combustion. Can cause breathing problems.
- **Carbon monoxide** – another product of incomplete combustion. Can cause brain damage or even death.

Maths skills

Percentages of greenhouse gas levels

Work out 92 per cent of the UK's 1990 greenhouse gas emissions. Is it a big decrease compared to current levels? Are we on track to meet it?

$774 \times \dfrac{92}{100} = 712$ million tonnes

The UK beat this target in 1996!

b Which atmospheric pollutants are caused by transport?

Finally, fossil fuels can cause even greater problems when things go wrong. In 2010, the Deepwater Horizon, a floating oil drilling rig in the Gulf of Mexico, exploded. Eleven people died in the explosion, which also blew a hole in the sea floor.

Nearly 10 million litres of oil gushed from the hole per day. This caused an environmental disaster. Many habitats and countless thousands of animals were destroyed.

Figure 1 The Deepwater Horizon disaster in 2010

Summary questions

1. Copy and complete using the words below:

   nitrogen spills carbon dioxide acid fossil greenhouse

   All fuels produce _____ when they burn. Some also produce sulfur dioxide and _____ oxides. These gases add to problems like _____ rain and the _____ effect. Using _____ fuels can also result in environmental disasters like oil _____.

2. **a** Use the table opposite of the UK's greenhouse gas emissions to draw a line graph.
   b Draw a dotted horizontal line at 92 per cent of 1990 levels to show the progress the UK is making.

3. Complete the following table using the information on these pages about crude oil.

Social impacts (how does this affect me?)	Economical impacts (how does this affect jobs, business and money?)	Environmental impacts (how does this affect our environment?)

links
For information about the greenhouse effect look back at 1.7 Maintaining our atmosphere.

Key points
- Fossil fuels are running out. They are non-renewable.
- Burning fossil fuels produces carbon dioxide (the main greenhouse gas), which can affect our climate.
- Fossil fuels can harm the environment in other ways, such as acid rain or oil spills.

Summary answers

1. carbon dioxide, nitrogen, acid, greenhouse, fossil, spills

2.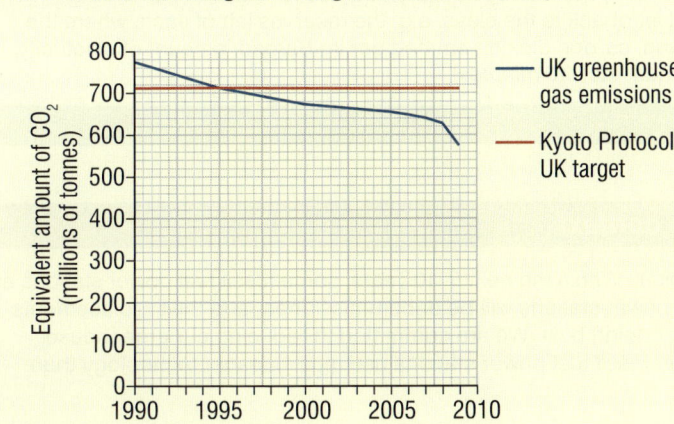

3.

Social impacts (how does this affect me?)	Economic impacts (how does this affect jobs, business and money?)	Environmental impacts (how does this affect our environment?)
I need crude oil for transport. I need crude oil for plastic products. I need crude oil for medicines.	The petrochemical industry provides thousands of jobs. The petrochemical industry is worth billions of pounds.	Burning crude oil can contribute to climate change. Burning fossil fuels can cause acid rain. Extracting and transporting crude oil can cause oil spills.

Unit 2, Theme 2 – My home

8.5 Generating electricity

Learning objectives

Students should learn:
- that fossil fuels release energy when burned
- how electricity is generated in fossil fuel power stations.

Learning outcomes

Most students will be able to:
- state that burning fossil fuels releases energy
- describe the stages in a fossil fuel power station.

Some students will also be able to:
- explain how the energy is transferred at each stage in a fossil fuel power station.

Specification link-up: Science B 3.4.2.3

- Know that fossil fuels (natural gas, oil and coal) release energy when they are burned, which can be used to generate electricity in our homes.
- Describe how electricity can be generated from fossil fuels …

Lesson structure

Starters

How fossil fuels are formed – Check students can recall how coal oil and gas were formed millions of years ago. Remind students that reserves will only be found in certain places where the geological conditions allowed the fields to develop. *(5 minutes)*

What do I know? – Students write down what they know as facts about fossil fuels and what their opinions are. Use their answers to identify misconceptions. Support students by giving them prompts, e.g. how and where are fossil fuels formed? Are they renewable? How long will supplies last? How are supplies used/transported? Extend students by asking them to suggest evidence that will back up their opinions. *(10 minutes)*

Main

- Demonstrate a steam generator: many school suppliers sell model engines that can be linked to flywheels or generators and possibly used to light an LED. Students should spot the similarities and differences between this and real power stations. Ask them to identify the energy changes taking place at each stage. Hint: to save time in the lesson, add freshly boiled water to the tank or the demonstration will take longer than expected.
- The e-on Energy World website has excellent interactive materials for this topic: www.eon-uk.com/EnergyExperience/557.htm discusses many aspects of different fuels. www.eon-uk.com/EnergyExperience/658.htm includes a turbine and generator activity. However, the whole website has excellent resources for use on a wide range of energy topics
- Students should draw diagrams for the processes occurring in the power stations. Extend students by asking them to identify stages where most energy is wasted. Some students will benefit from an unlabelled drawing to add labels to, or a card matching activity (matching what happens in each part of the power station or matching the energy change to the different stages).

Plenaries

Revisit the starter, what do I know? – Can students add more information? Have they changed their opinions? Have they more evidence to back up their opinions? Students should be supported by working as small groups. To extend students, ask them to suggest implications as well as identifying evidence for their opinion. *(5 minutes)*

Looking at the facts – Students use secondary sources to research some data about fossil fuels and feedback to the class, e.g. the reserves left of each; where the UK's reserves are found; carbon dioxide emitted per kg burned; amount of electricity generated per kg burned, etc. *(10 minutes)*

Support

- Provide a set of cards for students to sequence showing how a fossil fuel power station generates electricity. Also provide cards showing the energy changes for students to match to a diagram.

Extend

- It is very revealing to research the statistics about energy production worldwide – this is suggested as a brief plenary, but could be extended. For example, we import large quantities of gas and coal. What implications are there if Russia or the Middle East become unable or unwilling to supply us? How is gas stored? What is meant by 'security of supply'?

Science in context

This is extremely topical. Within the next 5–10 years, some fossil fuel power stations and many of our nuclear power stations will be decommissioned, and their replacements are not yet planned or being built. We are committed to reducing our greenhouse gas emissions. Newer fossil fuel power stations use much cleaner technology than previously.

Chapter 8 – Fuels and electricity

Further teaching suggestions

Demonstrate electromagnetic induction

Demonstrate electromagnetic induction. Check your equipment is sensitive enough before the demonstration. Use a transformer coil with as many turns as possible. (e.g. 1000 turns)

Connect a milliammeter to the coil – an ac milliammeter is best. Move the strongest magnet you have into and out of the coil. You should see the milliammeter flick one way when the magnet moves in, and the other way when it is pulled out of the coil. If you have a magnet with different poles on the flat faces of the magnet, a current will be generated when the magnet spins inside the coil.

Making an alternator

With a motivated group and some spare time, students could use a kit (e.g. Philip Harris Westminster kits) to make an alternator. Full instructions will be included with the kit. Students enjoy carrying this out, but think of this as a related, but optional extra.

Location, location, location

Use a map showing the location of different power stations in the UK to identify the one closest to your school.
- Shows coal-fired power stations; www.ukqaa.org.uk
- Downloadable data which includes a list of the power stations in the UK; www.decc.gov.uk

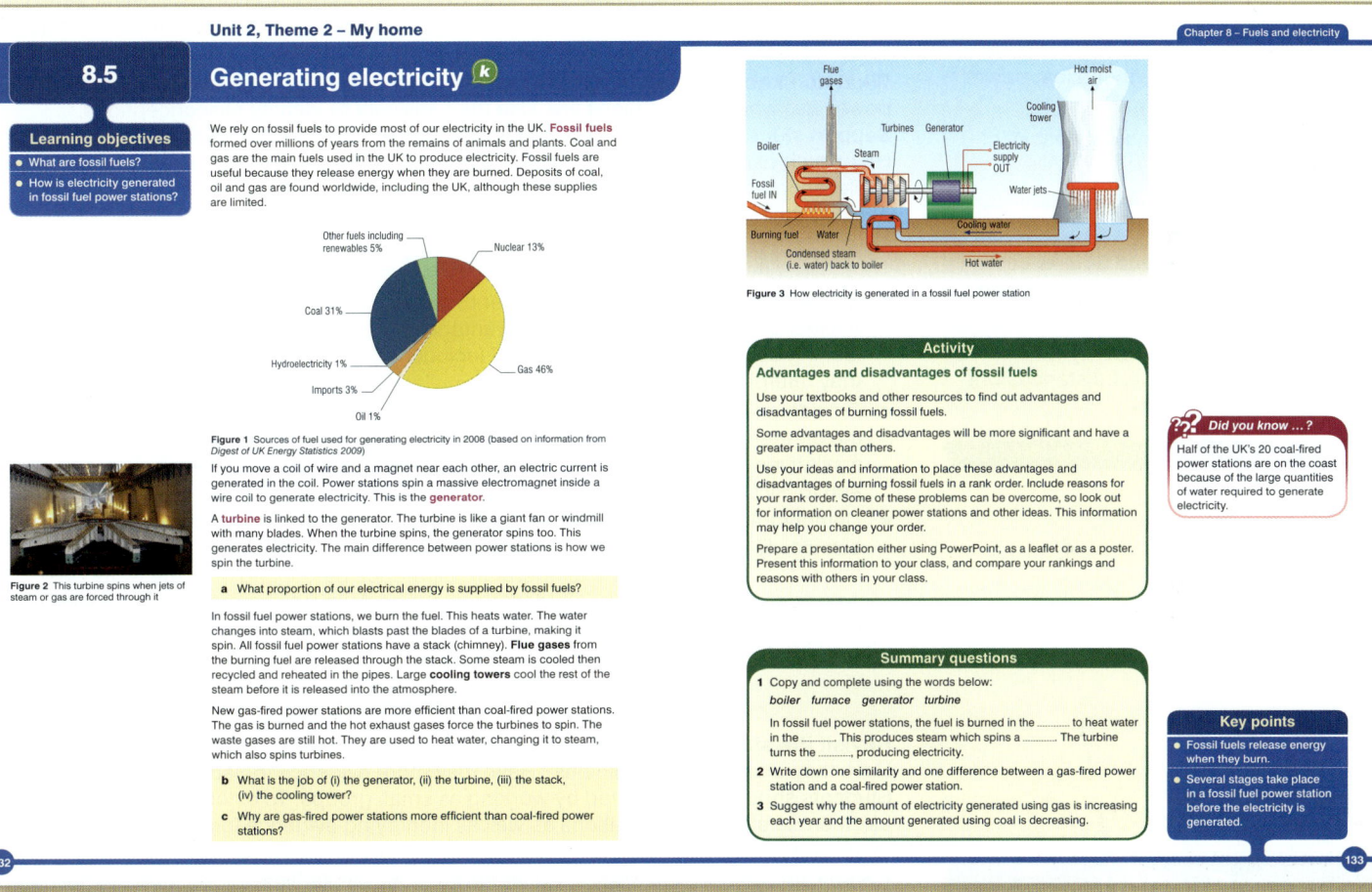

Summary answers

1 furnace, boiler, turbine, generator
2 Similarity: both heat water to change it to steam, which drives turbines; both burn fossil fuels.
  Difference: coal is less efficient/gas can be used in two ways to spin turbines.
3 Gas is more efficient to use so new power stations tend to be gas powered. Students may come up with sensible suggestions that may or may not true, but show an appreciation of the issues, e.g. we are producing less coal than gas; gas is cheaper than coal.

Answers to in-text questions

a 78%
b i Generator changes kinetic energy into electricity.
  ii Turbine forces the generator to spin.
  iii Stack releases pollutants from burning the fuel.
  iv Cooling towers release moist air from the surplus steam.
c Gas-fired power stations use the same gas to spin the turbines twice – once using the hot gases, when they are burnt and then using steam, created from the hot gases.

Unit 2, Theme 2 – My home

8.6 The nuclear alternative

Learning objectives

Students should learn:
- that nuclear fuels produce energy from nuclear fission
- how nuclear energy generates electricity
- some benefits and problems connected with the use of nuclear energy.

Learning outcomes

Most students should be able to:
- state that energy from nuclear fission generates electricity in nuclear power stations
- list some benefits and drawbacks of nuclear energy.

Some students should also be able to:
- explain similarities and differences between nuclear power stations and fossil fuel power stations
- explain why nuclear energy is beneficial in some circumstances but not in others.

AQA Specification link-up: Science B 3.4.2.3
- Explain how nuclear fuels … may be used as an alternative to fossil fuels.
- Know that nuclear fuels produce energy from nuclear fission.
- Explain the problems of using nuclear fuels (problems of radioactive emissions, disposal of waste) …

Lesson structure

Starters

Spot the difference – Provide data about the energy resources France and the UK use. Ask students to spot the differences. **France** – oil 1%; gas 4%; coal 5%; nuclear 78%; hydroelectricity 11%; other renewables 1%; **UK** – oil 1.5%; gas 35.5%; coal 38%; nuclear 19%; hydroelectricity 2%; other renewables 4%. *(5 minutes)*

What do I know, nuclear? – Students write down what they know as facts about nuclear power and what their opinions are. Use their answers to identify misconceptions. Support students by giving them some prompts, e.g. is there nuclear power in the UK? How much of our energy does it supply? Is it safe? Does it release carbon dioxide? Does it release radioactivity? Extend students by asking them to suggest evidence that will back up their opinions. *(10 minutes)*

Main

- Students can research nuclear power to prepare a PowerPoint presentation on nuclear energy, how it works, its advantages and disadvantages. Ensure students are looking out for reliable websites, and include facts rather than opinion. It is useful if they reference their sources too.
- Provide an unlabelled diagram of a nuclear power station. Ask students to label the different sections and explain what the role of each part is. Students should list similarities and differences between the different parts of the nuclear power station and fossil fuel power station.
- Demonstrate a simple heat exchanger. Arrange a rubber tube so that cold water can slowly trickle through it. Partially submerge the rubber tube in a beaker of hot water and measure the temperature difference before and after the water has passed through the tubing. Students can suggest ways to increase the temperature difference.
- Detailed knowledge of fission is not required.

Plenaries

Revisit the starter, what do I know, nuclear? – Can students add more information? Have they changed their opinions? Have they more evidence to back up their opinions? Students should be supported by working as small groups. To extend students, ask them to suggest implications as well as identifying evidence for their opinions. *(5 minutes)*

Comparing coal and nuclear – Ask students to list similarities and differences between the use of coal and nuclear power stations. Who can identify the most points? Some students may need support, e.g. type of fuel, effect on environment, treatment of waste, proportion used by the UK, what makes the turbines spin. *(10 minutes)*

Support
- Provide a set of cards for students to sequence showing how a nuclear power station generates electricity. Provide cards for a coal powered station so students can identify the similarities and differences more easily.

Extend
- It is very revealing to research the statistics about energy production worldwide, e.g. accident and fatality statistics.
- Investigate the most suitable materials used to construct nuclear power stations.

Science in context

Within the next 5–10 years, many of our nuclear power stations will be decommissioned, and their replacements are not yet planned. Either our electricity bills will rise dramatically, we may have to import electricity (which we already do to a certain extent), rely more on fossil fuels or have power cuts more frequently. Having pioneered nuclear power, many skilled workers required to build new power stations have gone as so little investment has been carried out in the UK.

Chapter 8 – Fuels and electricity

Further teaching suggestions

Nuclear fission reactions

Demonstrate chain reactions using dominoes or cassette tape boxes arranged so that knocking down each one will knock down two more (up to four layers). Highlight to students that knowledge about a chain fission reaction is not examined.

Simulations

Work through a simulation of nuclear power on the internet – there are several versions although you may need to register for these.

Unit 2, Theme 2 – My home

8.6 The nuclear alternative

Learning objectives
- What are nuclear fuels?
- How do nuclear fuels generate electricity?
- What are some benefits and problems connected with the use of nuclear energy?

Burning fossil fuels produces greenhouse gases. There are other ways to generate electricity without producing greenhouse gases. Nuclear power stations use **uranium** and **plutonium** as fuels. **Nuclear fission** reactions in the nuclear fuel release energy. Depending on the type of power station, a gas or liquid is piped through a heat exchanger. Here, water in a separate piped system changes to steam. The steam turns turbines, which spin generators, which generate the electricity.

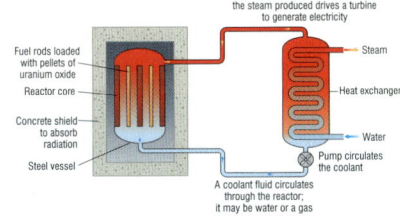

Figure 1 What happens in a nuclear power station

a Name **two** nuclear fuels.

Advantages of nuclear power

There are many advantages to using nuclear power. It is a reliable form of energy, and does not release greenhouse gases. There are currently plenty of supplies of nuclear fuels. Using nuclear power conserves our fossil fuel supplies.

Nuclear power is very safe. The only accident to cause deaths occurred nearly 25 years ago in Ukraine. Several hundred people died as a direct result of the accident. There was also a noticeable increase in thyroid cancers for several years afterwards. There have been close calls in USA and some minor accidents in the UK.

This compares with thousands of deaths per year while mining for coal in China alone, and over 170,000 deaths caused when the Banqiao hydroelectric dam in China collapsed in 1975.

Did you know …?
France generates 78 per cent of its electricity by nuclear power, producing about a third of the carbon dioxide per person per year compared to America.

Activity
Nuclear power
Prepare a leaflet explaining how a nuclear power station generates electricity.

b State **two** advantages of using nuclear power.

Disadvantages of nuclear power

There are some disadvantages to nuclear power. Many safety precautions are built into the design of nuclear power stations, so the cost of producing electricity is high.

It is very expensive to decommission nuclear power plants safely at the end of their working lives. Several of the UK's nuclear power plants will need to be decommissioned in the next ten years.

If there were to be a nuclear accident the economic and environmental damage would be enormous. It would take tens, if not hundreds, of years to recover from.

There are some radioactive emissions from nuclear power stations. However, emissions from coal-fired power stations are 100 times more radioactive.

The main drawback of nuclear reactors is the highly radioactive waste produced from the used fuel rods.

c State **two** disadvantages of using nuclear power.

Radioactive waste

Low-level radioactive waste is not dangerous to handle and is disposed of in landfill sites.

Medium-level waste is solidified in concrete or bitumen, then buried underground in special storage areas.

Highly radioactive waste has to be stored safely for years while the radioactivity dies away. Some can be recycled, but the rest is vitrified (changed into glass). After 50 years, the waste is about 1000 times less radioactive, but still needs to be safely stored for hundreds of years. Relatively small amounts of this highly radioactive waste are produced.

Figure 2 A nuclear fuel recycling plant at Sellafield

Figure 3 Nuclear waste is changed into glass (vitrified) and stored to stop radioactivity seeping into the surroundings

AQA Examiner's tip
Remember that nuclear fuel is non-renewable (like fossil fuels) but it does not give off any air pollution (unlike fossil fuels).

Summary questions

1 Copy and complete using the words below:
   rods   generators   exchanger   turbines   fission
   In nuclear power stations, nuclear _____ reactions cause fuel _____ to heat up. A heat _____ uses this energy to change water, in pipes, into steam. The steam spins _____, which turn _____ to produce electricity.
2 Write down one similarity and one difference between a nuclear power station and a fossil fuel power station.
3 Explain one benefit and one drawback of nuclear power on the environment.

Key points
- Nuclear power generates electricity using nuclear fission reactions.
- There are advantages and disadvantages in using nuclear power.
- Nuclear power is safe and does not produce greenhouse gases. It produces radioactive waste.

Summary answers

1 fission, rods, exchanger, turbines, generators
2 Similarity: e.g. both heat water to change it to steam, which drives turbines.
   Difference: e.g. fuel is coal/oil/gas in fossil fuel power stations, and plutonium/uranium in nuclear power stations.
3 Benefit: e.g. no greenhouse gases; fewer and smaller mines needed to access the fuel
   Drawbacks: some radioactive emissions; storage of radioactive waste

Answers to in-text questions

a uranium, plutonium
b reliable, does not release greenhouse gases, safe, produces ample supplies of fuel
c storage of radioactive waste, costs involved in setting up and decommissioning reactors

Unit 2, Theme 2 – My home

8.7 Renewable energy resources

Learning objectives

Students should learn:

- that renewable energy resources will not run out
- how different renewable energy resources generate electricity
- some benefits and problems connected with the use of renewable energy sources.

Learning outcomes

Most students should be able to:

- state that renewable energy resources will not run out
- describe the different types of renewable energy resource
- explain how renewable energy resources generate electricity.

Some students should also be able to:

- explain why renewable energy resources are beneficial in some circumstances but not in others.

Specification link-up: Science B 3.4.2.3

- Define the terms renewable and non-renewable in the context of energy resources.
- Explain how … renewable energy sources (wind, solar, hydroelectricity, wave, tidal, biomass and geothermal) may be used as alternatives to fossil fuels.
- Explain the possible problems of using … renewable energy resources (unreliability and possible effects on the environment).

Within this context, candidates should be able to use scientific data and evidence to discuss, evaluate or suggest implications of the following:

- the environmental impact over time of energy production by comparing the advantages and disadvantages of using alternative energy sources
- the economic impact of using alternative energy sources.

Support

- Write advantages and disadvantages of renewable energy sources on sheets of paper stuck round the classroom. In small groups, students use sticky notes to add the names of energy sources that they think these apply to. Discuss the answers with students as this exercise may throw up misconceptions.

Extension

- Students should consider why a mixture of renewable energy resources is best for the UK. They should be aware that energy can be used to generate electricity, but also for heating and for transport uses.
- They could research the statistics for different countries to explain why different energy sources are used in different places.
- Students could consider the capital costs involved and how long the payback time is for individual householders installing solar panels or wind turbines (payback is the years needed to recover the initial cost by savings on energy bills). Grants are available for some installations – students could consider why the government feels this is a useful move.

Lesson structure

Starters

Spot the renewable energy – Ask students if they have seen wind turbines/solar panels/solar cells/hydroelectric schemes/ground source heat pumps. Most students will have seen some of these either in real life or on the TV. (Remember that solar cells provide electricity from the Sun's light, but solar panels – often seen on roofs – absorb the Sun's heat to heat water directly. Ground source heat pumps use geothermal energy to heat water directly.) *(5 minutes)*

Energy in the news – Show a recent news clip about renewable energy (look for a clip in the last year of less than 5 minutes from the BBC or Sky news sites). Support students by asking them to consider how renewable energy impacts us in everyday life. Use this to explain why renewable energy is different from non-renewable energy, and how the energy source works. Extend students by asking them to consider problems that may be caused or solved by using the energy source described. They should also consider if the report is biased. *(10 minutes)*

Main

- Provide maps of your local area. Help students to identify the location of main towns, hills, rivers and other features. Based on your local geography, ask students to suggest what type of renewable energy would be most suitable for your area. Local weather data for the past couple of years would also be useful.
- This activity could be extended worldwide. Provide a data sheet stating the main source of renewable energy for European countries. Use maps or atlases to consider how the local geography contributes to the choice of renewable energy source.
- Build up a collection of news stories concerning our use of electricity and renewable energy sources (either from newspapers or online). Provide groups with a news story and ask them to consider how this relates to the features, advantages and disadvantages of renewable energy. Examples include the granting of licences for offshore wind farms.
- Students can prepare a PowerPoint presentation or poster on one renewable energy source and how it works. They should consider the environmental impact over time of energy production by comparing the advantages and disadvantages of using alternative energy sources.
- Students present their PowerPoint presentations to each other, perhaps in small groups so that all the energy sources are covered. Students should discuss a wider range of issues and be able to evaluate the more appropriate uses for each method.
- This website has particularly good notes, video clips and images. Students can use these to prepare notes or presentations on renewable sources of energy: www.darvill.clara.net/altenerg/.
- This website has excellent materials for all key stages: www.eon-uk.com/EnergyExperience/. There are activities, animations, information and quizzes. Before preparing your lesson, look though to find the most suitable activities.

Chapter 8 – Fuels and electricity

Plenaries

Where in the world? – Put up a large map of the UK or the world or Europe. Identify regions likely to have high use of hydroelectricity (e.g. Scandinavia, Scotland), solar power (e.g. equatorial regions), wind power (coastal regions), geothermal (e.g. Iceland). *(5 minutes)*

Match the resources – Students should match the renewable resource with the advantages and disadvantages of that source. To support students, provide a set of cards with the names of different renewable resources and some of the features of different resources. Students group cards together matching different features. To extend students, they put pairs of energy sources together and list at least one thing each pair of energy sources has in common. This can be repeated putting the resources into groups of three. List at least one thing each group of energy sources has in common. *(10 minutes)*

Science in context

The UK has strict targets to increase its production of electricity from renewable sources. Students will become increasingly familiar with wind farms and there have been some plans for the River Severn to have a tidal barrage. They will also see solar panels on house roofs – these heat water directly so reducing heating bills.

Unit 2, Theme 2 – My home

8.7 Renewable energy resources

Learning objectives
- Will renewable energy resources run out?
- How do different renewable energy resources generate electricity?
- What are the benefits and problems connected with the use of renewable energy sources?

We are surrounded by natural resources that can be used to generate electricity. As we have become more dependent on electricity, we need resources that will not run out – these are **renewable energy** resources.

Electricity is generated when a generator connected to a turbine spins. When wind, water or steam flows over blades connected to a turbine, electricity is generated. The energy source is free for most renewable energy sources. However, the structures built to generate the electricity can be very expensive for the amount of electricity generated.

a What is the difference between a renewable energy source and a non-renewable energy resource?

Burning biomass
Biomass power stations generate electricity when wood, poultry litter and straw are burned. The energy released changes water to steam which spins the turbines, generating electricity. One disadvantage is that carbon dioxide (a greenhouse gas) is emitted from the power stations. However, if plant matter is burned, the plants have absorbed carbon dioxide as they grew. The energy source is reliable and a single power station can generate large amounts of electricity. The cost of a power station is similar to the cost of a fossil-fuel power station.

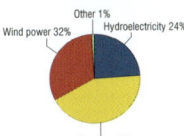

Figure 1 How our renewable energy was provided in 2008

Wind farms
Wind farms can be found near the coast, on hills or offshore. They use a free source of energy that is non-polluting. Large blades on wind turbines spin when it is windy generating electricity. To improve efficiency, turbines can be up to 100 m tall.

However, these turbines cannot produce energy if it is calm or very stormy. The noise they cause can be very disturbing and they are often very visible against the natural landscape.

Figure 2 The UK is planning to build more wind farms offshore

Hydroelectric power
Hydroelectric schemes use dams to trap water in reservoirs in mountainous areas. The water is released when needed through pipes in the dam wall. Turbines in the pipes spin as the water travels through them.

Hydroelectric dams produce electricity very quickly when needed to cope with surges in demand. In quiet periods, some hydroelectric schemes pump the water back up into the reservoir so it is ready for the next time it is needed. Hydroelectric schemes are reliable and non-polluting, but can only be sited in mountainous areas.

The biggest disadvantage with hydroelectric power is that, to trap water, local habitats have to be flooded. This often means people have to move from surrounding towns and villages.

Figure 3 Valleys in mountainous regions are flooded when hydroelectric schemes are developed

Tidal power
Tidal schemes use large barriers across the mouths of rivers. These are called tidal barrages. They force seawater through pipes when the tide goes in or out. Turbines in these pipes spin when the water is flowing, and electricity can be generated for about 40 per cent of the time. The schemes are reliable and can generate large amounts of electricity.

However, they are expensive to build, and there are only a limited number of places in the world where the schemes would work. The river flow and habitat around the estuary are damaged by flooding.

Geothermal power
Geothermal schemes are most successful in volcanic regions. Shafts are sunk very deep into the Earth's surface. Cold water is pumped down these deep shafts and there is enough heat transferred from hot rocks to change the water to steam. It escapes up another shaft to spin turbines in a power station. In the UK, some geothermal energy is used to heat water rather than produce steam.

Wave power
At the moment, there are no large-scale electricity supplies powered by waves, although the technology is developing. Some sites in Scotland have been identified for the first commercial-scale developments. These should be completed by 2020.

Although wave power is non-polluting, it is unpredictable and it has proved hard to harness the energy from the sea.

Solar power
Solar cells generate small amounts of electricity directly from sunlight. The cells are expensive for the amount of electricity generated. However, solar cells are useful in devices like calculators, which only use a small amount of electricity. They are also useful on satellites or rural bus stops, which are remote from other sources of energy. The cells can be useful in very sunny climates but have limited uses in the UK.

b Which types of renewable energy can generate electricity at any time when required?

c Which forms of renewable energy can generate a lot of electricity in the UK?

Figure 4 The world's first tidal scheme was in France. Only a few schemes have been built worldwide.

Did you know ...?
There has been a geothermal energy scheme in the UK since the 1980s. In Southampton, enough energy has been retained underground for useful amounts to be extracted and used to heat houses, shops, hotels and leisure centres. Electricity from the scheme will be used in Southampton port.

Figure 5 The Pelamis generator will start producing commercial wave energy in Scotland by 2020

Summary questions
1 Copy and complete using the words below:
   biomass geothermal hydroelectric solar tidal waves wind
   a These types of renewable energy can only be used in certain places: .... and ....
   b These forms of renewable energy can only be used at certain times: .... and ....
   c These types of energy depend on the weather: .... and ....
2 Write down one advantage and one disadvantage for each renewable energy source.

Key points
- Renewable energy sources will not run out.
- The main renewable energy sources are biomass, wind, hydroelectricity, tidal, geothermal, solar and wave power.
- There are advantages and disadvantages in these schemes.
- Renewable energy sources use free resources and do not produce greenhouse gases. They have an impact on the environment.

Further teaching suggestions

Solar cells
Solar cells kits can be used to investigate electricity produced by solar cells.

Windmill investigation
Falling water and hairdryers can be used to force toy windmills to spin. Students investigate how many blades are most effective. Use a glass tube pushed through the centre of a cork to allow the cork to freely spin on a supporting spindle. Cut slots into the cork and slip different numbers and shapes of plastic blades into the slots before using water or a hairdryer to spin the cork.

Answers to in-text questions
a Non-renewable sources run out if used, they cannot be replaced in a short time scale.
Renewable resources will not run out or can be regenerated in a short time scale.
b hydroelectricity, biomass, geothermal
c biomass, hydroelectricity, wind

Summary answers
1 **a** geothermal, hydroelectric (tidal)
  b tidal, solar (wind)
  c wind, solar

2 Biomass is reliable but creates greenhouse gases.
  The fuel source is free and non-polluting for solar/wind/hydroelectricity/waves/geothermal and tidal.
  Solar energy cannot be used at night.
  Wind power cannot be used if it is calm.
  Geothermal, tidal and hydroelectric can only be used in certain places.
  Wave power is too unreliable.

Unit 2, Theme 2 – My home

8.8 The National Grid

Learning objectives

Students should learn:
- how electricity gets to our homes
- how the National Grid might affect our health
- how the National Grid damages the environment.

Learning outcomes

Most students should be able to:
- explain the role of different parts of the National Grid
- describe some of the advantages and disadvantages of high voltage cables.

Some students should also be able to:
- explain in detail why electricity is transferred at very high voltages
- evaluate the evidence that there may be risks associated with the use of high voltage cables.

AQA Specification link-up: Science B 3.4.2.3

- Describe how electricity is distributed through the National Grid via high voltage cables.

Within this context, candidates should be able to use scientific data and evidence to discuss, evaluate or suggest implications of the following:
- environmental and health concerns arising from the distribution of electricity by pylons and high voltage cables.

Lesson structure

Starters

The impact of the National Grid – Show students pictures of the local area with parts of the National Grid visible. Students should identify any visual and environmental impact of the National Grid, e.g. that pylons and cables are visible because the surrounding vegetation has been cleared. *(5 minutes)*

What causes electric fields? – Students identify equipment in the room that they think has associated electric or magnetic fields. This includes computer monitors, mobile phones, wi-fi systems, etc. Support students by demonstrating the magnetic fields by showing their effect on a sensitive compass either when equipment is turned on or off. Extend students by asking them how they could tell if there was an associated magnetic field, and if they have some idea of which field would be the strongest. *(10 minutes)*

Main

- Use photographs and local knowledge to make students familiar with the different components of the National Grid – see if you can find out where your nearest power station is.
- In small groups, students research the impact of high voltage cables on communities in more detail. Some points to investigate are: are the clusters of children with leukaemia only found by high voltage cables? If not, other factors may be important. Roughly how many people are involved in any cluster? With small numbers the importance of the cluster may be overstated. Which organisations carried out the research and are they likely to be biased? Do any features of the studies make them more or less reliable? They should be aware that different sources may have a hidden agenda (e.g. a journalist and an Electricity Board representative will have different priorities).
- This can be extended to role-play with students playing the roles of concerned resident, scientist who has researched the phenomena and representative from the Electricity company. Alternatively, students could take part in a debate.
- Discuss how likely it is that the magnetic and electric fields actually cause an effect on people, with a brief analogy. If a lamp is turned on in a brightly lit room, it does not cause extra stress on our eyes – equate this to the fact that the magnetic fields from cables are about 1 per cent of the Earth's magnetic field. However, the magnetic fields in the cables constantly change due to the way that the electric current is generated, but the Earth's field is constant. Turning on a different coloured lamp may cause extra stress – even if it is a small percentage of our total load.

Plenaries

The route of electricity – Students list the components of the National Grid that the electricity flows through, working back from the consumer unit in homes to the power stations. Support students by giving them a list of components parts of the National Grid and ask them to order these. *(5 minutes)*

Where are our power stations? – Show a picture with the locations of the different power stations. Discuss why the power stations are distributed as they are (e.g. near a large supply of water, near population centres, near transport routes). Ask students what they think the most important factors are when locating power stations (e.g. to minimise the transport of electricity around the country by being close to a population centre, to minimise the transport of raw materials such as coal, to be close to raw materials). You will find that the choice of location will usually be due to a combination of several factors. *(10 minutes)*

Support
- Use a card matching exercise to link different components of the National Grid with their functions.
- Use images showing components of the National Grid in sequence, from the consumer box in the home to the power stations.
- Use magnets and iron filings or compasses to demonstrate what is meant by a magnetic field.

Extend
- Use a set of plotting compasses surrounding a coil of wire that forms part of a circuit to demonstrate the magnetic fields associated with electric currents.
- Discuss whether an effect linked with a factor means that the factor causes the effect (e.g. do electric lights cause the night to be dark? Does watching too much TV cause obesity?).

Science in context

People at work are often exposed to electric and magnetic fields. Less time is normally spent at work than at home, but the levels of exposure can be higher. This is true in many offices where equipment is more concentrated. There are no legally set limits on exposure so workers are not monitored. In the electricity generation and transmission industry, levels of exposure are much higher than normal so the companies set time limits to try and reduce the doses that their workers receive.

Did you know …?

Birds sitting on high voltage cables do not get electrocuted because there has to be a voltage difference (p.d.) for a current to flow. If both feet are on the same wire, there is no p.d. Problems come if someone stands on an earthed object and touches the cable, or something connects the cable to Earth by a conductor. Then there is a large p.d. between 0 V at the Earth's surface and the high voltage of the cable.

Unit 2, Theme 2 – My home

8.8 The National Grid

Learning objectives
- How does electricity get to our homes?
- Does the National Grid affect our health?
- Does the National Grid damage the environment?

The **National Grid** is used to deliver electricity throughout the country. **Pylons** are large metal towers that support the cables which carry electricity from **power stations** to homes and factories. Power stations supply the electricity at several thousand volts. Then **step-up transformers** increase this to even higher voltages.

Increasing the voltage lowers the current. A lower current means that less energy is wasted by the cables getting hot although the same power is supplied. This makes the National Grid more efficient with less energy transfer from overhead cables. **Sub-stations** near homes contain **step-down transformers** to reduce the voltage to a safer level, 230 V.

a Why is electricity transferred in the National Grid at high voltages?

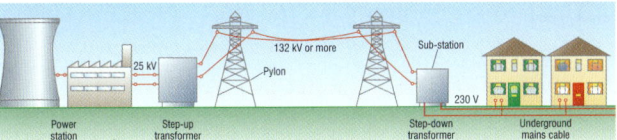

Figure 1 The National Grid

Is there a health risk?

Cancers in children may have many causes, such as a mother having X-rays or smoking during early pregnancy. Two separate studies have given us a cause for concern, although there is no proven link.

Children living within 100 m of high voltage cables were compared with other children. The scientists found that the children living near the cables were nearly twice as likely to suffer from leukaemia as children living further away (leukaemia is a cancer of the blood).

These studies compared thousands of children with cancer born over more than 25 years in two countries. In some cases, the cancers occurred in groups of people living near each other (a cluster). This made experts wonder if the cause was a factory, or pollution in the surrounding area.

b Explain why many people trust these studies.

Electric and magnetic fields

Electric and magnetic fields are places where electric or magnetic forces can be detected. These fields surround high-voltage cables as well as electrical equipment. They are weaker further from the equipment, and when the current or voltages are lower. Experts do not believe the high voltages from the cables are causing the cancers because:

- Electric fields from the high-voltage cables cannot pass through buildings.
- Magnetic fields due to high-voltage cables are about 100 times weaker than the Earth's magnetic field.

Figure 2 People living close to high-voltage cables may worry about the effects of electric and magnetic fields

- Experiments using electric and magnetic fields at similar levels did not cause changes to living cells.

c What are electric and magnetic fields?

Activity
Cable health risk
Write a leaflet for **one** of the situations below:
1 From an electricity company explaining to local residents why high voltage cables are not a risk to health.
2 From a pressure group of parents of children with leukaemia explaining why they think that high-voltage cables are a risk to health.

What are the environmental problems with maintaining the National Grid?

Pylons supporting high-voltage cables can be 50 m high and so stand out against the natural landscape. Some are painted to try to merge with the sky. The area surrounding the cables is kept free from trees and other structures. This is so the cables can be maintained safely and are less likely to be damaged.

In some conditions, power cables hum or crackle, which can disturb people living nearby. The high-voltage cables and pylons can be damaged by wind, ice and lightning. Damaged cables can cause power cuts as well as fire or electrocution.

d List **three** environmental effects caused by the National Grid.

Activity
Underground or overhead?
Discuss the advantages and disadvantages of burying electric cables underground. Decide what research is needed, and how you will present your findings. Keep a record of each website you use and decide how reliable it is. In small groups, pool your information and present it to the class.

Summary questions
1 Copy and complete using the words below:
   increase lines pylons reduce
   The National Grid is a network of .... and high voltage power .... Step-up transformers .... the voltage before it is carried by the power lines. Step-down transformers .... the voltage before it is supplied to homes.
2 Explain why the evidence does not prove that living near high voltage cables causes cancer.

Did you know …?
Our homes are full of equipment that has electric and magnetic fields. TV screens, computers, fluorescent lamps and microwave ovens all have electric and magnetic fields. Cordless phones have one of the strongest fields of anything in our homes. However, there is no proof that these fields cause any ill effects.

Figure 3 Trees near power cables are cut back to prevent damage and power cuts during storms

Key points
- Electricity is supplied through the National Grid.
- The National Grid includes power stations, sub-stations, pylons and cables.
- Electromagnetic fields surround high voltage cables.
- Some health risks, and damage to the environment, are increased near high voltage cables.

Summary answers

1 pylons, lines, increase, reduce

2 There are other possible causes that haven't been ruled out; the fields are too weak to cause measurable changes in cells.

3 Electric fields do not pass through buildings and magnetic fields are much smaller than the Earth's field; no discernable effect has been found when exposing cells to the different fields.

Answers to in-text questions

a To reduce energy losses.
b Results were repeated, carried out over a long time period and involved many people.
c places where electric and magnetic forces can be detected
d visual pollution, noise pollution, damage to the surrounding vegetation

Unit 2, Theme 2 – My home

Summary answers

1. **a** Petrol engines run more smoothly and quietly.
   b It won't freeze at high altitude.
2. Carbon dioxide (will turn limewater cloudy); water (will turn blue cobalt chloride pink).
3. **a** $C_{12}H_{26}$
   b Higher
   c Because it is a larger molecule and needs more energy to turn into a gas.
4. **a** pentane + oxygen $\longrightarrow$ carbon dioxide + water
   b $C_5H_{12} + 8O_2 \longrightarrow 5CO_2 + 6H_2O$
5. **a** Carbon dioxide – greenhouse gas – climate change
   Sulfur dioxide – acid rain
   Nitrogen oxides – acid rain
   Particulates – asthma
   Carbon monoxide – toxic
   b They have slowly decreased.
6. **a** The furnace is where the coal is crushed and burned.
   b The boiler is where water is heated and turned into steam.
   c The turbine is turned by jets of steam, capturing their kinetic energy.
   d The generator changes kinetic energy into electrical energy.
7. Similarities: e.g. both use heat to change water into steam; both use steam to turn turbines;
   Differences: e.g. coal-fired power stations burn coal but nuclear power stations use nuclear fuels (uranium and plutonium); coal-fired power stations have a furnace but nuclear power stations have a reactor, etc.
8. **a** The wind turns the blades that are connected to the generator. As the generator spins, electricity is generated.
   b Advantage e.g. renewable source of energy, non-polluting.
   Disadvantage e.g. no electricity generated on calm days, noise pollution, takes up large areas of land.
9. - power station
   - step up transformer
   - power cable
   - step down transformer
   - sub-station
   - house
10. Geothermal – C
    Hydroelectric – D
    Wave power – B
    Wind power – A

Kerboodle resources

- Practical: Burning fuels (8.2)
- Data handling skills: Comparing the energy content of various fuels (8.3)
- Extension: Cracking science (8.3)
- Interactive activity: Electricity generation (8.5)
- Revision podcast: Fuels
- Examination-style questions
- Answers to examination-style questions
- Test yourself: My home

Unit 2, Theme 2 – My home

Summary questions

1. Fuels are chosen because they have the right properties for a particular task.
   a Explain why many cars use petrol instead of diesel, even though diesel is more efficient.
   b Explain why kerosene is used as a fuel in jet planes.
2. When natural gas (methane) burns, what products are made? How could you prove it?
3. Dodecane is a hydrocarbon with 12 carbon atoms.
   a What is its chemical formula?
   b Would dodecane have a higher or lower boiling point than hexane (six carbon atoms)?
   c Explain the reasoning of your answer to part b.
4. Pentane is a hydrocarbon with the formula C_5H_{12}.
   a Write a word equation for the combustion of pentane.
   b Write a balanced symbol equation showing its complete combustion in oxygen. [H]
5. Our society relies heavily on fossil fuels for energy.
   a Fossil fuels produce pollutants when they burn. List five of these pollutants and explain the harm they cause.
   b How have the UK's greenhouse gas emissions changed over the past 20 years?
6. Explain what the role is for these parts of a coal-fired power station.
   a furnace
   b boiler
   c turbine
   d generator
7. Write down two differences and two similarities between a coal-fired power station and a nuclear power station.
8. **a** Explain how a wind turbine generates electricity.
   b Write down one advantage and one disadvantage of using wind turbines to generate electricity.
9. Put these parts of the National Grid in the order that electricity passes through them from a power station to the home.
   - step-down transformer
   - step-up transformer
   - power cable
   - power station
   - house
   - sub-station

10. Match the diagram of the energy source A–D to its name

Energy source	Diagram
Geothermal	
Hydroelectric	
Wave power	 A
Wind power	 B

Practical suggestions

Practicals	AQA	k	📖
Demonstrate fractional distillation of crude oil using CLEAPPS mixture (take care to avoid confusion with the continuous process in a fractionating column).		✓	✓
Test oil fractions for viscosity, ease of ignition and sootiness of flame.		✓	✓
Compare the energy content of different fuels – for example, by heating a fixed volume of water.	✓	✓	
Test the products of combustion of fuels to show that carbon dioxide is produced.		✓	✓
Investigate the effect of changing different variables on the output of solar cells (e.g. distance from the light source, the use of different coloured filters and the area of the solar cells).		✓	
Demonstrate a model water turbine linked to a generator.	✓		✓
Model the National Grid.		✓	

Chapter 8 – Fuels and electricity

Examination-style questions

There are many different ways to produce energy by burning fuels.

a Give examples of **two** renewable fuels. (1)
b Name **two** fossil fuels. (1)
c Name a fuel that can be used without burning it. (1)
d Complete a balanced symbol equation for the combustion of pentane (C_5H_{12}). [H] (4)

Match each type of energy source up with the correct disadvantage.

Energy source	Disadvantage
Fossil fuel	Waste stays dangerous for a long time
Nuclear fuel	Can be a problem distributing the electricity
Tidal	Stops fish swimming up river to spawn
Wave power	Produces greenhouse gases

(3)

The diagram shows the generation of electricity in a power station.

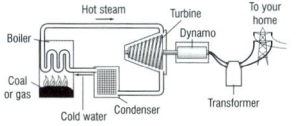

a Use the diagram to help you describe how to generate electricity in a power station. (4)
b The national grid uses cables to distribute electricity. The electricity that passes through these cables is at a high voltage. Why? (3)

A caravan owner did an experiment to find out which make of camping gas was the best value for money.

Make of gas	Cost in £	Time it lasts in days
Campagas	28	20
Gas-on-the-go	27	18
Porta-gas	30	24
Travelgas	19	15

a Calculate the cost per day for Campagas. (2)
b Explain which gas was the best value for money. (2)
c Choose **one** thing that the caravan owner could do to make the experiment more accurate.

Only do the experiment at the weekends	Only buy one kind of gas	Repeat the experiment	Time how long the gas lasts in seconds

(1)

5 Read this article from a local newspaper, then answer the questions.

Uproar over new pylons

Electricity pylons are used to distribute electricity all over the world yet in one Yorkshire town there have been mass demonstrations against a new line of them across the dales. Dr David Winter, spokesperson for Greenenergy said 'if everyone used alternative ways of generating their own electricity, then we would not have this problem'. Dr Winter himself has a solar panel and a wind turbine on the roof of his house. Another alternative to using pylons is underground cables. These are high voltage cables that are buried deep underground. However, underground cables sometimes get really hot which can cause damage to them. This is addressed by pumping water or oil along the cable to cool it.

a What are **two** advantages of having underground cables? (2)
b Explain why using alternative ways of generating electricity would be much better for the environment. (2)
c *In this question you will be assessed on using good English, organising information clearly and using specialist terms where appropriate.*

Suggest why people like Dr Winter have chosen to install solar panels and wind turbines on the roofs of their houses but other people do not. (6)

AQA Examination-style answers

1 a Any **two** from: biomass, wood, biodiesel, (allow ethanol) (2 marks)
b Any **two** from: coal, gas, peat, oil (crude) (2 marks)
c nuclear (1 mark)
d $C_5H_{12} + 8O_2 \rightarrow 5CO_2 + 6H_2O$
(1 mark for O_2)
(1 mark for CO_2)
(1 mark for H_2O)
(1 mark for the correct balancing)

2 Fossil fuel — Produces greenhouse gases
Nuclear fuel — Waste stays dangerous for a long time
Tidal — Stops fish swimming up river to spawn
Wave power — Can be a problem distributing the electricity
(All correct = 3 marks, 3 correct = 2 marks, 1 or 2 correct = 1 mark)

3 a Burning fuel produces heat. (1 mark)
This causes water to turn to steam. (1 mark)
The steam turns the turbine. (1 mark)
The turbine spins the generator. (1 mark)
b So current can be kept low. (1 mark)
So less heat / energy is lost. (1 mark)
So more electricity can get through. (1 mark)

4 a $28 \div 20$
= £1.40 (2 marks)
b Porta-gas, becaues it is the least cost per day. (2 marks)
c Repeat the experiment. (1 mark)

5 a Any **two** from:
Less of an eyesore
Do not take up space
Less dangerous to play / fish around (2 marks)
b No greenhouse gases produced
So no global warming
OR
No sulfur dioxide / nitrogen oxides produced
So no acid rain (2 marks)
c Marks awarded for this answer will be determined by the Quality of Written Communication (QWC) as well as the standard of the scientific response.

There is a clear, balanced and detailed suggestion as to why people like Dr Winter have chosen to install solar panels and wind turbines on the roofs of their houses but other people do not. The answer is well structured with minimal repetition or irrelevant points. There is an accurate, fluent and clear expression of ideas with only minor errors in the use of technical terms, spelling, punctuation and grammar. (5–6 marks)

There is some suggestion as to why people like Dr Winter have chosen to install solar panels and wind turbines on the roofs of their houses but other people do not, with some omissions. The answer shows some attempt at structuring and the ideas are expressed with reasonable fluency and clarity. There are some errors in the use of technical terms, spelling, punctuation and grammar. (3–4 marks)

There is a brief suggestion as to why people like Dr Winter have chosen to install solar panels and wind turbines on the roofs of their houses but other people do not. The answer is largely incomplete and may contain some valid points which are not clearly structured. It lacks fluency and/or clarity. It contains errors in the use of technical terms, spelling, punctuation and grammar. (1–2 marks)

No relevant content (0 marks)

Examples of points made in the response:
- Neither solar panels nor wind turbines release any polluting gases
- Cheaper as they are not paying for electricity from the National Grid
- Can be paid for putting electricity back into the National Grid
- Eyesore
- Initial outlay of cost is expensive
- Solar panels heat up water for heating and washing
- Solar panels do not work at night
- Wind turbines do not work when there is no wind
- So on a still night there will be no electricity or hot water.

Unit 2, Theme 3

My property

AQA Specification link-up: Science B 3.4.3

We use a range of electrical appliances in our homes on a daily basis and as a result many families have to budget carefully to cover the cost of electricity bills. Many appliances work using electromagnetic waves. Electrical and electronic engineers work to ensure that our appliances are efficient and that they are safe to use.

In this theme there are two contexts:

3.4.3.1 The cost of running appliances in the home

3.4.3.2 Electromagnetic waves in the home

Unit 2, Theme 3
My property

In Unit 2, Theme 3 you will work in the following contexts, covered in Chapter 9:

The cost of running appliances in the home

How can we find out how much our electricity bill is?
Many people are sent an electricity bill every three months. The bill uses readings from our electricity meter to calculate how much we owe. We can also work out which equipment adds most to the bill.

How quickly do we use electricity?
We are often told to cut down on the electricity we use to reduce greenhouse gas emissions. Different equipment doing the same job can use very different amounts of electricity. How can we compare the electricity used by different appliances so we can make the correct choices?

How can we choose the most efficient equipment?
Making clever choices in the shop saves pounds on running costs. All equipment transfers some of the electrical energy it uses into wasted forms. Buying energy-efficient equipment means we waste less of the electricity that we buy.

Why are electrical appliances so useful?
Electricity is useful because it can can flow through wires to different places. Appliances at home transfer electrical energy to many different forms of energy. Find out how we can show these energy transfers using diagrams and how these can be used to compare different devices.

Electromagnetic waves in the home

How do we use electromagnetic waves in our lives?
Electromagnetic waves transfer energy at the speed of light. We use different types of waves for communicating, cooking and medical uses.

Are electromagnetic waves harmful?
Different electromagnetic waves affect cells in our body in different ways. Used properly they are useful, in other situations they can be harmful. How can we use different electromagnetic waves while limiting any harm from them?

The cost of running appliances in the home

How can we find out how much our electricity bill is?

Activity:
- Students all use electrical equipment and this unit will help them to consider how their bill is prepared. It will also allow students to make decisions about how they use equipment so they can reduce the electricity they use.

How quickly do we use electricity?

Activity:
- Students can estimate the cost of the electricity they used in the past week.
- Students can rank the equipment they think used most electricity during that week.
- Provide students with a sample breakdown based on known prices, actual readings, known power ratings of equipment used.
- Compare answers, explaining the differences.

How do we choose the most efficient equipment?

Activity:
- Use an Argos catalogue or similar data source for students to compare the power of different models of specific appliances.
- Students should consider reasons why different models are produced (e.g. catering for different sized households).
- Prepare a daily bill based on a household using a sample of the most efficient and the least efficient equipment.

Possible misconceptions

Students do not usually have a reasonable idea of the costs of electricity. Asking students to read their electricity meter daily for a week is a useful exercise, particularly if they can suggest reasons why their readings may vary from day to day. Students could identify sources of wasted electricity in the school and suggest solutions.

Be aware that some students will use a prepayment meter.

Remind students that there are several factors affecting the bill, for example, the equipment used, and how we use it (e.g. if it is left on).

Electromagnetic waves in the home

Activity:
- Introduce the different members of the electromagnetic (EM) spectrum. Students could consider the types of radiation that the Sun emits (ultraviolet, visible and infrared are the most obvious forms).
- Students can list all the devices they have used that make use of this type of radiation.
- Students could group different types of radiation, e.g. by typical use (communication, medical) or by effect on health (cause burns; cause cancer), etc.
- In groups students discuss:
  – why electrical appliances are so useful
  – how we use EM waves in everyday life
  – Are electromagnetic waves useful or dangerous?

Possible misconceptions

Students may believe that all radiation is harmful. Remind them that the type, the intensity and time of exposure all play a part in the effect that radiation has on cells.

Students may not realise that cancers are caused over many years and that low-level exposure is less likely to cause harm than higher doses. Some people are more susceptible to harm, e.g. genetically, or because other factors also have an effect.

Part of my job is helping customers choose the most suitable equipment for their needs. It's often the most efficient equipment too as this will save money on their electricity bills. We can use efficiency ratings to save customers' money.

My job as an electronics engineer means that I design and install communications networks using electromagnetic waves daily. Our networks use fibre optic cables and phone lines as well as wireless connections. We may set up temporary systems to cover events such as the Olympic Games, or permanent systems linking households and businesses.

Some students may have friends or relatives suffering from cancer so be sensitive to this.

It is hard to imagine what EM waves are as they are not seen directly as waves, but are detected by their effect on different things. Ask students to describe how they visualise EM waves. This may help you identify some misconceptions. Remind students that EM waves travel at the speed of light only in a vacuum – the speed of these waves is less in gases, liquids and solids.

Remind students that visible (white) light is a continuous spectrum of different colours, and that the spectrum of all EM waves is continuous.

Unit 2, Theme 3 – My property

9.1 Energy

Learning objectives

Students should learn:
- that energy is measured in J and kJ
- how to draw and interpret Sankey diagrams.

Learning outcomes

Students should be able to:
- state the units of energy
- interpret simple Sankey diagrams
- draw their own Sankey diagrams.

Some students should also be able to:
- interpret more complicated Sankey diagrams.

AQA Specification link-up: Science B 3.4.3.1
- Know that energy is normally measured in joules …
- Draw and interpret Sankey diagrams that show the types of energy transferred by an electrical appliance.

Lesson structure

Starters

What is energy? – Students write down what they think energy is, using as many scientific terms as possible. *(5 minutes)*

What 'uses' the most energy? – Students write a list of activities ranked in order of most energetic first. They should include reasoning for their rank ordering. Support students by giving them an activity, e.g. putting a book on a shelf. Ask them to suggest a similar activity using more energy (e.g. putting two books on the shelf). *(10 minutes.)*

Main

- Discuss what is meant by energy and review the different forms of energy.
- Sketch a Sankey diagram on the board. Ask students to help you fill in details as you work through it.
- Provide students with squared paper and rulers and ask them to draw simple Sankey diagrams – be rigorous about drawing them correctly.
- An internet search engine will bring up sites with more Sankey diagrams to discuss. Search for: Sankey diagram GCSE.

Plenaries

What have I learned? – Students write down three key points they have learned together with one question they still have not understood. They hand these in for you to judge what to revisit next lesson. *(5 minutes)*

My MP3 player's Sankey diagram – Students sketch a Sankey diagram for their MP3 player, estimating the types and amounts of energy. Support students by asking them to identify the energy transfers and you sketch the Sankey diagram on the board, asking them for help completing the labels. *(10 minutes)*

Support

- Provide Sankey diagrams for students to label. Ask them to label either the forms, or the amounts of energy first and then to match these to complete the remaining information.

Extend

- Be aware that this work leads onto calculations of efficiency in 9.4 Efficiency. Students can carry out calculations of multistage energy transfers, e.g. a battery powered torch:

    chemical energy $\longrightarrow$ heating the surroundings + electrical energy

    electrical energy $\longrightarrow$ heating the surroundings + light energy

Further teaching suggestions

Model steam engine

Set up a model steam engine (scientific suppliers have models). Ask students to identify the different stages and draw descriptive Sankey diagrams for each stage. Combine this into one large Sankey diagram.

How many electrical devices do you use?

The idea of why electrical energy is so convenient can be reinforced by asking students to list the number of electrical devices they have used that day.

Science in context

This is extremely topical. We are very dependent on electricity now as it is clean (at the point of use) and convenient to use. Students really only realise our dependence on electrical energy during power cuts.

Chapter 9 – Using energy and radiation

Practical support

Energy transfers

Equipment and materials required: A selection of electrical equipment, e.g. hairdryer, lamp, radio, kettle.

Details
The input energy for each piece of equipment is electrical. Students turn on the equipment and list all the forms of output energy that are present. Draw their attention to wasted forms such as sound and heat.

Safety: Demonstrate the kettle. Water in the kettle can cause scalds; students may be scalded by steam. Be aware that students may plug in power cables and the ends of these can cause electric shocks if they are not plugged in to the equipment. Keep working area dry.

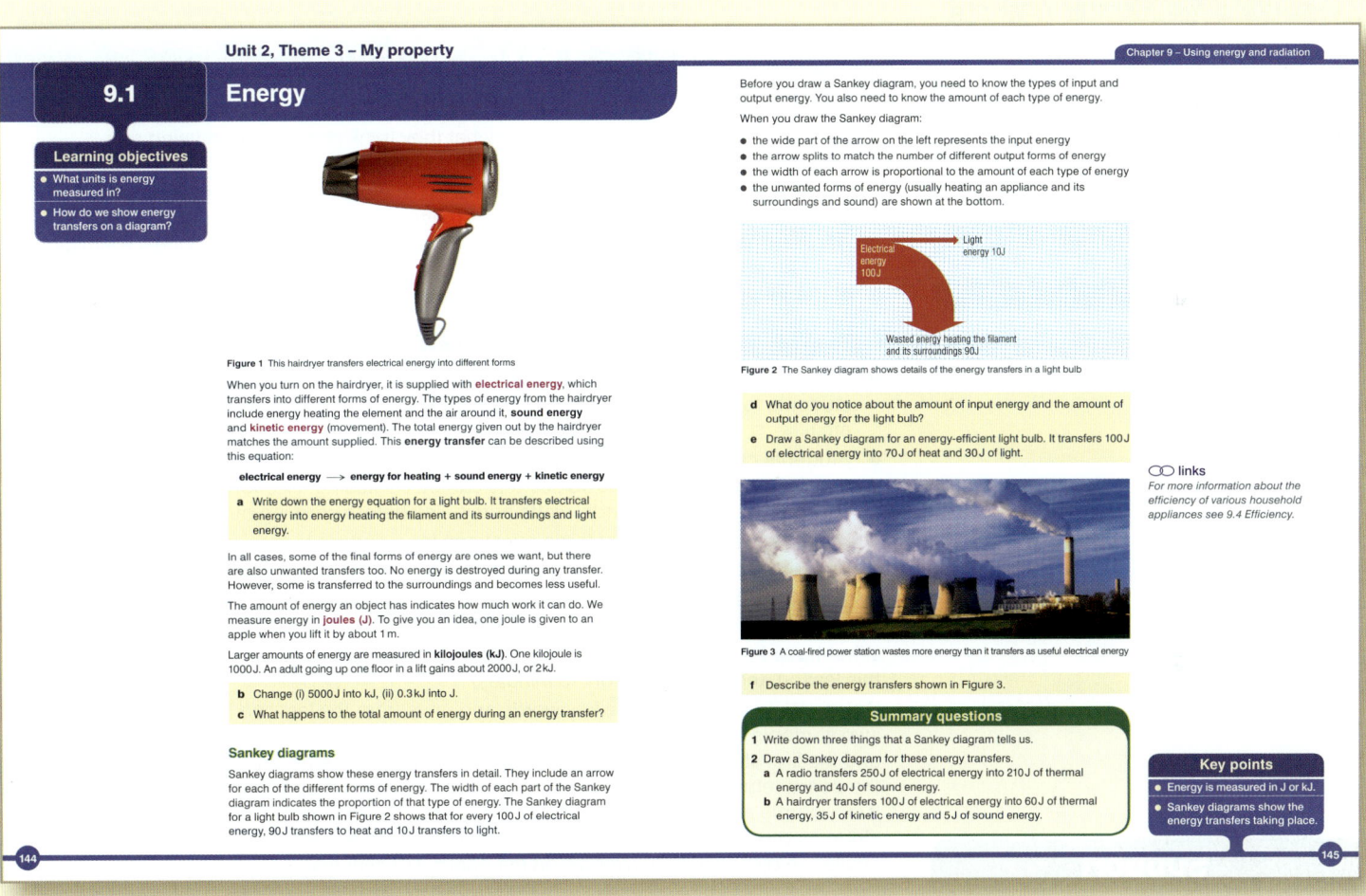

Summary answers

1 For example: types of energy involved, amount of each type of energy involved, proportion of each type of output energy.

2 a Similar shape to Figure 2 (Student Book) but arrow 25 units wide labelled 'electrical energy to start', becoming two arrows: one is 21 units wide labelled 'thermal energy' (210J) and the other is 4 units wide labelled 'sound energy' (40J).

b Similar shape to Figure 2 (Student Book) but arrow 10 units wide labelled 'electrical energy to start', becoming three arrows: the bottom one is 6 units wide labelled 'thermal energy' (60J), the middle one is 3.5 units wide labelled 'kinetic energy' (35J) and the top one is 0.5 units wide labelled 'sound energy' (5J).

Answers to in-text questions

a electrical ⟶ thermal + light
b i 5kJ   ii 300J
c The total amount of energy stays the same.
d total input energy = total output energy
e Similar shaped diagram, but the top arrow is wider (3 units wide) and the bottom arrow is narrower (7 units wide). The labels read light energy 30J and thermal energy 70J.

Unit 2, Theme 3 – My property

9.2 Electrical power

Learning objectives

Students should learn:
- what is meant by power
- that power is measured in W and kW
- how to calculate power.

Learning outcomes

Students should be able to:
- state that power = energy transferred per unit time
- state the units of power
- calculate power.

Some students should also be able to:
- rearrange equations to calculate power, potential difference, current, time or energy.

Specification link-up: Science B 3.4.3.1

- Calculate the power consumed by an electrical appliance using the formula:
  Power (watts) = potential difference (volts) × current (amps).
- Carry out simple calculations for different electrical appliances in the home using the formula:
  Power (kilowatt, kW/watts, W) = energy transferred (kilowatt-hour, kWh/joules) ÷ time (hours – h/seconds – s).
- Interpret information from energy labels on appliances and know why this is useful.

Lesson structure

Starters

What is power? – Students write down what they think power is, identifying any similarities or differences between power and energy. *(5 minutes)*

What uses most power? – Students write a list of electrical equipment ranked in order of most powerful first. They should include reasoning for their rank ordering. Support students by giving them an example, e.g. a ceiling light. Ask them to suggest a similar piece of equipment that is more powerful (e.g. a search light) or less powerful (e.g. a torch) – this does not need to be electrical equipment at this stage. *(10 minutes)*

Main

- Discuss what is meant by power and hand round a selection of equipment for students to examine the information on the base (e.g. hairdryer, bulbs, radio, kettle, etc.)
- Students carry out calculations based on one equation at a time – they will eventually use three equations for power, so this can be confusing. Reinforce the importance of converting units to seconds, watts and joules before carrying out the calculation.
- Students write out a set of three questions for a partner to try.
- Students could carry out the practical on calculating power (see 'Practical support').

Plenaries

What have I learned? – Students prepare a simple mind map using topics covered in this lesson and the last lesson. *(5 minutes)*

Splat – Write about 12 of the key words and units from the last two lessons on the board. Two students stand by the board. One student reads out a definition of one of the words, and the first student to 'splat' the correct word stays at the board. The other student at the board returns to their seat, and the questioner takes their place for the next round. It is a simple and engaging activity that involves students of all abilities. To support students provide them with a list of key definitions. To extend students increase the challenge by only using symbols on the board. *(10 minutes)*

Support

- Help students convert time into seconds – practise this with them.
- Help students to calculate W into kW, and kW into W – practise this with them.
- If students make reasonable progress, use a triangle to lay out the power equation to calculate values as required.

Extend

- Remind students to rearrange equations carefully – give them plenty of examples calculating different quantities.
- Give plenty of calculations with different units to practise converting these correctly.

Science in context

This is very much an everyday life issue – students will be aware, for example, that some equipment uses more energy than other equipment and that it is important to turn equipment off when they leave a room. This should help them realise that equipment for cooking and heating tends to be more powerful than equipment for lighting, sound and computing.

Further teaching suggestions

Use a plug-in energy consumption meter. Plug different devices into the monitor to demonstrate the power consumption of different devices. You can also plug an appliance in at the start of the lesson and monitor how much energy it has used by the end.

Practical support

Calculating the power

Equipment and materials required: Power pack or battery, connecting leads, voltmeter, ammeter, components such as bulb, motor.

Details

Connect the powerpack, ammeter and one component in series. Connect the voltmeter in parallel across the components. When the circuit is switched on, measure the voltage and current. Calculate the power for each component in turn.

Safety: Ensure that the voltage in each circuit does not exceed the power rating of different components.

Unit 2, Theme 3 – My property

9.2 Electrical power

Learning objectives
- What is power?
- What units is power measured in?
- How do we calculate power

Figure 1 These bulbs have different power ratings, and use electricity at different rates

The equipment in our homes transfers energy at different rates. An electric heater transfers energy more quickly than a lamp. We say the electric heater is more powerful than the lamp. **Power** measures how quickly energy is transferred. It is measured in **watts** (W) or kilowatts (kW), where 1 kW = 1000 W. If the power of a device is 1 W, then it transfers one joule of energy per second.

Some equipment like a tumble dryer is very powerful, but only used for short periods of time. Equipment like freezers are less powerful, but are left on all the time. The energy transferred by the freezer in a year can be higher than the energy transferred by the tumble dryer. This means you could save more money than you might think on electricity bills if you choose a freezer with a low energy rating.

Maths skills

Measuring power
power (W) = energy (J) ÷ time (s)

Worked example
A lamp transfers 1200 J in 20 seconds. What is its power?
Its power is 1200 ÷ 20 = 60 W.

Figure 2 This information tells you the power of this device

a What is the power of a heater that transfers 6 kJ in 6 seconds?
b How much energy does it transfer in 1 minute (60 seconds)?

Maths skills

Calculating electrical power
The power of electrical equipment can be calculated using:

power (W) = potential difference (V) × current (A)

For example, if a computer mouse charger is rated at 3.3 V and 0.3 A, what is its power?
Its power is 3.3 × 0.3 = 0.99 W

Electrical equipment usually has information about its energy usage printed on its base. The information is also printed on bulbs. It states the power rating, voltage and/or current used by the equipment when it is turned on.

The **potential difference** is often called the voltage. In the UK, our electricity is supplied at 230 V. In Europe and America, electricity is supplied at a different voltage so the equipment settings may need to be changed if an appliance is used abroad.

c Why doesn't equipment always use the same amount of power in America compared with the UK?
d When a kettle is plugged into the mains supply (230 V), its current is 10 A. What is the power of this kettle?

Practical

Calculating the power
Use a joulemeter to measure how much energy is transferred in one minute by low-voltage bulbs of different powers. Calculate the power of each bulb using your results and compare this with the theoretical amount. Repeat this with a motor or immersion heater.

Did you know...?
Electrical cookers use more power than most electrical devices in the home. Often they use their own circuit containing a 30 A fuse.

Summary questions

1 Complete these sentences:
   a Power measures how quickly .... is transferred.
   b A kilowatt is one .... watts.
   c Electricity is supplied at 230 .... in the UK.
2 Work out the power in each of these examples.
   a A radio transfers 250 J in 2 minutes.
   b A hairdryer uses a current of 9 A when it is plugged into the mains supply in the UK.
3 How much energy is transferred when a 100 W bulb is left on for 1 hour?

Key points
- Power is the rate at which energy is transferred.
- Power is measured in W or kW.
- Power is calculated using:
  power = energy ÷ time, or
  power = voltage × current.

Summary answers

1 a energy   b thousand   c volts
2 a 250 ÷ 120 = 2.1 W
   b 230 × 9 = 2070 W (2.07 kW)
3 100 × 60 × 60 = 360 000 J (or 360 kJ)

Answers to in-text questions

a 6000/6 = 1 kW (1000 W)
b 60 s × 1 kW = 60 kJ
c The UK power supply is at a different voltage to the American supply.
d power = 230 × 10 = 2300 W

Unit 2, Theme 3 – My property

9.3 Buying electricity

Learning objectives

Students should learn:

- that a unit of electricity is a kilowatt-hour
- how to calculate the units of electricity used
- how to calculate the cost of electricity used
- about the implications of running home appliances.

Learning outcomes

Students should be able to:

- understand that electricity is sold in units (kWh)
- calculate the electricity used when given in appropriate units
- calculate the cost of electricity sold
- identify some consequences of running electrical appliances
- understand that one unit is a kilowatt-hour
- convert units and joules
- calculate the electricity used.

Some students should also be able to:

- evaluate the implications of costs of running home appliances.

Specification link-up: Science B 3.4.3.1

- Interpret the readings taken from a domestic electricity meter and know that a unit of electricity = 1 kWh.
- Calculate the costs of using different electrical appliances using:
  Total cost = number of kilowatt-hours × cost per kilowatt-hour.

Within this context, candidates should be able to use scientific data and evidence to discuss, evaluate or suggest implications of:

- the costs of running home appliances.

Lesson structure

Starters

What's on standby? – Students write down the items they left on standby (equipment left on standby is on a low power mode, not fully on but still consuming energy) at home today. *(5 minutes)*

Look at the bill – Hand round two different electricity bills either from the same house for different seasons, or from different houses for the same time period. Students to write down at least five pieces of information that they can tell about the different usage from the two bills (e.g. if the cost per unit changed, which household used most electricity, etc). Support students by only asking them to look at one bill and find out answers to specific questions, e.g. cost per unit, the season of the bill, the units bought, etc. *(10 minutes)*

Main

- Practise converting W to kW (divide by 1000), and seconds to hours (divide by 3600).
- Introduce the unit kilowatt-hour – explain that joules are too small a unit of energy to be useful for everyday calculations.
- Use information from the first starter to introduce the idea that leaving items on standby costs money – take a typical example such as a TV. An internet search will give you data on standby powers.
- Explain that you can monitor your usage by checking the electricity meter. Readings are usually taken three months apart, and the difference between the two readings is the number of units (kWh) used. Electricity bills will show both sets of readings. Practise calculating how many units have been used and their cost from two sets of readings – students can find this surprisingly hard to grasp.
- Link the calculations to the electricity bills used in the starter: you could hand out a mock bill with blank sections for students to write in new readings and costs for the next bill that one of these families receives.
- Discuss the implications of using wasteful equipment or using equipment in a wasteful way (higher electricity bills and ultimately more pollution). However, only about a third of our electricity in the UK is used domestically, and the UK contributes about 2% of the world's carbon dioxide so individual action will have little direct impact.

Plenaries

Turn it off – Students think of a slogan to use in school to encourage people to turn off items left on unnecessarily. *(5 minutes)*

My bill yesterday night – Students list the items they were directly responsible for using last night (e.g. between 5.00 pm and 11.00 pm) and approximately how long these were on for. Provide a list of approximate powers for these items in kW, so students can calculate the amount they added to the electricity bill. *(10 minutes)*

Support

- Students will struggle to convert units. Provide all data in hours and kilowatts for initial calculations.
- Students should try converting watts to kilowatts (divide by 1000).
- Help them recognise times such as 30 minutes as 0.5 hours.
- Remind them that answers for costs are likely to be in pence, not pounds.

Extend

- The challenge for this topic comes in familiarity with different units: students should be given a lot of practise converting W and kW, hours and seconds.
- Link ideas in more complex questions. Use the preceding spreads to help with calculations of power using current/voltage readings, then calculating total costs involved.

Science in context

This is extremely relevant; students may hear comments at home about the cost of electricity and be aware of their electricity bills. Each home has an electricity meter that many students are likely to have access to if required.

Chapter 9 – Using energy and radiation

> ## Further teaching suggestions
>
> ### Read your own electricity meter
> Students may be able to read their own electricity meters. It can be interesting to do this daily for a week or two, especially if the weather is different throughout that period.
>
> ### Electricity bills
> Bring in samples of real electricity bills – it can be useful to compare the different charges from different companies.

Unit 2, Theme 3 – My property

9.3 Buying electricity

Learning objectives
- What is the unit of electricity?
- How do we calculate the Units of electricity used?
- How do we calculate the cost of electricity used?
- What are the implications of running home appliances?

Figure 1 This electricity meter shows the Units used since it was installed

AQA Examiner's tip
Remember to include the correct units with your answer.

All homes have an electricity meter to measure the amount of electricity used. Readings are taken or estimated four times a year. It is inconvenient to charge customers for electricity in joules. One joule is a tiny amount of energy. We are billed in Units instead. 1 Unit = 1 kilowatt-hour (kWh). It is the energy used if something with the power of 1 kW (1000 W) is switched on for 1 hour, or if a 100 W light is on for 10 hours.

Units (kWh) = power (kW) × time (h)

a How many Units are used if a 2 kW electric fire is on for 5 hours?

The extra number of Units used since the last bill is used to calculate the bill. The number of Units used is the difference between this reading and the last reading. Different companies charge different amounts per Unit used. The cost of the electricity is calculated using:

cost of electricity = Units used × price per Unit
(pence)   (kWh)   (pence per kWh)

The total bill usually includes a quarterly charge. This has to be paid whether you use electricity or not.

b An electricity meter reads 34578. Three months ago it read 33235. How many Units have been used?

c If each Unit costs 14p, what is the cost of electricity used?

Figure 2 This electricity bill gives information on electricity usage and costs

We can all make changes to reduce the size of our electricity bills. There are three main ways this can happen:

- We can use equipment more efficiently. This includes:
  – not leaving computers and TVs on standby
  – not leaving phone chargers on all night
  – only boiling the amount of water you need in a kettle.
- We can buy equipment that wastes less electricity. This includes:
  – using energy-efficient bulbs instead of filament bulbs
  – using equipment with a better efficiency rating, e.g. washing machines and freezers.
- We can use equipment less. This includes:
  – turning lights off when we leave the room
  – hanging clothes out to dry instead of using a tumble dryer.

Some savings make a lot of sense. Freezers and fridges are not powerful but they are on all the time. This means inefficient models can use quite a lot of electricity over a year. Electric heaters and tumble dryers are powerful so they add Units to the electricity bill quite quickly. Reducing the time that they are in use makes sense.

d Write down **one** way to reduce your electricity bill for each of the three main methods. Use different examples to the ones in the text.

e If you only had one energy-efficient bulb to use in your home, where would you use it? Explain.

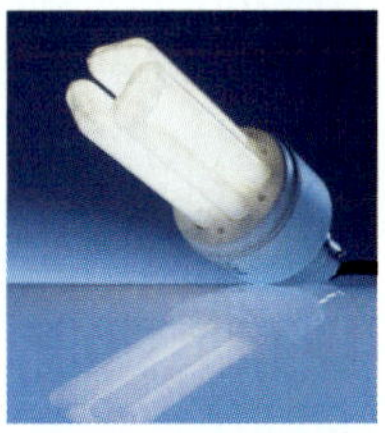

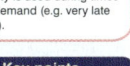

Figure 3 Use energy-efficient bulbs to reduce the electricity used compared with filament bulbs

Electricity is a clean energy source at the point of use. However, pollution is caused by power stations where electricity is generated. If all households used less electricity, the demand for electricity would fall, generating less pollution. However, the impact of a single household on the amount of electricity generated is too small to measure. Concentrate on the benefits of saving money at home instead!

Figure 4 Reducing our electricity bills helps to reduce demand for electricity

f What is the main benefit to each household of reducing electricity usage?

?? Did you know ...?
Electricity bills usually contain a quarterly charge, paid regardless of usage. This covers administrative costs of the account.
Some bills charge a different rate for higher usage, or if electricity is used during times of low demand (e.g. very late at night).

Summary questions
1 Explain why we buy electricity in kWh instead of J.
2 a How much electricity did Joel use? He had a 100 W light on for 4 hours, watched TV (500 W) for 2 hours and used his computer (750 W) for 2 hours.
  b Each Unit costs 15p. How much did Joel add to the electricity bill?
3 Write a paragraph explaining several ways that electricity bills can be reduced.

Key points
- Electricity is sold in Units of kWh
- Units are calculated using power (kW) × time (h).
- The electricity bill is based on the Units used and price per Unit.

> ## Summary answers
>
> 1 kWh is the equivalent of 3 600 000 J – so the numbers are more straightforward to deal with.
>
> 2 a Usage is 4 × 0.1 + 2 × 0.5 + 2 × 0.75 = 2.9 units
>   b 43.5 pence
>
> 3 Use the ideas in the Student Book. Stick to the main point of reduced use, more efficient/less powerful equipment and more efficient habits.

Answers to in-text questions
a 10 Units
b 1343 Units
c 18802p (£188.02)
d for example, turning settings lower/choosing less powerful equipment/using timer switches
e Aim to put the bulb in a room where it will be on for longest, e.g. a hallway.
f reduced bills

Unit 2, Theme 3 – My property

9.4 Efficiency

Learning objectives

Students should learn:
- what is meant by efficiency
- how efficiency is calculated
- the implications of efficient use of home appliances.

Learning outcomes

Most students should be able to:
- calculate the efficiency of simple devices
- describe some places where energy is wasted in devices
- identify some consequences of running electrical appliances efficiently.

Some students should also be able to:
- interpret Sankey diagrams to calculate efficiency
- explain features that make some devices more efficient than others
- evaluate the implications of running home appliances efficiently.

AQA Specification link-up: Science B 3.4.3.1

- Explain the meaning of the term *efficiency* when applied to simple energy transfers in electrical applications, and give reasons for the energy losses in appliances.
- Calculate the efficiency of an appliance using the equations:
  efficiency = useful energy out/total energy in
  effficency = useful power out/total power in

Within this context, candidates should be able to use scientific data and evidence to discuss, evaluate or suggest implications of
- the efficiency of different appliances used in the home
- the costs of running home appliances.

Lesson structure

Starters

What is efficiency? – Students describe what they think efficiency is. Support students by asking them to think about an example of very inefficient or very efficient equipment. Extend students by asking them to explain how you would compare efficiencies. *(5 minutes)*

Energy transfers – Students write down energy transfer equations for several simple devices, including all forms of wasted energy. Use this to introduce the concept of efficiency qualitatively, e.g. efficiency = useful output ÷ input. This means in a lamp, the energy transfer equation is: electrical (input) ⟶ heating surroundings (waste output) + light (useful output). You can express the efficiency as light output ÷ electrical input. *(10 minutes)*

Main

- Develop the Energy transfers starter with more examples. Put numerical values to each type of energy in each energy transfer equation discussed and use these to calculate the efficiency (see 'Practical support').
- Ask students to draw Sankey diagrams to represent the differences, as well as carrying out calculations. This can lead on to interpretations of more Sankey diagrams for different equipment.
- Point out that efficiency = useful energy output ÷ energy input, and also efficiency = useful power output ÷ power input.
- Consider cost implications of using less efficient electrical appliances.

Plenaries

Energy companies and efficiency – Electricity companies have an obligation to help customers reduce their carbon emissions, e.g. nPower sent all its customers unsolicited energy efficient bulbs. All students identify one advantage and one disadvantage of this action. The Green Party opposed this. Students should discuss whether nPower will achieve what it intended and suggest reasons why the Green Party may have opposed this action. *(5 minutes)*

What changes could I make? – Students list 3–5 changes they could make to use equipment more efficiently at home, and rank these in order of impact. *(10 minutes)*

Support

- Students may be confused when calculations involve efficiency as a percentage and need support with this.
- Efficiency is often shown as a percentage. It is never more than 100% (this can be because the student has the calculation the wrong way up). It can also be expressed as a decimal (e.g. 0.1) in which case it is never more than 1. This is simpler initially, but may not be the way it is shown in a question.

Extend

- In their answers, students should aim to explain the physics behind the facts. Many white goods now have energy ratings – students could discuss how reliable these features are, why they are needed and how manufacturers have improved efficiency. Changes in other equipment mean increased energy usage (e.g. digital radios compared with conventional radios) and batteries are very inefficient.

Did you know ... ?

Electrical devices can be very efficient, but the efficiency of the production of electricity in power stations ranges between roughly 30% and 60%. You can tell some devices are inefficient by feeling the warmth they produce – this includes the wiring and plugs.

Science in context

Manufacturers increasingly give information about the efficiency of electrical items. This idea can also be linked to the energy efficiency certificates that properties for sale require (although this is not just linked to electrical items but also is linked to insulation, etc.)

Practical support

Comparing bulbs

Equipment and materials required: Lamp fitted with an energy efficient bulb, lamp fitted with a filament bulb, temperature sensors, light sensor, clamp stand.

Details

Switch on the two lamps. Leave them for a few minutes to come to full brightness. Use the light meter to set up the temperature sensor next to each lamp in a position where the light output is the same. Leave the equipment for at least 15 minutes. Read the temperature sensors.

The efficiency of a 100W filament bulb is usually given as about 4%, and the efficiency of an energy efficient bulb is approximately 25%. This means that a 20W energy efficient bulb gives out a similar amount of light as a 100W filament bulb.

Safety: The bulbs will become hot. Do not look directly at the bright lights. Take care when using mains electricity.

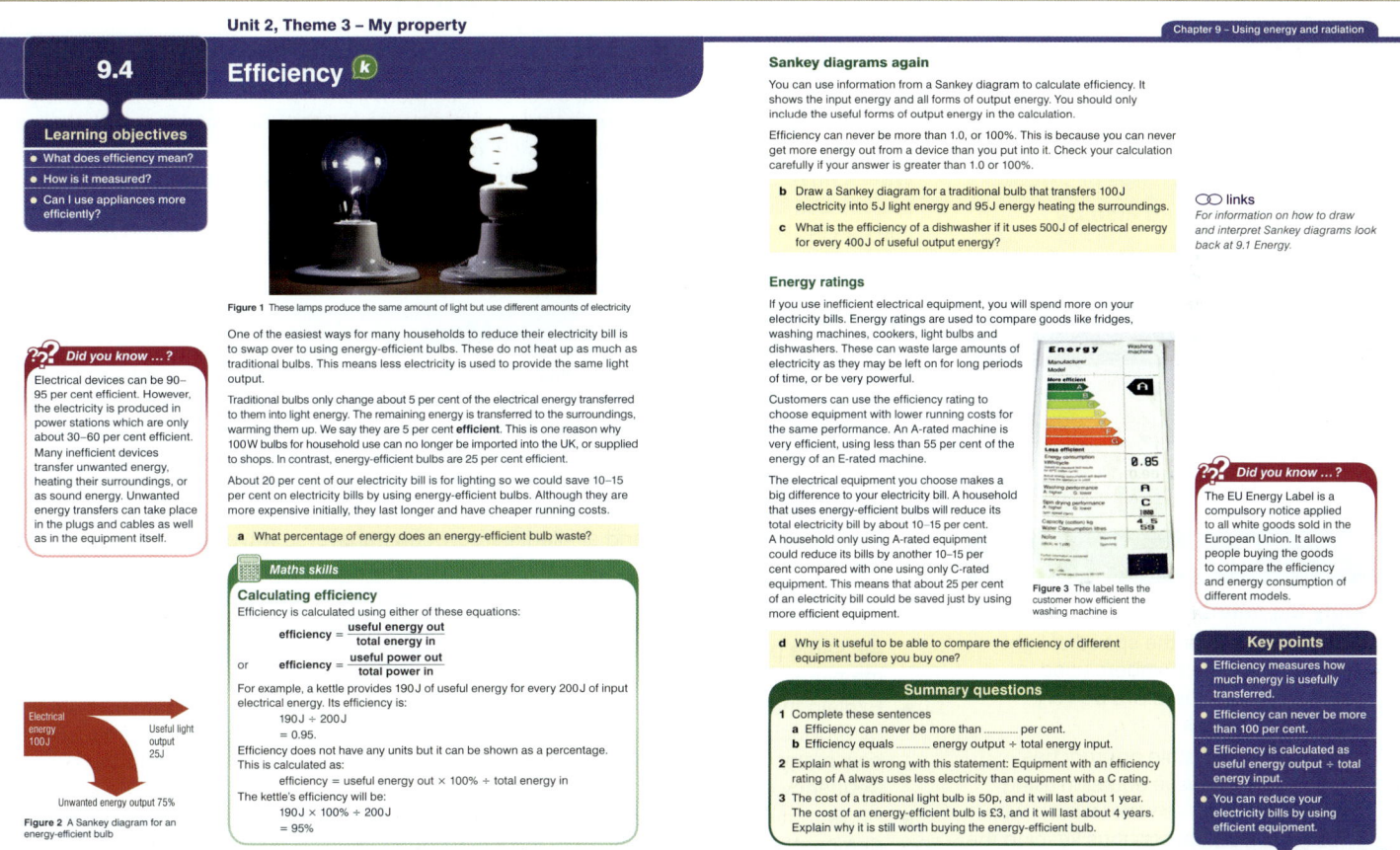

Further teaching suggestions

Efficiency versus cost

Students calculate the lifetime cost savings of buying more expensive, but more efficient equipment such as freezers and washing machines. The Argos catalogue has information on a variety of white goods or you can refer to a comparison website for white goods.

Summary answers

1. **a** 100   **b** useful.
2. The statement can be true in many cases, but is only always true if both devices are used in the same way (e.g. for the same amount of time) and if their energy output is the same.
3. The energy-efficient will save money on the electricity bills.

Answers to in-text questions

a 75%

b This is an arrow which is 10 units thick on left splitting into two arrows, top one is 0.5 units wide and bottom one is 9.5 units wide. Label arrow in left as '100J electricity input'; label top (thinner) arrow as '5J useful light output' and bottom part of arrow as '95% wasted energy output'.

c 80%

d More efficient equipment uses less energy so you can save on your electricity bills.

Unit 2, Theme 3 – My property

9.5 How fast do waves travel?

Learning objectives

Students should learn:
- that waves transfer energy from place to place
- what is meant by frequency and wavelength
- how to calculate wave speed.

Learning outcomes

Most students should be able to:
- describe how waves transfer energy from place to place
- define what frequency is and the units it is measured in
- calculate wave speed
- link frequency and energy of a wave.

Some students should also be able to:
- use the wave equation to calculate wavelength and frequency.

AQA Specification link-up: Science B 3.4.3.2
- Know that electromagnetic radiation travels as waves and moves energy from one place to another.
- Know that the number of waves per second produced by a source is called the frequency and is measured in hertz (Hz).
- Use the equation: velocity (m/s) = frequency (Hz) × wavelength (m).

Lesson structure

Starters

What is a wave? – Students write down what they think is a wave. Students can give examples of different waves. More able students should aim to write down what the properties of waves are and what distinguishes them from non-waves. (You are looking for points like: they transfer energy; the energy moves but the material doesn't; there are cyclical movement/peaks or toughs, etc.). *(5 minutes)*

Waves have different speeds – Students give examples of waves that move at different speeds and how they can tell that this happens. They may need prompts, e.g. how you can tell light and sound travel at different speeds? (Thunder and lightning reach us at different times). How can you tell that water waves travel at different speeds? (By watching the crest of water waves move in different depths of water). *(10 minutes)*

Main

- Demonstrate several different waves, e.g. waves along a slinky spring (either longitudinal when the person vibrating one end of the spring moves their hand parallel to the spring), or transverse (when the person vibrating one end of the spring moves their hand perpendicular to the spring), and water waves. In all of these examples students can see the movement of energy from one place to another, even though the material does not move.
- Identify the differences and similarities between transverse waves and longitudinal waves.
- Identify the features specific to transverse waves (peaks, troughs and wavelength) on a labelled diagram. Ensure students realise the wavelength is the distance between the same point on each wave cycle, e.g. between peaks, or between troughs. It is measured in metres.
- Explain that frequency is the number of cycles per second – you can demonstrate this by producing waves on the slinky spring or using a dipper in a ripple tank (measure how many waves are produced in 10 seconds, and divide this by 10 to find the frequency in Hertz).
- Show that increasing the frequency shortens the wavelength and vice versa. This demonstration leads nicely on to the wave equation,

    wave speed = wavelength × frequency.

- Initially carry out simple calculations using water waves and sound waves (that travel at 340 m/s), then move onto calculations involving the EM waves travelling at 3×10^8 m/s.

Plenaries

Electromagnetic waves – List the seven members of the EM spectrum and their frequencies. Divide the class into seven groups. Each group works out the wavelength of one member. Each group prepares an A4 page including the frequency, wavelength and name of the member of the EM spectrum, which can be used as a visual aid for the next few lessons. *(5 minutes)*

Comparing waves – Students prepare a table stating two or three differences and similarities between transverse waves and longitudinal waves. *(10 minutes)*

Support

Simplify the maths
- EM waves travel at the speed of light (3×10^8 m/s). The numbers used in calculations will therefore be very large or very small.

    Here are two possible approaches:
    1. Introduce the wave equation using examples such as water waves, which travel at slow speeds thus keeping the numbers involved in the calculation simple. This will help students visualise exactly what is going on.
    2. Refer to the speed as 300 million metres per second. Carry out calculations using 300, rather than 300 000 000. The factor of a million can then be written in to the answer (e.g. 0.03 million hertz; 6 million metres)

Extend

- Encourage students to carry out calculations using powers of ten correctly, involving waves with wavelengths of micrometres ($\times 10^{-6}$), nanometres ($\times 10^{-9}$) or frequencies of kilohertz ($\times 10^3$) and megahertz ($\times 10^6$).
- Vary the units (giving students information as km, mm, etc.).

?? Did you know ... ?

Nothing can travel faster than the speed of light in a vacuum. At these speeds, the mass of tiny moving objects like subatomic particles increases so much that they can't accelerate any more. This is one effect of Einstein's Theory of Relativity.

Further teaching suggestions

Animations to demonstrate wave speed

Where possible use animations to demonstrate wave speed – it will help students to spot features of the wave and enable you to discuss how the frequency and wavelength are related.

Science in context

Students will be familiar with the sight of water waves, e.g. if a stone is dropped in a puddle, or at the seaside. All waves including light waves, sound waves and seismic waves have a particular speed which can vary in different media.

Unit 2, Theme 3 – My property

9.5 How fast do waves travel?

Learning objectives
- What do waves transfer from place to place?
- How do we describe waves?
- How do we calculate the speed of waves?

Figure 1 When a stone falls into a pond, waves transfer energy as they travel across the water's surface

AQA Examiner's tip
Remember that all the waves in the electromagnetic spectrum travel at the same speed. They just have different wavelengths and, therefore, frequencies.

We are surrounded by **waves** in nature. These include water waves, sound and light as well as seismic waves.

Electromagnetic waves are waves that include radio, microwave, infrared, visible, ultraviolet, X-rays and gamma waves.

Many waves need a material to travel through, such as a solid, liquid or gas. Electromagnetic waves can also travel through a vacuum.

All waves have features in common:
- They are a regularly repeated **disturbance** that passes through the material.
- They all transfer energy from one place to another, but the material doesn't travel with the energy.

a Write down **two** things that all waves have in common.

Transverse waves cause disturbances at right angles to the direction in which the energy travels. Water waves are one example of transverse waves. The surface of the water moves up and down but the energy travels along the surface to the edge of the pond.

The wavelength is the distance between one peak and the next, or between one trough and the next. It is measured in metres.

The frequency is the number of waves produced each second, or the number of waves passing a certain point each second. Frequency is measured in hertz. One hertz is a frequency of one wave per second.

b The frequency of one wave is 12 Hz. How many waves are produced each second?

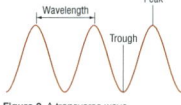

Figure 2 A transverse wave

All waves travel at a certain speed. This depends on how long each wave is (its wavelength) and how many waves are produced each second (its frequency).

Maths skills

Wave calculations

The velocity of a wave is calculated using:

velocity (m/s) = wavelength (m) × frequency (Hz)

What is the velocity of a sound wave? Its frequency is 1000 Hz and its wavelength is 0.34 m.

The velocity of the sound wave = 1000 × 0.34
= 340 m/s

All electromagnetic waves travel at 300 million m/s. This means that:
the frequency of any electromagnetic wave is 300 million ÷ wavelength.
the wavelength of any electromagnetic wave is 300 million ÷ frequency.

For example, calculate the frequency of a radio wave. Its wavelength is 1000 m.

The frequency = 300 million ÷ 1000
= 0.3 million Hz

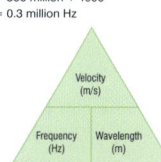

Figure 3 The wave speed triangle

c What is the speed of a water wave? Its wavelength is 1.5 m and its frequency is 0.5 Hz.

d What is the wavelength of a microwave? It is a member of the electromagnetic spectrum, and its frequency is 10 000 million Hz.

Summary questions

1 Complete these sentences choosing the correct word:
 a Water waves/**microwaves** are an example of electromagnetic waves.
 b Water waves are an example of **transverse**/longitudinal waves.
 c Wave speed equals frequency **multiplied**/divided by wavelength.
2 Which of these waves is travelling faster? A wave with a frequency of 1000 Hz and a wavelength of 0.02 m, or a wave which travels 25 m in 2 s?
3 Sketch and label a transverse wave with a wavelength of 2 m.

Did you know...?
Nothing can travel faster than the speed of light in a vacuum. The mass of objects, however tiny, increases so much at speeds near this that they can't accelerate any more.

Key points
- All waves transfer energy without transferring matter.
- Waves are regular disturbances passing through matter.
- Wave speed = wavelength × frequency.

Summary answers

1 **a** microwaves **b** transverse **c** multiplied

2 The first wave. Its speed is 1000 Hz × 0.02 m = 20 m/s and the second wave's speed is
$$\frac{25\,m}{2\,s} = 12.5\,m/s$$

3 The diagram should be the same shape the diagram of a transverse wave in the Student Book. Students should label the distance between troughs or between peaks as 2 m.

Answers to in-text questions

a For example, transfer energy; cause regular disturbances; do not transfer material.
b 12
c 0.75 m/s
d 0.03 m

Unit 2, Theme 3 – My property

9.6 Electromagnetic waves

Learning objectives

Students should learn:
- the types of wave that make up the EM spectrum in sequence
- how the frequency and energy of different members relate to their uses
- the implications of using members of the EM spectrum.

Learning outcomes

Most students should be able to:
- state the waves that make up the EM spectrum in sequence
- explain at least one use for each member of the EM spectrum
- link the properties of the waves with their uses.

Some students should also be able to:
- evaluate the implications of using EM radiation.

AQA Specification link-up: Science B 3.4.3.2

- Know the order of the electromagnetic spectrum, from radio waves (low frequency/long wavelength) to gamma rays (high frequency/short wavelength).
- Know that the higher the frequency of the wave the higher the energy.
- Name and describe the uses of the different types of electromagnetic waves used in our homes:
  (a) Radio waves – TV and radio
  (b) Microwaves – mobile phones, satellite TV and cooking
  (c) Infrared – remote controls for TV and DVD players
  (d) Visible light – fibre optic cables
  (e) UV – sun beds.

Within this context, candidates should be able to use scientific data and evidence to discuss, evaluate or suggest implications of:
- how the uses of different types of waves depend on their properties.

Lesson structure

Starters

What do you know? – Write down the waves that make up the EM spectrum. Students write down one fact about each member – share these to find out prior knowledge/misconceptions. *(5 minutes)*

What is communication? – Brainstorm different methods of communication. Students can group methods of communication as ones with a direct physical link (e.g. a letter) and indirect (e.g. mobile phone call). Introduce the idea that the fastest communications involve waves. Extend students by discussing whether communication has to involve physical movement of material or not, and what affects the speed of transmission. They rank the methods in order of speed. *(10 minutes)*

Main

- Ensure students can see a diagram of the EM spectrum. It is reasonably easy for students to visualise waves (e.g. wavelengths of radio waves are metres/km; of microwaves, cm; of infrared, mm; of visible light, tenths of millions of a metre).
- Remind students that all EM waves travel at the speed of light. Use the wave equation and typical wavelengths to calculate frequencies within each group.
- Explain that the energy carried by the waves depends on the frequency. Radio waves have the lowest energy in the EM spectrum.
- Stress that the waves form a continuous spectrum. For example, very long microwaves are also considered to be very short radio waves with similar properties and uses. Within each grouping, there will be a wide range of wavelengths with different associated uses linked to these.
- Different uses of radio waves: tune into different radio stations – remind students that as well as national broadcasts, there are very local broadcasts for emergency services, etc.
- Different uses of microwaves: these are used in mobile phone networks and satellite TV networks because microwaves can travel through the Earth's atmosphere (however the signal can be blocked by large quantities of water – demonstrate this by surrounding a mobile phone with plastic bags full of water then trying to get a signal). Wi-fi connections in schools use microwaves. Microwave ovens cook food.
- Use a TV remote control to demonstrate infrared (you can also demonstrate that the signal reflects off a wall). Show thermal images of houses and people.
- Students will be familiar with fibre optic lamps and Christmas trees. Be warned – optical fibres look flexible, but are made of glass that will snap if bent.
- You can also explain that some materials fluoresce (when they absorb UV from a UV lamp, these materials glow as they re-emit the radiation as visible light).
- There are very many more uses of EM radiation than are covered in the specification. You could set a competition for students to find out as many reasonably common uses for each member of the spectrum as possible. This can become very factual and dry – try to get students involved in investigating the uses and it becomes more interesting and memorable.

Support

- Use a very kinaesthetic approach, demonstrating radio, TV, mobile phones and remote controls throughout the lesson.
- Provide students with a simple diagram showing the seven members of the EM spectrum. Add labels as the lesson progresses indicating typical uses, wavelength, frequency, energy. Aim to keep words to a minimum, but include images.

Extend

- Set a project to research the uses of one member of the EM spectrum in detail. Within each member of the EM spectrum, the uses vary depending on the wavelength (e.g. radio broadcasts on shortwave, long wave, medium wave and VHF have different uses linked to their properties).

Chapter 9 – Using energy and radiation

Plenaries

Splat – Write key words from the lesson on the board, including uses and properties. Two students stand at the board. Another student asks them a question. Whoever 'splats' the correct answer stays at the board and is joined by the questioner. The other person returns to their seat and someone else asks a question. *(5 minutes)*

What did you know then? – Students write down three questions that they could not answer at the start of the lesson, but can answer now. Swap these with a partner and compare answers. Students could then each prepare a question with the answer they think is right. The teacher uses these to reinforce points and correct misconceptions. *(10 minutes)*

Answers to in-text questions

a Travel at speed of light; carry energy; absorbed/transmitted/reflected differently by different materials.
b They are not absorbed by the atmosphere.
c They are not absorbed or reflected by the atmosphere.
d More infrared is emitted by warmer objects like a person.
e This is when light (repeatedly) reflects off the inside surface (of a glass fibre).

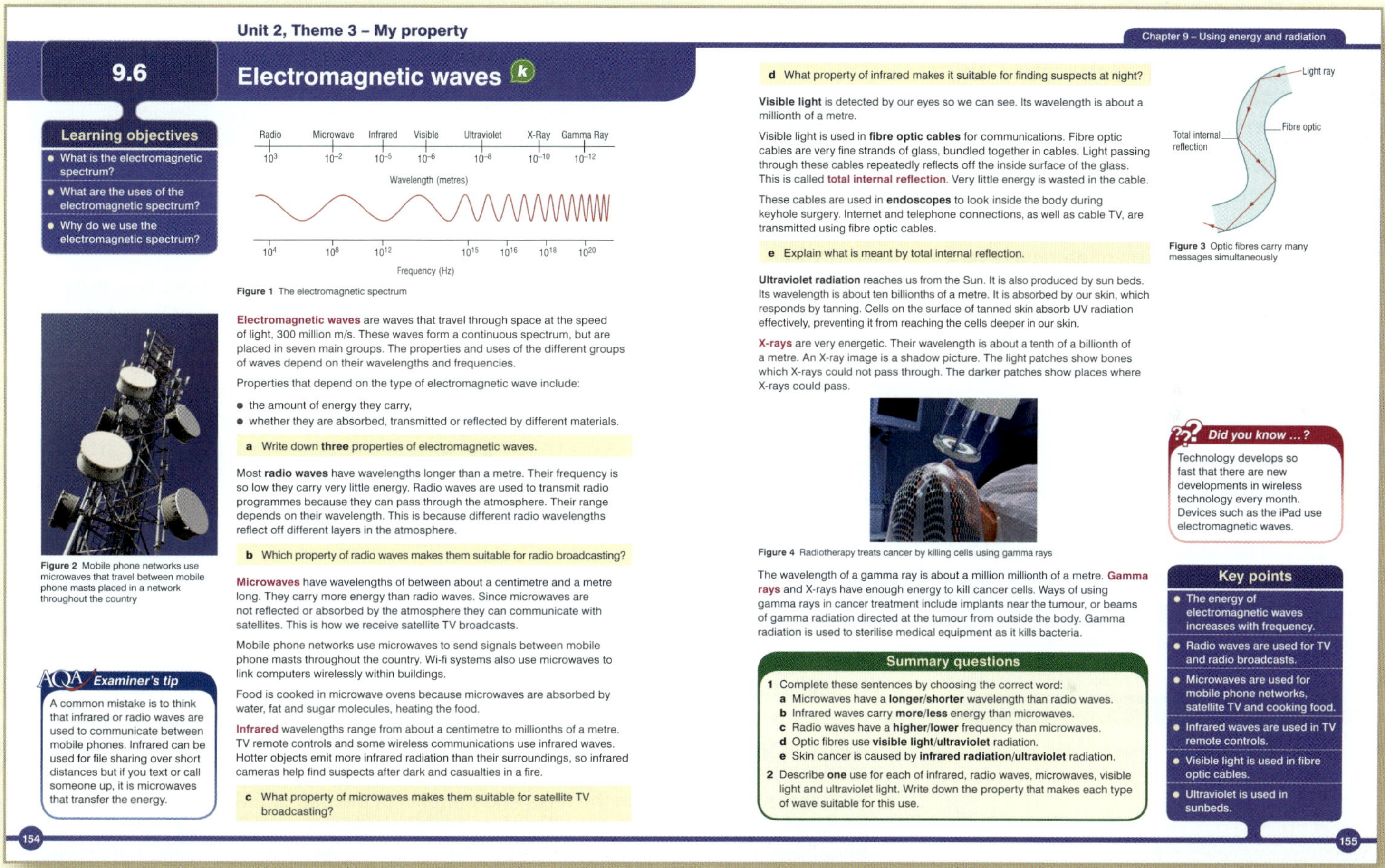

Unit 2, Theme 3 – My property

9.6 Electromagnetic waves

Learning objectives
- What is the electromagnetic spectrum?
- What are the uses of the electromagnetic spectrum?
- Why do we use the electromagnetic spectrum?

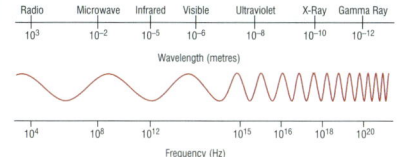

Figure 1 The electromagnetic spectrum

Electromagnetic waves are waves that travel through space at the speed of light, 300 million m/s. These waves form a continuous spectrum, but are placed in seven main groups. The properties and uses of the different groups of waves depend on their wavelengths and frequencies.

Properties that depend on the type of electromagnetic wave include:
- the amount of energy they carry,
- whether they are absorbed, transmitted or reflected by different materials.

a Write down **three** properties of electromagnetic waves.

Most **radio waves** have wavelengths longer than a metre. Their frequency is so low they carry very little energy. Radio waves are used to transmit radio programmes because they can pass through the atmosphere. Their range depends on their wavelength. This is because different radio wavelengths reflect off different layers in the atmosphere.

Figure 2 Mobile phone networks use microwaves that travel between mobile phone masts placed in a network throughout the country

b Which property of radio waves makes them suitable for radio broadcasting?

Microwaves have wavelengths of between about a centimetre and a metre long. They carry more energy than radio waves. Since microwaves are not reflected or absorbed by the atmosphere they can communicate with satellites. This is how we receive satellite TV broadcasts.

Mobile phone networks use microwaves to send signals between mobile phone masts throughout the country. Wi-fi systems also use microwaves to link computers wirelessly within buildings.

Food is cooked in microwave ovens because microwaves are absorbed by water, fat and sugar molecules, heating the food.

AQA Examiner's tip
A common mistake is to think that infrared or radio waves are used to communicate between mobile phones. Infrared can be used for file sharing over short distances but if you text or call someone up, it is microwaves that transfer the energy.

Infrared wavelengths range from about a centimetre to millionths of a metre. TV remote controls and some wireless communications use infrared waves. Hotter objects emit more infrared radiation than their surroundings, so infrared cameras help find suspects after dark and casualties in a fire.

c What property of microwaves makes them suitable for satellite TV broadcasting?

d What property of infrared makes it suitable for finding suspects at night?

Visible light is detected by our eyes so we can see. Its wavelength is about a millionth of a metre.

Visible light is used in **fibre optic cables** for communications. Fibre optic cables are very fine strands of glass, bundled together in cables. Light passing through these cables repeatedly reflects off the inside surface of the glass. This is called **total internal reflection**. Very little energy is wasted in the cable.

These cables are used in **endoscopes** to look inside the body during keyhole surgery. Internet and telephone connections, as well as cable TV, are transmitted using fibre optic cables.

e Explain what is meant by total internal reflection.

Ultraviolet radiation reaches us from the Sun. It is also produced by sun beds. Its wavelength is about ten billionths of a metre. It is absorbed by our skin, which responds by tanning. Cells on the surface of tanned skin absorb UV radiation effectively, preventing it from reaching the cells deeper in our skin.

X-rays are very energetic. Their wavelength is about a tenth of a billionth of a metre. An X-ray image is a shadow picture. The light patches show bones which X-rays could not pass through. The darker patches show places where X-rays could pass.

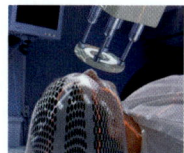

Figure 4 Radiotherapy treats cancer by killing cells using gamma rays

The wavelength of a gamma ray is about a million millionth of a metre. **Gamma rays** and X-rays have enough energy to kill cancer cells. Ways of using gamma rays in cancer treatment include implants near the tumour, or beams of gamma radiation directed at the tumour from outside the body. Gamma radiation is used to sterilise medical equipment as it kills bacteria.

Summary questions

1 Complete these sentences by choosing the correct word:
 a Microwaves have a **longer/shorter** wavelength than radio waves.
 b Infrared waves carry **more/less** energy than microwaves.
 c Radio waves have a **higher/lower** frequency than microwaves.
 d Optic fibres use **visible light/ultraviolet** radiation.
 e Skin cancer is caused by **infrared radiation/ultraviolet** radiation.
2 Describe **one** use for each of infrared, radio waves, microwaves, visible light and ultraviolet light. Write down the property that makes each type of wave suitable for this use.

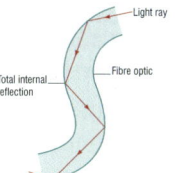

Figure 3 Optic fibres carry many messages simultaneously

Did you know ...?
Technology develops so fast that there are new developments in wireless technology every month. Devices such as the iPad use electromagnetic waves.

Key points
- The energy of electromagnetic waves increases with frequency.
- Radio waves are used for TV and radio broadcasts.
- Microwaves are used for mobile phone networks, satellite TV and cooking food.
- Infrared waves are used in TV remote controls.
- Visible light is used in fibre optic cables.
- Ultraviolet is used in sunbeds.

Summary answers

1 a shorter b more c lower
 d visible light e ultraviolet
2 Infrared thermal cameras are used to find people because they emit more thermal radiation than their surroundings; radio waves are used for communications because they are not absorbed by the atmosphere; microwaves are used to cook food as they are absorbed by molecules heating it up; visible light is used to look inside the body as it can travel long distances in optic fibres; UV light causes a skin tan so is used in sunbeds.

Further teaching suggestions

Preparing a PowerPoint
Students prepare a short PowerPoint (e.g. limited to four slides) based on uses and properties of one member of the EM spectrum per group. As a plenary, students can present these presentations to the class.

Diffraction
Discuss diffraction, which refers to the amount of spreading by waves round gaps and past obstacles. Radio waves diffract a lot so they easily pass round hills and buildings; microwaves diffract less so the beam spreads less. This means that microwaves can be received using smaller dishes than radio waves.

Science in context
Electromagnetic waves are used in a wide variety of familiar contexts, e.g. mobile phones, satellite and terrestrial TV, remote controls, fibre optic cables used for telephone communications, etc.

Did you know ...?
Many new wireless communication devices will use combinations of microwaves, infrared and visible radiation to communicate successfully.

Unit 2, Theme 3 – My property

9.7 Dangers of radiation

Learning objectives

Students should learn:
- the dangers posed by different waves in the EM spectrum
- the uses of X-rays and gamma rays in medicine
- the implications of using shorter wavelength members of the EM spectrum.

Learning outcomes

Most students should be able to:
- describe the effect of EM radiation on cells
- describe medical uses of gamma rays and X-rays
- describe some of the risks and precautions linked to the use of EM radiation
- discuss the benefits and drawbacks of using EM radiation for different purposes.

Some students should also be able to:
- evaluate the implications of using EM radiation for medical purposes
- evaluate the possible risks of low levels of radiation from various sources.

AQA Specification link-up: Science B 3.4.3.2

- Know that X-rays and gamma rays are not usually used in the home as they can damage the body, but can be used in medicine for X-rays and radiotherapy.

Within this context, candidates should be able to use scientific data and evidence to discuss, evaluate or suggest implications of the following:
- how the uses of different types of waves depend on their properties
- the dangers of using electromagnetic waves for various purposes e.g. sun beds, mobile phones, microwave cookers.

Lesson structure

Starters

X-ray images – Show students a selection of different X-ray images. Point out features such as breaks in bones or tumours, etc. Explain why the denser tissues are white (these absorb X-rays more effectively). *(5 minutes)*

What's in the tin? – Hand round a tin of food with no label. Ask all students how they can tell what is inside without damaging the tin. Link this idea to the use of X-rays in hospitals to investigate patients without operating on them. Ask more able students for reasons why you may want to leave the tin undamaged. Use their ideas to lead a brief discussion on why it is important to have ways of investigating patients without operating on them. *(10 minutes)*

Main

- Remind students of the main reasons why EM radiation can be harmful (it can be absorbed by cells, either killing them, or damaging them causing them to mutate, or even heating them).
- Explain radiation is most harmful if it is a high-energy form such as gamma radiation, or if it is in high doses or over a prolonged period. These factors all increase the chances of harmful mutations.
- Explain to students that the risk of exposure to radiation may be out weighed by the potential benefit it brings, especially as we are exposed to radiation naturally.
- Discuss how to assess risk and remind students that we tend to understate familiar risks and overstate unfamiliar risks. Examples of risky activities include crossing the road, smoking, etc.
- Students should research the risks and benefits of the use of EM radiation in different situations (either give the class the same topic, or allocate one topic per group) Situations may include: using mobile phones, having dental X-rays, using sun beds, using microwave cookers, taking a transatlantic flight, etc.
- Ask students to prepare a PowerPoint (perhaps with four or five slides only to limit time). Remind students to reference their sources of information. Include slides, e.g. to introduce the application, to explain the benefits, to explain the risks/disadvantages, to evaluate the argument and come to a conclusion.

Plenaries

What type of radiation am I? – Provide a set of statements describing several different members of the EM spectrum and harmful effects on the body. Students match the correct member to each statement. Extend students by asking them to write these statements themselves for three or more members of the EM spectrum and test these on their classmates. *(5 minutes)*

What did you find out? – Students explain their PowerPoint presentation to another group. If groups have worked on different topics, students can move round and find out quickly about several different situations. *(10 minutes)*

Support

- Select several suitable websites in advance. The cancer research site explains several issues clearly, for example.
- Provide a template or writing frame for the PowerPoint presentation.

Extend

- There is a lot of scope for discussion of complex issues. Being able to analyse risk successfully is a skill that students can work on. Also being selective with data from different sources and ranking these in terms of reliability will help students make successful assessments.

Further teaching suggestions

Uses of EM radiation

Students each write down one use for each member of the entire EM spectrum. If there is time, combine these as a table on the board. You could make this a quick competition with students working in teams. Add marks for correct uses and deduct marks for incorrect uses.

RMIVUXG

Students can try and write a mnemonic to remember all the members of the EM spectrum in order. If you are short on time, give the students one to remember, e.g. Rabbits Make Interesting Very Unusual Xmas Gifts.

Did you know …?

Fluoroscopes were X-ray machines used in shoe shops into the 1970s. Before the dangers of too much exposure to X-rays was known, these machines used X-rays to see how well the shoes fitted children's feet.

Unit 2, Theme 3 – My property

9.7 Dangers of radiation

Learning objectives
- Can electromagnetic waves be dangerous?
- How do we use X-rays and gamma rays?
- Why are X-rays and gamma rays not used in the home?

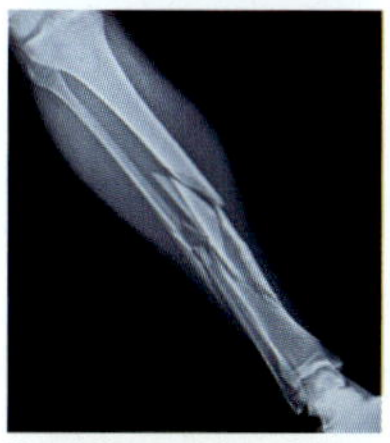

Shorter wavelengths

As the wavelength of electromagnetic waves gets shorter, the energy carried by the waves increases. The waves are absorbed by our cells. Higher energy waves have a bigger effect on our bodies and can cause harm.

a Why can shorter electromagnetic waves cause harm?

X-rays and gamma rays pass through a patient's body. They are used to diagnose and treat different medical problems without operating on a patient.

X-rays

X-rays are mainly used in a hospital or dentist's surgery because too much exposure can be harmful, and their use must be controlled. If an X-ray is needed, the parts of the body that are not being X-rayed are shielded.

b What property of X-rays means they are not suitable to use at home?

Gamma rays

Cancer cells can grow rapidly and stop nearby organs from working properly. Gamma rays and X-rays have enough energy to kill cells so doctors use these to target the cancer cells. It is important that nearby healthy cells are not damaged. So treatment is designed to shield these cells or expose them to a lower dose of radiation.

c How are healthy cells protected during cancer treatment?

When X-rays and gamma rays are absorbed, they can damage or kill cells. Damaged cells may mutate and become cancerous. This means the use of X-rays and gamma rays must be controlled. To reduce harm, doctors limit how often a person is treated. They aim to use as low a dose as possible. This reduces the chance of long-term damage.

Different patients and different cells are more vulnerable. Unborn babies and children are vulnerable because they are growing so their cells are dividing rapidly. Damage in children and adults is harmful if a vital organ like the heart, kidney or liver stops working properly. When a patient is treated, lead screens shield areas that are not being treated.

Workers in X-ray and radiotherapy departments must also be protected. They are frequently exposed to small doses of radiation.

d Why can X-rays and gamma rays be harmful?

UV radiation

Ultraviolet radiation is less energetic than X-rays but still causes harm. A tan is the way skin cells try to protect themselves from exposure to UV radiation. Doctors are worried because there are more skin cancer sufferers nowadays. Skin cancer can take years to develop. Doctors think over-exposure to the Sun during overseas holidays and in sunbeds caused damage that is now turning cancerous in some people.

UV from the Sun is very intense in hotter countries. People can protect themselves by covering up, staying inside during the early afternoon and using sun cream. Some regular exposure to sunlight is important, however. It helps us to produce vitamin D in our bodies, which is essential for the formation of bones and teeth.

e Why are some people more likely to suffer from skin cancer after holidays in hot countries?

Figure 3 Microwaves from mobile phones are unlikely to be harmful

Microwaves

Microwaves are not very energetic but may possibly still cause harm. Harm may be caused from a high dose over a short period of time. Microwave ovens cook food because molecules in cells absorb microwaves and heat up. Damaged microwave ovens may let some microwaves seep out around the door seals. This is unlikely to be harmful as the dose is extremely low.

Some people are worried that low doses of microwaves over many years from mobile phones may cause harm. So far, studies suggest the risk, if it exists, is tiny. The radiation emitted by mobile phones is too low to cause cancers directly. Mobile phone use has increased dramatically but the incidence of brain tumours has not. However, cancer takes years to develop and the way the studies have been carried out has not convinced everyone.

f Why may some people be unaware that ultraviolet waves are harming them?

Did you know …?

Mobile phones are linked to an increased risk of death. This is because car drivers making or receiving calls are more likely to have accidents because they are distracted.

Activity

Are mobile phones dangerous?

Investigate the evidence that mobile phones may cause tumours or cancers. Present your data as a poster suitable to be displayed in school.

Summary questions

1 Complete these sentences choosing the correct word:
   X-rays   gamma radiation   ultraviolet waves   microwaves

   .... can be used to treat cancer. .... and .... are too dangerous to be used in the home. .... are given out by mobile phones.

2 Write down three pieces of advice that a person who works in an X-ray department could follow to reduce the risk of harm to patients from X-rays.

3 Write down three pieces of advice that a mobile phone user should follow to reduce any possible risk of harm from microwaves.

Key points

- Microwaves and ultraviolet radiation can be harmful.
- Hospitals use X-rays to produce shadow pictures of bones and gamma rays are used to treat cancer.
- X-rays and gamma rays are harmful in large doses so their use is controlled.

Figure 1 X-rays reveal damage to bones

Figure 2 X-rays and gamma rays can only be used by authorised workers in controlled areas

---

Science in context

EM waves are used in a wide variety of familiar contexts, e.g. in vision; medical uses of X-rays and radiotherapy. Many students will identify with the use of mobile phones, though it is unlikely that any will use sunbeds as the use of sunbeds by under-18s is restricted by law.

Activity

Are mobile phones dangerous?

The mobile phone debate continues to be unresolved. There is no biological reason for increased cancer risk but cancers take years to develop. Very few studies surveyed people who have used phones since being teenagers or younger. Mobile phones emit less radiation than before, but usage has increased dramatically particularly with younger people. Methodology for the various surveys has been flawed – there is scope for HSW discussion on the best way to carry out the research.

Answers to in-text questions

a More energetic so absorbed by body and damage/kill cells.
b They are very energetic and can penetrate into cells.
c They are shielded or exposed to less radiation.
d They damage cells and can cause cancer/cell mutations.
e Exposure to intense UV from the sun that their skin was not prepared for.
f The damage may take years to become obvious.

Summary answers

1 gamma radiation, gamma radiation, X-rays, microwaves
2 Check if the X-ray is needed; use as low a dose as possible; shield the body using lead.
3 Turn off the phone if not in use; don't make long calls; look for a phone that emits low levels of microwaves.

Unit 2, Theme 3 – My property

Practical suggestions

Practicals	AQA	k	📖
Construct a model house. Use sensors and data logger to measure temperatures with and without various types of insulation.	✓		
How the efficiency of an electric motor varies with the load.	✓		
Candidates read the electricity meter at home on a daily or weekly basis (with permission from their parents). They could then look for trends in usage and try to explain these, e.g. in terms of weather conditions.	✓		✓
Demonstrate the use of an electrical joulemeter to investigate the energy transferred by low-voltage lamps of different powers, and by low-voltage motors and low-voltage immersion heaters.	✓		
Investigate the efficiency of low-voltage bulbs – invert in water and measure temperature change.	✓		
Use a class set of skipping ropes to investigate frequency and wavelength.	✓		
Carry out traditional experiments with a slinky spring.	✓		✓
Carry out traditional investigations using ripple tanks, including the relationship between depth of water and speed of wave.	✓		

Kerboodle resources

- Theme map: My property
- Bump up your grade: Which is the best buy? (9.4)
- How Science Works: Efficiency of electrical appliances used in the home (9.4)
- Maths skills: Velocity = frequency × wavelength (9.5)
- Revision podcast: Electromagnetic spectrum (9.6)
- Viewpoints: UV sunbeds (9.7)
- Interactive activity: Electrical power consumption
- Examination-style questions
- Answers to examination-style questions
- Test yourself: My property

Unit 2, Theme 3 – My property

Summary questions

1. Draw a Sankey diagram to show this energy transfer: a lamp transfers 10 J of electrical energy into 2 J of light energy and 8 J of heating the lamp and its surroundings energy.
2. The power of a kettle is 2000 W. It is used for 15 minutes each day. The power of a fridge is 25 W. It is on for 24 hours. Which appliance uses more energy?
3. Calculate the missing values in the table.

Power in watts	Current in amps	Voltage in volts
45	0.9	
	1.3	14
62	8	

4. These readings were taken from an electricity meter.
   21 June
   | 2 | 6 | 3 | 2 | 4 | 2 |
   21 September
   | 2 | 6 | 4 | 8 | 5 | 6 |
   a How many Units of electricity were used?
   b How much did this cost if each Unit cost 15p?
5. a A 100 W light bulb is 4 per cent efficient. How much energy does it waste each second?
   b A person changes their light bulbs for energy-efficient light bulbs, which are 24 per cent efficient. Explain how this may change their electricity bill.
6. Calculate the frequency of a microwave. Its wavelength is 10^{-3} m (0.001 m). The speed of light is 300 000 000 m/s.
7. Use the equation: velocity = frequency × wavelength fill in the table.

Velocity (m/s)	Frequency (Hz)	Wavelength (m)
330	30	
20	8	
	19	5

8. Match the type of electromagnetic wave to its use.

Type of wave	Use
Radio	In mobile phone networks
Microwave	In TV remote controls
Infrared	To broadcast programmes

9. Explain why microwaves are used to communicate with satellites.
10. Match the type of electromagnetic wave to its use.

Type of wave	Use
Ultraviolet	To kill cells in a tumour
X-ray	In sun beds
Gamma ray	To identify if a patient has cancer

11. Explain why gamma rays and X-rays are not suitable to use in the home.
12. Write down three properties of all electromagnetic waves.
13. What precautions should workers in X-ray department take to reduce their risk from X-rays?

Summary answers

1. The diagram is an arrow that splits. It is 10 units wide on the left labelled 'Electrical 10 J' and splits into two arrows facing to the right, one is 2 units wide and labelled 'light 2 J' and the other is 8 units wide and labelled 'wasted energy (or energy heating the lamp and its surroundings) 8 J'.
2. Kettle: 2000 × 15 × 60 = 1800 kW
   Fridge: 25 × 24 × 60 × 60 = 2160 kW
   The fridge uses the more electricity.
3.

Power in watts	Current in amps	Voltage in volts
45	0.9	50
18.2	1.3	14
62	8	7.75

4. a 264 856 − 263 242 = 1614 Units
   b 1614 × 15p = 24 210 pence or £242.10
5. a It wastes 96 J per second.
   b The bill will be reduced because less electricity is wasted as unwanted heat.
6. 300 000 000 ÷ 0.001 = 300 000 000 000 Hz
7.

Velocity in m/s	Frequency in Hz	Wavelength in m
330	30	11
20	2.5	8
95	19	5

Chapter 9 – Using energy and radiation

Examination-style questions

1. Label each of the types of energy on this Sankey diagram for a television.

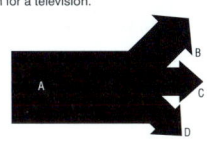

(1)

2. Match the type of electromagnetic wave with its use.

Type of wave	Use
Infrared	Security lights
Gamma	Mobile phone
Microwave	Sterilisation
Ultraviolet	Toaster

(3)

3. Choose words from the box to complete the sentences. Words can be used more than once or not at all.

> energy   frequency   matter
> second   wavelength

Waves transfer .... from place to place without any .... being transferred. If a wave has a small .... it has a high .... This means that more .... is being transferred per .... (3)

4. Use the power label from a microwave oven to answer the questions. Show your workings as part of your answer.

> Microwave oven (Household)
> Model number: 196136
> Serial number: 7168301403
> Manufactured: November 2008
> Input: 1100 W   Output: 700 W

  a. Calculate how many units the microwave oven would use if left on for 15 minutes. (4)
  b. Calculate how much this would cost if the price per unit is 12 pence. (2)
  c. Calculate the efficiency of the microwave oven. (3)

5. Read the article below from an internet site about skin cancer and then answer the questions.

> **Cancer-beds outlawed for under 18s**
>
> An expert committee that makes recommendations to the World Health Organization has suggested putting a ban on people under the age of 18 from using sunbeds.
>
> It made its decision following a review of research which concluded that the risk of melanoma – the most deadly form of skin cancer – was increased by 75% in people who started using sunbeds regularly before the age of 30.
>
> The Sunbed Association (TSA) supports a ban on under-16s, but argues there is no scientific evidence for a ban on young people aged 17 or 18.

  a. *In this question you will be assessed on using good English, organising information clearly and using specialist terms where appropriate.*
  Explain the advantages and disadvantages to a person using a sun bed. (6)
  b. What could the government do to warn young people about the risks of using sunbeds? (1)

Examination-style answers

1. A: electricity
   B, C and D: light, sound and heat (in any order) (1 mark)

2.
Type of wave	Use
Infrared	Security light
Gamma	Sterilisation
Microwave	Mobile phone
Ultraviolet	Toaster

(All 4 correct = 3 marks, 3 correct = 2 marks, 2 correct = 1 mark, 1 correct = 0 marks.)

3. energy, matter, wavelength, frequency, energy, second

(All 6 correct = 3 marks, 4 or 5 correct = 2 marks, 2 or 3 correct = 1 mark)

4. a $1100\,W = 1.1\,kW$
      15 minutes = 0.25 hours
      $1.1 \times 0.25 = 0.275\,kWh$ (unit not needed for the mark) (4 marks)

   b 0.275×12 (allow error carried forward from 5a)
      = 3.3 pence   (unit not needed for the mark) (2 marks)

   c Efficiency = useful energy ÷ total energy × 100
      = 700 ÷ 1100 × 100
      = 64%   (unit not needed for the mark) (3 marks)

5. a Marks awarded for this answer will be determined by the Quality of Written Communication (QWC) as well as the standard of the scientific response.

   There is a clear, balanced and detailed explanation of the advantages and disadvantages to a person using a sun bed. The answer is well structured with minimal repetition or irrelevant points. There is an accurate, fluent and clear expression of ideas with only minor errors in the use of technical terms, spelling, punctuation and grammar.
   (5–6 marks)

   There is some explanation of the advantages and disadvantages to a person using a sun bed, with some omissions. The answer shows some attempt at structuring and the ideas are expressed with reasonable fluency and clarity. There are some errors in the use of technical terms, spelling, punctuation and grammar.
   (3–4 marks)

   There is a brief explanation of the advantages and disadvantages to a person using a sun bed. The answer is largely incomplete and may contain some valid points which are not clearly structured. It lacks fluency and/or clarity. It contains errors in the use of technical terms, spelling, punctuation and grammar. (1–2 marks)

   No relevant content (0 marks)

   Examples of points made in the response:
   - Ultraviolet waves
   - Small wavelength
   - High frequency
   - Lots of energy
   - Can give you skin cancer
   - Look good
   - Get a tan.

   b Adverts, leaflets, school talks (1 mark)

---

8. Radio – to broadcast programmes
   Microwave – in mobile phone networks
   Infrared – in TV remote controls

9. Microwaves can pass through the atmosphere – they are not reflected or absorbed by it.

10. Ultraviolet – in sun beds
    X-ray – to identify if a patient has cancer
    Gamma ray – to kill cells in a tumour

11. Gamma rays and X-rays are very energetic and can be harmful to cells.

12. Electromagnetic waves travel at the speed of light in a vacuum; they carry energy from one place to another; they can all travel in a vacuum. Students could also include: they are all transverse waves; they can be reflected and refracted.

13. Reduce the time they are in the X-ray room when X-rays are being taken; stand behind the protective glass shield.

Unit 2 – My family and home

AQA Examination-style answers

1 a CDAB   (4 correct = 3 marks, 2 or 3 correct = 2 marks, 1 correct = 1 mark)

b It does not involve the brain, so it is faster, which means any damage is lessened.   (3 marks)

2 parents
characteristics
offspring
generations   (4 marks)

3 a Both carry a dominant gene which masks the other gene.   (1 mark)

b The genotype of both parents is Aa (this could be shown as the headings on the Punnett square and gains one mark).
Punnett square filled in correctly (one mark can be awarded even if genotype is incorrect as long as the Punnett square is filled in according to the genotype given by the student).
25% (one mark can be awarded for the correct interpretation of an incorrect Punnett square)   (3 marks)

c Less oxygen carried around body.
So breathlessness / tiredness / lack of concentration.
OR
Blood cells might get stuck in capillaries.
Can lead to stroke.   (2 marks)

4 a Burnbright: 25 (%)
Lighteze: 4 (J)
Sureshine: 25 (%)   (3 marks)
(unit not needed for the mark).

b Burnbright
It needs the most energy.   (2 marks)

c Waste energy
As heat   (2 marks)

5 a A – boiler
B – turbine
C – generator
D – pylon   (All 4 correct for 2 marks, 2 or 3 correct = 1 mark)

b $CH_4 + O_2 \longrightarrow CO_2 + 2H_2O$
(1 mark for each correct symbol, with the reactants and products being on the correct side and 1 mark for the correct balancing)

6 a Any **three** from:
Use different amounts of sand to mix up mortar.
Hang different masses from it OR drop a known mass from different heights onto it OR drop different masses onto it from the same height OR stand different masses on a block of it across a gap.
Any correct fair test reference.
Repeat for accuracy.
Compare OR find which one is the best.   (3 marks)

b Correct reference related to strength.
The fact that strength / braking mass went up then down as more sand is added.
The best mass of sand to add is 15 g.   (2 marks)

c Repeat
For accuracy
OR
Collect more results for values between 10 and 20 g
For precision   (1 mark)

AQA Examination-style questions

1 a The letters represent the parts involved in a reflex action. Put them in the correct order in the flow chart.
.... → .... → .... → ....
A: Relay neuron
B: Motor neuron
C: Receptor cell
D: Sensory neuron   (3)

b Explain the advantage of having a reflex action instead of a normal response. [H]   (3)

2 Choose words from the box to complete the sentences.

characteristics   generations   offspring   parents   years

Selective breeding involves selecting .... with desired ...., crossing them then selecting the best .... This process is repeated over several ....   (4)

3 Sickle-cell anaemia is a common disease around the equator. A person with sickle-cell anaemia has red blood cells that are a different shape to normal. A husband and wife from Kenya wanted to start a family together but they were worried that their children might inherit the disease they were both carrying. Both of the parents had inherited one sickle-cell recessive gene and one normal dominant gene.

a Explain why neither parent shows the symptoms of sickle-cell anaemia.   (1)

b Use a Punnett square to find the probability of these parents having a child with sickle-cell anaemia. Use **A** for the normal dominant gene and **a** for the recessive disease gene. [H]   (3)

c Explain a problem there might be for someone suffering from sickle-cell anaemia.   (2)

4 A shop owner wanted to put the most efficient light bulb on display. She used a joulemeter to find the amount of energy that went into each bulb in a given time. She also measured the amount of light energy given off by each bulb in a given time.

a Complete her table of results

Name	Energy needed by the bulb in joules	Light energy given from the bulb in joules	Efficiency (%)
Burnbright	32	8	
Lighteze	20		20
Sureshine	22	5.5	

   (3)

b Explain which bulb would be the most expensive to leave on.   (2)

c Explain why the bulbs are not 100 per cent efficient.   (2)

5 a The diagram shows a fuel-burning power station. Label the parts A to D

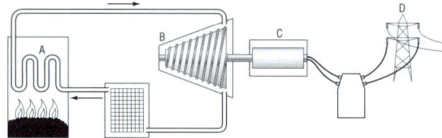

   (2)

b Methane is a gas that can be burnt in a power station to generate electricity. Give a balanced symbol equation for the combustion of methane. [H]   (4)

7 a Any **two** from:
Need more electrical energy.
Running out of fossil fuels.
No pollution from nuclear power station.
Better safety in power plants.
Do not need as big an area of land as some other types of power plant.
More jobs.   (2 marks)

b Marks awarded for this answer will be determined by the Quality of Written Communication (QWC) as well as the standard of the scientific response.

There is a clear, balanced and detailed argument giving the reasons for and against having a nuclear power plant built close to a community. The answer is well structured with minimal repetition or irrelevant points. There is an accurate, fluent and clear expression of ideas with only minor errors in the use of technical terms, spelling, punctuation and grammar.   (5–6 marks)

There is some argument giving the reasons for and against having a nuclear power plant built close to a community, with some omissions. The answer shows some attempt at structuring and the ideas are expressed with reasonable fluency and clarity. There are some errors in the use of technical terms, spelling, punctuation and grammar.   (3–4 marks)

There is a brief argument giving the reasons for and against having a nuclear power plant built close to a community. The answer is largely incomplete and may contain some valid points which are not clearly structured. It lacks fluency and/or clarity. It contains errors in the use of technical terms, spelling, punctuation and grammar.   (1–2 marks)

Unit 2 – My family and home

End of Unit 2 questions

Mortar is made from cement, water and sand.

a A builder used sand, water and cement to make up some mortar. He used $10\,cm^3$ of water and 20g of cement. He wanted to do an experiment to find how the amount of sand affects the strength of the mortar.
Give an outline of an experiment he could do to find the best composition for his mortar. (3)

b The table shows a list of results that the builder found when he did his own more accurate.

Amount of sand in grams	Breaking mass in kilograms
5	10
10	20
15	35
20	30
25	26

c Write a conclusion based on the builder's results. (2)
d Suggest and explain what the builder could have done to make his experiment more accurate. (1)

Read the newspaper article below about the opening of a coal mine and then answer the questions.

Today the community of Lowfield was picketing the entrance of one of five sites identified by the government as possible locations for a nuclear power plant. Over 200 residents from the local community have decided to form an action group to oppose the site.

The leader of the action group Josephine Lauder said 'There is no way this community is going to allow itself to be blighted so that the government can provide the country cheap electricity. We think the cost to our community is simply too high.' Miss Lauder's statements were in response to a report yesterday on a regional news programme. The report stated that not only will house prices dramatically decrease in the area but there is also scientific evidence that the power station can have a devastating effect on a community living in that area.

In response, the site manager looking at the location, Steven Ashman, adamantly denied the allegations made in the report and by Miss Lauder, and defied them to show him evidence that would back up their allegations. He also explained that the proposed plans would have an overwhelming positive effect on the community.

e State **two** reasons why the government wants to build more nuclear power plants. (2)
f *In this question you will be assessed on using good English, organising information clearly and using specialist terms where appropriate.*
The site manager, Steven Ashman, said that the power station would have a positive effect but people like Miss Josephine Lauder do not want to have a nuclear power plant built in their community. Give the reasons for and against having a nuclear power plant built close to a community. (6)

161

Bump up your grades

Students on the C/D borderline will need to make sure they can explain rather than just describe. So a D grade student would not use the word 'because' in their answer as often. Encourage your students to go that step further and explain why things happen.

C grade students should know the electromagnetic waves in order of frequency and be able to explain why one wave is more dangerous than another. They should be able to describe the difference between a reflex and a normal response. Knowledge of Punnett squares is needed. Knowledge of word equations for the conversion of limestone into various useful products is an advantage at this level. Quite often students at this border find the mathematical equations difficult so plenty of practise would give the student more confidence.

Grade F students should be able to draw and label an animal cell, give some uses of limestone and list some examples of renewable energy resources.

Grade A students should complete balanced equations for neutralisation reactions and the combustion of hydrocarbons. Also, an in-depth understanding of the various bodily processes explored in this unit would need to be shown; the best way for this is to find past exam questions for the students to practise on.

No relevant content (0 marks)

Examples of points made in the response:
- Increased traffic
- Worried about safety
- Eyesore
- More noise
- Destroy local habitat
- More jobs
- More money coming into the local community.

Unit 3, Theme 1

Improving health and wellbeing

AQA Specification link-up: Science B 3.5.1

People working in the medical professions use their knowledge to diagnose and treat disease and illness, or to research new ways of treating disease. In both cases, medical scientists need to be aware of how the healthy body works as well as what may cause the body to become unhealthy.

In this theme there are three contexts:
- 3.5.1.1 The use (and misuse) of drugs
- 3.5.1.2 The use of vaccines
- 3.5.1.3 The use of ionising radiation in medicine

The use (and misuse) of drugs

Most drugs are legal; they are deemed to be safe to use. Other drugs, which can damage your body even in very small amounts, have been made illegal by governments. These drugs may not only harm the person taking them, but can indirectly harm others around them. Testing for illegal drugs is increasingly being carried out in the workplace, to improve the health and safety of employees.

Medical drugs are prescribed by doctors to improve quality of life, by helping to cure or prevent disease. However, some medical drugs can cause unpleasant side effects. Many of these drugs are extremely damaging to health if an overdose is taken – often with irreversible effects.

Recreational drugs have no medical benefits. They are taken purely for personal enjoyment. Not all recreational drugs are illegal; however these drugs can still damage the body. For example, there is evidence that respiratory and circulatory disorders are linked to the misuse of tobacco and alcohol.

What are medical drugs?

Activity:
- Ask students to list all the medical drugs they think you would find in a standard first aid box. Share the lists in groups of four. The group should then try to sort the drugs into groups, based on their use and the information on the packaging. Examples include allergy relief, pain relief, or to reduce swelling.

What are recreational drugs?

Activity:
- Provide small groups of students with two body outlines. Students are to label the effects that smoking, or drinking alcohol, have on the body. They should include both long- and short-term effects. The images could be made body-sized, to make sharing of class results more visual.
- Ask students to make a list of all the recreational drugs they can think of. They should divide the list into two groups – those which are legal and those which are illegal.

Possible misconceptions

Many students believe that legal drugs cause no harm. This point should be discussed when studying the damaging effects that tobacco and alcohol have on the body. This is also the case with medicinal drugs. Tranquillisers are a good example of a medical drug that people can become addicted to.

Some students also believe that you cannot become addicted to a substance by only taking it a couple of times. Students should be aware that drugs such as cocaine are highly addictive and for some individuals taking the drug once or twice can lead to addiction. They should also be aware that many addicts do not initially recognise that they are suffering from addiction.

The use of vaccines

When a microorganism enters the body and causes disease, antibodies are made against it. These antibodies fight off the disease and a person recovers. Some of these antibodies will remain in the body to fight off the microorganism if it returns, preventing the microorganism from causing the same disease again – this is known as immunity.

Medical scientists have used the concept of immunity to develop vaccines. They can trick the body into developing immunity using a dead or weakened microorganism that is unable to cause disease. Vaccinations have helped to reduce the occurrence of certain diseases in many parts of the world.

There are occasional scares about the safety of some vaccines, which can result in people choosing not to be vaccinated. This often results in an outbreak of disease. Some people also believe that vaccines overload our immune system, making it less able to react to other diseases such as meningitis, AIDS and cancer. Others are concerned about possible side effects of vaccines, although these are usually mild and not life threatening. Doctors and medical professionals believe that these concerns are unfounded, and recommend that vaccinations are the safest and best way of protecting yourself from infectious diseases.

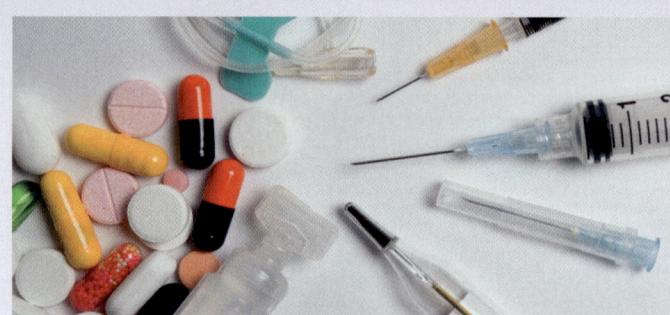

Unit 3, Theme 1

Improving health and wellbeing

In Unit 3, Theme 1 you will work in the following contexts, covered in Chapters 10 and 11:

The use (and misuse) of drugs
What are medical drugs?

Medical drugs improve the quality of our lives by preventing, treating or curing diseases. These include antibiotics, such as penicillin, anti-inflammatory drugs like aspirin, and painkillers like paracetamol.

What are recreational drugs?

Recreational drugs are taken for an individual's personal enjoyment. They have no medicinal benefits. In fact many cause more harm than good. Tobacco and alcohol are examples of legal recreational drugs. People often take them to relax. However, if they are taken over a long time they seriously damage your health.

The use of vaccines
What is a vaccination?

Medical scientists have developed vaccines against a number of diseases caused by bacteria and viruses. This results in a person developing immunity, and prevents them getting the disease. Vaccination is the simplest, most efficient and cost-effective way of preventing life-threatening infections in the community. Common vaccines include tetanus, polio and measles.

How do our bodies prevent pathogens causing disease?

You come into contact with pathogenic microorganisms every day. Thankfully, they do not always make you ill. The skin acts as a barrier preventing the entry of the majority of these microorganisms. If the skin is damaged it must seal itself as quickly as possible. Your platelets play a major role in this process.

The use of ionising radiation in medicine
How is ionising radiation used in hospitals?

Medical professionals can diagnose and treat certain diseases using ionising radiation. For example, X-rays can detect broken bones, and gamma radiation can be used to treat some types of cancer. Gamma radiation is also used in the diagnosis of some medical disorders. There are many other medical uses of radiation that you will learn about in this theme.

How is exposure to ionising radiation monitored?

Although ionising radiation has many medical uses, it is also potentially harmful to people who work with it every day. Radiation protection supervisors, medical staff, engineers and radiographers all work together to ensure that the received dose of radiation does not exceed safe limits. Workers who are exposed to radiation wear a film badge. This records the received dose of ionising radiation.

Improving health and wellbeing

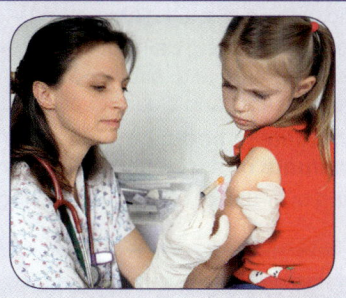

Matching patients to treatments by screening their genetic makeup is the goal of some drugs companies. The prescribing of drugs has always tried to tailor treatment to reduce side effects, but generally this relied very much on trial and error. Personalised medicine tries to predict treatment response or prevents disease before symptoms appear.

Drugs are chemical substances. They work by altering the chemical reactions that take place inside the body. If the body gets used to these changes, it may become dependent on a drug. This is known as addiction. Many highly addictive drugs are illegal in the UK; however, tobacco and alcohol are widely available to over-18s. Some medical drugs are also addictive – however, their usage is carefully monitored by doctors and medical professionals.

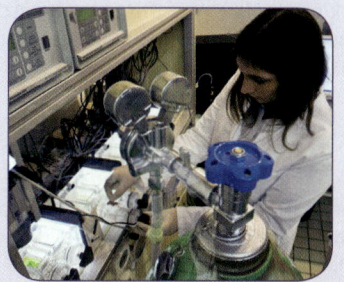

Before a drug can be prescribed by doctors it has to be carefully researched and monitored. Part of this process involves the new drug being tested on animals to ensure it is safe to use. Testing new drugs on animals is extremely controversial. Some people believe it is cruel, and a violation of the animal's rights. However, many people believe these tests are very important, as the drugs may have the potential to save human lives.

What is a vaccination?

Activity:

- Ask students to try to write a definition for the term 'immunity'. They should first try the task individually. Students should then share their definition with a partner or small group to try to improve their definition. They should try to make their definition as specific as possible.
- Ask students to make a list of the vaccinations they have received. Can they remember why they had them, and at what age? Students could then produce a list of the arguments used for and against vaccinations.

How do our bodies prevent pathogens causing disease?

Activity:

- Provide students with a large image of a person. In groups ask them to label areas where microorganisms could enter the body. They should then try to explain how the body naturally prevents this from occurring.
- Provide students with a series of situations that could increase the likelihood of microorganisms entering the body, such as a through an open cut, during sexual intercourse or while eating. Students should suggest steps they could take to reduce the risk of this occurring.

Possible misconceptions

Some students believe that vaccines contain small doses of an infectious disease. This is not the case – even a small number of microorganisms can lead to a disease. Vaccines contain dead or weakened versions of a pathogen.

Not all microorganisms are harmful – remind students of everyday food sources, such as yoghurts, which contain microorganisms that aid digestion and are therefore beneficial to health.

The use of ionising radiation in medicine

Ionising radiation was discovered just over 100 years ago. Some of the potential medical applications were quickly realised. Unfortunately, the potential health hazards were not identified so quickly.

Since those early days, a wide range of medical applications for ionising radiation has been identified. Medical professionals diagnose and treat certain diseases, such as cancer, by using ionising radiation. However, ionising radiation also has the potential to cause cancers and so its use must be carefully controlled. People who work with radiation wear film badges to monitor their exposure and ensure that safe limits are not exceeded.

How is ionising radiation used in hospitals?

Activity:

- Split students into small groups. Each group is provided with a large sheet of paper and a marker pen. The word 'Radiation' is written in large letters in the centre of the paper. Students complete a 'graffiti' exercise – adding as many words or terms, related to the word radiation, as possible in a short period of time (for example, 90 seconds). Each group then passes its paper onto the next group, adding to the previous graffiti added. Once groups have added to all pieces of paper, groups feed back to the class on the contents of their sheet, allowing you to correct misconceptions and discuss relevant points.
- Ask students to make a list of the ways they think ionising radiation can be used in hospitals. Visual prompts could be supplied to support students, such as images of a CT scanner, an X-ray, and gamma camera images of medical tracers.

How is exposure to ionising radiation monitored?

Activity:

- Show students an image of a radiation film badge. Ask students to hypothesise on what it might be and why people who work with radioactive isotopes are required to wear one at all times.

Possible misconceptions

Many students have difficulty understanding that if ionising radiation can cause cancer, it can also be used to treat cancer. Time is required to deal with this issue. Discuss with students how a cancerous tumour can form and the effect of ionising radiation on living tissue. The key point to draw out is that cancerous tissue is more affected by ionising radiation than healthy tissue.

Some students also have the misconception that 'all radiation is dangerous'. While this is true to an extent, it should be stressed to students that the risks can be minimised and controlled. Great care is taken to ensure that the exposure to sources of ionising radiation is within safe limits.

Unit 3, Theme 1 – Improving health and wellbeing

10.1 Medicines

Learning objectives

Students should learn:

- that medical drugs relieve symptoms, cure or prevent diseases
- that analgesic drugs reduce pain and anti-inflammatory drugs reduce swelling, which may also relieve pain
- that the over-use of some medical drugs may lead to long-term health problems.

Learning outcomes

Most students should be able to:

- state what a medical drug is
- distinguish between an analgesic and an anti-inflammatory drug
- explain that some drugs relieve symptoms but do not cure a disease
- describe how the over-use of symptom-relieving drugs can damage health.

Some students should also be able to:

- explain how aspirin, paracetamol and blood pressure-reducing tablets treat the symptoms of some diseases.

Support

- Provide students with a list of conditions, such as headache, sprained ankle and hay fever, and a list of drugs that can be bought over the counter to treat these conditions. Then they match the condition to the drug used to treat it.

Extend

- Ask students to produce a list of questions to ask their local pharmacist about their job. These could include their education and career path, and how they use science every day as a part of their job.
- Students could find out how different treatments available help to lower high blood pressure.

AQA Specification link-up: Science B 3.5.1.1

- Know that disease may be treated with medicines that contain useful drugs (e.g. penicillin is an antibiotic, aspirin is anti-inflammatory).
- Know that some medicines, including painkillers, help to relieve the symptoms of disease, but do not provide a cure (e.g. aspirin, paracetamol, treatments for high blood pressure, anti-depressants and sleeping pills).

Within this context, candidates should be able to use scientific data and evidence to discuss, evaluate or suggest implications of the following:

- the issues caused by the over-use of symptom-relieving drugs.

Lesson structure

Starters

What are the differences between a medical and a recreational drug? – Working in pairs, students make a list of the differences between medical and recreational drugs. Examples of paracetamol and cocaine could be quoted to start the discussion. *(5 minutes)*

What are these drugs used for? – Show students a range of everyday treatments (e.g. for headaches, insect bites, period pain and minor sports injuries). Ask them to study the packaging to identify the drug and say what conditions they can be used to treat. To support students, provide them with labels of the names of the drugs and conditions they could be used for. Ask students to put the correct cards with the appropriate drug. To extend students, ask them to try to name an alternative drug or treatment for each condition. For example, instead of using eye drops for sore eyes they could suggest bathing eyes in cold water. *(10 minutes)*

Main

- Discuss with students what they understand by the terms 'painkiller' and 'analgesic'. What conditions can these drugs be used to 'treat'? Explain simply the difference between drugs that treat or manage the symptoms of a condition and drugs that provide a cure. (Immunisation and the prevention of the spread of diseases can be mentioned here, but do not need to be discussed in detail as this will be covered later in the topic.)
- Introduce students to 'over the counter' medicines. Ask students to produce a list of the drugs found in their first-aid cupboard at home. What are they used for? Discussion could be stimulated by showing a range of products and their packaging. Be aware that many students will refer to a generic drug by its brand name – for example, Nurofen (ibuprofen) or Panadol (paracetamol). Explain the difference between a generic name for a drug and the brand name it may be sold under.
- Students should then research some common conditions that can be treated by 'over the counter' medicines. Which drugs could they take to ease these conditions? To support students provide them with a list of common ailments that would be dealt with using painkillers or anti-inflammatory drugs. Students can use information they find from the internet/drug packets/worksheets, or the Student Book, to complete the following table:

Condition	Drug used	How drug works
e.g. swollen ankle	e.g. aspirin	e.g. reduces swelling

- Ask students to feedback their findings to the rest of the class.
- Working in small groups, students should discuss and suggest what they think the long-term risks of over-using 'over the counter' drugs may be (many students will believe they cannot be harmful because they do not have to be prescribed by a doctor). Ensure that students do not just focus on the effects of an overdose – were there any other side effects they read earlier in the drug packaging?

Answers to in-text questions

a a drug that relieves symptoms, cures, or prevents a disease
b an anti-inflammatory drug – reduces swelling and pain so also analgesic
c analgesic – pain relieving. (Paracetamol can also be used to treat a fever [anti-pyretic].)
d It can cause heart attacks, and other circulatory diseases.
e a drug that can be bought without a prescription

Chapter 10 – The use (and misuse) of drugs

Plenaries

True or false –
- Paracetamol can be taken to cure a cold. [F]
- High blood pressure can be reduced by taking more exercise. [T]
- Aspirin reduces pain by reducing swelling. [T]
- Paracetamol is an anti-inflammatory drug. [F]
- 'Over the counter' drugs have no harmful effects on health. [F]

(5 minutes)

Treating high blood pressure – Working in pairs, students take on the role of a doctor and a patient with high blood pressure. The student taking the role of the doctor should explain to the patient about changes they should make to their diet and current exercise routine. They should explain how this will lower the patient's blood pressure. The patient should question the doctor on any aspect that was not well explained. The roles are then reversed and an alternative treatment explored. To support students, provide them with key words or phrases to include in their explanation. To extend students, ask them to provide a written guide to high blood pressure treatments, aimed at a person with little scientific knowledge. *(10 minutes)*

Science in context

Pharmacists and doctors need to be aware of the wide range of drugs that are available to treat a particular disease or condition. Understanding the way in which these drugs work allows the pharmacist or doctor to determine which treatment to prescribe. Many drugs are available to buy without prescription – it is important for everyone to be aware of the way in which these common drugs work in order to be able to make an informed choice about how to treat a minor condition, e.g. a cold or a muscle sprain.

Unit 3, Theme 1 – Improving health and wellbeing

10.1 Medicines

Learning objectives
- What is a medical drug?
- What is the difference between a painkiller and an anti-inflammatory drug?
- What are some of the side effects of taking 'over the counter' medicines?

links
For more information on antibodies see 10.2 Antibiotics.

Medical drugs

When you feel ill, doctors can prescribe drugs to make you feel better. These are known as **medical drugs**. These are legal, and help improve quality of life by curing, preventing or treating a disease.

Some drugs work by killing the microorganism that has made you ill. These microorganisms are known as **pathogens**. Antibiotics work in this way. Other drugs work by relieving the symptoms of an illness, like painkillers, sleeping tablets, high-blood-pressure tablets and antidepressants. These drugs do not provide a cure.

a What is a medical drug?

Aspirin

Aspirin is a painkiller, often taken to relieve headaches. Drugs that reduce pain are also known as **analgesics**. Aspirin is also an **anti-inflammatory** – that means it reduces swelling – which in turn reduces pain. Headaches are caused when capillaries in the brain become inflamed. Aspirin does not kill pathogens.

?? Did you know ...?
Aspirin has been developed from the bark of the willow tree. For thousands of years, ancient peoples used willow bark for pain relief.
From the Cinchona tree, we get quinine, which is used to treat malaria.

b What type of drug is aspirin?

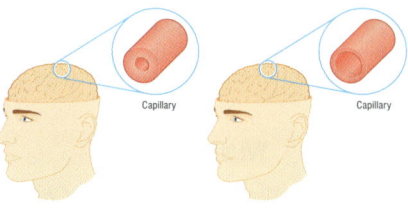

Figure 1 Action of aspirin

Paracetamol

This drug is used as a painkiller. Like aspirin, it treats the symptoms of a disease but does not kill pathogens.

When you hurt yourself, your body releases a hormone-like substance that makes you feel pain. Paracetamol reduces the production of this chemical, decreasing the pain that you feel.

c What type of drug is paracetamol?

Treatments for high blood pressure

Having high blood pressure puts strain on your heart and blood vessels. This increases the risk of heart attacks and other circulatory diseases.

If a person's blood pressure is slightly high, it can often be reduced by eating a healthier diet and exercising more. If it is dangerously high, medication will be prescribed. A range of drugs is available to reduce a person's blood pressure temporarily. These drugs act in a number of ways including:

- by reducing the amount of water in the blood
- by widening arteries, allowing blood to flow more freely.

These drugs do not provide a cure, they only relieve symptoms. Patients must take them constantly. If they stop, their high blood pressure will return.

d Why is high blood pressure dangerous?

'Over-the-counter' drugs

Many medicines are available 'over the counter'. This means people can buy them without consulting a doctor and getting a prescription. Many of these treatments are used to relieve aches, pains and itches. Others can prevent or cure ailments, such as athlete's foot. Some over-the-counter drugs help to manage recurring problems, such as migraines or period pain.

Generally, if you follow the instructions carefully, these drugs are very beneficial. However, long-term use or an overdose can have serious side effects. Interactions with other drugs or supplements can cause unexpected complications. People with other medical conditions, or those who are pregnant, should always seek the advice of a doctor before taking over-the-counter drugs.

e What is an 'over-the-counter' drug?

A number of problems can be caused by the over-use of medical drugs. The most severe conditions include kidney failure and liver damage. Anti-inflammatory drugs can also cause stomach ulcers, or problems with your intestines.

Many of these drugs cause serious complications if large quantities are taken in a short period of time – an overdose. This could result in a coma, or even death. Even if an overdose is not fatal, long-term complications can result. The long-term use of many drugs can also lead to addiction.

?? Did you know ...?
Analgesics are often used to reduce the symptoms of a headache. However, over-use of these drugs can trigger even more frequent headaches.

Figure 2 Many drugs can be bought from supermarkets or chemists to treat common ailments. These are known as 'over the counter' medicines, and do not need to be prescribed.

Summary questions

1. Copy and complete the following sentences using the words below:
   inflammatory   pain   analgesic   symptoms   cure   arteries   swelling

   Some drugs treat the .... of a disease, but do not provide a .... Aspirin is an anti-............. It is a drug that can relieve .... Paracetamol is an ...., as it can reduce .... Drugs that treat high blood pressure often widen ....

2. Explain why aspirin is described as an analgesic and an anti-inflammatory drug.

3. What are the potential health risks associated with over-use of paracetamol?

Key points

- Some medical drugs improve a person's health by curing, or preventing disease. Others relieve the symptoms of disease but do not cure it.
- Analgesic drugs reduce pain. Anti-inflammatory drugs reduce swelling, which may also relieve pain.
- Over-use of symptom-relieving drugs may lead to addiction, and can cause kidney and liver damage.

Further teaching suggestions

Role of a pharmacist
Ask a pharmacist to come into the class to talk about their job and their role in helping people to choose the correct 'over the counter' medicine.

Active ingredients
Many students will be unaware of the active ingredients that are packaged as part of a branded drug. Provide a range of commonly available treatments and ask students to study the packaging, and name the active ingredient(s) in each one. Students could then compare the prices of branded drugs against generically available (often 'own-brand') products. Do students feel there is any advantage to buying the more expensive version of the drug?

Summary answers

1. symptoms, cure, inflammatory, swelling, analgesic, pain, arteries
2. Aspirin reduces the swelling. This often reduces pressure on nerves, reducing pain.
3. Addiction, kidney and liver damage, and recurring headaches.

Unit 3, Theme 1 – Improving health and wellbeing

10.2 Antibiotics

Learning objectives

Students should learn:
- that antibiotics are drugs that kill bacteria
- MRSA is an example of an antibiotic-resistant bacterium
- that doctors should not over-prescribe antibiotics, as this can increase the number of antibiotic resistant bacteria
- how antibiotic resistant strains of bacteria develop. **[HT only]**

Learning outcomes

Most students should be able to:
- describe how antibiotics can treat disease
- describe how over-prescribing antibiotics has led to antibiotic-resistant bacterial strains
- discuss what MRSA is and why this is a problem in hospitals
- explain the financial implications to the NHS of increasing numbers of antibiotic-resistant bacteria.

Some students should also be able to:
- explain how antibiotic-resistant strains of bacteria develop **[HT only]**
- explain why antibiotics are no longer widely prescribed for mild throat infections. **[HT only]**

AQA Specification link-up: Science B 3.5.1.1

- Know that disease may be treated with medicines that contain useful drugs (e.g. penicillin is an antibiotic, aspirin is anti-inflammatory).
- Know that most bacteria, but not viruses, may be killed by antibiotics.
- Describe how some bacteria develop resistance to, or may not be easily treated by, antibiotics (e.g. MRSA). Pathogens mutate spontaneously, producing resistant strains.
- Explain how resistant strains develop: **[HT only]**
   a antibiotics kill individual pathogens of the non-resistant strain
   b individual resistant pathogens survive and reproduce, so the population of the resistant strain rises
   c now, antibiotics are not used to treat non-serious infections such as mild throat infections in order to slow down the rate of development of resistant strains.

Within this context, candidates should be able to use scientific data and evidence to discuss, evaluate or suggest implications of the following:
- the misuse of antibiotics, resulting in bacterial resistance and increased costs to the NHS.

Lesson structure

Starters

Matching pairs – Students match a list of drugs to their uses. This should revisit work covered previously on analgesics and anti-inflammatory drugs, as well as introducing antibiotic drugs. *(5 minutes)*

What are bacteria? – Project an image of a typical bacterial cell. This should only contain a cell wall and cytoplasm (containing genetic material). Students should study this image and contrast with their knowledge of animal cells (plant cell structure is not required for this specification). What are the similarities and differences? These could be presented in a comparison table. To support students, provide a 'tick-box' type table, for them to select which features are present in which type of cell. An image of an animal cell could also be projected alongside the bacterial cell. To extend students, revise the features of plant and animal cells from previous work, and ask students to produce their comparison table looking at three types of cell. Are any features specific to any of the cells? *(10 minutes)*

Main

- If not already covered in the starter activity, discuss briefly what is meant by 'bacteria' – single-celled organisms (microorganisms) that have no nucleus.
- Ask students to discuss together if they have ever been prescribed antibiotics – what condition was this for? Can they remember what the antibiotic was called? Do they know if it cured their condition? Explain to students that antibiotics work by killing bacteria; however, they have no effect on viruses or most fungi. Discuss also that penicillin is actually the name given to a family of antibiotics; these are the most commonly prescribed.
- Carry out the practical activity 'antibiotics' in the 'Practical support' box.
- Explain simply how the over-prescription of antibiotics can increase the risk of bacterial resistance, and how this has cost implications for the NHS (the evolution of antibiotic-resistant bacteria is covered in more detail in 3.5 Evolution. (Higher Tier only – explain how bacteria can develop antibiotic resistance).
- Organise students to hold a discussion on the treatment of flu. Questions that could be posed to students include: Why can't we treat flu with antibiotics? What chemical-based drugs can we take to help deal with the symptoms of flu? Why are some people who have severe flu prescribed antibiotics? [Secondary bacterial infections can occur due to an impaired immune system.]

Plenaries

Antibiotic definition – Ask students to write no more than two sentences explaining how an antibiotic works and to illustrate their answer with an example. *(5 minutes)*

Testing bacterial susceptibility – Ask students to describe, in a series of numbered steps, how microbiologists can determine which antibiotic they should use to treat a bacterial infection in the laboratory. *(10 minutes)*

Support

- Provide a series of cards listing a range of bacterial, fungal and viral conditions and a range of possible treatment drugs. Students need to select which drugs could be recommended for each of the conditions. Each drug card should note whether it is used to treat the condition or the symptoms of the condition.

Extend

- Ask students to produce an A5 flyer that could be given to doctors, reminding them why they should not over-prescribe antibiotics.

Practical support

Antibiotics

Equipment and materials required: Agar plates impregnated with different species of harmless bacteria, discs of different antibiotics, tea tree oil, washing-up liquid, toothpaste.

Details

Provide students with agar plates impregnated with different species of harmless bacteria. Ask students to act as microbiologists in hospital laboratories – they are trying to find out which antibiotic(s) will act on the bacteria provided, grown from a patient's blood samples. Students should be taught to use aseptic technique during this activity. Ask students to place discs of different antibiotics onto the agar plates. After a few days, they should measure the clear zone around each disc – the zone of inhibition. The bigger the zone the more effective the antibiotic has been. As an alternative, the practical could have been set up in a previous lesson, and students could spend time in this lesson measuring the zones of inhibition. The activity can be extended to investigate the antibiotic effect of a range of substances – examples could include tea tree oil, washing-up liquid and toothpaste.

Safety: Follow aseptic techniques. Do not exclude oxygen when incubating. Dispose of plates following CLEAPSS guidance.

10.2 Antibiotics

Unit 3, Theme 1 – Improving health and wellbeing

Learning objectives
- What is an antibiotic?
- Why should doctors be careful not to over-prescribe antibiotics?
- What is MRSA?
- How do antibiotic-resistant strains of bacteria develop? [H]

What are antibiotics?
Antibiotics are drugs that kill *bacteria*, but do not damage the cells in your body. They have *no* effect on viruses or many fungi. Some antibiotics may have side effects that damage animal cells.

Penicillin
Penicillin is an antibiotic. It is used to treat a range of conditions caused by bacteria, including ear, nose and throat infections.

Penicillin was discovered by accident! Back in 1928, pharmacologist Alexander Fleming was growing bacteria on agar plates. One day he forgot to seal one of the plates, leaving it open. When he returned he found a mould (called *Penicillium notatum*) growing.

He noticed that where the mould was growing the bacteria had died – something in the mould had killed them. The substance that had killed the bacteria was extracted and named penicillin.

a What do antibiotic drugs do?

b Name some common conditions penicillin can be used to treat.

The photo of an agar plate in Figure 2 shows where bacteria have been killed by penicillin. Notice the 'halo' around the penicillin disc. It is called the **zone of inhibition**.

c How can you tell from the appearance of an agar plate if bacteria have been killed by an antibiotic?

There are several different types of antibiotic. Each kills a different species or range of species of bacteria. To identify the type of bacteria that is making you ill, doctors may send samples, such as blood or urine, to be tested at public health laboratories in hospitals. Scientists grow the bacteria in the samples on agar plates so that they can be identified. The best antibiotic to kill the microorganism will then be prescribed to the patient.

Practical
Antibiotics
Test the sensitivity of harmless species of bacteria to particular antibiotics using antibiotic rings. You could also investigate other materials that may have antibiotic properties – these could include tea tree oil, washing up liquid and toothpaste, for example.

Figure 1 This is the mould *Penicillium* growing on an orange

Figure 2 Penicillin on an agar plate

Antibiotic resistance
Bacterial infections have been treated with antibiotics ever since penicillin was discovered and developed. Since then antibiotics have saved millions of lives.

However, unfortunately, bacteria can spontaneously **mutate**, and these mutations can lead to some strains becoming **resistant** to antibiotics. This means that many types of antibiotic will no longer kill them. They are often referred to as 'super bugs'.

How bacteria become resistant [Higher]
When antibiotics are used to treat an infection, they will kill individual pathogens that do not have antibiotic resistance. However, resistant pathogens will survive. These will then reproduce, increasing the population of the resistant strain. Failure to complete a course of antibiotics 'because you feel better' also enables the more resistant strains to survive.

To try to slow down the rate of development of resistant strains, doctors no longer prescribe antibiotics for non-serious infections, such as mild throat infections.

MRSA
MRSA is a bacterium that is resistant to many antibiotics. This means that MRSA infections are very difficult to treat and can be fatal. It is a particular problem in hospitals, because seriously ill patients are more likely to pick up this infection. However, the spread of MRSA can be limited through good hygiene practices.

d What does 'antibiotic resistance' mean?

Over-prescribing antibiotics
Many people visit their doctor with minor illnesses, such as coughs and colds. In the past, doctors would nearly always prescribe antibiotics to treat these conditions. However, many of these conditions are caused by viruses – and so the antibiotics have no effect on the illness. The widespread use of antibiotics has increased the rate of development of antibiotic-resistant bacteria. These bacteria survive and reproduce, causing an increase in their numbers.

This creates extra costs for the National Health Service:
- more nurses and doctors are needed to treat new infections caused by the drug-resistant bacteria
- more staff are required to control outbreaks of these infections, and improve hygiene standards
- research into new drugs is expensive
- the new drugs themselves are expensive.

e Why shouldn't antibiotics be used routinely to treat minor infections?

Activity
Infection control
The spread of MRSA and other 'superbugs' can be prevented in hospitals by strict hygiene procedures. Write a set of rules for medical staff, to help stop the spread of MRSA in hospital wards.

Figure 3 MRSA bacteria

AQA Examiner's tip
It is a mistake to think that your body becomes resistant to the antibiotics, so they stop working. It is the bacteria which cause the disease inside your body that may become resistant to the antibiotics.

Key points
- Antibiotics are drugs that kill bacteria.
- The over-prescribing of antibiotics has increased the number of antibiotic resistant bacteria. This has significantly increased costs for the NHS.
- Spontaneous mutations can result in bacteria developing antibiotic resistance. When antibiotics are taken, the non-resistant strains are killed. The resistant strains survive and reproduce, increasing their population. [H]

Summary questions
1. Copy and complete using the words below:
   antibiotic   bacteria   MRSA   penicillin   resistant

   .... is an .... drug. It works by killing .... Some bacteria are now becoming .... to antibiotics. One example is ....

2. Tonsillitis is a bacterial infection. It causes swelling in the tonsils, making the throat very sore. If you have it:
   a. how can taking antibiotics make you better?
   b. how can taking aspirin help relieve the symptoms?

3. How do antibiotic-resistant strains of bacteria develop? [H]

links
For information about how bacteria evolve to become resistant to antibiotics look back at 3.5 Evolution.

Answers to in-text questions

a They kill bacteria.

b ear, nose and throat infections

c There is a ring around the disc where the bacteria have been killed.

d Certain types of bacteria are no longer vulnerable to antibiotics.

e Minor infections are often caused by viruses, so antibiotics will have no effect. Some infections will clear up without the use of antibiotics. The widespread use of antibiotics has increased the rate of development of antibiotic-resistant bacteria.

Summary answers

1. penicillin, antibiotic, bacteria, resistant, MRSA

2. **a** The antibiotics kill the bacteria that are causing the disease.

   b Aspirin reduces the swelling in the tonsils, reducing pain in the throat.

3. Spontaneous mutations can result in a bacteria developing antibiotic resistance. When antibiotics are taken the non-resistant strains are killed. The resistant strains survive and reproduce increasing their population.

Activity

Infection control
Students could begin by discussing the steps that they think all hospital staff should take to prevent the spread of infectious diseases. This could include what they should wear and everyday procedures they should follow.

MRSA should then be introduced. Basic information is provided in the Student Book. Students should research techniques that are used by nurses to prevent the spread of MRSA. Key terms that could be entered into a search engine are – 'MRSA prevention', 'MRSA and hospitals' and 'MRSA spread'. You may also wish to direct students to www.mrsainfection.org, which links to many other webpages.

Students should analyse the information they have found to produce a standard hygiene procedure for nurses and other hospital staff to follow, to help prevent the further spread of MRSA.

Unit 3, Theme 1 – Improving health and wellbeing

10.3 Approving a new drug

Learning objectives

Students should learn:
- that new drugs are tested on animals to assess their effects on living organisms
- that new drugs are tested on human volunteers, before being given to patients
- the advantages and disadvantages of testing drugs on animals and humans.

Learning outcomes

Most students should be able to:
- state that new drugs are tested chemically, then on animals, healthy volunteers and finally patients
- state some benefits and drawbacks of testing new drugs on animals and humans.

Some students should also be able to:
- explain the purpose of each stage of testing a new drug
- evaluate the usefulness of animal and human testing.

Specification link-up: Science B 3.5.1.1
- Know that before a medicine can be used for treating a disease it undergoes extensive clinical trials.

Within this context, candidates should be able to use scientific data and evidence to discuss, evaluate or suggest implications of the following:
- the issues of testing new drugs on animals and humans.

Lesson structure

Starters

Continuum activity – Students line up along an imaginary line with 'yes' at one and and 'no' at the other in response to questions about testing new drugs on animals and humans. Individual students are asked to explain why they are standing in a particular place. This works best if started with sweeping black-and-white statements, such as 'All animal testing should be banned' before more subtle statements, such as 'The health of a human is worth more than the life of a mouse'. *(5 minutes)*

Common medicines – Ask students to list at least five medicines which they have used or heard of; each medicine listed should be used to treat a different condition. To support students, provide a series of images to prompt thought. These should be differentiated according to the needs of the group. For example, you could provide images of common conditions, or images of some common medicines students may have been exposed to. To extend students, ask them to explain how each of the medicines they have named benefits humanity. Ask them to put the medicines in order of importance and justify their decisions. *(10 minutes)*

Main

- Pose the question 'How can we know that medicines are safe?' as an opener on this topic.
- Provide the steps involved in a clinical trial in a jumbled order, as a cut-and-stick activity. The labels provided at the side of the flow chart in the Student Book could be added as an extension for more able students.
- Without any prompting, ask the class to vote on whether they are for or against animal testing of new medicinal drugs. Record the results on the board.
- Split the class into smaller groups – half of the groups are to be in favour of animal testing, and half against. Give each group ten minutes to research and discuss the arguments for, or against, testing new drugs on animals. Each group should produce a bullet pointed list of their findings, which can be collated to form a class summary sheet.
- Hold a class discussion on the students' research. Ask the class to vote again. Have their opinions changed in light of their discussions? Explain that by carrying out their own research, they are now able to make an informed decision.

Plenaries

How do we know medicines are safe? – Pose the question 'How do we know that medicines are safe?' again. Students compete in groups to find as many answers as possible. *(5 minutes)*

Continuum activity – Return to the continuum activity from the start of the lesson. Make the same statements again and see if students' opinions have changed. To support students, reuse the images from the start of the lesson to reinforce the statements being made. To extend students, also ask them to do it in the role of the stakeholder they took on earlier in the lesson. *(10 minutes)*

Support

- Further (support) work on the advantages and disadvantages of testing drugs on animals: Provide students with a pack of cards containing statements about animal testing e.g. short life cycle, potential suffering. Students categorise into 'arguments for' and 'arguments against' and write a sentence explaining their classification of each argument.

Extend

- Students could be introduced to the concept of placebos and double-blind tests. Why are these important in clinical trials? Students could be asked to set up their own clinical trial on the effect of caffeine on heart rate. The test could involve students drinking caffeinated and decaffeinated coffee, and their pulse could be measured to study heart rate.

Answers to in-text questions

a Medicines and Healthcare products Regulatory Agency
b To test for side effects.

Practical support

Testing a new antacid

Equipment and materials required: Eye protection, hydrochloric acid (0.4 mol/dm³), burette, pipette filler, 25 cm³ volumetric pipette, 100 cm³ conical flask, phenolphthalein indicator, clamp and stand, white tile, funnel, pestle and mortar, selection of antacids.

Details

Fill the burette with hydrochloric acid, crush one dose of antacid, dissolve it in 25 cm³ distilled water in the conical flask and add a few drops of phenolphthalein. Slowly titrate the acid into the antacid solution; trial runs will be necessary to get an idea where the end-point is. Compare the amount of acid one dose of each antacid could neutralise.

Safety: Wear eye protectin. CLEAPSS Hazcard 47A Hydrochloric acid – corrosive. CLEAPSS Hazcard 32 Phenolphthalein – irritant.

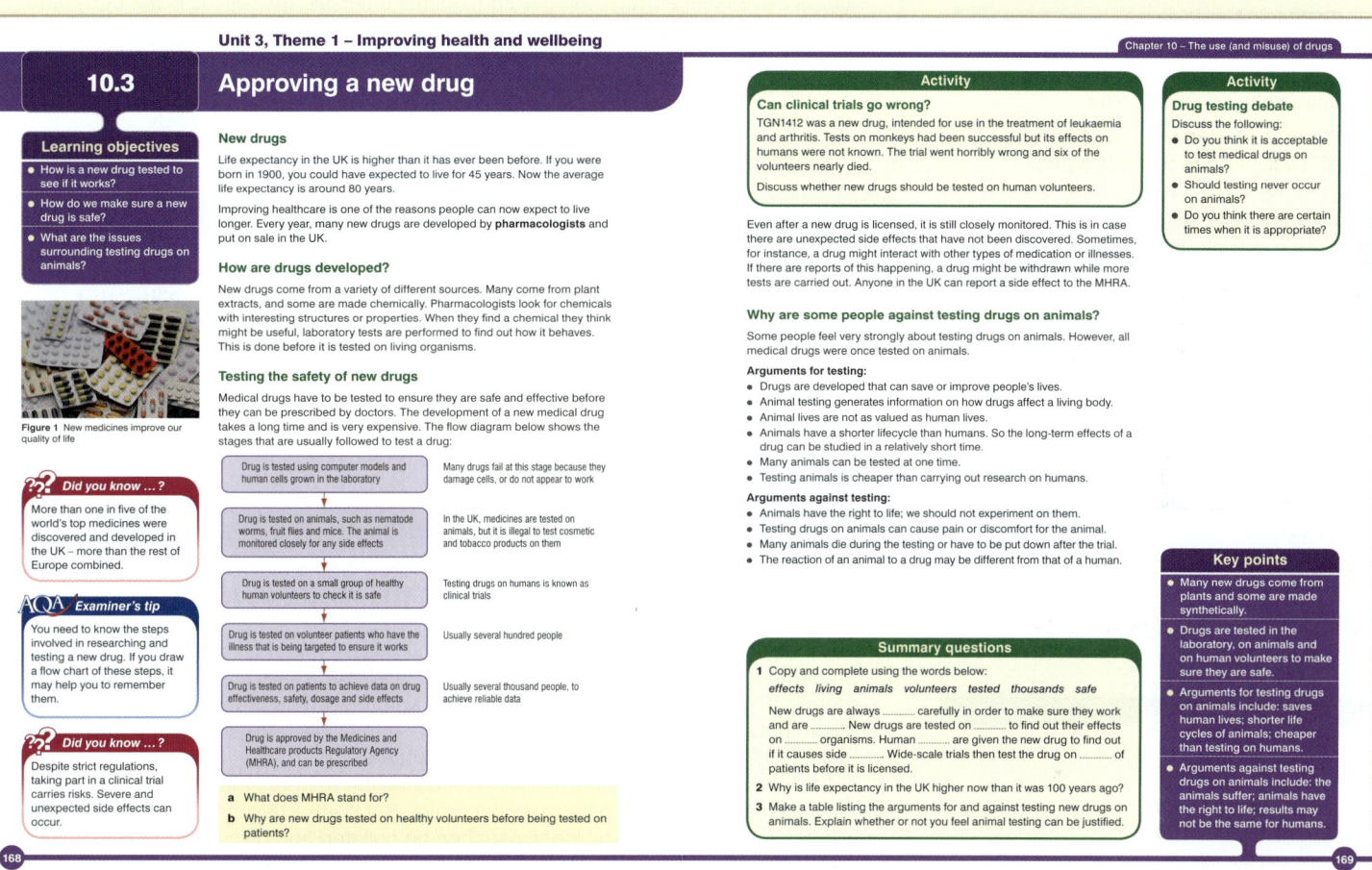

Summary answers

1 tested, safe, animals, living, volunteers, effects, thousands
2 Drugs have been developed that help to treat illnesses which were previously life threatening. Other possible answers could refer to quality of diet/medical advances/better quality accommodation/improved sanitation/less manual labour.
3

Arguments for include:	Arguments against include:
Can save human lives	Animals have a right to life
Provides information on effect on a living body	Testing would be against the will of the animals
Animal lives are not as valuable	Can cause suffering to animals
Animals have a shorter life cycle	Many animals may die as a result
Cheaper than testing on humans	Animal reactions are often different to those of a human

Activity

Drug testing debate

Using the photo in the Student Book ask students how this makes them feel. Explain that the animal is participating in drug trials for a potentially life saving drug. Does this change their feelings? Hold a class discussion on medical drug testing on animals. Using this information (and any activity in this box) ask students to write a newspaper article about drug testing on animals. Alternatively, students write a presentation for a court to persuade them to ban, or allow testing on animals.

Can clinical trials go wrong?

This activity could be introduced by students watching TV clips of the news, or reading newspaper articles about the clinical trial on TGN1412. Students should be asked whether or not they would be prepared to take part in a clinical trial. Divide the class into groups based on their opinion. Each group should list as many supporting arguments for their viewpoint as possible, and should justify their answers. Each group should then present their case. Has anyone in the class changed their opinion?

Unit 3, Theme 1 – Improving health and wellbeing

10.4 Recreational drugs

Learning objectives

Students should learn:
- that drugs are chemicals that affect the body in a beneficial or harmful way
- that recreational drugs are taken for a person's personal enjoyment, and have no medical benefits
- that people can become addicted to drugs, and suffer withdrawal symptoms if they attempt to stop taking them.

Learning outcomes

Most students should be able to:
- list examples of legal and illegal drugs
- state harmful effects some drugs have on the body
- explain what is meant by the terms 'addicted' and 'withdrawal symptoms'.

Some students should also be able to:
- explain how some drugs cause the changes that occur in the body
- evaluate the impact of legal and illegal drugs.

AQA Specification link-up: Science B 3.5.1.1

- Give examples of recreational drugs that may harm the body (alcohol, nicotine, antidepressants, amphetamines, barbiturates, heroin, cocaine and cannabis).
- Know that some people may become dependent on, or addicted to, recreational drugs because the drug changes some of the chemical processes in the body, and they suffer withdrawal symptoms without them (e.g. nicotine in tobacco).

Within this context, candidates should be able to use scientific data and evidence to discuss, evaluate or suggest implications of the following:
- the impact of legal (alcohol and tobacco) and illegal drugs on the body
- the issues caused by the over-use of symptom-relieving drugs.

Lesson structure

Starters

Name some drugs – Ask students to make a list of ten drugs. Which are the most dangerous? To support students, provide them with some anagrams of common drugs. To extend students, ask them to justify why they have positioned the drugs in a certain order. For example is heroin more dangerous than cannabis because you can die immediately from taking it? *(5 minutes)*

Legal drugs? – Provide students with a list of drug names, to sort into legal and illegal categories. Then ask them to work in pairs discuss why the government has made some drugs illegal. *(10 minutes)*

Main

- Begin by defining what is meant by the term 'drug' and the difference between medical and recreational drugs. Explain that in this lesson students are going to focus on recreational drugs.
- Ask students to make lists of the reasons why people might take recreational drugs. Students should work by themselves initially, and then share their ideas in small groups. These ideas can then be discussed with the whole group. Parallels can be drawn between why people do some sports – especially adrenaline sports (to get a 'buzz') –and why some people take recreational drugs.
- Ask students to work in pairs to make a list of all the dangers they can think of which are associated with taking recreational drugs. Set them a target to try and come up with ten reasons. Students' ideas can then be collated and discussed.
- Ensure that the following terms have been covered during the class discussions: addictive, stimulant, depressant, hallucinogen and withdrawal symptoms.
- Using the internet/leaflets/Student Book, students produce an information sheet about alcohol, nicotine, antidepressants, heroin, cocaine, amphetamines, barbiturates and cannabis. The sheet needs to be divided into eight sections to represent each of the eight drugs they are researching. Each section must include information on whether the drug is legal, a stimulant, depressant or hallucinogen, how addictive it is, harmful effects on the body and any known positive effects of the drug, e.g. cannabis is used by some people who sufferer with multiple sclerosis. Students are to summarise the information so that it fits onto one side of A4 paper. It should be written in a style that could be handed out to year 10 students in a PSHE lesson. Or they could do a PowerPoint presentation – each member of the group could look at one drug then some students can prepare a slide show and others prepare the presentation. (This is an opportunity for team work.)
- Individuals report their findings to the class. The key information can then be recorded in a large table on the board and shared with the class.
- Using the whole class results, students are to swap their own information sheet with a partner. This is to be peer assessed in terms of its content, readability and presentation. Any key missing information should be added.

Plenaries

Associated dangers of taking drugs – Ask students to write a list, in pairs, of reasons why drugs should be illegal, e.g. damaging to health, can stimulate crime, cause driving accidents, fights, etc. *(5 minutes)*

Student descriptions – Give students the opportunity to write their own descriptions of drug addiction and withdrawal symptoms, then in pairs, share with the group. To support students, provide them with some key words or phrases that they must include in their definitions. To extend students, ask them to illustrate their definitions with examples of withdrawal symptoms. *(10 minutes)*

Support

- Provide students with an information-sheet template to complete the main lesson task. They can use this template to record their research findings on how alcohol, nicotine, antidepressants, amphetamines, barbiturates, heroin, cocaine and cannabis cause their effects on the body, and the associated dangers. For some students the template could be partially filled in so they have to research fewer drugs.

Extend

- Ask students to choose one of the five drugs: heroin, cocaine, amphetamines, barbiturates or cannabis and find out in detail using secondary sources how it causes its effects on the body.

Chapter 10 – The use (and misuse) of drugs

Further teaching suggestions

The cannabis debate

Should cannabis be legalised? In groups, students research and debate the reasons for and against legalising cannabis. Give students different roles to play in the debate, e.g. a politician, a paramedic, a member of the police, a teenager, a parent whose child died as a result of a drug overdose and a multiple sclerosis sufferer who uses cannabis.

What support can recovering addicts receive?

Students research differing treatments available to recovering drug addicts, e.g. use of methadone, counselling and rehabilitation centres.

Drug awareness campaign

Students draw or write a storyboard for an advert to be shown on television as part of a drug awareness campaign. The advert should be aimed at teenagers.

Answers to in-text questions

a They alter the chemical reactions that take place inside the body.
b Their body becomes used to the changes that the drug has made. They crave the drug and think they can't survive without it.
c It affects the nervous system and damages the liver.
d They can damage the body and can kill even in small quantities.
e **three** from: barbiturates, heroin, amphetamines, cannabis, cocaine, etc. (Remember barbiturates and amphetamines are not necessarily illegal – they can be prescribed drugs.)

Unit 3, Theme 1 – Improving health and wellbeing

10.4 Recreational drugs

Learning objectives
- How do drugs affect the body?
- What are recreational drugs?
- What is drug addiction?

How do drugs affect us?

Drugs are chemical substances that affect the way our bodies work. Some have a helpful effect when we are ill. However, many can seriously harm our health and even cause death. **Recreational drugs** are taken for a person's enjoyment. They have no medical purpose.

Drugs work by altering the chemical reactions that take place inside the body. If the body gets used to these changes, it may become dependent on a drug. If this happens to someone, they become **addicted**. They start to crave the drug and feel that they can't survive without it.

If addicts attempt to stop taking a drug they suffer **withdrawal symptoms**. This happens because the body is no longer being provided with a chemical that it is used to having. It takes a long time for the body to get used to not having the chemical and for it to start working normally again. The symptoms can be very unpleasant and make it even harder to give up.

Every drug produces different amounts and types of withdrawal symptoms. Physical withdrawal symptoms include: headaches, sweating, palpitations, muscle tension, sickness and diarrhoea. Emotional withdrawal symptoms include: feeling panicked or irritable, being unable to sleep or concentrate and depression.

a What sort of effects do drugs have on the body?
b What happens if a person becomes addicted to a drug?

Legal drugs

Some drugs that can harm our body are legal. These include alcohol, tobacco and prescription drugs, such as antidepressants and barbiturates.

- Many people think **alcohol** is perfectly safe, but it can seriously damage your body. It affects your nervous system and damages your liver.
- Most people realise **smoking** is dangerous, but around one in five of the population smoke. Cigarettes contain a drug called nicotine, which is very addictive. This makes it very difficult to stop smoking once you have started. Smoking seriously increases your risk of cancer and lung and heart diseases.
- **Antidepressants** are prescribed by doctors to help relieve depression. They provide patients with short-term benefits. However, if people use them for a long time, it can lead to addiction.
- **Barbiturates** are sedatives or tranquillisers and are present in some sleeping tablets prescribed by doctors. They are highly addictive and people can easily take an overdose. Long-term use can lead to depression.

Figure 1 These are examples of legal drugs

c What harmful effects does alcohol have on your body?

Illegal drugs

Some drugs that can damage your body, even in very small amounts, have been made illegal by the government. Despite this, many young people will be offered these drugs and some may want to experiment with these deadly chemicals. Others may feel pressurised by their friends to join in.

Illegal drugs can be grouped together by their effect on the body. There are three main groups:

1 **Stimulants** – make you feel more alert, awake and generally happier. They work by speeding up the nervous system
2 **Depressants** – reduce feelings of stress and panic, which can make you feel more relaxed. They work by slowing down the activity of the nervous system.
3 **Hallucinogens** – alter what you see and hear. They work by interfering with normal brain function.

d Why are some drugs illegal?
e Name **three** groups of illegal drugs.

Illegal drugs can increase the risk of some diseases of the lungs and heart. Smoking cannabis, for instance, can increase the risk of lung disease. Cocaine tightens up blood vessels in the heart, which increases the risk of heart disease. One recent study of people admitted to Accident and Emergency units for chest pains found that 10 per cent of them had cocaine in their blood.

The table below shows some examples of illegal drugs and their effects on the body.

Drug	Type	Harmful effect on the body
Barbiturates, e.g. dolls	Depressant	Addictive; hallucinations; heart attack
Heroin	Depressant	Addictive; risk of coma
Amphetamines, e.g. speed	Stimulant	Addictive; memory loss; increased blood pressure
Cocaine	Stimulant	Addictive; aggression; brain damage; damage to soft tissues of the nose
Cannabis	Hallucinogen	Addictive; mental health damage; bronchitis; lung cancer

Activity

Is this 'cool'?

This woman has died from taking drugs.

Discuss the following:
a Why do people take drugs?
b What short-term and long-term risks are involved in drug taking?
c What effects can drugs have on your body and on your behaviour?
d Photographs like the one above are used by the government and health professionals to try to shock young people and stop them from taking drugs. Do you feel that campaigns like this work?
e How would you inform people of the dangers of taking drugs?

Summary questions

1 Copy and complete using the words below:
   addicted   withdrawal symptoms   unpleasant   drugs

   Chemicals that affect the way your body works are called .... If you take them too often you may become .... When addicts stop taking drugs, they suffer .... These can be very .... and make it harder to give up.

2 State **four** reasons why people may be tempted to take drugs.
3 When is a barbiturate a legal drug?

Key points

- Drugs are chemicals that affect the body in a helpful or harmful way.
- Recreational drugs are taken for personal enjoyment. They have no medical benefits.
- People can become addicted to drugs and suffer withdrawal symptoms if they try to stop taking them.

Activity

Is this 'cool'?

In addition to using the Student Book, students could be shown the Leah Betts video 'Sorted' as a stimulus for discussing the issues raised in this activity. Students should be asked to explore their response to the material. This could be carried out independently, in groups or as a class activity. Students should be encouraged to look at the issues from a scientific, as well as personal perspective. Students should be asked to justify why they believe in a particular viewpoint. For example, if a student feels that shock tactics from an advertising campaign are not effective, they should be asked to give reasons why they feel this. The activity could be extended further by asking students to produce their own campaign to try to persuade young people not to take drugs. This could take the form of a television advertisement, a poster, a website or a leaflet.

Summary answers

1 drugs, addicted, withdrawal symptoms, unpleasant
2 curiosity, peer pressure, stress, pain relief, etc.
3 When it is prescribed by a doctor as a sleeping tablet.

Unit 3, Theme 1 – Improving health and wellbeing

10.5 Tobacco

Learning objectives

Students should learn:

- that tobacco smoke contains tar, nicotine and carbon monoxide
- that smokers are more likely than non-smokers to suffer from respiratory and circulatory diseases
- that tobacco smoke contains carbon monoxide, which reduces the oxygen-carrying capacity of the blood.

Learning outcomes

Most students should be able to:

- state some harmful effects of smoking
- name the main components of tobacco smoke
- describe how tar, nicotine and carbon monoxide can damage a person's health.

Some students should also be able to:

- explain in detail how tobacco smoke can cause disease.

AQA Specification link-up: Science B 3.5.1.1

- Know that tobacco smoke contains substances that cause diseases of the respiratory and circulatory systems.
- Know that tobacco smoke also contains carbon monoxide, which reduces the oxygen-carrying capacity of the blood.

Within this context, candidates should be able to use scientific data and evidence to discuss, evaluate or suggest implications of the following:

- the impact of legal (… tobacco) and illegal drugs on the body
- the link between smoking and respiratory and circulatory diseases.

Lesson structure

Starters

Smoking anagrams – Provide students with anagrams of some key words associated with smoking, e.g. tar, nicotine, high blood pressure and cancer. *(5 minutes)*

Why is smoking harmful? – Small groups of students complete a 'graffiti' exercise. Smoking is written in large letters in the centre of a piece of paper. Each group then adds as many words or terms to the paper as possible in a short period of time (for example 3 minutes). Posters should then be compared and misconceptions and relevant points can then be discussed. To support students, provide them with some key areas to organise their ideas. For example, components of tobacco and health. To extend students, ask them to define the key terms on their poster such as carbon monoxide and tobacco. *(10 minutes)*

Main

- If the 'Why is smoking harmful?' starter was not completed, ask small groups of students to discuss the dangers to their health that smoking causes. These ideas can be noted in bullet-point format on a large sheet of paper, and then fed back to the class.
- The 'smoking machine' demonstration (see 'Practical support' opposite for details) can be used to look at two of the main components of tobacco smoke – tar and carbon dioxide (and, by the implication of incomplete combustion, carbon monoxide). These components, along with the presence of nicotine in a cigarette, can then be linked to the diseases they cause.
- Show students the photograph of a diseased lung in the Student Book – discuss how this differs from a healthy lung, and what has caused these changes.
- Show students a slide/diagram of ciliated cells. Explain their role in keeping the respiratory tract clean, and how they are damaged by smoking.
- Select data and/or graphs showing the percentage of a population who smoke, against the incidence of lung cancer within that population. Does the data show a link? This exercise could be repeated by looking at data for the incidence of heart disease and low birth-weight babies. Students should be asked to draw their own conclusions – is the evidence compelling, or is further research required? At what point does research evidence become accepted fact?
- Show students a range of cigarette packets. Student should be asked to look at the government warning labels they contain. Hold a discussion, to gauge whether or not students think these warnings would persuade them not to buy cigarettes. Do they think that the legal age for buying cigarettes (18) is low enough? Can they think of any better ways that the government could use advertising campaigns to prevent young people from smoking?
- Discuss the reasons why it is hard to give up smoking. Nicotine patches and chewing gum could be used as a stimulus to the discussion. Revise the concept of addiction and withdrawal symptoms.

Plenaries

Tobacco components – Ask students to match a list of tobacco smoke components (tar, nicotine and carbon monoxide) to the diseases and health problems they cause. *(5 minutes)*

Smokers' cough – Ask students to write a *short* paragraph explaining why smokers cough a lot in the mornings. To support students, provide them with key words, such as cilia and mucus to include in their explanation. To extend students, ask them to extend their answer to explain what would happen if a smoker didn't cough to clear their lungs. *(10 minutes)*

Answers to in-text questions

a Because tobacco contains nicotine, an addictive drug.
b Chemicals in the tar.
c For example, bronchitis, emphysema, lung cancer, stroke and heart attacks.
d They sweep mucus up the trachea, so they can be swallowed down the oesophagus into the stomach.
e It carries oxygen around the body.

Support

- Provide students with cards listing the components of tobacco smoke, their effect on the body, and why they are dangerous. Students need to match the statements.

Extend

- Ask students to carry out some research into Richard Doll – the scientist who first proved there was a link between smoking, lung cancer and an increased risk of heart disease. Students could write a short account detailing his work in this field.

Practical support

Components of tobacco smoke

Equipment and materials required: The smoking machine – should consist of cigarette smoke being drawn through a U-shaped tube containing cotton wool, and a test tube containing limewater, hand pump, eye protection.

Details
Connect the smoking machine to a hand pump, and light the cigarette. The pump should draw the smoke through the cotton wool and the limewater. The cotton wool will turn brown as tar is collected, and the limewater will turn cloudy due to the presence of carbon dioxide in the cigarette smoke (avoid skin contact).

Safety: Must be carried out in a fume cupboard. Use gloves when disposing of cotton wool into a sealed plastic bag. See CLEAPSS Guide L195 'Safer chemical, safer reactions'. CLEAPSS Hazcard 18 Limewater – irritant.

Summary answers

1. Tar – Contains chemicals that cause cancer
   Nicotine – Addictive and makes the heart beat faster
   Carbon monoxide – Lowers the oxygen-carrying capacity of the blood
2. £2007.50
3. They have to cough up the mucus from their lungs, which has built up overnight while they have slept.

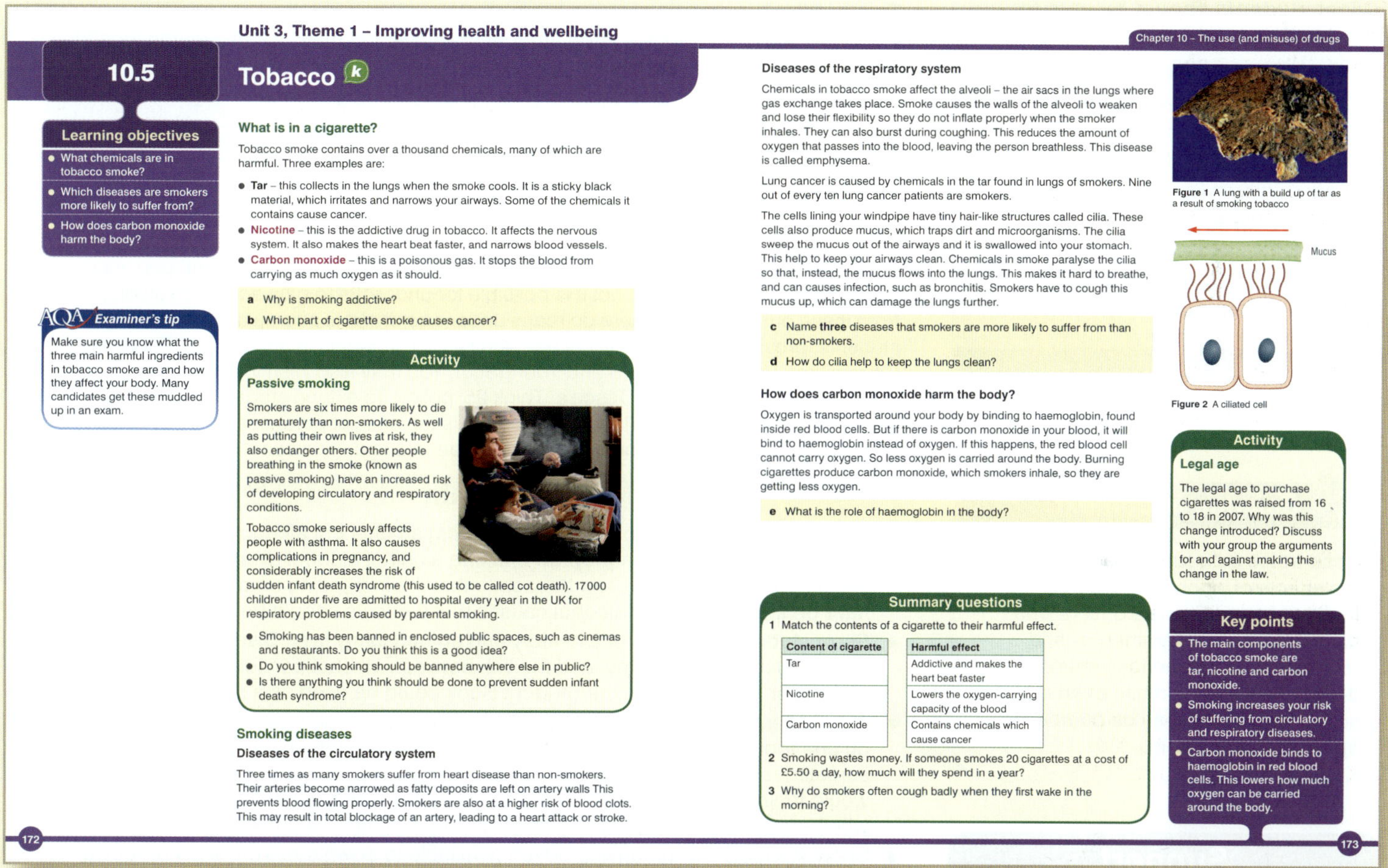

Unit 3, Theme 1 – Improving health and wellbeing

10.5 Tobacco

Learning objectives
- What chemicals are in tobacco smoke?
- Which diseases are smokers more likely to suffer from?
- How does carbon monoxide harm the body?

What is in a cigarette?
Tobacco smoke contains over a thousand chemicals, many of which are harmful. Three examples are:

- **Tar** – this collects in the lungs when the smoke cools. It is a sticky black material, which irritates and narrows your airways. Some of the chemicals it contains cause cancer.
- **Nicotine** – this is the addictive drug in tobacco. It affects the nervous system. It also makes the heart beat faster, and narrows blood vessels.
- **Carbon monoxide** – this is a poisonous gas. It stops the blood from carrying as much oxygen as it should.

a Why is smoking addictive?
b Which part of cigarette smoke causes cancer?

AQA Examiner's tip
Make sure you know what the three main harmful ingredients in tobacco smoke are and how they affect your body. Many candidates get these muddled up in an exam.

Activity
Passive smoking

Smokers are six times more likely to die prematurely than non-smokers. As well as putting their own lives at risk, they also endanger others. Other people breathing in the smoke (known as passive smoking) have an increased risk of developing circulatory and respiratory conditions.

Tobacco smoke seriously affects people with asthma. It also causes complications in pregnancy, and considerably increases the risk of sudden infant death syndrome (this used to be called cot death). 17 000 children under five are admitted to hospital every year in the UK for respiratory problems caused by parental smoking.

- Smoking has been banned in enclosed public spaces, such as cinemas and restaurants. Do you think this is a good idea?
- Do you think smoking should be banned anywhere else in public?
- Is there anything you think should be done to prevent sudden infant death syndrome?

Smoking diseases
Diseases of the circulatory system
Three times as many smokers suffer from heart disease than non-smokers. Their arteries become narrowed as fatty deposits are left on artery walls. This prevents blood flowing properly. Smokers are also at a higher risk of blood clots. This may result in total blockage of an artery, leading to a heart attack or stroke.

Diseases of the respiratory system
Chemicals in tobacco smoke affect the alveoli – the air sacs in the lungs where gas exchange takes place. Smoke causes the walls of the alveoli to weaken and lose their flexibility so they do not inflate properly when the smoker inhales. They can also burst during coughing. This reduces the amount of oxygen that passes into the blood, leaving the person breathless. This disease is called emphysema.

Lung cancer is caused by chemicals in the tar found in lungs of smokers. Nine out of every ten lung cancer patients are smokers.

The cells lining your windpipe have tiny hair-like structures called cilia. These cells also produce mucus, which traps dirt and microorganisms. The cilia sweep the mucus out of the airways and it is swallowed into your stomach. This help to keep your airways clean. Chemicals in smoke paralyse the cilia so that, instead, the mucus flows into the lungs. This makes it hard to breathe, and can causes infection, such as bronchitis. Smokers have to cough this mucus up, which can damage the lungs further.

c Name **three** diseases that smokers are more likely to suffer from than non-smokers.
d How do cilia help to keep the lungs clean?

How does carbon monoxide harm the body?
Oxygen is transported around your body by binding to haemoglobin, found inside red blood cells. But if there is carbon monoxide in your blood, it will bind to haemoglobin instead of oxygen. If this happens, the red blood cell cannot carry oxygen. So less oxygen is carried around the body. Burning cigarettes produce carbon monoxide, which smokers inhale, so they are getting less oxygen.

e What is the role of haemoglobin in the body?

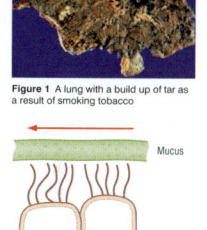

Figure 1 A lung with a build up of tar as a result of smoking tobacco

Figure 2 A ciliated cell

Activity
Legal age

The legal age to purchase cigarettes was raised from 16 to 18 in 2007. Why was this change introduced? Discuss with your group the arguments for and against making this change in the law.

Summary questions
1. Match the contents of a cigarette to their harmful effect.

Content of cigarette	Harmful effect
Tar	Addictive and makes the heart beat faster
Nicotine	Lowers the oxygen-carrying capacity of the blood
Carbon monoxide	Contains chemicals which cause cancer

2. Smoking wastes money. If someone smokes 20 cigarettes at a cost of £5.50 a day, how much will they spend in a year?
3. Why do smokers often cough badly when they first wake in the morning?

Key points
- The main components of tobacco smoke are tar, nicotine and carbon monoxide.
- Smoking increases your risk of suffering from circulatory and respiratory diseases.
- Carbon monoxide binds to haemoglobin in red blood cells. This lowers how much oxygen can be carried around the body.

Further teaching suggestions

Smoking and pregnancy
Students research the reasons why women who smoke have babies with a lower birth weight, and the long-term potential health risks associated with this.

Don't smoke campaign
Students could produce a poster/leaflet/PowerPoint presentation aimed at teenagers to persuade them not to start smoking.

Giving up
Students in small groups could be given one 'giving up smoking' aid (e.g. patches, chewing gum, nasal spray). They need to find out how it is meant to help (e.g. something to hold, something to chew), what chemicals it contains and how these help a person to give up smoking. Does the product carry any health risks? Can they find any data on its success rates? Students report their findings to the class.

Science in context

Tobacco smoke has a negative effect on the health of the population. This not only affects the people who choose to smoke, but also as a result of people breathing in 'second-hand' smoke. People need to be aware of the dangers of smoking to their own health and to people around them. This consideration led to the Smoke-free (Premises and Enforcement) Regulations 2006. People working in the NHS are regularly involved in the treatment of medical conditions caused by smoking. Scientists are also carrying out research into the link between tobacco smoke and sudden infant death syndrome.

Unit 3, Theme 1 – Improving health and wellbeing

10.6 Alcohol

Learning objectives

Students should learn:
- that alcohol is a drug that affects the nervous system
- that alcohol can damage the liver and the brain
- that guidelines exist for the safe use of alcohol.

Learning outcomes

Most students should be able to:
- describe the immediate effects alcohol has on the body
- describe the long-term effects alcohol has on health
- evaluate strategies to discourage young people from abusing alcohol.

Some students should also be able to:
- explain what is meant by the term 'alcohol tolerance'.

Specification link-up: Science B 3.5.1.1

- Know that alcohol affects the nervous system by slowing down reactions (loss of self-control) and causes long-term damage to the liver and brain.

Within this context, candidates should be able to use scientific data and evidence to discuss, evaluate or suggest implications of the following:

- the impact of legal (alcohol ...) and illegal drugs on the body.

Lesson structure

Starters

True or false? –
- People who have drunk a large amount in an evening may still be over the legal drink-drive limit the following morning. [T]
- Alcohol is a legal drug and is therefore not harmful to your health. [F]
- Alcohol contains the drug ethanol. [T]
- A pint of beer contains one unit of alcohol. [F]
- Drinking alcohol damages your liver. [T]

(5 minutes)

Where is alcohol available? – Students make a list of the types of outlets where alcohol is available to purchase – for example, supermarket, off licence, newsagent, pub, etc. Students discuss whether or not it is possible for under-18s to gain access to alcohol from these outlets. If not, where do many under-age drinkers source alcohol? To support students, provide them with a list of age-verification schemes to think about. These include CitizenCard, Validate, ProofGB and YoungScot. You may also wish to include retailers' voluntary schemes, such as Tesco's 'Think 25' policy. To extend students, they could further discuss the wide availability of alcohol. Is the ease of purchase (and hence its social acceptability) a factor that encourages young people to drink alcohol? *(10 minutes)*

Main

- Provide students with a pack of cards stating an effect of alcohol on the body. Students have to imagine they are drinking a unit of alcohol every 30 minutes; they should sort the cards to show the action that the increasing amount of alcohol would have on their body. How alcohol causes these effects should be discussed.
- Show students a picture of a healthy and an unhealthy liver. Discuss the action of the liver on alcohol in the body, and the link between alcohol consumption and cirrhosis of the liver. The photos in the Student Book could be used as a stimulus.
- Ask students to perform a range of simple tasks (walk along a straight line, balance on one foot for 30 seconds, touch their nose, read a sign in the distance), which are sometimes used by the police to assess whether a person has been drinking. How would their responses be different if they had been drinking? Why? Students discuss whether or not this is a reliable test for the level of alcohol consumption.
- Show students a drink-driving advert (available to download from www.dft.gov.uk/think/). Why is it important not to drink and drive? Discuss the drink-drive limit – what does this mean in terms of pints of beer and glasses of wine? Students could be given empty glasses for a range of drinks such as beer, wine and vodka. They could then fill them with water to the level they think equates to one unit.
- Explain to students what is meant by the term 'alcohol tolerance' and how this increases the risks of alcohol-related conditions and alcoholism.
- Students should use the prompts in the 'Activity' box to discuss the pressures put on younger people to start drinking – and why the social problem of binge drinking has become such an issue in the UK.

Plenaries

How many units? – Show students a variety of drinks. Students should estimate how many units of alcohol are found in each. *(5 minutes)*

Dangers of alcohol – Encourage students to write a list of as many of the problems or conditions caused by drinking alcohol in large quantities as they can think of. For each problem or condition listed, they should define it as a short- or long-term effect of alcohol. To support students, provide them with some key terms, which they could use in their lists. To extend students, ask them to extend their lists further by including social problems associated with excessive alcohol consumption. *(10 minutes)*

Science in context

Road traffic police need knowledge of the effects of alcohol on the body so that they can recognise drivers who are over the limit. Paramedics and other emergency services need to recognise if a person has been drinking so that the person can be offered the appropriate treatment.

Support

- Students could produce a poster, explaining the dangers of drinking to excess, to reinforce how alcohol damages the body.

Extend

- Discuss what makes '1 unit' of alcohol – how is this actually calculated? Students could be given data on a range of drinks (e.g. different strength beers), along with their alcohol content and volume. They can then calculate how many units of alcohol each drink contains.
- Students discuss how a driver might drink more than the drink-drive limit, without realising it.

Chapter 10 – The use (and misuse) of drugs

Further teaching suggestions

Alcohol and reactions
Students could analyse data on human reaction times, with and without the presence of alcohol. Students link this to the risks associated with drinking and driving.

Drink driving test
Students could write a short procedure that would be used by traffic police to determine whether or not a person had been drinking and driving. This would include details of the tests they could carry out and the expected result for a person who had been drinking alcohol.

Alcohol and pregnancy
Students can look at and discuss data linking the consumption of alcohol during pregnancy with birth defects. This could lead to a class debate on how they would feel about themselves/a partner drinking during pregnancy.

Legal drinking ages
As well as discussing the issues highlighted in the Student Book surrounding licensing laws, happy hours and 'alcopops', the discussion could be extended to look at legal drinking ages. Some people believe that the legal drinking age should be increased from 18 to 21 (as in the USA); others believe it should be lowered to 16 (as in Italy). Students need to decide which statement they agree with, giving reasons to support their argument.

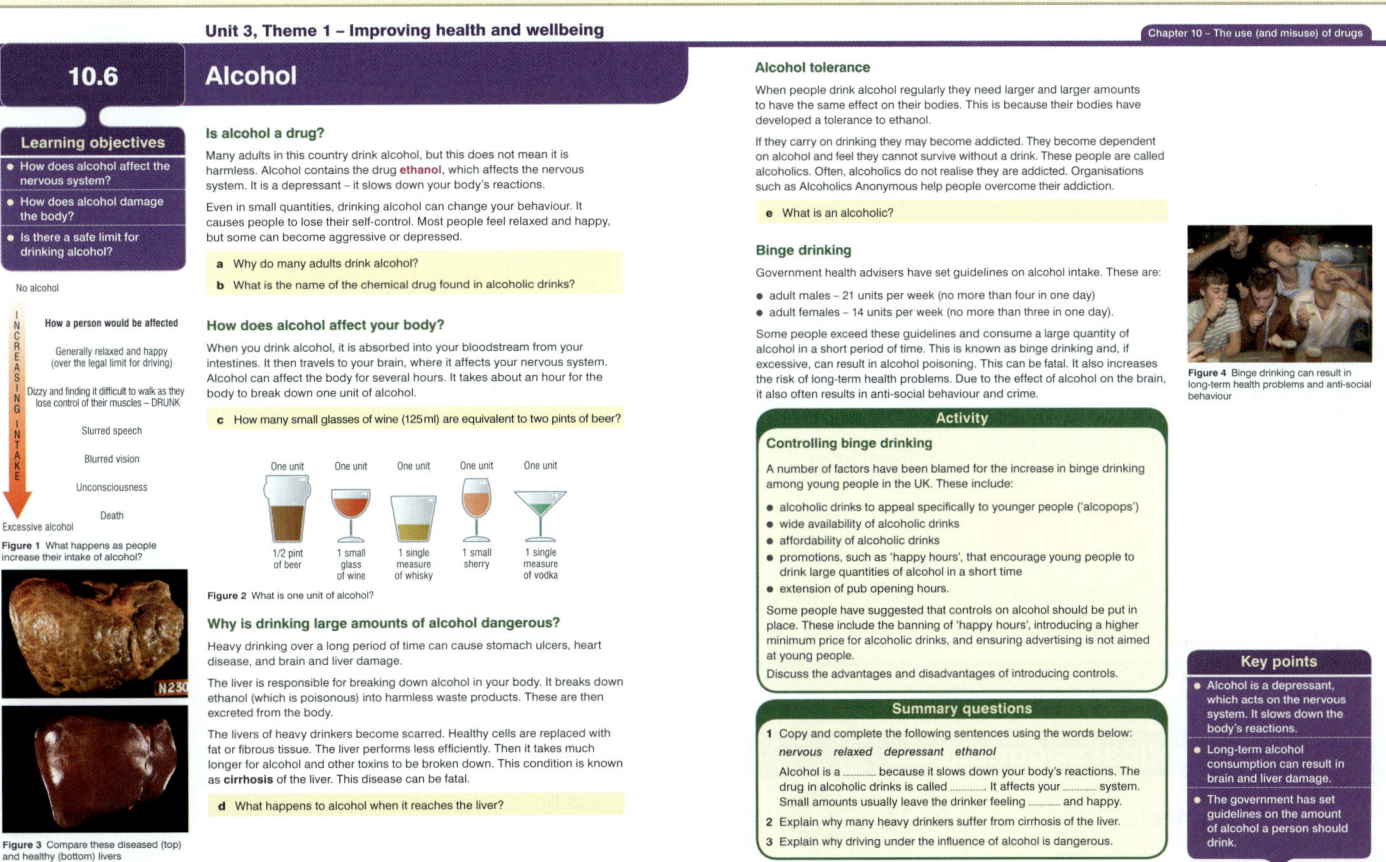

Summary answers

1 depressant, ethanol, nervous, relaxed
2 Their livers have had to work extremely hard breaking down alcohol. As a result they become scarred. Healthy cells are replaced by fat or fibrous tissue.
3 Drinking alcohol even in small amounts slows a person's reactions making it harder for them to respond rapidly to situations as they are driving. With larger amounts of alcohol a person's vision may be blurred and they may have difficulty in coordinating their muscles.

Answers to in-text questions

a To make them feel relaxed and happy.
b ethanol
c four small glasses of wine
d Ethanol is broken down into waste products.
e a person who is addicted to alcohol

Activity

Controlling binge drinking
Students should discuss the detrimental effects of excessive alcohol consumption. This could be looked at from a number of viewpoints:

- The health effects on the individual.
- The effects of increased crime and anti-social behaviour on society.
- The cost to the economy of lost working days and poor health.
- The costs to the NHS of treating alcohol-related conditions.

Students should discuss why younger people, in particular, become involved in binge drinking – and if there are any steps the government could take to help dissuade people from excessive alcohol intake.

Unit 3, Theme 1 – Improving health and wellbeing

Summary answers

1 **a** A – 3
   B – 2
   C – 1

   b analgesic and anti-inflammatory drugs

2 viruses, flu, tonsillitis, antibiotics, prescribed, resistant

3 **C** Drug is tested using computer models and human cells.

   B Drug is tested on animals. The animal is monitored closely for any side-effects.

   E Drug is tested on a small group of healthy human volunteers to check it is safe.

   F Drug is tested on volunteer patients who have the illness that is being targeted to ensure it works

   A Drug is tested on patients to achieve data on drug effectiveness, safety, dosage and side effects

   D Drug is approved by the Medicines and Healthcare products Regulatory Agency (MHRA), and can be prescribed.

4 **a** A recreational drug is taken for pleasure; a medical drug is taken to treat a condition.

   b See below.

Stimulant	Depressant
cocaine	barbiturates
amphetamines	heroin

   c Many recreational drugs are addictive.

5 **a** ciliated cell

   b To move mucus away from the lungs.

   c The cilia become paralysed, preventing them from sweeping the mucus away from the lungs.

6 **a** ethanol (alcohol)

   b the liver

   c For example, brain damage, liver damage, heart disease, stomach ulcers

Unit 3, Theme 1 – Improving health and wellbeing

Summary questions

1 **a** Select an example for each type of drug, below, by matching the letters and numbers.

Type of drug	Name of drug
A Painkiller (analgesic)	1 Penicillin
B Anti-inflammatory	2 Aspirin
C Antibiotic	3 Paracetamol

   b Which two types of drug, from the examples above, only treat the symptoms of a condition?

2 Copy and complete the following sentences, using the words below:

   antibiotics resistant viruses flu
   tonsillitis prescribed

   Some infections are caused by ...., such as .... Bacterial infections, such as ...., can be treated using .... These have to be .... by a doctor. The over-use of antibiotics in the past has led to some bacteria, such as MRSA, becoming .... to these drugs.

3 Before a drug can be prescribed it has to undergo clinical trials. Reorganise the following steps of drug testing into the correct order.

   A Drug is tested on patients to achieve data on drug effectiveness, safety, dosage and side effects

   B Drug is tested on animals. The animal is monitored closely for any side effects.

   C Drug is tested using computer models and human cells.

   D Drug is approved by the Medicines and Healthcare products Regulatory Agency (MHRA), and can be prescribed.

   E Drug is tested on a small group of healthy human volunteers to check it is safe.

   F Drug is tested on volunteer patients who have the illness that is being targeted to ensure it works.

4 **a** What is the difference between a recreational drug and a medical drug?

   b Copy and complete the following table, using these drugs:
   barbiturates cocaine heroin amphetamines

Stimulant	Depressant

   c Why do people find it difficult to stop taking drugs?

5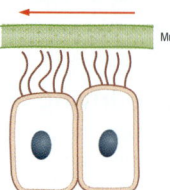

   a What is the name of this type of cell?
   b What is its job in the respiratory system?
   c How can this type of cell be damaged by smoking?

6

   a What is the name of the drug in this picture?
   b Which organ breaks down this drug?
   c State two medical conditions that may be caused by the long-term use of this drug.

Practical suggestions

Practicals	AQA	k	📖
Use pre-inoculated agar in Petri dishes to evaluate the effect of disinfectants and antibiotics.	✓	✓	✓
Investigate the effects of drugs (caffeine-based drinks, sleeping pills, alcohol) on Daphnia heartbeat rate.		✓	

Kerboodle resources

- Theme map: Improving health and wellbeing
- Animation: Bacterial resistance to antibiotics (10.2)
- WebQuest: Smoking timeline (10.4)
- Support: What's the harm in that? (10.4)
- Interactive activity: Drugs
- Practical: Antibiotics and bacteria (10.2)
- Examination-style questions
- Answers to examination-style questions

Chapter 10 – The use (and misuse) of drugs

Examination-style questions

Choose words from the box to answer the questions.

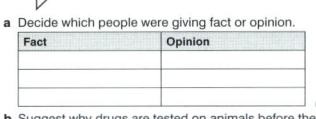

a Give the **two** legal medicines that are taken as a painkiller. (2)
b Which drug causes harm to the liver? (1)
c Which is the addictive drug in tobacco smoke? (1)
d Penicillin is an example of which type of medicine? (1)
e Which are the **two** illegal drugs? (2)

2 Many microorganisms can cause diseases.
a Give examples of **two** diseases that are caused by a virus. (2)
b Describe how viruses make us feel ill. (1)
c Explain the role of platelets in stopping us from getting diseases. (2)
d Explain how our body becomes immune to a disease. (3)

3 The table shows the effects of different types of performance-enhancing drugs used by sports men and women. Use the table to answer a–d.

Performance-enhancing drug	Effect
Diuretic	Makes the person urinate
Lean mass builder	Grows muscle and reduces body fat
Painkiller	Allows a person to compete past their usual pain threshold
Sedative	Overcomes nervousness
Stimulant	Increases alertness

a Explain the type of drug taken by a snooker player. (2)
b Which type of drug is caffeine? (1)
c Which type of drug might be taken by a wrestler to help him meet weight restrictions? (1)
d Which type of drug might be taken by a weight lifter? (1)

4 Read these opinions about testing medicines on animals and then answer the questions.

A diabetic: 'The insulin I have to take daily was tested on animals.'

A bypasser: 'Drugs should be tested on prisoners who have committed serious crimes.'

A government official: 'In Britain, the law states that if a drug is to be used for medicine, it has to be tested on at least two different types of live mammal.'

A Peta spokesperson: 'The animals are routinely killed after they have been used in experiments.'

A news reporter: 'Drugs testing on human tissue samples in a test tube and experiments carried out using computer models are both alternatives to animal testing.'

a Decide which people were giving fact or opinion.

Fact	Opinion

(2)

b Suggest why drugs are tested on animals before they go on sale. (1)
c Put the steps involved in research carried out in laboratories on drugs before they go on sale in the correct order. (4)

A	B	C	D	E
Animals	Cells	Healthy humans	Tissues	Unhealthy humans

(4)

5 A study was completed in Bangladesh in 2004 to find if the number of years a person went to school for had any effect on them taking up smoking.
The graph shows the percentage of men and women smokers and how much schooling they received.

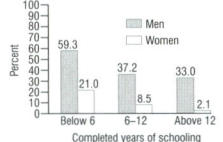

In this question you will be assessed on using good English, organising information clearly and using specialist terms where appropriate.

a Write a conclusion based on the results in the graph. (6)
b Explain why some people find it hard to give up smoking. (2)

Examination-style answers

1 a aspirin
   paracetamol (2 marks)
  b alcohol (1 mark)
  c nicotine (1 mark)
  d antibiotic (1 mark)
  e cannabis
   heroin (2 marks)

2 a any **two** from: (2 marks)
   flu / measles / mumps / rubella / polio
  b cause cell damage (1 mark)
  c they form a barrier / scab (1 mark)
   So no microbes can get in (1 mark)
  d When our white blood cells encounter
   a bacterium or virus (1 mark)
   they make antibodies against it. (1 mark)
   These antibodies can be produced quickly
   on the next encounter with the same organism. (1 mark)

3 a sedative (1 mark)
   So they have a steady aim. (1 mark)
  c diuretic (1 mark)
   So they are lighter. (1 mark)

4 a
Fact	Opinion
Diabetic	Bypasser
Government official	
Peta spokesperson	
News reporter	

(All correct = 2 marks, 3 or 4 correct = 1 mark)

b To make sure they are safe. (1 mark)
c BDACE (4 or 5 correct = 4 marks, 3 correct = 3 marks, 2 correct = 2 marks, 1 correct = 1 mark)

5 a Marks awarded for this answer will be determined by the Quality of Written Communication (QWC) as well as the standard of the scientific response.

There is a clear, balanced and detailed conclusion which is well structured with minimal repetition or irrelevant points. There is an accurate, fluent and clear expression of ideas with only minor errors in the use of technical terms, spelling, punctuation and grammar. (5–6 marks)

There is an attempt at a conclusion which shows some structure and the ideas are expressed with reasonable fluency and clarity. There are some errors in the use of technical terms, spelling, punctuation and grammar. (3–4 marks)

There is a brief attempt at a conclusion which is largely incomplete and may contain some valid points which are not clearly structured. It lacks fluency and/or clarity. It contains errors in the use of technical terms, spelling, punctuation and grammar. (1–2 marks)

No relevant content (0 marks)

Examples of points made in the response:
- There are more male smokers than female in all categories
- The less schooling the more likely you are to smoke
- Quoting of any of the data to help prove their point, e.g. women with more than 12 years of schooling are the least likely to smoke with only 3.9 % of them smoking
- Any manipulation of the data, e.g. Men with less than 6 years of schooling are almost twice as likely to smoke than men with more than 12 years OR women with less than 6 years of schooling are 41 % less likely to smoke than men in the same category
- A plausible explanation, e.g. people who stay in education for longer might learn about the health implications of smoking so are less likely to take it up.

b Smoking is addictive (1 mark)
 therefore people usually suffer withdrawal symptoms (1 mark)

Unit 3, Theme 1 – Improving health and wellbeing

11.1 Harmful microorganisms

Learning objectives

Students should learn:
- that bacteria and viruses are two groups of microorganism
- that harmful microorganisms are referred to as pathogens
- how bacteria and viruses cause disease.

Learning outcomes

Most students should be able to:
- name some diseases caused by microorganisms
- describe how bacteria and viruses cause disease.

Some students should also be able to:
- explain in detail how bacteria and viruses cause disease.

Specification link-up: Science B 3.5.1.2

- Name some diseases caused by bacteria (tuberculosis, cholera, typhoid) and viruses (influenza, measles, mumps, rubella, polio).
- Know that some types of bacteria and viruses make us feel ill when they reproduce rapidly in the body (bacteria, by producing toxins and viruses, by causing cell damage).

Lesson structure

Starters

Matching diseases – Put a list of diseases caused by microorganisms on the board or give students a list of diseases. Then ask students to sort the diseases into a table with the headings 'bacteria' and 'virus'. *(5 minutes)*

Microorganism spider diagrams – Ask students to write down as many facts about microorganisms that they can remember. To support students provide them with a list of key words to include in their diagrams. To extend students, ask them to identify specific features of bacteria and viruses. *(10 minutes)*

Main

- Show students a photograph of a 'simple' bacterium and a virus at a magnification that allows the main features to be seen (the photos in the Student Book would be ideal). Note the different features they possess and differences in their size. Students can use these photos to produce a list of similarities and differences between bacteria and viruses. These could be summarised in a table.
- Show students photographs of a range of pathogenic viruses to illustrate the very different appearances of viruses and the wide range of diseases they cause.
- Explain how bacteria and viruses that have entered the body cause disease by damaging and destroying cells, and producing toxins.
- Show students animations/diagrams of bacteria and viruses replicating. The Student Book includes these diagrams. This information could be summarised by students in a format suited to the students' learning style. Examples could include flow charts, concept maps, posters or in a text-based format.

Plenaries

Viral replication – Provide students with a pack of cards containing all the steps involved in viral replication. Students sort the cards into the correct order. *(5 minutes)*

Bacterial replication – Ask students to sketch a simple diagram to show how bacteria replicate. To support students, provide them with diagrams of the stages of bacteria replication for them to organise into the correct order. To extend students, ask them to give an indication of the elapsed time at each stage of replication. *(10 minutes)*

Support

- Provide students with a diagram of a simple bacterium and a virus to label.
- Then provide them with a series of pictures of bacteria and viruses that they can match to their labelled reference organisms.

Extend

- Students can analyse graphs showing the population of bacteria replicating within an organism. It is important for them to recognise the incubation period, the rapid increase in population size and factors that can limit the size of the population.

Further teaching suggestions

How do viruses cause disease?
Produce a cartoon strip showing one way in which a virus enters the body, replicates and causes disease. This task could be extended to include specific details of one particular type of virus.

Treatments for diseases caused by microorganisms
Students could research the range of treatments that doctors can prescribe to treat diseases caused by bacteria and viruses. They need to state how each method treats the condition.

Observing bacteria
Show students a range of bacteria on prepared mircroscope slides. They should be able to recognise the difference in shape between cocci (spherical), bacilli (rod-shaped) and vibrio (comma-shaped) bacteria.

Chapter 11 – Modern medicine

Answers to in-text questions

a microorganisms that cause disease
b microbiologists
c Bacteria have a cell wall; viruses have a protein coat. Viruses are much smaller than bacteria. Bacteria have many genes whereas viruses only have a few.
d They enter your cells and tell the nucleus to make lots of copies of themselves. The cell bursts, releasing many viruses which then infect other cells.
e Microorganisms reproduce rapidly.

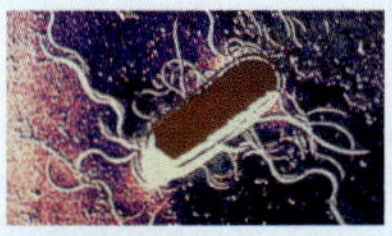

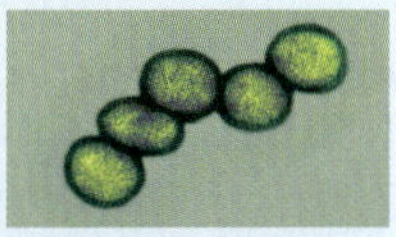

Science in context

Microbiologists study microorganisms to identify what is causing a plant or animal disease. They also use this information to decide which type of drug or treatment should be administered, and in some cases what steps can be taken to prevent a disease spreading. Incidences of plant diseases have huge agricultural implications.

Summary answers

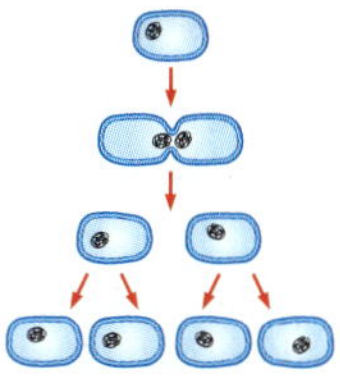

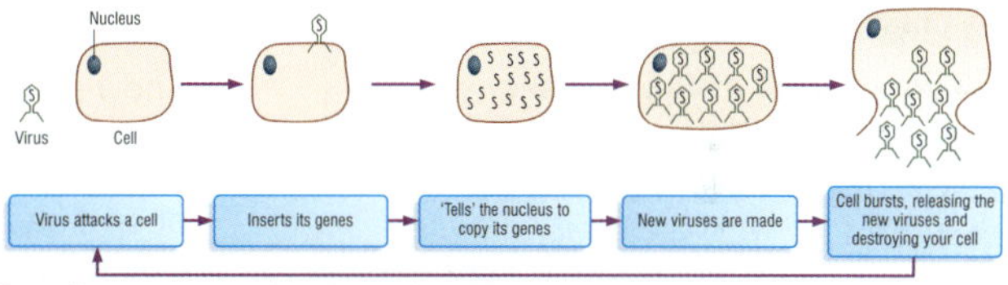

1 pathogens, bacteria, viruses, fungi, mumps, tuberculosis

2 a 160, 640.

   b a line graph showing an exponential curve as the number of bacteria double every 20 minutes.

   c The bacteria double in number every 20 minutes. Soon there are thousands of bacteria attacking the person's cells.

Unit 3, Theme 1 – Improving health and wellbeing

11.2 How are diseases spread?

Learning objectives

Students should learn:

- that pathogens can enter the body in a number of ways
- that infected people can spread disease
- that unhygienic conditions can lead to diseases spreading.

Learning outcomes

Most students should be able to:

- state how pathogens can enter the body
- describe how diseases can be spread by infected people
- describe how unhygienic conditions can lead to the spread of diseases
- list some ways in which disease can be prevented.

Some students should also be able to:

- explain in detail, steps which can be taken to control the spread of a disease.

Specification link-up: Science B 3.5.1.2

- Know that pathogens can enter the body through wounds, the respiratory system, the digestive system and by sexual transmission, as a result of unhygienic conditions or contact with infected people.

Lesson structure

Starters

Pathogen entry – Give students an outline of the human body. On this picture, students label sites where pathogens can enter the body. *(5 minutes)*

Unhygienic conditions – Project an image of an unhygienic individual – for example, messy hair hanging into food, which is being eaten with dirty hands. In pairs, students should identify possible sources of contamination. To support students, supply the image to students on paper, for them to circle the possible sources of contamination. To extend students, for each source of potential contamination, ask them to suggest a way it could be prevented. *(10 minutes)*

Main

- Show students a photograph of *Clostridium difficile* (causes sickness and diarrhoea). These are infectious bacteria that spread easily through unhygienic practises and conditions. Discuss with students – what is *Clostridium difficile* and what symptoms does it cause? Have any students had these types of symptoms? How is it spread? Could its spread be prevented? This links together the two key points of the lesson – how infectious diseases are spread, and unhygienic conditions.
- Split the class into two halves. One half will research how an individual can prevent themselves from catching an infectious disease. Good examples, which would be easy to locate via an internet search engine, include measles and flu. The other half of the group will research how being hygienic helps to prevent diseases from spreading. Good examples to study include cholera and food poisoning.
- Each half should be split into several smaller groups. For larger groups, different groups could study different diseases – however, the means of infection and control methods will be consistent. Each group should present their findings to the rest of the class, e.g. via PowerPoint presentations or A2 posters. (The Student Book is an ideal starting point for the research, but other sources should be made available, e.g. the internet, health books, microbiology textbooks.)
- Key ideas from the each group should be summarised by all students. This could take the form of a concept map, table, bullet pointed summary or in a 'cartoon' format.

Plenaries

Preventing disease spreading – Put two lists on the board or give students lists of ways diseases can be spread and methods of prevention in a muddled order. Students match the way a disease can be spread to a method of prevention. *(5 minutes)*

Disease outbreak – Ask students to imagine that there has been an outbreak of an infectious disease. Students have to prioritise what are the three most important steps that should be employed to prevent the disease spreading. They need to have reasons to justify their decisions. To support students, provide them with five suggestions of steps they could take. They need to decide which three are the most important. To extend students, ask them to consider the idea of quarantine. Is it morally acceptable for a government to quarantine people, or even communities, to prevent the spread of contagious diseases? Students could discuss the pros and cons of this action for individuals, and for the population as a whole. *(10 minutes)*

Support

- Provide students with a set of cards containing photos of products you can use to protect yourself from infections (face mask, condom, plaster, etc.) and descriptions of how they reduce the risk of disease transmission. Students need to match the appropriate cards.

Extend

- Write a list of rules to be displayed in a commercial kitchen about basic hygiene procedures that must be followed. These need to be detailed enough to prevent any customers getting food poisoning due to unclean work areas, but short and simple enough to be clear and well understood by the workforce.

Answers to in-text questions

a Through cuts in the skin (injury and bites) and through the digestive, respiratory and reproductive systems.
b Through coughing and sneezing – droplet infection (also by touching contaminated surfaces and transferring to mucous membranes, e.g. by rubbing eyes).
c By drinking clean water, cooking food properly, covering food, washing your hands and covering cuts and grazes.
d By landing on an infected surface, e.g. animal faeces, and then transferring these pathogens to the food.

Chapter 11 – Modern medicine

Further teaching suggestions

Controlling disease outbreaks
Students could research a real life disease outbreak (e.g. Asian Bird Flu) or natural disaster (e.g. Asian tsunami, flooding of New Orleans) and the steps that were taken to control the potential disease outbreak.

Swine flu
Provide students with the media campaign (adverts, poster and information booklet) used to tackle swine flu. They then need to analyse the science behind the campaign, whether they thought it was effective, and how they think it could be improved if a similar situation was to arise again.

The role of charities
Students could look into the work of Médicins sans Frontières, or the Red Cross. They need to find out the role these charities play in disaster zones, and what type of aid they distribute to survivors.

Unit 3, Theme 1 – Improving health and wellbeing

11.2 How are diseases spread?

Learning objectives
- How do pathogens enter the body?
- How do diseases spread?
- How can you protect yourself from disease?

In order to cause harm, pathogens have to enter our bodies. This can happen in four main ways:
- through cuts in the skin – from injury, or insect/animal bites
- through the digestive system – when you eat and drink
- through the respiratory system – when you breathe through your mouth and nose
- through the reproductive system – during sexual intercourse.

You are more likely to become unwell if large numbers of microorganisms enter your body. This is likely if you have contact with an infected person, or if you are exposed to unhygienic conditions.

a Name **four** ways pathogens can enter your body.

How can you prevent yourself catching a disease from an infected person?

By covering your mouth and nose. When somebody sneezes or coughs, tiny drops of liquid are released into the air – **droplet infection**. Colds and flu are spread by this method. In 2009, many people wore protective masks to try to reduce the spread of swine flu.

By not touching. Some diseases are **contagious**. They are spread by touching infected people. Some can even be spread by touching objects an infected person has touched. Mumps and chicken pox can be spread in this manner.

By using protection. Body fluids are exchanged during sexual intercourse. Syphilis and gonorrhoea are examples of sexually transmitted diseases. Using condoms can help prevent diseases, including HIV, being transferred in this way.

By not sharing needles. People who inject medicines or illegal drugs should never share needles. Diseases can be passed on in blood on the needle. HIV and hepatitis can be spread in this way.

b How are colds and flu spread?

How being hygienic can help prevent diseases spreading

Cook food properly. Some animals contain bacteria that could cause food poisoning. These include some species of E. coli and Salmonella. However, if the foods are cooked properly, the bacteria will be killed. You can also prevent the transmission of bacteria into cooked food by keeping raw and cooked food separate.

Drink clean water. Untreated water can contain microorganisms that cause diseases like cholera and typhoid. This may happen after flooding if sewage contaminates the fresh water supply. If water might be infected, you must boil it or use sterilisation tablets.

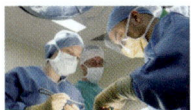

Figure 1 What steps are taken to prevent the spread of disease in an operating theatre? Do these protect the surgeon or the patient?

Figure 2 When someone sneezes, you can catch a disease through droplet infection

? Did you know … ?
Flies are often found on animal dung. They can then land on your food, spreading microbes all over it!
Restaurants and food outlets need to make sure all food is covered to prevent this happening.

Protect yourself from animal bites. Mosquitoes spread malaria. By biting an infected person, and then someone else, the mosquito can pass a disease on. You can protect yourself from malaria by taking anti-malarial drugs, and using insect-repellent sprays.

Wash your hands. To reduce your risk of infection from microorganisms, wash your hands regularly. This is essential for employees in businesses involving the handling of foods.

Cover cuts and grazes. To prevent infection, you should thoroughly clean a cut. It should then be covered with a plaster, so microorganisms cannot enter your body.

c State **five** ways you can help prevent the spread of infections.

d How can insects transfer pathogens onto food?

An important part of many jobs is to ensure the health and safety of employees and clients. This includes ensuring that work premises are hygienic.

Restaurants are monitored by environmental health officers to ensure premises are clean, and that food is stored and cooked properly. This helps prevent the spread of microorganisms such as Salmonella, which could cause food poisoning.

Lifeguards are trained to check water purity in swimming pools and to use chemicals to ensure that water is clean. This prevents the spread of waterborne diseases.

It is very important that you maintain good personal hygiene at home to help prevent diseases spreading. This includes washing hands thoroughly before preparing and eating food. You should also ensure that you cover your mouth when you cough or sneeze. Ideally you should use a tissue, which should immediately be placed in a bin.

Activity
How can we prevent an epidemic?

'Swine flu' (Influenza H1N1) affected many countries across the world in 2009. Generally, it was a mild condition requiring painkillers, and a few days' rest. However, it was fatal for several thousand people around the world. To try to prevent its spread in the UK, the government:

1 Made the disease symptoms widely known – via the TV, radio and internet.
2 Promoted good hygiene procedures – for example, by encouraging the use of anti-viral handwashes.
3 Advised ill, elderly and pregnant people to avoid crowded places.
4 Shut schools where there were outbreaks.
5 Set up a swine flu helpline where symptoms could be checked over the phone.
6 Developed a vaccine against the virus that causes swine flu.
Write a magazine article explaining how these steps reduced the spread of the disease.

AQA Examiner's tip
Make sure you read the question properly. There is usually a clue about the way the disease in question is transmitted. If you are asked for two ways to stop a disease being transmitted, make sure you do not give the same answer twice, e.g. 'cover mouth when sneezing' and 'cover mouth when coughing'.

Summary questions
1 Copy and complete the table:

Method of transmission	Diseases spread in this way
Shared needles	
Touch	
Droplets	
Sexual intercourse	

2 Why must meat be thoroughly cooked before you eat it?
3 Why do diseases spread more easily in:
 a highly populated areas
 b countries with poor hygiene?

Key points
- Pathogens enter the body through wounds (injury or bites), the digestive system, the respiratory system or during sexual intercourse.
- Diseases are spread by contact with an infected person, or exposure to unhygienic conditions.
- You can help to protect yourself from diseases by avoiding infected people, using condoms during sexual intercourse and by being hygienic.

Summary answers

1

Method of transmission	Diseases spread in this way
Shared needles	HIV, hepatitis
Touch	Mumps, chicken pox
Droplets	Colds, flu
Sexual intercourse	Syphilis and gonorrhoea (or HIV and hepatitis)

2 To kill any microorganisms that could be present in the meat.

3 a More people come into contact with each other, increasing the risk of diseases being spread.
 b Poor hygiene leads to the survival of more microorganisms, therefore leading to a greater risk of an infection spreading.

Science in context

Hygiene is extremely important in the catering industry (from factory level, through to chefs and waiting staff) to ensure that customers are not exposed to the pathogens that cause food poisoning. Hygiene and disease-prevention mechanisms are essential in hospitals to prevent diseases spreading between patients and to staff.

Activity

How can we prevent an epidemic?
Lots of useful information to support this activity is available on the www.direct.gov.uk/swineflu website, or from the NHS. Many newspapers also hold relevant articles in their archives – locate via a search engine.

Unit 3, Theme 1 – Improving health and wellbeing

11.3 Body defence mechanisms

Learning objectives

Students should learn:

- that platelets are involved in clotting blood and forming scabs
- that there are two types of white blood cell – phagocytes and lymphocytes
- how white blood cells deal with pathogens.

Learning outcomes

Most students should be able to:

- list the steps involved in forming a scab
- describe how phagocytes engulf pathogens
- describe how lymphocytes make antibodies.

Some students should also be able to:

- explain in detail the role of white blood cells in immunity.

AQA Specification link-up: Science B 3.5.1.2

- Describe how platelets help to form a barrier to infection through a cut.
- Describe how white blood cells help to defend against pathogens.
- Describe how antibodies in the blood provide immunity to certain diseases.

Lesson structure

Starters

White blood cells – Ask students to sketch a white blood cell, labelling as many features as possible. Revision of KS3 knowledge. *(5 minutes)*

Role of blood in the body – Provide students with the following statements. For each one they should state if it is true or false. Review of prior knowledge.

- Red blood cells fight disease. [F]
- White blood cells do not have a nucleus. [F]
- Platelets are involved in blood clotting. [T]
- White blood cells are smaller than red blood cells. [F]
- Scabs prevent microorganisms entering the body. [T]

To support students, allow them access to the Student Book so they can find the answers. To extend students, ask them to add as many blood facts to their list as they can remember. *(10 minutes)*

Main

- Ask students to discuss in pairs how your skin heals after it has been cut.
- Then show students an animation, or a sequence of pictures (these are in the Student Book) of blood clotting, so that they can see how a scab is formed. This information can be summarised as a flow chart.
- Show students a slide/photograph of white blood cells. Introduce the fact that there are two types – lymphocytes and phagocytes. Students need to recognise the differences in the appearance of these types.
- Split the class into pairs. One student in each pair should use the Student Book to produce a series of plasticine models to explain how lymphocytes deal with disease. The other student should use the Student Book to produce plasticine models to explain how phagocytes fight disease.
- Each student then teaches the other (using their models) how their white blood cells deal with disease.
- Discuss with students how immunity can be provided via the production of antibodies.

Plenaries

Blood clotting – Provide students with a pack of cards showing the main steps involved in clotting the blood and a scab forming. Students can then sort cards into the correct order. *(5 minutes)*

Antibody production – Ask students to draw a series of diagrams showing how lymphocytes destroy microorganisms, using antibodies. To support students, provide them with a series of diagrams for them to annotate. To extend students, ask them to do the same again for a different microorganism. This will demonstrate that they understand antibody specificity. *(10 minutes)*

Support

- Provide students with a jumbled version of the flow diagram of blood clotting from the Student Book. Ask them to match the diagram to the correct annotation and then sequence them into the correct order.

Extend

- Ask students to show their plasticine models to the whole class and explain how lymphocytes and phagocytes defend the body. They need to explain any limitations or inaccuracies in their models.

Chapter 11 – Modern medicine

Further teaching suggestions

Haemophillia research
Research haemophilia – a blood-clotting disorder – to find out the problems that occur if your blood cannot clot. This will provide an introduction to inherited genetic disorders.

Blood tests
When you are ill doctors sometimes perform blood tests. Find out what is meant by the term 'white cell count'. What happens to the number of your blood cells when you are ill? How can monitoring your white cell count let doctors know that you are recovering from a disease?

Childhood diseases
Students could research a common childhood disease and its effect on the body. They can then explain in detail how the body would fight this disease.

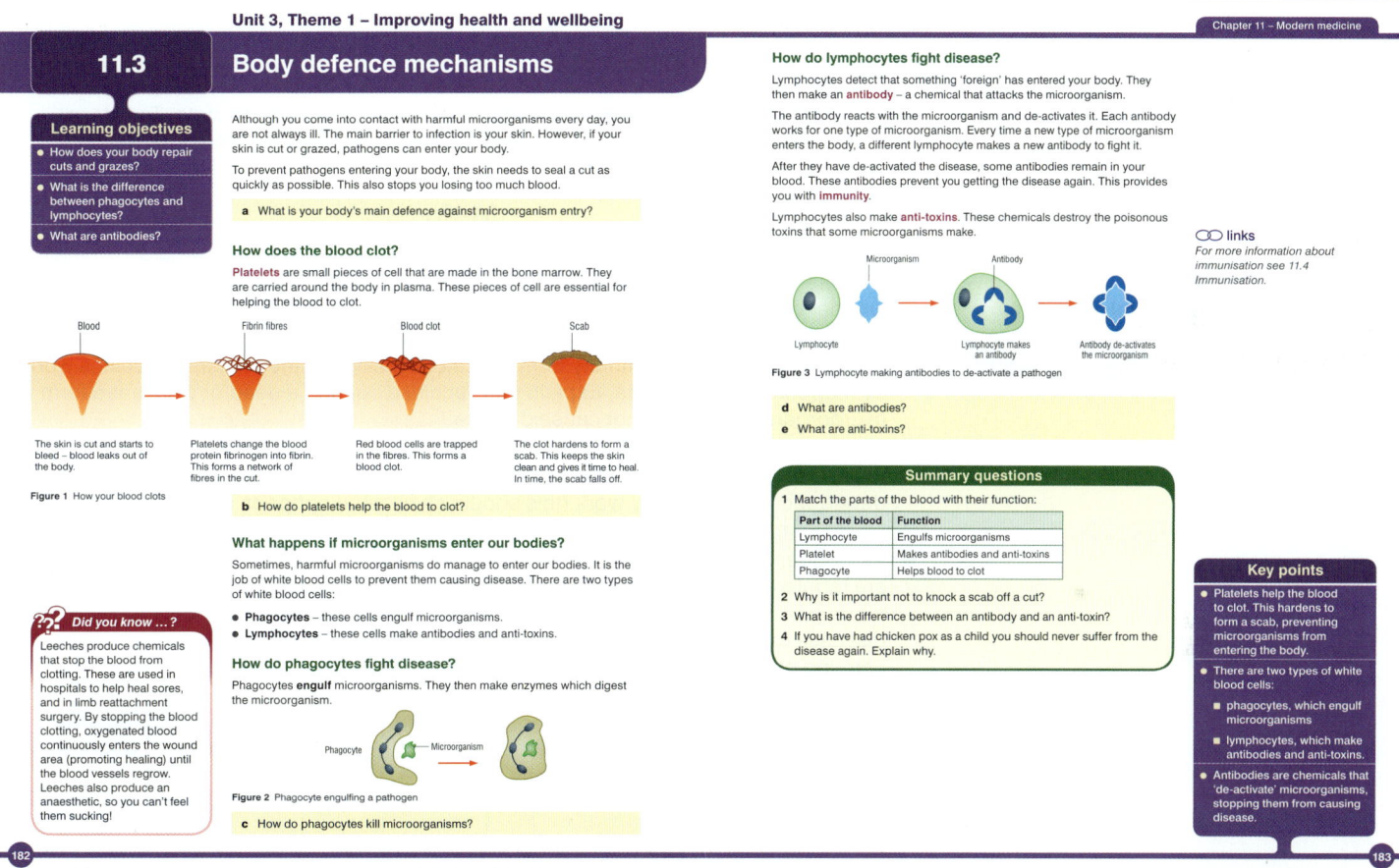

Summary answers

1. Lymphocyte – Makes antibodies and anti-toxins

   Platelet – Helps blood to clot

   Phagocyte – Engulfs microorganisms

2. You will open up the cut allowing microorganisms to enter.

3. Antibodies deactivate microorganisms; anti-toxins destroy the toxins that some microorganisms make.

4. Antibodies against the chicken pox virus will remain in your blood. If the virus enters your body, it will be destroyed before it has the chance to cause disease.

Answers to in-text questions

a skin

b By changing fibrinogen into fibrin. This forms a network of fibres that trap red blood cells, forming a clot.

c They engulf the microorganism, and then destroy it using enzymes.

d chemicals that attack microorganisms

e chemicals that destroy the toxins that some microorganisms make

Unit 3, Theme 1 – Improving health and wellbeing

11.4 Immunisation

Learning objectives

Students should learn:
- that immunisation can protect you from some diseases
- that immunisation works by your body developing an immunity to a disease
- how your body develops immunity.

Learning outcomes

Most students should be able to:
- name some diseases that you can be immunised against
- describe simply how the body develops immunity
- explain how immunisations trigger immunity, without making a person ill.

Some students should also be able to:
- explain the difference between artificial and natural immunity.

AQA Specification link-up: Science B 3.5.1.2
- Describe how white blood cells help to defend against pathogens.
- Describe how antibodies in the blood provide immunity to certain diseases.
- Explain how vaccination protects humans from infection.

Lesson structure

Starters

Vaccinations – Ask students to write a list of all the diseases they have been immunised against. This could include travel vaccinations for those who have travelled outside Europe. *(5 minutes)*

How do lymphocytes fight disease? – Ask students to draw and annotate a diagram to explain how lymphocytes fight off microorganisms (revision of last lesson 11.3). To support students, provide them with diagrams to sequence and annotate. To extend students, ask them to extend their answer to explain how phagocytes fight disease. *(10 minutes)*

Main

- Students look at the list of immunisations they should have already received in their life. Has anybody had any extras – why? (e.g. for foreign holidays, if they were born in another country). Why is it important to have these injections?
- Recap how white blood cells (lymphocytes) fight disease by producing antibodies and how this leads to immunity from a disease. Introduce the idea that vaccines work by triggering this response. Explain that vaccines contain a dead or weakened form of microorganism that can not cause disease.
- Provide students with information on how Jenner developed the smallpox vaccine. Students use this information to produce a cartoon strip to explain how vaccines are developed, and how they work (this should show what happens inside the blood to cause immunity).
- Ask students to carry out some research into common diseases that children are regularly immunised against. These could include measles, mumps, rubella, tuberculosis and polio. This should demonstrate how unpleasant and deadly the diseases can be, and how suffering can be prevented by a simple injection. Whilst carrying out their research they could complete the following table.

Disease	Type of microorganism that causes the disease	Symptoms	Age immunised against disease	Vaccine contains
e.g. Measles				

- Discuss with students how national immunisation programmes not only protect the individual who has been immunised, but other people in the population who have not been immunised. This would include, for example, those with medical conditions that prevent them from receiving the immunisation.

Plenaries

Immunisation true or false? – *(5 minutes)*
- Immunisations are always given as an injection. [F]
- Immunisations can give you the actual disease. [F]
- Vaccines contain dead or weakened microorganisms. [T]
- People require extra immunisations to visit some countries. [T]
- Phagocytes in your blood produce the antibodies required for immunity. [F]

Immunisation definition – Ask students to write their own definitions for the terms 'immunity' and 'immunisation'. To support students, provide them with key words or phrases to be included in their definition. To extend students, ask them to also define the word vaccine. *(10 minutes)*

Support
- Provide students with a flow diagram explaining how immunisation works in a number of steps. Students need to rearrange the diagram into the correct order.

Extend
- Ask students to research why booster immunisations are required for the prevention of some diseases. They should then feedback this information to the class.

Further teaching suggestions

Advantages of immunisation

Students produce a leaflet that could be given out by health visitors to help convince parents that they should immunise their children. It needs to state the advantages for the child and the health of the population as a whole, and how an immunisation works.

Flu jabs

Ask students to research why people over the age of 65 are invited to receive a flu jab every year. Why are they not protected from one year to the next? (Annual flu strain varies each year as flu rapidly mutates – scientists need to develop a vaccine for the new strain.)

Activity

Chicken pox vaccination

The internet can be used to find out additional information on immunisation. A useful website is the NHS direct website. When you have entered the site, type 'immunisation' into the search facility.

11.4 Immunisation

Learning objectives
- How do you gain immunity to a disease?
- Why is immunity important?
- How do immunisations work?

When a pathogen enters your body, lymphocytes make an antibody against it. After the antibodies have destroyed the pathogens causing the disease, some remain in the body. If the same type of microorganism enters your body again, the antibodies will destroy it before it can cause disease. This is called **immunity**. It prevents you suffering from the same disease again.

Figure 1 The effects of chicken pox

Did you know …?
People often refer to immunisations as giving you artificial immunity. If you suffer from a disease and develop your own immunity to it, this is normally called natural immunity.
Both types of immunity develop as a result of your body producing antibodies. The major advantage of artificial immunity is you never have to suffer from the disease to start with!

Examiner's tip
A vaccine contains a small amount of the disease. The MMR jab contains small amounts of three diseases. If someone decided not to go for the MMR but had the three injections separately then each of the three injections would not contain a small amount of the MMR jab. Each of the three injections would contain a different disease.

Activity
Chicken pox vaccination

Scientists have developed a vaccine against chicken pox, but do not routinely administer it. This is because in most cases chicken pox is a mild illness, and nearly 90 per cent of the population will develop immunity naturally. The immunisation is only usually offered to siblings of children with certain cancers, or of children who have had an organ transplant.
- Do you think the chicken pox immunisation should be routinely offered to every child? Carry out research using the Internet into the risks associated with chicken pox, shingles, and the vaccine, to make an informed decision.

a What is meant by the term 'immunity'?

What is an immunisation?

Through studying immunity, medical scientists have developed **immunisations**. These are also referred to as vaccinations. Immunisations can protect you against some diseases caused by microorganisms. They are the simplest and most cost-effective means of preventing life-threatening infections in a population.

Are your immunisations up to date? Look at the table below:

Child's age	Disease immunised against
2, 3 and 4 months	Polio, diphtheria, tetanus, whooping cough, Hib meningitis and meningitis C
About 13 months	Measles, mumps and rubella (MMR)
3–5 years	MMR, polio, diphtheria, tetanus and whooping cough
10–14 years	Tuberculosis (TB)
12–13 years	Cervical cancer (girls only)
13–18 years	Polio, diphtheria and tetanus

b Name **three** diseases you should be vaccinated against while you are at secondary school.

How do immunisations work?

Immunisation involves a **vaccine** being inserted into your body. This normally takes the form of an injection but some vaccines can be taken by mouth.

Most vaccines contain dead microorganisms, or microorganisms that have been weakened so that they can no longer cause disease. The microorganisms do not make you ill but still trigger your white blood cells (lymphocytes) to make antibodies.

The antibodies destroy the microorganisms. Some antibodies remain in your body. These will be able to fight off the pathogen quickly if it enters your body again, preventing it from causing disease. You are now immune.

c What does a vaccine normally contain?

Diseases the whole population should be immunised against

Polio

Polio is a disease that affects your nervous system. It can damage your nerves, and lead to permanent paralysis of parts of your body. In very severe cases it can cause death. The polio vaccine is not always injected as it can be given by mouth. It is often provided on a sugar lump as it doesn't taste very nice!

Tuberculosis (TB)

You may recently have been injected with the BCG vaccine. This protects you against tuberculosis (TB). TB is a disease which affects your lungs and causes breathing problems. In very severe cases it can cause death.

Measles, mumps and rubella

Just after their first birthday, children are given an immunisation called MMR. This single injection protects them against measles, mumps and rubella.

As well as being unpleasant, measles and mumps can cause permanent damage. In some cases, measles can cause deafness and mumps can make men infertile. In extreme cases, both diseases can be fatal.

Rubella is generally less unpleasant. However, if a woman gets the disease when she is pregnant, the unborn baby may be born deaf or blind. In some cases it can even die.

d Why should children have the MMR immunisation?

Summary questions

1. Copy and complete using these words:
   vaccines   immunity   diseases   microorganisms
   The spread of infectious .... can be prevented by the use of .... These work by introducing dead or weakened .... into the body. Your body builds up a defence to this infection – this is known as ....
2. Which diseases would a 15 year old have been immunised against?
3. Why do some parents choose not to have their children immunised against some diseases?

links
For information about how lymphocytes destroy microorganisms look back at 11.3 Body defence mechanisms.

Figure 2 This child is being given the polio vaccine

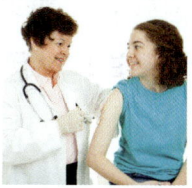

Figure 3 The BCG vaccine is injected into the arm. This protects you from tuberculosis.

Key points
- When a pathogen enters your body, lymphocytes make antibodies against it. Once these antibodies have got rid of the disease they remain in your body. This gives you immunity to the disease.
- Immunity prevents you suffering from the same disease again.
- Immunisations are doses of dead or weakened microorganisms. These are given by injection or by mouth. They trigger your body to produce antibodies. This gives you protection against a disease.

Summary answers

1. diseases, vaccines, microorganisms, immunity
2. choice from: polio, diphtheria, tetanus, whooping cough, Hib meningitis, meningitis C, measles, mumps and rubella
3. Some vaccines have risks associated with them. Some parents choose not to expose their children to this risk.

Answers to in-text questions

a Immunity to a disease means that if you come into contact again with that pathogen, you will be able to fight it off without getting ill.
b TB, polio, diphtheria, for example
c Vaccines contain dead or weakened microorganisms.
d To stop them getting the diseases measles, mumps and rubella, which would make them ill and in rare cases could kill.

Unit 3, Theme 1 – Improving health and wellbeing

11.5 Vaccination issues

Learning objectives

Students should learn:
- that scientists study data to check that immunisation is working
- some common side effects of immunisation, including fever and swollen glands
- that some people have concerns over safety, and choose not to be immunised.

Learning outcomes

Most students should be able to:
- name some common side effects of immunisation
- describe some advantages and disadvantages of immunisation
- describe how changes in the number of people receiving an immunisation affects the number of cases of a disease.

Some students should also be able to:
- analyse data to explain how changes in the number of people receiving an immunisation affects the number of cases of a disease.

Specification link-up: Science B 3.5.1.2

Within this context, candidates should be able to use scientific data and evidence to discuss, evaluate or suggest implications of the following:
- the value to individuals and populations of being vaccinated against diseases, including concerns about side-effects and effects on the immune system
- how the occurrence of diseases has changed as a result of increased use of vaccinations.

Lesson structure

Starters

How do microorganisms enter the body? – Ask students to make a list of the five main ways microorganisms enter the body. (Revision of earlier work in this chapter.) *(5 minutes)*

Advantages and disadvantages of immunisation – Ask students in small groups to write advantages and disadvantages of immunisation on sticky notes. To support students, provide them with a list of some of the factors surrounding immunisation for students to sort into advantages and disadvantages. To extend students, ask them to add scientific evidence where possible to justify the advantage or disadvantage stated. *(10 minutes)*

Main

- Divide the board into two halves – advantages and disadvantages of immunisation. If the starter has been completed above, ask students to take it in turns to put their reasons for being immunised on the board. Each idea should be discussed, drawing out the scientific evidence to support each argument.
- Students should have identified that the main reason for being immunised is to save lives and to prevent suffering from infectious diseases. Explain to students that scientists need to constantly monitor the occurrence of diseases and the number of immunisations that have been administered, to check the effectiveness of a vaccine.
- Show students a graph of the rate of vaccination uptake for a particular disease, against the number of cases of this disease (for example, the Student Book contains data for the occurrence of tuberculosis and measles). Students should analyse the data, making conclusions about the effect of a vaccine on the health of a population.
- Students should complete the activity task on whooping cough. This requires more detailed analytical skills to explain the peaks and troughs in the graph.
- Repeat the initial task on disadvantages of immunisation. Students' ideas should be grouped together under the subheadings – concern over safety, concern over side effects and concern over overloading the immune system. Each subheading should then be studied in turn.
- Students should be aware that there are occasional scares over the safety of vaccines, but often this is a result of media hype and is not supported by scientific evidence. Remind students that vaccines, like all drugs, have to undergo vigorous safety tests before they can be routinely administered.
- Ask students to discuss with a partner any side effects they may have experienced after receiving an immunisation. Share experiences with the class, emphasising that these are generally very mild and many people do not experience any side effects at all.
- Revise with students how microorganisms can enter the body. Explain that this is how we build up a natural immunity, and that people are constantly exposed to microorganisms. It should be emphasised that there is no firm evidence that vaccines overload the immune system.

Plenaries

Side effects of vaccines – Ask students to produce a list of five side effects they may experience as a result of an immunisation. *(5 minutes)*

Removing concerns – The following three statements are given by people as reasons for not having an immunisation. Ask the students to use scientific reasons to try to convince others that immunisation is the best way to protect against infectious diseases.
1 Vaccines are not safe.
2 Immunisation causes side effects.
3 Immunisation overloads the immune system.

To support students, supply a list of reasons. Students need to select the most appropriate reasons for each of the statements. To extend students, supply them with scientific data relevant to each of the three statements. They need to think about how the data could be presented to support a coherent argument for the safety of immunisation. *(10 minutes)*

Answers to in-text questions

a People were being immunised against TB and measles, so fewer contracted the disease.
b concerns over safety, side effects and overloading the immune system
c eating, putting their hands and objects in their mouths

Support

- Provide students with more graphs to enable them to analyse obvious trends between the number of vaccines administered, and the number of cases of a disease.

Extend

- Students can be provided with more data to analyse referring to the use of vaccines and outbreaks of disease. For example, they could study data on the occurrence of measles amongst the UK population before and after the alleged link between the MMR vaccine and autism in 1998.

Chapter 11 – Modern medicine

Further teaching suggestions

Should a child be routinely immunised?
Students could be asked to discuss in pairs whether or not they would have a child of their own immunised against common childhood illnesses. Their discussion should take into account the potential risks of the immunisation, and the health benefits to both the child and the wider population.

MMR debate
In groups, students research the reasons why a child should be immunised against measles, mumps and rubella, and the reasons why some parents have chosen not to have their children vaccinated. The discussion needs to study how the incidence of these diseases has changed over time, with a varying proportion of immunised children in the population. This can be contrasted with the potential risk of complications from receiving the vaccine. A search engine will provide many possible sources of information – students will need a degree of direction to find relevant information.

Unit 3, Theme 1 – Improving health and wellbeing

11.5 Vaccination issues

Learning objectives
- How do we know that immunisations work?
- Do immunisations have side effects?
- Why do some people choose not to have immunisations?

The Government recommends that all children should have a number of immunisations during their childhood. These will protect them from many childhood diseases, and provide long-term immunity. If you travel abroad, other immunisations are sometimes recommended. This is especially true if you are visiting less-developed or tropical countries.

How do we know that immunisations work?
During the period 1971–2000, the population of the UK increased. During this time, tuberculosis and measles injections were given to large numbers of the population. Despite the rise in population, the number of cases of these diseases decreased.

a Explain why the number of cases of TB and measles fell during the period 1971–2000.

Why do some people choose not to be immunised?
Immunisations are an effective way of protecting yourself against many life-threatening conditions. Vaccines have to undergo many stages of testing, like all medical drugs. This ensures they are safe before they can be routinely administered to the population. However, not everyone chooses to be immunised. This may be because:

1. There are occasional scares about the safety of some vaccines.
2. Concerns over possible side effects.
3. Some people believe vaccines overload our immune system. They think this makes the immune system less able to react to other diseases such as meningitis, AIDS and cancer.

b State **three** reasons why people may choose not to be immunised.

What are the common side effects of immunisation?
Most people suffer few or no side effects from an immunisation. Any side effects are normally very short lived, and can easily be treated with a painkiller. The side effects that can be experienced include:
- fever
- sickness and/or diarrhoea
- swollen glands
- a small lump at the site of the injection, which may last for a few weeks
- irritability.

Severe reactions to immunisations are very rare.

Figure 1 Number of cases of measles and TB in the UK 1971–2000

AQA Examiner's tip
When answering questions referring to statistical data you should give figures from the graph to support your analysis.

links
For information on the many stages of testing that vaccines have to undergo look back at 10.3 Approving a new drug.

Do vaccines overload the immune system?
There is no evidence to suggest that having a number of vaccines, even several at a time, overloads the immune system.

Our immune systems are constantly being challenged by many different microorganisms. From the moment a baby is born, it is exposed to numerous bacteria and viruses on a daily basis. This is especially true whilst eating, and through putting their hands and objects into their mouth. Scientific studies have estimated that vaccines occupy less than 0.1 per cent of a child's immune system.

c Name **two** ways that babies naturally increase their exposure to bacteria and viruses.

Activity

Are vaccines effective?
Whooping cough is a disease which can cause long bouts of coughing and choking. This can make it hard to breathe. It can be fatal to babies younger than 1 year old. Children are now routinely vaccinated against whooping cough.

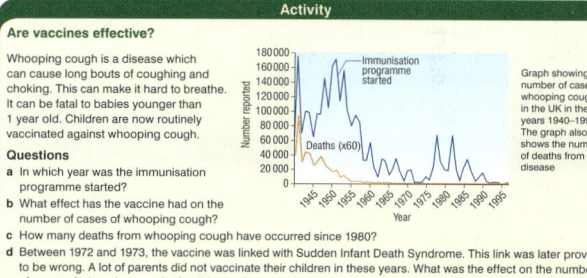

Graph showing the number of cases of whooping cough in the UK in the years 1940–1997. The graph also shows the number of deaths from this disease.

Questions
a. In which year was the immunisation programme started?
b. What effect has the vaccine had on the number of cases of whooping cough?
c. How many deaths from whooping cough have occurred since 1980?
d. Between 1972 and 1973, the vaccine was linked with Sudden Infant Death Syndrome. This link was later proved to be wrong. A lot of parents did not vaccinate their children in these years. What was the effect on the number of cases of whooping cough after this time?
e. Suggest why the number of deaths from whooping cough was decreasing even before the vaccination programme was started.

Summary questions
1. Sort the following statements into reasons for and against immunisations:
   - BCG leaves a scar
   - Many lives are saved
   - May get a temperature
   - People don't suffer from infectious diseases
   - Injections may hurt
   - Side effects of disease may last for years, e.g. paralysis/deafness
2. Why do scientists need to continually monitor the population for side effects caused by receiving immunisations?
3. Some people cannot have immunisations, due to other medical conditions. Explain why immunising others helps to prevent these people from getting an infectious disease

Key points
- Scientific data provides evidence that immunisations have decreased the number of cases of infectious diseases.
- Most people suffer no side effects from immunisation. If they do, they are generally very mild.
- Some people choose not to have immunisations because they are worried about side effects, the safety of vaccines and overloading their immune system.

Activity

Are vaccines effective?

a 1953

b Since the vaccination programme started, the number of deaths from whooping cough has dropped dramatically.

The number of cases dropped from around 150 000 to less than 5000. Between about 1976 and 1986, there was an increase in the number of cases of whooping cough. The number of cases then fell again after this period.

c None

d The number of cases of whooping cough increased again. However, there were no deaths due to increased medical advances and therefore people no longer died from the disease.

e The numbers of deaths from whooping cough were decreasing due to advances in medical treatment, e.g. the widespread use of antibiotics.

Summary answers

1. Reasons for immunisation: many lives are saved; people don't suffer from infectious diseases; side effects of disease may last for years, e.g. paralysis/deafness.

   Reasons against immunisation: may get a temperature; BCG leaves a scar; injections may hurt.

2. To confirm (and reaffirm) that vaccines are safe to use.

3. If most of the population are immune to a disease, there will be few people who suffer from the disease. This means that a person who was not well enough to be immunised against a disease is unlikely to come into contact with an infected person. Therefore they would be unlikely to catch the disease.

Unit 3, Theme 1 – Improving health and wellbeing

11.6 Medical uses of X-rays

Learning objectives

Students should learn:
- why ionising radiation is dangerous
- that X-rays are high energy waves in the electromagnetic spectrum
- that X-ray images can be formed using photographic film
- how some different diseases can be identified using X-rays.

Learning outcomes

Most students should be able to:
- define what an X-ray is and why they are hazardous to health
- describe how an X-ray image is made
- name some medical conditions that can be diagnosed using X-rays.

Some students should also be able to:
- explain how the density of a material affects its absorption of X-rays.

Support

- Provide students with a series of diagrams and annotations to explain how an X-ray image is produced. Students should sequence the diagrams into an appropriate order.

Extend

- Provide an X-ray image of a healthy lung and a diseased lung. Students should study the images carefully, highlighting any differences between them. Students should then explain why the different tissue types produce slightly different X-ray images.

AQA Specification link-up: Science B 3.5.1.3

- Know that X-rays are examples of transverse waves.
- Know that X-rays are a form of electromagnetic radiation.
- Understand that ionising radiation kills living cells and because of this can be used to treat cancer.
- Describe the characteristic properties of X-rays (penetration) that enable them to be used to diagnose medical disorders.
- Know that the use of high-energy radiation can be dangerous and needs to be monitored.
- Explain why people who work with radiation wear film badges and why these are monitored regularly to check the levels of radiation absorbed.

Lesson structure

Starters

True or false – Students select true or false for each of the following statements:
- X-rays are dangerous. [T]
- X-rays can pass easily through dense materials. [F]
- X-rays can be used to detect some cancers. [T]
- X-rays can cause burns. [T]
- X-rays are low-energy electromagnetic waves. [F] *(5 minutes)*

The electromagnetic spectrum – Students should be asked to produce a diagram of the electromagnetic spectrum, highlighting the position of X-rays within the spectrum. To support students, provide a partially completed copy of the spectrum, with labels to add to the appropriate positions. To extend students, ask students to add information to the spectrum, detailing wavelengths (long to short), frequencies (low to high) and energy transmitted (low to high). *(10 minutes)*

Main

- Using a projection of the electromagnetic spectrum, the different waves within the spectrum should be briefly revisited from previous knowledge and the main uses of the different waves revised. Include that these are transverse waves.
- Discuss with students – which waves are dangerous to humans? Students should discuss the potential health risks of UV radiation (skin cancer) and gamma rays (cancers, radiation burns). X-rays are found in between these two waves – students can therefore predict the potential health risks of receiving X-rays.
- Using the Student Book as a prompt, ask students to produce a short cartoon strip, explaining how X-rays can be used to produce an image of internal structures in the body. Ensure that students note that the more dense a structure, the greater the absorption of X-rays.
- Project images of a range of medical conditions (obtainable via a search engine). Good examples include a compound fracture, broken bone, TB, lung cancer and a blood clot. Ask students to explain how X-rays can be used to diagnose these conditions.
- Discuss with students if X-rays are potentially harmful to human health and why are they used to diagnose medical conditions in people who are already unwell? Students should compile a list of the advantages to the patient and the disadvantages of having X-rays rather than invasive medical procedures.
- Extend the discussion to include people in hospitals who work with radiation on a daily basis – radiographers. How are the risks controlled? Discuss briefly the use of film badges to monitor received radiation doses, and the use of lead shielding to protect workers from unnecessary exposure to ionising radiation.

Plenaries

Clothing for X-rays – Ask students to explain why, when receiving an X-ray, patients are asked to remove jewellery, watches, belts and other metallic objects, but do not need to remove any items of clothing. *(5 minutes)*

Producing an X-ray image – Students should produce a flow chart, detailing the steps involved in producing an X-ray image. To support students, provide visual prompts to help students sequence their flow charts. To extend students, they should be asked to ensure that they include information on how X-rays are affected by materials of different density. *(10 minutes)*

Chapter 11 – Modern medicine

Further teaching suggestions

Investigating X-rays

Students could carry out research into how X-rays can be used to diagnose medical conditions in the digestive system that would not show up on a 'normal' X-ray. The concept of the 'barium meal' should be included in this research.

X-ray risks

Students could carry out further research into the damage caused by X-rays. This could include a quantification of the risk associated with a single X-ray, or a course of X-rays when treating a medical condition.

Science in context

X-rays are used in hospitals to diagnose a number of medical conditions. Radiographers need to be aware of the health risks associated with X-rays to ensure they stay safe while at work. Doctors need to be aware of the conditions that can be diagnosed using X-rays, e.g. fractures, some cancer tumours, and how to recognise these conditions from an X-ray image.

Unit 3, Theme 1 – Improving health and wellbeing

11.6 Medical uses of X-rays

Learning objectives
- What is an X-ray?
- How are X-ray images formed?
- How can we use X-rays to identify some illnesses?

links
For information on the electromagnetic spectrum look back at 9.6 Electromagnetic waves.

You may have had an **X-ray** photograph, perhaps to identify a broken bone. Have you ever asked yourself how an image of the inside of your body is made?

X-rays (a form of **ionising radiation**) were discovered over 100 years ago. Over time, scientists have discovered that ionising radiation can be very helpful to us, but it also kills living cells. Many of the early researchers working with X-rays developed forms of cancer.

Doctors and dentists use X-rays to see inside the body of a patient, without the need for an operation. This removes the chance of a patient developing an infection.

a Why do doctors take X-ray images?

What are X-rays?

X-rays are high-energy transverse waves. They are just a small part of a large family of waves known as the **electromagnetic spectrum**.

  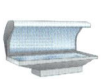

Radio | Microwaves | Infrared radiation | Light | Ultraviolet radiation | X-rays and gamma radiation

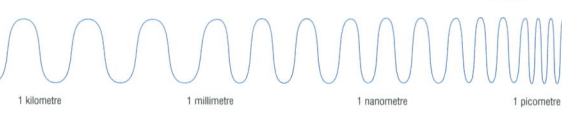

1 kilometre | 1 millimetre | 1 nanometre | 1 picometre

Wavelength
(1 nanometre = 0.000 001 millimetres, 1 picometre = 0.001 nanometres)

Figure 1 The position of X-rays within the electromagnetic spectrum

AQA Examiner's tip
Instead of just writing that X-rays can be used to detect bone damage, you may need to explain why. Your explanation should include the word 'absorbed'.

X-rays can penetrate through some materials. This means they can pass through them, without being absorbed. Dense materials absorb X-rays; this includes bones and teeth. The greater the density of the structure, the more absorption occurs. Modern X-ray imaging systems allow doctors to see fine detail in an X-ray image. This allows them to diagnose a wide range of medical conditions, including tumours and chest infections like pneumonia.

b What is an X-ray?

c State **three** medical conditions that can be diagnosed using X-ray images.

How are X-ray images produced?
- Photographic film is placed behind the part of the patient being investigated.
- The X-ray generator is placed in front of the patient. The patient is exposed to X-rays.
- The X-rays penetrate through soft tissues like skin and muscle. They are absorbed by denser structures, such as bones and teeth.
- The X-rays that penetrate through the patient expose the film.
- The image is then developed. Regions of the film that were exposed to X-rays show up black. Areas of film that were not exposed to X-rays, because they were absorbed, are white.

Images produced in this manner are known as 'shadow pictures'. The image is formed because an X-ray shadow is produced behind dense material in the body.

d Name some parts of the body which absorb X-rays well.

Activity

Using X-rays to diagnose conditions

Doctors and radiologists use X-rays to diagnose a wide range of medical conditions. Carry out some research using the internet into which conditions can be diagnosed using this technique and those that cannot.

How can we protect ourselves from the dangers of X-rays?

As X-rays are ionising radiation, they are known to cause cancer. However, the risk to your health of having a few X-rays is tiny.

Radiographers have to accept working with X-rays as part of their job. They are at risk of receiving high doses of this radiation. It is important that they protect themselves from this danger.

Lead (because it is very dense) absorbs X-rays, so a lead screen is placed between the radiographer and the patient. Radiographers also wear film badges. These measure the dose of radiation received.

e How do radiographers monitor their exposure to X-radiation and protect themselves from the X-rays?

Summary questions

1 Copy and complete using these words:
soft   waves   dense   inside

X-rays are high energy .... They are used to make images of the .... of the body. Only .... parts of the body can be imaged, as X-rays will pass through .... tissues.

2 Small children can have X-rays taken sitting on their parent's lap. Why does the parent need to wear a lead apron?

3 Explain why X-ray images are known as 'shadow pictures'. (Hint – think about how shadows are made using light.) Use a diagram to help with your explanation.

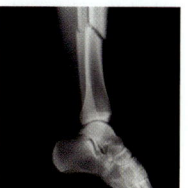

Figure 2 X-ray image of a broken leg

Figure 3 This radiographer wears a film badge to monitor her exposure to radiation

links
For more information on how film badges work see 11.7 What is radioactivity?

Key points
- Ionising radiation can treat cancer as it kills cells.
- X-rays are high energy, transverse, electromagnetic waves.
- 'Shadow pictures' are images of the inside of the body, which are produced using X-rays.
- Radiographers and medical professionals need to be protected from the harmful effects of X-rays.

Summary answers

1 waves, inside, dense, soft

2 To protect the parent from unnecessary exposure to X-rays.

3 Dense materials absorb X-rays. The photographic film is exposed by regions, which X-rays pass through. Regions of the film that are not exposed are in the shadow of dense structures – hence the term 'shadow images'. The contrast between exposed and unexposed regions allows analysis of the X-ray image.

Answers to in-text questions

a To take an image of internal structures in the body, without the need for surgery.

b a high-energy electromagnetic wave

c broken bones, cancers and pneumonia, for example

d bones and teeth, for example

e Monitor: through wearing a film badge.
Protect: through lead shielding.

Unit 3, Theme 1 – Improving health and wellbeing

11.7 What is radioactivity?

Learning objectives

Students should learn:
- the different properties and characteristics of radiation
- that the use of radioactivity needs to be monitored
- how a film badge can be used to monitor radioactivity.

Learning outcomes

Most students should be able to:
- describe the different forms of radioactivity
- state the properties of the different forms of radioactivity
- link the properties of the different forms of radiation with their effects
- explain how film badges distinguish between doses of the different forms of radiation received by workers.

Some students should also be able to:
- evaluate the implications of the care of health workers that use ionising radiation.

Specification link-up: Science B 3.5.1.3

- Know that gamma rays are examples of transverse waves.
- Know that gamma rays are a form of electromagnetic radiation.
- State the characteristics and properties of the three main types of nuclear radiation emitted continuously by radioactive sources (alpha particles, beta particles and gamma rays).
- Know that the use of high-energy radiation can be dangerous and needs to be monitored.
- Know that people who work with radiation wear film badges and that these are monitored regularly to check the levels of radiation absorbed.
- Be able to describe the construction of the film badge.

Within this context, candidates should be able to use scientific data and evidence to discuss, evaluate or suggest implications of the following:
- care of health workers that use ionising radiation as part of their everyday work.

Lesson structure

Starters

Source-radiation-effect – Explain to students that sources give out radiation that can be detected away from the source. Ask for examples, e.g. a torch is a source of light radiation that is detected by our eyes. Introduce the idea that atomic nuclei can be a source of radioactivity but in this case, small particles are thrown out of the atom. Ask them for suggestions about how to detect radioactivity. *(10 minutes)*

Main

- Remind students that the radioactivity comes from certain atoms. If you are exposed to a radioactive source, you do not become radioactive unless you become contaminated with some of the radioactive material.
- Not all elements are radioactive. Not all atoms of radioactive elements are radioactive. Students may have heard of isotopes, i.e. atoms of the same element with different numbers of neutrons in their nuclei. Many elements have radioactive isotopes and at least one stable isotope.
- Demonstrate radioactive sources if these are available – remember that you cannot use a student as a helper. This demonstration is valuable and is simple to do.
- Use a Geiger counter or similar ratemeter to demonstrate the activity of the sample. You will also need the sources, samples of lead, aluminium, card and other materials of different thicknesses and protective equipment. Change the settings as appropriate for the activity levels of your samples. (See 'Practical support'.)
- The dosage from the samples is unlikely to be harmful but you must demonstrate correct handling procedures. Use tongs to handle samples, and keep students away from the sources. Keep sources in their container if they are not being used and limit the time of exposure.
- Demonstrate that there is a background count at all times. This is naturally occurring radioactivity from our surroundings. Show how the reading varies as different sources are moved closer to and away from the data logger. Use different absorbers between the source and the ratemeter to compare the penetrating power of sources.
- Remind students that radioactivity must reach cells in order to harm them. Alpha particles must be inside the body to harm cells as they can't penetrate far. However, they are very damaging inside the body as they are very ionising. Gamma radiation is hard to contain as it is so penetrating. However, it is less ionising. You can't say which is more dangerous – it depends on the situation and the energy of the radiation which can vary widely.
- Discuss the film badge in detail in this lesson, having introduced it in the previous lesson.

Plenaries

Role play: am I at risk? – Students can work in small groups. They should act out a scenario of a worker, safety officer and employer. The worker routinely forgets to wear his/her badge – the safety officer explains the implications of this. Another scenario is a worker whose employer does not provide the badge. The worker and safety officer explain the implications of this. Support students by giving them a prompt sheets to work though either scenario in groups, or with three students acting this out for the class. *(10 minutes)*

Support

- Use Lego™ pieces for visual impact. Make up a nucleus using roughly equal numbers of two different colours of Lego™ representing neutrons and protons. Demonstrate alpha decay by removing a particle made from two pieces of each colour. Represent beta decay by substituting an extra proton and removing a neutron.
- Beware of students thinking that beta radiation comes from the outer electrons in the atom – radioactivity is from the nucleus.

Extend

- Explain the impact on the nucleus of emitting an alpha particle (the mass drops by four, the charge drops by two so the remaining element is two lower in the periodic table. If a beta particle is emitted, the mass is unchanged and a neutron changes into a proton. The new element is one higher in the periodic table. Gamma radiation is emitted with alpha or beta radiation but not on its own.

Practical support

Penetrating power

Equipment and materials required: Geiger–Müller tube, ratemeter (and possibly high voltage power supply), large plastic tray, long tweezers, radioactive sources, set of absorbers (paper, card, plastic, aluminium of various thicknesses and lead plates).

Details: Position the GM tube in the tray and switch it on. Bring the sources close to the tube window (and above the tray) and the ratemeter should count. If you can find a ratemeter that clicks, the demonstration is a lot more fun.

Add a mount to position a source in place. Between the source holder and GM tube position a holder to hold the absorbers. Make sure that the GM tube is less than 10 cm from the source holder or the alpha particles will not reach. Turn on the GM tube and then mount an alpha emitter in the holder and note the count rate. Position a paper absorber between the source and the GM tube and note the count rate. Test the beta source with paper, plastic and then aluminium plates. Test the gamma with aluminium and then various thicknesses of lead.

Safety: Do not handle radioactive sources until you have had a training session from the school Radiation Protection Supervisor. Under *no* circumstances allow the students to handle the source. Always handle the sources with tongs away from the trunk of your body to minimise exposure.

Science in context

This may not seem directly relevant, but in fact we are surrounded by radioactivity from natural sources, e.g. the Sun and rocks. The Earth has been more radioactive in the past and we have evolved in this environment.

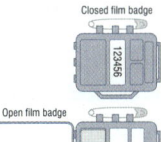

Summary answers

1 inside, nucleus, weak
2 beta – it can penetrate through plastic but not through metal
3 Alpha radiation is very ionising but it cannot penetrate far into the skin so it can't reach vulnerable cells unless it is breathed, eaten or injected in. Beta and gamma radiation are less ionising so are less likely to be absorbed and cause damage but are more likely to penetrate into the body.

Did you know …?

Gamma rays are absorbed by the atmosphere because, although it is not a dense medium, it is deep.

Answers to in-text questions

a alpha, beta, gamma
b It loses or gains an electron (from its outer shell).
c i alpha, beta, gamma   ii gamma, beta   iii beta   iv alpha
d To limit the number of cells damaged /reduce cancer risk.
e To monitor their exposure over time.

Unit 3, Theme 1 – Improving health and wellbeing

11.8 Uses of ionising radiation

Learning objectives

Students should learn:
- that radioactivity is used to diagnose and treat cancer
- how radioactivity is used in tracers, radiotherapy, gamma cameras
- some of the ethical issues of treatment.

Learning outcomes

Most students should be able to:
- state different ways that radioactivity is used to treat cancer
- state that treatment should aim to kill cancer cells and not damage or kill healthy cells
- describe how gamma cameras and tracers can be used in diagnosis
- discuss and evaluate the ethical implications of cancer treatment using ionising radiation.

Specification link-up: Science B 3.5.1.3

- Know that some medical imaging equipment involves the use of gamma rays, which can be detected using a gamma camera.

Within this context, candidates should be able to use scientific data and evidence to discuss, evaluate or suggest implications of the following:
- the advantages and disadvantages of using ionising radiation for the diagnosis (including medical tracers) and treatment of diseases
- ethical issues that may need to be considered by doctors and patients before the treatment of cancers with ionising radiation.

Lesson structure

Starters

How big will it get? – Explain to students that once we stop growing, our cells divide and die at the same rate. Cancer cells divide more quickly than they die. Students start with the number 1, then double it (2) then double it (4), etc. To extend students, ask them how many doublings are needed to reach a million. Support students by helping them to double the numbers eight or nine times just to get the picture of how the growth gets more rapid. *(5 minutes)*

What do you know about cancer treatment? – Ask students to write down three facts they know about cancer treatment. Some may be well informed if their relatives have or have had cancer, but this is an opportunity to correct some misconceptions. *(10 minutes)*

Main

- There are many excellent sites written for patients, which are accessible to students. These include the Cancer Research website.
- Explain to students that each patient needs an individual plan as different cancers need different treatments and each patient reacts individually to treatment.
- Explain that gamma radiation is used as alpha and beta radiation will not penetrate far enough into the body. Internal radiotherapy can use beta or gamma sources as the implant is close to the tumour.
- Radioactive sources become less radioactive over time: their activity halves during a period of time called a half life. Sources used for tracers and internal radiotherapy have fairly short half lives, e.g. hours or days. This limits the dose of radioactivity the patient is exposed to. Sources used for external radiotherapy will have very long half lives (many years) so the equipment does not need recalibrating frequently.
- Ethical issues can be debated here, for example on the most appropriate treatment for a patient and whether everyone should receive the same treatment; whether money should be allocated for research into treatments, for screening healthy people or to educate people on lifestyle choices that reduce the risk of cancer. Students could consider if a screening programme offers good value for money especially considering the effect of false positives.

Plenaries

Why do we use radiation? – Students summarise at least three points that make radiation suitable for use in diagnosing and treating cancer. *(5 minutes)*

Spider diagram – Ask students to prepare a spider diagram to link the ideas from this lesson and the previous one. Support students by giving them a list of key words and ask them to link pairs of these in sentences before linking them within the diagram. *(10 minutes)*

Support

- Reinforce the idea that cells divide rapidly to form tumours – students may think the cells themselves grow. They may also think someone is radioactive from treatment – the cells of the patient are not radioactive, but radioactive material inside them will be (e.g. from a tracer or internal radiotherapy).

Extend

- The treatment of cancer is very individualised. Students can investigate other diagnostic techniques and treatments that are available. There are also sophisticated ways to ensure that the beams are targeted at tumours. This whole area is being actively researched and there is a lot of information that students can investigate.
- Students should debate ethical issues. How should a doctor treat a pregnant woman for cancer? Is aggressive cancer therapy the best option for an elderly patient? Are patients capable of choosing their best treatment plan?

Further teaching suggestions

Cancer treatment notes

Students could research and prepare a set of notes either for a patient, or for a member of the medical staff explaining the treatments, risks and precautions during cancer treatment. You can then compare the students work with some actual NHS information leaflets. Alternately use the NHS leaflets as stimulus material.

Unit 3, Theme 1 – Improving health and wellbeing

11.8 Uses of ionising radiation

Learning objectives
- How is radioactivity used to treat patients with cancer?
- What are the ethical issues that should be considered when using radioactivity to treat patients with cancer?

Gamma cameras

One way to investigate a patient who may have cancer uses **gamma cameras** and tracers. A tracer is a radioactive substance that is injected into the patient. The tracer travels through the bloodstream, and collects in the patient's bones or other tissues.

More of the tracer collects in regions where the cells are more active. This is where cancerous cells are repairing themselves. The tracer in the patient's tissues gives out gamma radiation. More intense gamma radiation is given out from places where the tracer collects. After a couple of hours, the patient passes through the centre of the gamma camera (see Figure 1).

The gamma camera detects where there is most radioactivity. This shows where the cancer is. Then doctors use this information to decide what treatment to give the patient.

There are some risks from using tracers and gamma cameras. The dose of radiation is equivalent to about 200 X-rays. This may be dangerous if the patient is a baby, a child or pregnant. The cells in very young children are more vulnerable to the effects of ionising radiation.

However, patients will only have a scan if doctors think they may have a serious health problem. The greatest risk is that cancer is not diagnosed and treated correctly.

a Write down one benefit and one problem for a patient having a gamma camera scan.

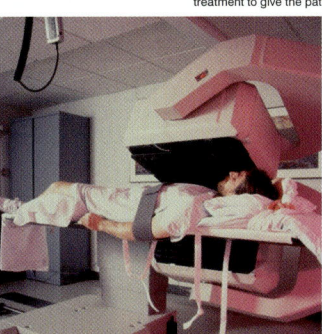

Figure 1 A gamma camera creates a 3-D image of the patient

Radiotherapy

Cancer cells can be killed using ionising radiation like X-rays or gamma rays. This is called **radiotherapy**. If cancerous cells are damaged or destroyed, they cannot divide so the tumour stops growing. Radiotherapy is also used after an operation to kill any remaining cancer cells. Cancerous cells are targeted to receive a large dose of radiation. Healthy cells are protected so that they receive a much lower dose.

In **external radiotherapy** ionising radiation is beamed at the tumour from outside the body. Doctors use several beams from different directions. This reduces the dose that healthy cells receive. Treatment takes a few minutes but is repeated over several days. Patients receiving external radiotherapy do not retain radioactivity as the radiation passes through them.

In **internal radiotherapy** a sample of radioactive material is placed next to the tumour inside the body. The sample emits gamma or beta radiation directly to the tumour. Cells further away receive a smaller dose. Less radiation is needed to treat the patient this way. Patients are radioactive during treatment because some radioactive material stays inside them.

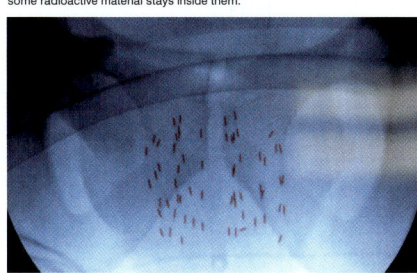

Figure 2 Brachytherapy uses radioactive pellets placed next to the tumour inside the patient

Ethical issues

Patients and doctors have to decide whether or not to use ionising radiation to diagnose or treat cancer. In many cases, a patient may be unwell during treatment but recover and live for many years afterwards. If the cancer can be diagnosed before it is advanced, patients are less likely to need surgery and they are more likely to be cured.

However, in some cases, the cancer may not be suitable for radiotherapy. Some patients may reject treatment if they are unlikely to recover or are worried about long-term damage from the treatment.

b What is radiotherapy?

Did you know ...?
One source of radioisotopes used for medical treatment is from nuclear reactors in nuclear power stations.

Summary questions

1 Copy and complete using the words below:
internal   tracers   tumours
  .... are formed when cells divide uncontrollably.
  .... are used with gamma cameras to diagnose cancer.
  .... radiotherapy is used to treat cancer inside the body.
2 Explain **two** advantages of diagnosing cancer using a gamma camera.
3 Prepare a leaflet to explain to patients how a gamma camera helps to diagnose cancer and why it is used in hospitals.
4 Find out more about the different uses of tracers.

Key points
- Radiotherapy is used to treat cancer.
- Gamma cameras and tracers can diagnose cancer.
- Radiotherapy is not suitable for treating all cancers.
- Patients and doctors need to balance the harm caused by receiving ionising radiation against the chances of successfully treating cancer.

Summary answers

1 tumours, tracers, internal

2 for example, no need for surgery, can diagnose small tumours within organs and tissues

3 Include description of the use of gamma cameras with tracers; explain that ionising radiation is too harmful to use outside a controlled setting; explain that it helps doctors spot places where cancer may be developing.

4 For example, tracers may be used in hospitals to investigate blood flow through different organs; in industry they can investigate the flow of oil through an engine, or track pipelines and detect leaks.

Science in context

Be aware that many students will know people who have been, or are being treated for cancer. Students are very likely to be aware of the illness, but not be very well informed.

Answers to in-text questions

a Cancer may be detected or ruled out ; the patient receives a relatively large dose of ionising radiation.

b Radiotherapy is treatment using ionising radiation to kill cells.

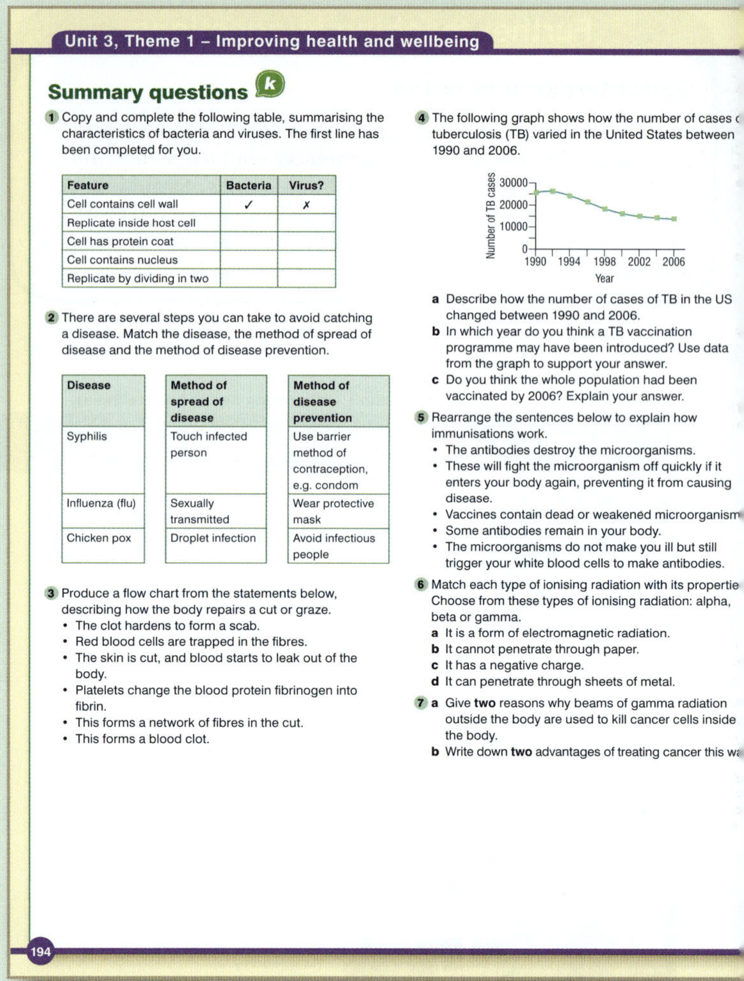

Unit 3, Theme 1 – Improving health and well being

Summary answers

1

Feature	Bacteria	Virus?
Cell contains cell wall	✓	✗
Replicate inside host cell	✗	✓
Cell has protein coat	✗	✓
Cell contains nucleus	✗	✗
Replicate by dividing in two	✓	✗

2

Disease	Method of spread	Prevention
Syphilis	Touch infected person	Use barrier method of contraception, e.g. condom
Influenza (flu)	Sexually transmitted	Wear protective mask
Chicken pox	Droplet infection	Avoid infectious people

Syphilis → Sexually transmitted → Use barrier method of contraception, e.g. condom
Influenza (flu) → Droplet infection → Wear protective mask
Chicken pox → Touch infected person → Avoid infectious people

3 The skin is cut, and blood starts to leak out of the body.
↓
Platelets change the blood protein fibrinogen into fibrin.
↓
This forms a network of fibres in the cut.
↓
Red blood cells are trapped in the fibres.
↓
This forms a blood clot.
↓
The clot hardens to form a scab.

4 a Between 1990–1992, no change (slight rise in cases)
Between 1992–2000, steady decrease in number of cases
Between 2000–2006, a steady decrease in the number of cases, but at a lower rate than 1992–2000.
 b 1990–1992 – it was after this point the number of cases decreased (the gradient becomes negative)
 c No, there were still many cases of TB in 2006.

5
- Vaccines contain dead or weakened microorganisms.
- The microorganisms do not make you ill, but still trigger your white blood cells to make antibodies.
- The antibodies destroy the microorganisms.
- Some antibodies remain in your body.
- These will fight the microorganism off quickly if it enters your body again, preventing it causing disease.

6 a gamma
 b alpha
 c beta
 d gamma

7 a Gamma radiation kills cells; gamma radiation can penetrate inside the body, etc.
 b There is no need for an operation; the patient is exposed to radiation for a shorter time, etc.

Kerboodle resources

- Revision podcast: Vaccines (11.4)
- Bump up your grade: Vaccines – Good or bad? (11.5)
- Interactive activity: Radiation (11.6)
- On your marks: Medicines and medical treatment
- Examination-style questions
- Answers to examination-style questions
- Test yourself: Improving health and well being

Examination-style questions

Your body has a defence system to get rid of any pathogens that enter.

Choose the correct word from each box to complete the sentences.

a There are two types of .... that help to get rid of disease. (1)

| red blood cells | platelets | white blood cells |

b These are the .... that make antibodies that stick to the disease ... (1)

| lymphocytes | pathogens | plasma |

c ... and the .... that engulf anything with an antibody attached to it. (1)

| microbes | phagocytes | virus |

2 The graph shows the number of measles cases per year in Britain between 1996 and 2008.

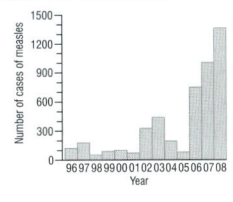

a In which year was the number of measles cases the lowest? (1)
b How many cases were there in 2003? (1)
c A triple vaccination called MMR is given to children under two.
What are the other two diseases the vaccination protects against? (2)
d Some parents want to give their child three separate vaccines rather than the triple vaccine MMR. Suggest one reason why this might increase the number of measles cases. (1)

3 Radiographers can see inside people's bodies using X-rays.
a Put these sentences in the correct order to describe how X-ray images of broken bones are produced.
A The X-rays that penetrate through the patient expose the film.
B The part of the patient affected is exposed to X-rays.
C Photographic film is placed behind the part of the patient being investigated.
D X-rays penetrate through soft tissue like skin and muscle. They are absorbed by denser structures such as bones and teeth.
E The image is then developed. Regions of film that were exposed to X-rays show up black. Areas of the film that were not exposed to X-rays because they were absorbed are white. (4)
b What type of wave is an X-ray? (1)
c Give the name and use of another electromagnetic wave that can be used in hospitals. (1)

4 A film badge can be worn to monitor exposure to ionising radiation. The diagram shows a type of a film badge.

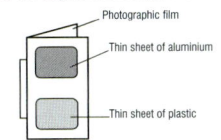

a In this question you will be assessed on using good English, organising information clearly and using specialist terms where appropriate.
Describe how the film badge is used to detect the amount and type of radiation someone has been exposed to. (6)
b Suggest what disease someone who is exposed to too much radiation could get. (1)

4 a Marks awarded for this answer will be determined by the Quality of Written Communication (QWC) as well as the standard of the scientific response.

There is a clear, full and detailed description of how the film badge is used to detect the amount and type of radiation someone has been exposed to. The answer is well structured with minimal repetition or irrelevant points. There is an accurate, fluent and clear expression of ideas with only minor errors in the use of technical terms, spelling, punctuation and grammar. (5–6 marks)

There is some description of how the film badge is used to detect the amount and type of radiation someone has been exposed to, with some omissions. The answer shows some attempt at structuring and the ideas are expressed with reasonable fluency and clarity. There are some errors in the use of technical terms, spelling, punctuation and grammar. (3–4 marks)

There is a brief description of how the film badge is used to detect the amount and type of radiation someone has been exposed to. The answer is largely incomplete and may contain some valid points which are not clearly structured. It lacks fluency and/or clarity. It contains errors in the use of technical terms, spelling, punctuation and grammar. (1–2 marks)

No relevant content (0 marks)

Examples of points made in the response:
- Film goes darker on exposure to radiation
- Can detect how much gamma behind the aluminium window
- Can detect how much beta and gamma behind the plastic window
- So can work out how much beta there was by comparing both photographic films
- Alpha can only travel a few cm in air and cannot get through the thin plastic sheet
- But alpha does not penetrate the skin so is not harmful outside the body.

b cancer OR radiation sickness (1 mark)

Examination-style answers

1 a white blood cells (1 mark)
  b lymphocytes (1 mark)
  c phagocytes (1 mark)

2 a 1998 (1 mark)
  b Any number between 420 and 480 (1 mark)
  c mumps
    rubella (2 marks)
  d May choose not to have particular vaccine
    OR
    May forget to have that vaccine
    OR
    May catch measles while waiting for vaccine. (1 mark)

3 a CBDAE (All correct = 4 marks, 4 correct = 3 marks, 3 correct = 2 marks, 1 or 2 correct = 1 mark)
  b transverse (1 mark)
  c Gamma imaging: sterilisation OR cancer cure OR medical
    OR
    Infrared waves: heat lamps
    OR
    Visible light: any sensible use
    OR
    Ultraviolet: sterilisation (1 mark)

Unit 3, Theme 2

Making and improving products

AQA Specification link-up: Science B 3.5.2

Scientists seek to produce new products from naturally occurring starting materials such as metals, rocks and minerals by physical or chemical change. These products affect our quality of life and well being.

When developing and making new products scientists have to be concerned about the effect of doing so on the environment.

In this theme there are three contexts:
- 3.5.2.1 Uses of electroplating
- 3.5.2.2 Developing new products
- 3.5.2.3 Selective breeding and genetic engineering

Uses of electroplating

Activity:
- Students would have heard the term 'gold plated' before, but may not be aware of the role of electricity in the process. Explore existing knowledge by showing the class a gold plated item (audio connectors, jewellery, printed circuit board) and asking students to suggest how and why the gold was added.
- Identify different metals used as a base metal and as a plating metal. Ask students for their ideas why some metals tend to be plated, and why other metals are used to make the object.

Possible misconceptions

Students may be confused by how the electrical current seems to flow through a liquid. Remind them that this is due to movement of ions, not electrons.

Help students follow the flow of electrons through the electrolysis circuit. The electrons flow between the power source, the electrodes and the ions at the electrodes, but not through the electrolyte.

Unit 3, Theme 2

Making and improving products

In Unit 3, Theme 2 you will work in the following contexts, covered in Chapters 12 and 13:

Uses of electroplating

Metals have many useful properties, making them tremendously important to us. However, some properties of certain metals make them less useful. For example, iron is prone to rusting and nickel can cause allergies. Electroplating is a way of using one metal to coat another. This can cover up a metal surface to stop it rusting, or to stop it coming into contact with skin.

Electroplating uses a process called electrolysis. This is when a compound is split up using an electric current. During electroplating, the object to be plated is put into a solution containing metal ions. The electric current makes the ions stick to the object and turn into atoms of the plating metal. In this way, the new metal builds up in layers.

Electroplating can be hazardous. Large electric currents can be dangerous and the chemicals used are often poisonous. People who electroplate metals need to minimise the risks they are exposed to. Safety equipment such as thick gloves and face masks are used to prevent workers from being harmed.

Developing products

There are lots of new materials in everyday use because of the hard work of materials scientists. We now have:
- plastics that change colour as the temperature changes
- glasses, jewellery and clothes that change colour when they are exposed to light
- metals that always bend back to their original shape, or change shape when heated
- paints for cars that can heal their own scratches
- superconductors that conduct electricity without resistance and can make trains levitate.

Research into these materials is ongoing. Every year, more and more products that improve our quality of life are made.

Selective breeding and genetic engineering

Fifty years ago, the human population was about 3 billion. Since then, it has more than doubled to about 6.8 billion and it is continuing to rise rapidly. As the number of people on the planet continues to increase, more food must be produced. Farmers and agricultural scientists try to solve this problem with selective breeding and genetic engineering.

Selective breeding has been used for hundreds of years to make farm animals grow faster and larger. Only the animals with desired characteristics are allowed to breed. One of the problems of this is that it takes a long time to improve an organism through breeding. Another problem is inbreeding. When the breeding population is too small, genetic disorders can develop.

Genetic engineering involves changing the DNA of an organism to give it new or better characteristics. For example:
- crops that are resistant to disease or can grow in colder weather
- fruit and vegetables that contain extra vitamins
- bacteria that produce human insulin to treat diabetes patients.

Techniques like cloning and tissue culturing are other forms of biotechnology aimed at improving human life. Soon it may be possible to grow replacement organs for people, using just a few cells.

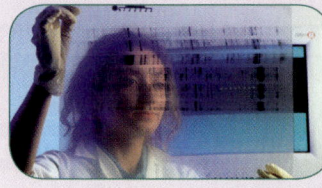

Developing products

Activity:
- Introducing students to smart materials is a perfect opportunity to work directly from contexts and every effort should be made to acquire samples of memory metal, thermochromic and photochromic materials. Students can work together on a project to develop a product that incorporates thermochromic or photochromic properties. Approaching this as a project also enables the teacher to include elements of the 'Making products' theme. Given data about costs of materials, students should be able to calculate the total cost of the new product they have made. Can they develop an innovative product that will sell for more than it costs to make? Potential projects include:
- thermochromic mug designs
- photochromic jewellery
- thermochromic stickers for mixer taps.

Possible misconceptions

Thermochromic pigments normally change from colourless to a coloured state. By painting them on top of existing colours, manufacturers give the illusion of changing between those colours.

Making and improving products

These earrings are made of nickel that has been electroplated with gold. The nickel is cheap and strong, but it is not very attractive and some people are allergic to it. Covering them with gold makes them look more attractive and prevents people developing allergies. It also helps the earrings last longer, as gold is less reactive than nickel and will not corrode.

Thermochromic pigments change colour when the temperature changes. For example, special inks have been developed that show when an egg has been cooked for the correct time. Customers can choose whether they want soft-, medium- or hard-boiled eggs, then buy eggs with a stamp that turns black after 3, 4 or 7 minutes respectively. Thermochromic road safety signs could be used in future to warn motorists about frosty weather conditions.

Superconductors are used in my MRI scanner. They are part of a very powerful electromagnet that is used to produce medical images. Without superconductors, it would be harder to detect brain tumours and other medical conditions. The superconductors allow electricity to go through them with no resistance. This means I can use very high currents to make strong magnetic fields. The superconductor needs to be very cold first, so liquid helium that is extracted from the atmosphere is used.

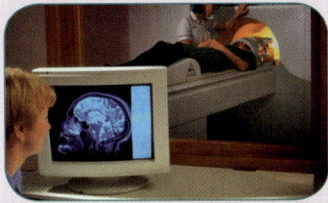

Researchers in Japan have added a gene to mice that makes them glow green. The gene was originally from a type of jellyfish that glows when exposed to blue light. The researchers did this in order to test different methods of genetic engineering. This kind of experiment improves scientists' understanding of genetic engineering. Greater understanding can lead to more ways to cure genetic diseases. Mice are not the only animals to have been genetically engineered in this way. Scientists have also created glowing pigs, monkeys, fish and even a rabbit. This type of research raises many ethical issues.

Selective breeding and genetic engineering

Activity:

- Selective breeding, genetic engineering, GM foods and cloning are terms students are likely to recognise from previous work. As a taster, the topic could be introduced through a news article, which is shared with the class. After reading the article, students can take part in the following activities.
- Ask students to discuss what they think genetic engineering is. In groups of four they should come up with five key words, or a diagram that describes this process. The ideas should be fed back to the class and discussed.
- Ask students to come up with three statements about cloning, which will be shared and discussed as a class. One statement should be about a fact they are sure of. One statement should state a fact that they are unsure if they understand, and one statement should mention something they would like to find out about cloning.

Possible misconceptions

Cloning a human would make a baby, not an adult. Students sometimes need reminding that cloning an organism is not the equivalent of a biological photocopy.

Unit 3, Theme 2 – Making and improving products

12.1 Electrolysis

Learning objectives

Students should learn:
- that electrolysis involves chemically breaking down substances by using electricity, and is used to electroplate metals
- that electrolysis involves the movement of charged particles called ions between electrodes within an electrolyte.

Learning outcomes

Most students should be able to:
- define electrolysis
- name and describe the equipment used in electrolysis
- explain electrolysis in terms of the movement of ions.

AQA Specification link-up: Science B 3.5.2.1

- Describe the process of electroplating as the application of a metal coating to a metallic or conducting surface by electrolysis.
- Know that electrolysis involves the movement of charged particles in an electrolyte.
- Know that the cathode is the negative electrode and the anode is the positive electrode in an electrolysis cell.
- Understand that charged particles are called ions and that ions are atoms which have either lost or gained an electron.

Candidates should be able to use scientific data and evidence to discuss, evaluate or suggest implications of the following:
- knowledge of charged particles to explain the electroplating of metal objects.

Lesson structure

Starters

Atoms and ions – Review students' understanding of atomic structure as a foundation to understanding ionisation. To support students, have them label a simple diagram of an atom, reminding them about the charges on each particle. To extend students, have them also name common ions from their formulae, such as SO_4^{2-} and Cl^-. *(5 minutes)*

Demonstration – Show students video clips and images of the electrolysis of bauxite. Review their knowledge of the extraction of aluminium from Chapter 2. *(10 minutes)*

Main

- Students can make a mind map of the types of chemical reactions they have met in science lessons before. More able students may be able to give specific examples. To support students, show them pictures or video clips of chemical changes (such as colour change, explosions etc) demonstrating ways that chemicals can react with each other.
- A good visual example of electrolysis with straightforward chemistry is the electrolysis of molten lead bromide (see 'Practical support'). It is easy to see the pure lead that has formed. This demonstration is probably better suited to small groups, as it requires a fume cupboard. Use this as an opportunity to introduce students to the names of the equipment involved. Copper chloride solution could also/alternatively be electrolysed.
- Make a large-scale model of electrolysis to demonstrate the electrolysis of molten ionic compounds. A detailed understanding of this is not needed, but it can provide a starting point for understanding the behaviour of aqueous solutions. Prepare cards with names of ions on them and sticky-tac on the backs. Use a large diagram of a standard electrolysis experiment, either on a whiteboard projector or overhead, and physically move the ions around, showing the products at the anode and cathode. This could easily be modified and used as a student demo, with students (as-particles) re-enacting the process themselves.
- Return to diagrams of atomic structure to explain why the loss or gain of electrons changes the overall charge of a particle. Use this to introduce various metal ions, asking the class to suggest how many electrons the ion would need to pick up at the cathode to regain metallic form.
- Ensure students understand that of the ions involved in electroplating, the metal ions are always deposited at the cathode (negative electrode).

Plenaries

Labelling – Students can label diagrams of electrolysis experiments. To support students, provide them with the labels. To extend students, get them to write in the ions present in the electrolyte and suggest what happens at the electrodes. *(5 minutes)*

Demonstrating understanding – Working in small groups, get students to plan (bullet points and sketch diagram will do) a procedure to plate gold into a nickel earring, using gold cyanide as an electrolyte. *(10 minutes)*

Answers to in-text questions

a Electroplating means coating a metal all over with a very thin layer of another metal.
b An ion is a charged particle. Losing electrons makes positive ions, gaining electrons makes negative ions.
c The anode is the positive electrode, the cathode is negative.
d Because it has gained or lost electrons.

Support

- Try to provide as much experience of the process as possible through demonstrations and examples. Focus on the names of the equipment.

Extend

- Use a variety of examples of metals and electrolytes to help students get a grasp of writing electrode equations.
- Provide students with a photograph of a brush plating process and ask them to explain how it works.

Practical support

Electrolysis of water

Equipment and materials required: Eye protection, distilled water acidified with dilute H_2SO_4, large beaker, electrodes, leads, dc power source, test tubes.

Details
Wearing eye protection, connect the electrodes to the power source and suspend them in the acidified water. Collect the resulting gases under water. Point out the difference in volumes and ask students which gas they think is which. Test accordingly for H_2 and O_2 with lit/glowing splints.

Safety: Wear eye protection. CLEAPSS Hazcard 98A Sulfuric acid – corrosive.

Electrolysis of molten lead bromide

Equipment and materials required: Fume cupboard, eye protection, crucible, clay triangle, tripod, mat, ammeter, Bunsen burner, lead(II) bromide, carbon electrodes mounted in rubber bung, dc power source and leads, tin tray.

Details
Perform in fume cupboard. Wearing eye protection, set up electrodes in the lead bromide – they need to be deep enough for a current to pass through, but not deep enough to cause a short circuit when pure lead forms at the bottom of the crucible). Use the ammeter to demonstrate that lead bromide conducts electricity when molten. Heat strongly, adjusting the voltage to give a current of about 1.5 A. Run the electrolysis for 10–15 minutes, pointing out the bromine gas formed at the anode. After 15 minutes, carefully pour away the molten lead bromide to reveal a bead of pure lead in the crucible.

Safety: Perform in a fume cupboard. Wear eye protection. CLEAPSS Hazcard 57A Lead(II) bromide – toxic. Dispose of waste following CLEAPSS guidelines. CLEAPSS suggest using zinc chloride as a safer alternative to lead(II) bromide – see CLEAPSS guidance.

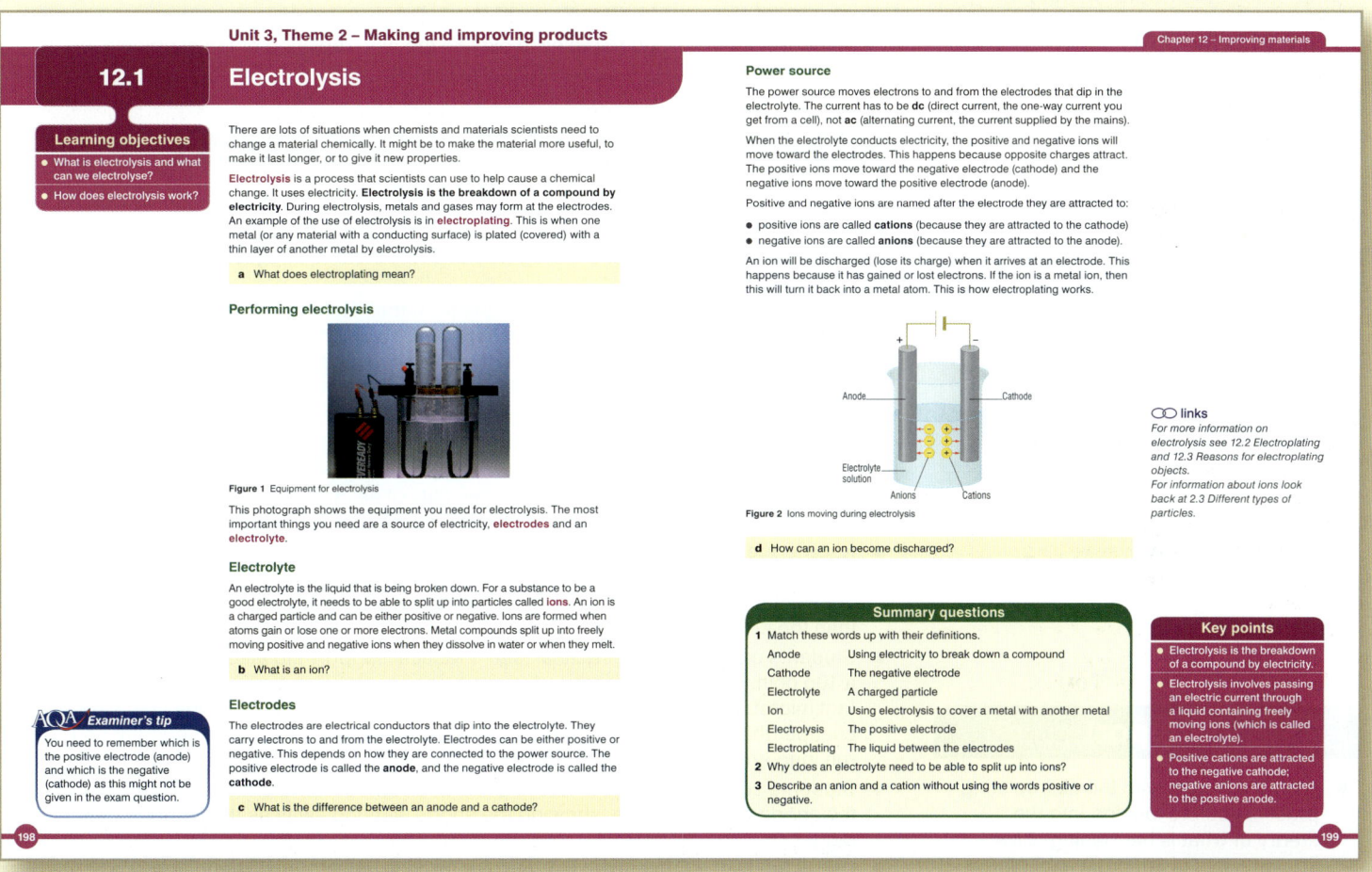

Summary answers

1. Anode – The positive electrode
   Cathode – The negative electrode
   Electrolyte – The liquid between the electrodes
   Ion – A charged particle
   Electrolysis – Using electricity to break down a compound.
   Electroplating – Using electrolysis to cover a metal with another metal.

2. Because electrolysis only works if the particles are charged.

3. Cation: ions (usually metal ions) that will move toward a cathode during electrolysis
   Anion: ions (usually non-metal ions) that will move toward an anode during electrolysis

Unit 3, Theme 2 – Making and improving products

12.2 Electroplating

Learning objectives

Students should learn:
- that metals are electroplated by using electrolysis to coat them with a thin layer of another metal]
- how metal ions move from the anode to the cathode during electroplating
- what happens at the electrodes in electroplating. [HT only]

Learning outcomes

Most students should be able to:
- state that electroplating is the coating of objects with other metals
- state that the object being electroplated should be the cathode, and the anode should be made of the electroplating metal.

Some students should also be able to:
- describe the process at the anode and cathode in terms of electron movement and ionisation
- write balanced equations for the process at each electrode. [HT only]

AQA Specification link-up: Science B 3.5.2.1

- Describe the process of electroplating as the application of a metal coating to a metallic or conducting surface by electrolysis.
- Know that the article to be electroplated is made the cathode and immersed in an aqueous solution containing ions of the required metal. The anode is usually a bar of the metal used for plating. During electrolysis metal is deposited on the article as metal from the anode goes into solution.
- Understand that charged particles are called ions and that ions are atoms which have either lost or gained an electron.
- Be able to complete simple equations to show the process at the cathode and anode
  a $M^{n+} + ne^- \longrightarrow M$
  b $M \longrightarrow M^{n+} + ne^-$ [HT only]

Candidates should be able to use scientific data and evidence to discuss, evaluate or suggest implications of the following:
- knowledge of charged particles to explain the electroplating of metal objects.

Lesson structure

Starters

Electroplated coins – Present students with some 'copper' coins and magnets. Ask them for ideas why the coins are magnetic and ask them to suggest why the coins are not 100% copper. *(5 minutes)*

Review – Assess what students remember from their last lesson by giving them an unlabelled diagram of an electrolysis experiment. To extend students, include reference to the ions in the electrolyte and require them to write a sentence explaining each feature they label. To support students, provide the labels as a list for them to choose from. *(10 minutes)*

Main

- Use a large diagram to return to the principles of electrolysis learned last lesson. Ensure students are still confident with the concept of electrostatic attraction.
- Knowledge of the electrochemical series is not needed. However, it will help if students understand that not all ions are discharged at the electrodes. Explain that the electrolyte is chosen carefully, so that it is easier for the metal ions to discharge than the non-metal ions.
- Students can perform a practical (see 'Practical support') to electroplate copper onto a 10p coin. Discuss the reasons for using copper plate on a coin in the first place [antimicrobial, relatively unreactive]. Why use a steel core? [makes the coin stronger, cheaper than pure copper]. If a 1p coin is used as the anode, the experiment could be run long enough to strip the copper plate, revealing the steel underneath.
- Students could produce a series of pictures showing how the metal ions move from the anode to the cathode, showing the anode wearing away as the cathode gets bigger.
- Use video clips of animations to help reinforce explanations of what happens at the electrodes.

Plenaries

Key word 'taboo' – In groups of three, students are given lists of key words they must describe to each other without using particular words (i.e. describe 'electrolyte' without using the word 'ion'). A point is scored for each member of the group when a word is correctly guessed. One student describes to another whilst the third keeps score. *(5 minutes)*

Mind maps – working individually, students make a web of links between key words associated with electroplating and electrolysis. They can explain why they are linking particular words together by writing a few words on the connecting lines. Students can then pair up or form larger groups to share their ideas. To extend students, provide them with extra words to link into their webs, such as 'economy' or 'society'. To support students, give them a pre-made web, asking them to explain why different words are connected. *(10 minutes)*

Support

- Start the lesson with the eletroplating practical before looking into the theory of what is happening. Give students lots of concrete experience of electroplating so they can understand what is going on. Stick to examples where a clear colour change can be seen (i.e. plating nickel onto copper).

Extend

- Students can perform an investigation into how various factors affect the amount of metal electroplated. These factors could be electrolyte concentration, distance between electrodes and current. Students can compare the mass of the electrodes before and after the experiment.

Chapter 12 – Improving materials

Practical support

Electroplating a coin

Equipment and materials required: Eye protection, 10p coins, copper strips (or 'copper' coins), crocodile clips, leads, dc power source (6V is adequate), 1 mol/dm³ copper sulfate solution, 100 cm³ beakers, sandpaper. Adding a lamp/ammeter to the circuit can help demonstrate the flow of current.

Details

Rub the surface of the copper source with sandpaper to remove some of its oxide coating. Set up the equipment as shown in the Student Book. Apply the current and allow the process to proceed until the coin looks well covered (about 5 minutes). The coins should be rinsed well before handling.

Safety: CLEAPSS Hazcard 27C Copper sulfate – harmful. Wash hands after using copper sulfate.

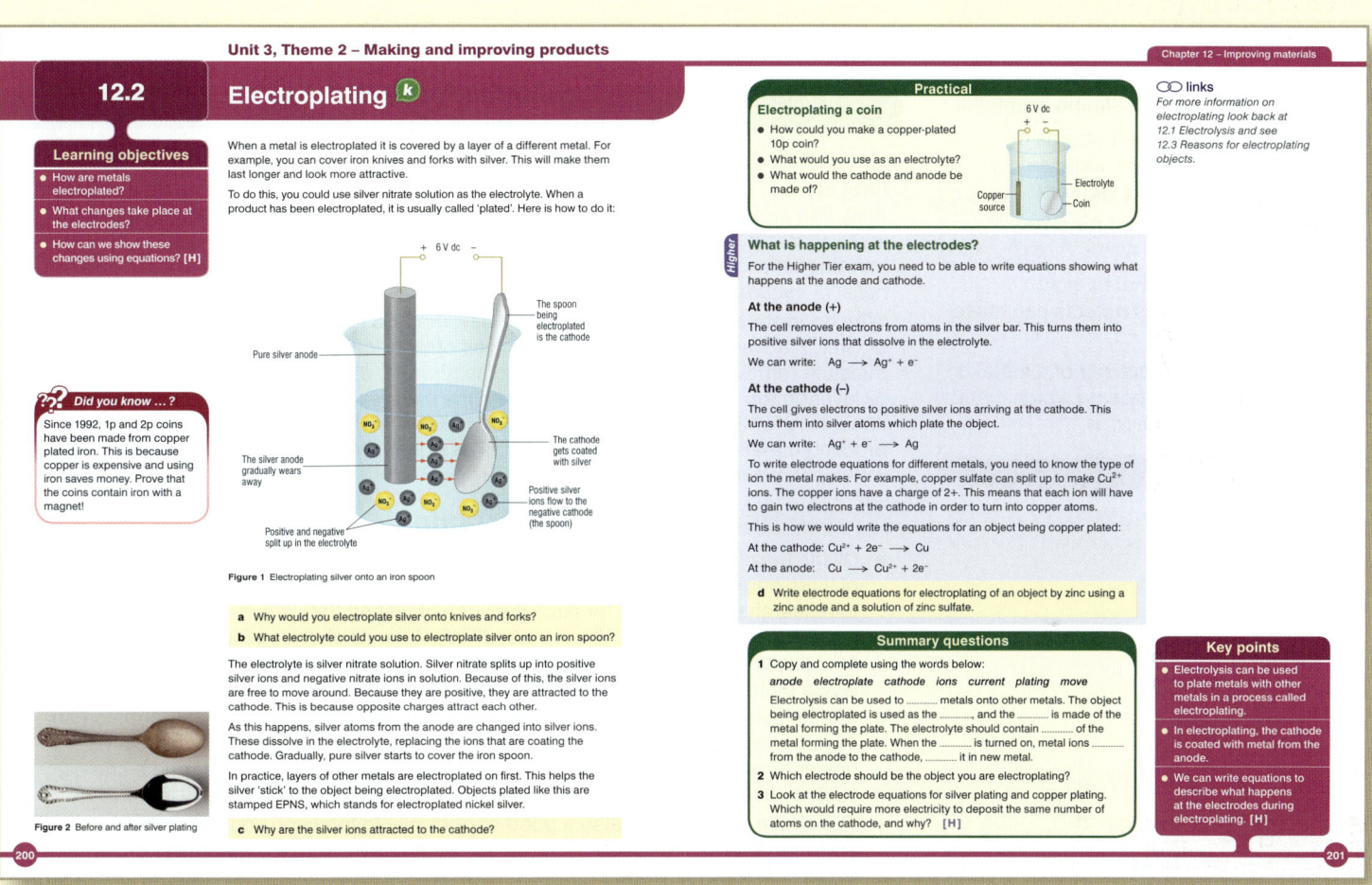

Further teaching suggestions

Plated or not?

Ask students to suggest ways to tell if an object has been electroplated. This can lead to a discussion about destructive and non-destructive testing methods.

Summary answers

1. electroplate, cathode, anode, ions, current, move, plating
2. cathode
3. Copper plating would require more energy than silver plating, as two electrons are transferred for each copper atom deposited at the cathode; but only one electron is needed to deposit each silver atom.

Answers to in-text questions

a. Reduce corrosion and improve appearance.
b. silver nitrate (or silver cyanide, which is mentioned on the next spread).
c. Because the silver ions are positively charged and the cathode ions are negatively charged.
d. Cathode: $Zn^{2+} + 2e^- \longrightarrow Zn$
   Anode: $Zn \longrightarrow Zn^{2+} + 2e^-$

Unit 3, Theme 2 – Making and improving products

12.3 Reasons for electroplating objects

Learning objectives

Students should learn:
- why metals are electroplated
- why electroplating can be hazardous and the safety precautions needed to protect workers in industrial electroplating.

Learning outcomes

Most students should be able to:
- state reasons why metal objects are electroplated, giving simple examples
- describe some of the hazards of electroplating and list some safety precautions taken by workers.

Some students should also be able to:
- explain, using examples, how electroplating metal objects can make them more useful
- explain how the reactivity of a metal is important when choosing which metal to electroplate with.

AQA Specification link-up: Science B 3.5.2.1

- Give reasons for electroplating metals (prevention of corrosion, decoration).
- Name some household objects that are electroplated to prevent corrosion.
- Explain why nickel jewellery is electroplated with precious metals (prevention of allergies or for decoration).

Candidates should be able to use scientific data and evidence to discuss, evaluate or suggest implications of the following:

- the potential risks to employees in the electroplating industry
- the suitability of different metals for electroplating items.

Lesson structure

Starters

First ideas – Present students with an assortment of nails, some of them new iron, some rusty iron and some zinc plated. Discuss the conditions that bring about rusting and invite students to explain why zinc plating is useful in this scenario. To extend students, introduce the idea of sacrificial protection and look at it in another context, such as ship hulls. *(5 minutes)*

Why do metals rust? – Students can review the KS3 topic of rusting. Either use pictures or (well in advance) pre-prepared examples of nails left in different conditions: air with moisture removed, air and water, air and salty water, oil, boiled distilled water, water with oil layer on top, etc. Present students with the pictures; they must deduce the conditions necessary for the nails to rust. To extend students, invite them to suggest ways of rust-proofing metals. To support students, just use dry air, air plus water and air plus salt water. *(10 minutes)*

Main

- Divide the class into three groups to research different uses of electroplating: chrome plating, tin plating and gold plating. Provide either printed material or use the internet to research the following for each instance of plating they are investigating. What metal is being covered? Why is it used? What are its shortcomings? What are the useful properties of the electroplated metal? What electrolyte is used? Are other metal layers plated on first (and why)? Are there alternatives to electroplating in this way?
- After initial research, students can form groups of three with one member being an expert in each type of plating. Each group then prepares a presentation, leaflet or poster on electroplating. These can then be assessed against appropriate criteria, such as use of technical vocabulary, presentation, scientific accuracy, etc.
- The above activities are also a good opportunity to focus on social, economic and environmental issues. Each group of three could be given a different aspect upon which to base their work.
- Review hazard symbols with the class, drawing attention to the hazardous nature of industrial electroplating. Students could match up hazard symbols to various activities and materials.
- Return to the electroplating/electrolysis activities in the previous lesson (or if these were not used, either demonstrate one now or use video clips). After making a list of the materials used, students can write a risk assessment for the procedures.

Plenaries

Hazard match-up – Get students to match up various hazard symbols to their meanings. *(5 minutes)*

Summarising – present students with headings 'metal object', 'metal electroplated' and 'reasons'. Students complete summaries with as many examples as they can. To support students, provide them with answers to put in the right place. To extend students, this could be an opportunity for peer assessment, with pairs of students reviewing each other's notes, presentations or leaflets, etc. and using the table as a means of collating the examples they have learned. *(10 minutes)*

Support

- Introduce the idea of plating to modify a material's properties. Start with a list of different metallic objects and get students to assign desirable properties to them, such as cutlery, motorbikes, jewellery, etc. For extra support, students can choose these from a list.

Extend

- Get students to suggest electroplating projects and provide reasons why it would be beneficial to plate an object with a particular metal.

Further teaching suggestions

Conducting a safety inspection at an electroplating workshop

Theme the lesson around a safety inspection at an electroplating workshop/factory. Examine the different stages of the process, from initial treatment of items to be electroplated to disposal of waste electrolyte. At each stage, highlight the hazards and risks involved, inviting students to suggest how to minimise the risk of harm.

Unit 3, Theme 2 – Making and improving products

12.3 Reasons for electroplating objects

Learning objectives
- Why do people electroplate metals and conducting surfaces?
- What precautions are taken to keep electroplating safe?

You would be surprised by the number of objects in your home that have been electroplated. Kettles, irons, parts of cars, jewellery, tin cans and even the coins in your pocket may have been electroplated.

This is normally done because another metal has more useful properties than the one the object is made of. Adding a very thin layer of the more useful metal gives the object many of its properties.

a Name **three** household items that are electroplated.

Why metals are electroplated
Preventing corrosion

Figure 1 Parts of motorcycles are chrome plated

Many of the metal objects we use are made from iron or its alloy steel. This is because iron is cheap, easy to hammer into shape, strong and dense. The disadvantage of using iron is that it rusts easily. By electroplating the object, we can protect it from rusting but still use iron and benefit from iron's useful properties. A metal is used that is less reactive than iron, usually chromium. Chromium is used because it is hard, shiny and rust resistant.

Another use of electroplating to stop corrosion is in tin cans. Most cans of food are made from steel for the reasons given above. However, if the steel rusts the food will be ruined. Tin is quite unreactive, so it's an ideal choice to store food inside. A very thin layer of tin is used on the inside of the can. Cans plated like this are called tinplate steel. If a can gets dented, the tin layer can break. This exposes the steel, which can then start to corrode.

Figure 2 Tin cans are actually tin-plated steel

??? Did you know …?
Girls are about ten times as likely to be allergic to nickel as boys.

b Which metals are used to electroplate iron in order to stop it rusting?

Reducing the effects of allergies
Nickel is used in a lot of cheap jewellery, tools and fastenings on clothes. However, millions of people around the world are allergic to it. **Nickel allergy** can cause blisters, rashes and even sores on skin that comes into contact with the metal. By plating nickel jewellery with another unreactive metal, jewellers can keep using nickel without it harming the wearer. Usually, a metal like silver or gold is used as they are attractive and do not react with water.

c What problems can nickel jewellery cause?

Figure 3 Electroplating can protect people from allergies like this

Decoration
Silver plating was described in the previous spread. Making objects like cutlery out of pure silver would be too expensive. Silver plating gives us the benefits of silver at a fraction of the cost.

Electroplating safely
Many of the electrolytes used in electroplating are very harmful. For instance, an electrolyte often used in silver plating is silver cyanide. Cyanides are very toxic and can kill you within seconds if you inhale or even touch them. As well as posing a danger to those working as electroplaters, the toxic wastes may damage the environment.

There are lots of hazards in the workplace for an electroplater. Chemical burns, respiratory problems and poisoning are just some of the risks faced in electroplating. Working directly with electricity is also hazardous.

Electroplating companies use many precautionary measures to reduce the risk of staff being harmed. Safety boots, breathing masks, gloves, eye protection and chemical-resistant clothing are used every day. There are also strict rules about disposing of any chemical waste.

d Name a poisonous chemical used in electroplating.

Figure 4 Electroplated nickel silver

Summary questions

1 Copy and complete using the words below:
carefully hazardous chromium allergic nickel corroding safety tin

.... is used to electroplate many iron items to make them rust resistant. .... is used to plate iron in cans of food, to stop the iron .... Cheap .... jewellery is electroplated with metals like gold or silver to prevent .... reactions. Electroplating can be .... Workers must use .... equipment at all times, and waste must be disposed of ....

2 Produce a summary table about electroplating using the information from this spread. Use the headings 'Item that is electroplated', 'Metal used', 'Reason for electroplating'.

3 List some of the hazards involved in electroplating, and describe the precautions taken to reduce the risks.

 links
For information on the properties of metals, look back at 7.4 Metals for construction.

Key points
- Objects are electroplated to make them resistant to corrosion, prevent allergic reactions and make them more attractive.
- The chemicals used by electroplaters are hazardous. Many safety procedures are in place to reduce the risk of harm.

Summary answers

1 chromium, tin, corroding, nickel, allergic, hazardous, safety, carefully

2

Item that is electroplated	Metal used	Reason for electroplating
Iron kettles, car/motorcycle parts	Chromium	Hard, shiny, resists corrosion
Iron cans	Tin	Resists corrosion
Nickel jewellery	Gold/silver	Nickel allergy

3 Poisonous electrolytes – gloves, goggles, masks
Electrocution/burns – rubber clothing / insulated gloves
Environmental pollution – control of waste disposal

Answers to in-text questions

a three from:
kettles, irons, parts of cars, jewellery, tin cans, coins
b tin and chromium
c rash and blisters – allergic reaction
d cyanide

Unit 3, Theme 2 – Making and improving products

12.4 Developing smart materials

Learning objectives

Students should learn:
- that smart materials change their properties according to their environment
- that smart materials can be used in medicine, car design and spectacles.

Learning outcomes

Most students should be able to:
- list some uses of smart materials
- describe the properties of a named smart material and explain its uses.

Some students should also be able to:
- evaluate how a smart material's properties make it suitable for a particular task compared with a traditional material.

AQA Specification link-up: Science B 3.5.2.2

- Give examples of new products and suggest uses for them:
  - smart (self-healing) paints – a coating that heals its own scratches when exposed to sunlight
  - smart materials – substances that are able to change their properties in response to the environment
  - chromic materials – thermochromic, photochromic – materials that change their colour according to changes in light or temperature.

Within this context, candidates should be able to use scientific data and evidence to discuss, evaluate or suggest implications of the following:
- the advantages and disadvantages of modern products compared with traditional products.

Lesson structure

Starters

What do you think? – Present pairs of students with the names of the different smart materials that will be used this lesson. Give them a minute's talking time, then each pair feeds back. What do they think the names mean? *(5 minutes)*

How old? – Get students to consider the order in which we started using various materials by sequencing a selection of cards with the names of the materials. Use materials such as leather, cotton, nylon, neoprene, silk, iron and rubber. To support students, provide pictures with the names of materials. To extend students, get them to consider how more recent materials improved upon older ones, and what they replaced. *(10 minutes)*

Main

- Start by demonstrating the behaviour of cornflour paste. Students should see that the more force is applied to it, the more viscous it becomes. It can actually be made to shatter when stuck with a hammer. Use this to introduce the ideas of smart materials as materials that have different properties depending on their environment.
- Students can make a sample of smart slime. The slime works in a similar way to the cornflour demonstration. The more shear force is applied, the more the material resists movement (see 'Practical support').
- Show students a video clip of a scratch shield paint advertisement.
- Set up a series of activities that allow students to get to know types of smart material and get some practical experience of working with them. Each activity station can be equipped with a research task either from a laptop or from supplied printed information. All materials are readily available through most science education suppliers.
- Nitinol wires – students observe different conformations after dipping in hot and cold water, and by passing a low electrical current through the wire.
- Thermochromic paints – students can make their own thermochromic paint by mixing zinc oxide into a paste with a little water (see 'Practical support'). Zinc oxide changes colour from white to yellow when heated strongly. Alternatively, a thermochromic paint set can be purchased for students to experiment with. They can paint onto plastic cups and observe the colour changes upon pouring in hot water.
- A tube of photochromic paint could be used along with a badge maker to make UV warning badges.
- If there is not enough time to rotate through all of the above activities in one lesson, students could remain at one station, becoming experts in that smart material. They can then feed back to the rest of the class.

Plenaries

What did I learn? – Students can write a sentence about each experiment they did, linking it to a real-life use of smart materials. *(5 minutes)*

Buy me! – Students make an advertisement for a product made from a smart material of their choice. To support students provide examples for them to write persuasively about. To extend students, require them to explain why the product they are advertising is better than one made from non-smart materials. *(10 minutes)*

Answers to in-text questions

a Memory metals can be made to change shape by changing their temperature.
b Self-healing paints can either flow to fill scratches or form crosslinks over the scratched area.
c Thermochromic: changes colour with temperature
  Photochromic: changes with light
d Prevention of burning, protecting eyes from bright light, warning about UV levels.

Support

- Start with the slime practical and get students to experiment with its properties to show how they change under different amounts of force.

Extend

- Get students to research the reasons why smart materials have such dynamic physical properties.

Chapter 12 – Improving materials

Practical support

Making a thermochromic pigment

Equipment and materials required: Zinc oxide, distilled water, boiling tube, 50 cm³ beaker, mat, Bunsen burner, stirring rod, spatula, eye protection.

Details
Mix two spatulas of zinc oxide with a few drops of water in the beaker with a stirring rod. Use the rod to apply the resulting paste to the inside of a boiling tube. Heat the tube gently at first to evaporate the water, then heat strongly and observe the colour change.

Safety: Wear eye protection. CLEAPSS Hazcard 108A Zinc oxide.

Making smart slime

Equipment and materials required: 20 cm³ 3.2% borax (hydrated sodium borate) solution, 20 cm³ PVA glue, food colouring, 2 plastic cups, disposable pipette, stirring rod.

Details
Add a few drops of food colouring to the PVA glue. Then starting dropwise, add the borax solution to the mixture. Stir thoroughly, continuing to add borax until there is no liquid left and only slime remains.

Safety: Wear eye protection. CLEAPSS Hazcard 14 Borax – toxic. 3.2% solutions of borax do not have a hazard classification.

Unit 3, Theme 2 – Making and improving products

12.4 Developing smart materials

Learning objectives
- What is a smart material?
- How do smart materials improve our lives?

Smart materials are materials that can change in response to their surroundings. More and more products are using smart materials because of their useful properties. Smart materials are used in clothes, medical devices, cars and many products you find in the home.

Memory metals

Examples of smart materials are the shape memory alloys, or 'memory metals'. Memory metals are mixtures of metals that are 'set' into a particular shape when they are formed. They can be designed to have one shape when they are cold, and a different shape when they are hot. They can also be designed to spring back into their original shape if they get bent. They have lots of uses:

- **Spectacle frames** – they can be twisted and bent around without being permanently changed.
- **Braces** – the metal braces are 'set' to be in the correct positions for teeth and then attached to uneven teeth. Body heat makes the brace slowly contract back into its original shape. This corrects the teeth.
- **Medicine** – bundles of thin wire tubes, called stents, are put into narrowing blood vessels. Body heat makes them expand, holding the vessel open properly.
- **Robots** – memory metal can make 'muscle wires'. These expand and contract like human muscles when heated by a current.

a How can memory metals be made to change back to their original shape?

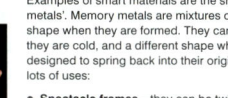
Figure 1 Memory metal lets this frame spring back into shape

Figure 2 Body heat is making this stent expand and hold a blood vessel open

Self-healing paints

Another group of smart materials are self-healing paints. They were first designed by car manufacturers, to reduce the appearance of scratches. Self-healing paints can be used on any painted surface that might get scratched. They work by releasing chemicals into the scratch after it is made. Some self-healing paints work like a very thick resin, slowly filling the scratch to make it disappear. They can make a scratch on a car door disappear within a week. Heat can speed the process up and warm water can help heal the scratch within a few minutes.

Figure 3 Self-healing paint can make scratches disappear

Other self-healing paints are helped along by light. A new type of paint is based on a protein found in lobsters. Ultraviolet (UV) light makes chemicals in the scratch join together in chains, closing the scratch up.

b How can self-healing paints repair scratches?

Chromic materials

Chromic materials have been around for a long time. They can change colour in different conditions. Materials that change colour at different temperatures are called thermochromic. Materials that change colour as a response to light are called photochromic.

c What is the difference between photochromic and thermochromic materials?

Thermochromic materials are usually painted in two layers. The top layer is thermochromic, and becomes transparent when heated. This reveals the layer underneath. Thermochromic materials can be used for:

- **Smart packaging** – if an item has been stored at the wrong temperature, the label will change colour so the user knows. This is useful for packaging medical supplies because certain medicines can be destroyed if they get too warm.
- **Safety materials** – spoons and bottles for babies can be made of thermochromic plastic. This changes colour if the food is too hot.

Photochromic materials have been used to make spectacle lenses for many years. In sunlight, photochromic lenses become tinted like sunglasses. When the wearer moves out of the light, the tint disappears. Other uses for photochromic materials are:

- **Rear-view mirrors in cars** – they darken at night so car headlights from behind don't dazzle the driver
- **Light detectors** – photochromic materials can absorb UV light, changing colour if the wearer has been in the sun for too long.

d Describe three ways chromic materials can stop people getting harmed.

Figure 4 This spoon changes colour if the food is too hot

Figure 5 A photochromic lens is only dark in bright sunlight

AQA Examiner's tip
Make sure you know which applications are in the specification. These will not be given in the question, so you will have to remember them.

Summary questions

1 Copy and complete using the words below:
   temperatures scratches change conditions heat colour memory healing

   Smart materials can .... in response to different .... Chromic materials change .... as a response to light or .... A .... metal can change shape at different .... Self-............ paints are able to fill in .... on paintwork.

2 Make a summary table of the information on this page. Use the headings 'Type of smart material', 'Uses', 'How we benefit'.

3 Write a paragraph to describe the impact smart materials have had on medicine. Remember to consider social, environmental and economic implications.

Key points

- Smart materials bring a wide variety of new properties to products. They can change depending on their environment.
- Smart materials include memory metals, self-healing paints, thermochromic materials and photochromic materials.

Summary answers

1 change, conditions, colour, heat, memory, temperatures, healing, scratches

2

Type of smart material	Uses	How we benefit
Memory metal	Braces, glasses, stents, robotics	Better medical treatment, more advanced robots
Self-healing paint	Cars, any paintable surface	Less need to repaint, so saves resources
Thermochromic paints	Kitchenware, packaging	Safety – avoiding burning. Getting information about storage conditions
Photochromic pigments	Glasses, car rear-view mirrors, UV detectors	Better glasses, safer cars, better protection from the sun

3 Mark scheme:
- Name of smart material used
- Describe how it improves quality of life.
- Explain potential damage to the environment from developing any new products.
- Note job creation in new areas, design, production and potential economic impact of people living longer.

Unit 3, Theme 2 – Making and improving products

12.5 Superconductors

Learning objectives

Students should learn:
- what is meant by a superconductor
- some applications of superconductors
- the advantages and disadvantages of superconductors.

Learning outcomes

Most students should be able to:
- understand what a superconductor is
- understand some uses of superconductors
- list some advantages and disadvantages of using superconductors.

Some students should also be able to:
- explain how superconductors work
- explain in more detail the advantages and disadvantages of using superconductors.

AQA Specification link-up: Science B 3.5.2.2

- Give examples of new products and suggest uses for them:
  – superconductors – substances whose resistance becomes almost zero at low temperatures, which reduces energy losses.
- Within this context, candidates should be able to use scientific data and evidence to discuss, evaluate or suggest implications of the following:
  - the advantages and disadvantages of modern products compared with traditional products.

Lesson structure

Starters

Using superconductors – Show a short clip of a Maglev train from Japan – explain that these trains are able to levitate because of strong electromagnets, which can run even when the current is turned off. This is one use of superconductors. *(5 minutes)*

Reviewing electromagnets – Extend students by getting them to write down what they know about electromagnets. You may want to give them some key words. Support students by giving them a series of statements and check they understand/ remember these points (e.g. electromagnets only work when a current flows through them; electromagnets are stronger if there are more coils, an iron core and a large current; electromagnets can be turned on and off). You could briefly demonstrate the electromagnet to remind students how they work. *(10 minutes)*

Main

- Superconductors have been around for about a century. Their use in everyday applications is limited because of the need to work at very low temperatures (e.g. some students may be familiar with liquid nitrogen if it has been applied to warts).
- Many materials are able to superconduct – these include elements as well as ceramic compounds. www.superconductors.org is a periodic table showing superconductors.
- Show videos of MRI scanners being used and Maglev trains. There might be topical news stories concerning superconductors that may be suitable.
- Obtain a list of instructions from hospitals for patients undergoing an MRI scan. Discuss the reasons for the various precautions (in particular that there should be no metal objects in the room).
- Encourage students to research one aspect of superconductors, e.g. a use, or research, or materials. Students prepare a poster showing the results of their research.

Plenaries

Summarising superconductors – Extend students by asking them to write down a definition of what makes superconductors different to normal conductors, a use, an advantage, and a disadvantage. They could compare these with a partner. Support students by providing them with a summary set of points to put into categories (e.g. uses, advantages). Students can add extra points to the list. *(5 minutes)*

Are superconductors important? – Ask students to vote yes or no and provide reasons for their decisions. At the end of the session, take the vote again to see if anyone has changed their mind. This is more useful after a session where students have carried out independent research so extra detail will be included in the discussion. *(10 minutes)*

Support

- There is a lot of science here related to previous work. This includes circuits, resistance, magnets and electromagnets. Set up a couple of demonstrations for students. One includes a circuit with a variable resistor and a bulb to show that more resistance in the circuit leaves less energy for the bulb to light. The other is a circuit making an electromagnet to remind students that the coil of wire becomes magnetic when the current flows.

Extend

- Students could research superconductors on several different themes.

  This is an area of active research with companies trying to develop materials that are superconductors at room temperature.

  Superconductors have some very specialist applications that students could investigate.

Chapter 12 – Improving materials

Further teaching suggestions

The How Stuff Works website (www.howstuffworks.com) has many links and includes information clearly presented for all abilities of student. There are videos and links to additional information that may be suitable for higher attaining students to carry out independent research.

Science in context

Superconductors have medical uses. Some students may know people who have had MRI scans and may also have seen levitating trains.

Did you know …?

Use the How Stuff Works website to find more detailed information on the discovery of superconductors. Go to www.howstuffworks.com and search under superconductivity. Look at 'what is superconductivity?' and press the link to 'A Teacher's Guide to Superconductivity for High School Students'.

Mercury is a liquid at room temperature, but becomes a solid at temperatures of about −40 °C.

Unit 3, Theme 2 – Making and improving products

12.5 Superconductors

Learning objectives
- What is a superconductor?
- What do superconductors do?
- What are some advantages and disadvantages of superconductors?

Some metals and **ceramics** have no **resistance** to electricity at very low temperatures. This is a very special electrical property called **superconductivity**. Resistance is a measure of how hard it is for an electric current to flow in any material. A current continues to flow in a superconductor even when the power supply is turned off because it has no resistance. Some superconducting wires have carried currents for years after the supply was turned off.

Superconductivity only occurs at temperatures so cold that the wires must be dipped in liquid helium or liquid nitrogen. Scientists are developing new materials that are superconductors at higher temperatures. So far, no material is a superconductor at room temperature.

Figure 1 A superconductor can be made from ceramics. Ceramics are compounds heated to very high temperatures and then cooled. The superconducting material shown above is flexible, and can be moulded. When it has been heated it becomes hard, and stays in the required shape.

a What is the resistance of a superconductor?
b Why don't we see superconductivity in ordinary circuits?

Superconductors are used in electromagnets. An **electromagnet** is a coil of wire in a circuit. When the current flows in the wire, the electromagnet produces a magnetic field. If the current is turned off, the magnetic field disappears. If a superconductor is used as an electromagnet it stays magnetized even when the power supply is turned off.

Any wire with resistance transfers some electrical energy into energy heating the wire and its surroundings. In superconductors, the current can travel any distance with no energy wasted in the wire.

c How does an electromagnet made from superconductors behave differently from other electromagnets?
d How is the energy wasted in a superconductor different from that in normal conductors?

Figure 2 Two sets of magnets in the track levitate the train and force it to move

Where are superconductors used?

The uses of superconductors are limited by the need to cool wires to very low temperatures.

Maglev trains in Japan use magnets to make the train levitate and move. The train has superconducting magnets in its undercarriage, and there are electromagnets in the track's sidewalls. These sets of magnets repel each other so the train floats above the track. This reduces friction so the train can travel at up to 500 km per hour.

Another set of electromagnets in the track and train make it move. When the train passes coils of wire in the track's sidewalls, magnetic fields are created. These interact with the train's magnets to pull the train along at high speeds.

The superconducting magnets work even when their electricity supply is turned off. This way energy is saved, although it can be expensive to cool the magnets to the low temperatures needed.

e What are the advantages of Maglev trains?

Medical scanning

Magnetic resonance imaging (MRI scan) uses pulses of radio waves to look inside your body. They build up a very detailed 2-D or 3-D image of tissues to investigate injuries and illnesses. Very strong magnets are needed for clear images – up to 40 000 times stronger than the Earth's magnetic field. The most common way to make magnets small enough and strong enough is to use superconductors.

The superconducting wire coil surrounds the tube that patients pass through. The coil is bathed in liquid helium at incredibly low temperatures. The helium is insulated in a container similar to a vacuum flask. This means the patient does not feel cold and the helium does not absorb energy from the surroundings. By using superconductors, the amount of electricity needed to run the MRI scanner is much lower than using other methods.

f Why are superconductors used in MRI scanners?

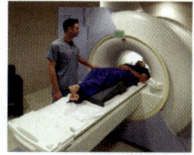

Figure 3 Superconductors surround patients as they pass through the MRI scanner

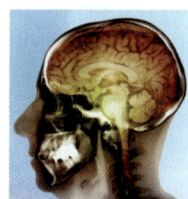

Figure 4 MRI scans are very detailed

Did you know …?

Superconductivity was discovered by a Dutch physicist, Heike Onnes, in 1911. He investigated how different metals behaved at very low temperatures and used a mercury wire in liquid helium.

Summary questions

1 Copy and complete using these words:
electricity   energy   resistance
Superconductors have no _____ at very low temperatures. This means that there are no _____ losses when the current flows. The amount of _____ needed to make a very strong electromagnet using superconductors is quite low.

2 Write down **two** advantages and **two** disadvantages of using superconductors in circuits.

3 Find out one other use of superconductors and write a paragraph explaining how the superconductor is used.

4 Some power stations use superconductors in their generators. These generators are about half the size of normal generators and over 99 per cent efficient. Why do you think some governments are funding research into superconducting generators?

Key points
- Superconductors have no electrical resistance at very low temperatures.
- They can make very strong electromagnets.
- They are used in MRI scanners and Maglev trains.

Summary answers

1 resistance, energy, electricity

2 Advantage: e.g. electromagnet stays magnetised when the circuit is off; there is no energy wasted in a superconductor.
Disadvantage: e.g. only some materials are superconductive; they are only superconductors at low temperatures.

3 Other uses include particle accelerators for nuclear research; some mine detection equipment (SQIDs); e-bombs which disable the enemies electronic equipment using an intense blast of electromagnetic radiation.

4 E.g. more efficient energy generation means less energy resources such as fossil fuels are used; it means fewer greenhouse gases produced; it means a country doesn't need to rely as much on imports of fuels; it may mean cheaper electricity as there is less wastage.

Answers to in-text questions

a zero ohms
b Not all materials are superconductors and superconductors only work at low temperatures.
c It stays magnetised after the current is turned off; there are no energy losses.
d No energy is wasted in superconductors (in resistors, energy is wasted heating the metal and its surroundings).
e They levitate above the tracks so less energy is lost as friction; they use superconducting electromagnets so the power supply can be turned off, saving energy; they are very fast.
f Superconductors are used as they produce very strong, small electromagnets.

Unit 3, Theme 2 – Making and improving products

12.6 Advantages and disadvantages of modern products

Learning objectives

Students should learn:
- the advantages and disadvantages of smart materials
- the advantages and disadvantages of superconductors
- the advantages and disadvantages of chromic materials.

Learning outcomes

Most students should be able to:
- understand some advantages of modern products and materials
- understand some disadvantages of modern products and materials.

Some students should also be able to:
- explain in detail some advantages of modern products and materials compared with traditional ones.

AQA Specification link-up: Science B 3.5.2.2

Within this context, candidates should be able to use scientific data and evidence to discuss, evaluate or suggest implications of the following:
- the advantages and disadvantages of modern products compared with traditional products.

Lesson structure

Starters

Reviewing modern products – Students write down three facts about superconductors, smart paints, smart materials and chromic materials. Pool ideas to check understanding and knowledge. *(5 minutes)*

Modern products – Get students to pair up equivalent modern and older products and give reasons why the modern products are now used more commonly. Support students by providing them with a list of products made from materials developed at different times to group as modern products or older products. Extend students by asking them to come up with their own list. *(10 minutes)*

Main

- This is very much a rapidly developing topic. Students should use the internet and materials from manufacturers to research the advantages and disadvantages of these materials.
- Students should prepare a PowerPoint presentation (maximum six slides) looking in detail at one type of modern material. They should identify the materials used previously and suggest advantages and disadvantages.
- Students should look at each other's presentations, so they understand the advantages and disadvantages of several materials. A tick sheet can be prepared to summarise the main points for the range of modern materials.
- Include a display of products made from different modern materials – as well as their older alternatives if appropriate – for students to handle and discuss.

Plenaries

One more use – Suggest several modern materials (e.g. self-healing paint, thermochromic paints) and ask students to suggest one or two uses for these materials not yet covered in the lesson. Strange and impractical uses are fine – thinking outside the box is how some great scientific discoveries began (e.g. self healing paint and reusable writing boards in Africa; thermochromic paint painted on a person's forehead to monitor a patient's fever)! *(5 minutes)*

Summarising modern materials – Extend students by getting them to draw a mind map identifying the advantages and disadvantages of each of the different types of modern material. Support students by providing them with a summary set of points to put into categories (e.g. disadvantages, advantages). Students can add extra points to the list. *(10 minutes)*

Support
- Try to include as many real examples of products that use modern materials as you can. Where possible, include actual objects made from modern and older materials so students can handle these and spot the differences. There is a real generation gap – many students will not recognise products as 'new' because they will have been familiar with these for several years.

Extend
- Students could research modern materials in different contexts such as food packaging, medical uses and electronic equipment.
- Students should realise that some products are designed for a certain job, some are 'spin-offs' from space research or other research programmes, while others are just from entrepreneurs recognising alternative uses.

Chapter 12 – Improving materials

Further teaching suggestions

Researching different uses

The How Stuff Works website (www.howstuffworks.com) has many links and includes information clearly presented for all ability students e.g. on smart windows, automotive finishing (self-healing paint), superconductors, thermochromic materials, etc. There are videos and links to additional information that may be suitable for more able students to carry out independent research. You could ask students to suggest different uses for these materials explaining their choices. Encourage them to think of as wide a selection as possible.

Science in context

Many students will have braces or spectacle frames that use modern materials. Modern packaging uses new materials.

Unit 3, Theme 2 – Making and improving products

12.6 Advantages and disadvantages of modern products

Learning objectives
- What are the advantages and disadvantages of smart materials?
- What are the advantages and disadvantages of superconductors?
- What are the advantages and disadvantages of chromic materials?

Smart materials

Many young people wear dental braces to straighten their teeth. Nowadays, some people use 'invisible' braces which are more comfortable and less noticeable than metal braces. However, these braces are expensive and need to be replaced every two weeks. They can only be used for teeth that are not too far out of place. The invisible braces are made from a modern plastic material that is transparent and strong.

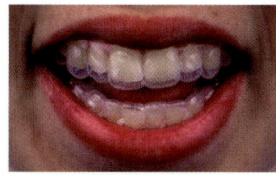

Figure 1 Modern braces can be removable and invisible

a Write down **two** advantages and **two** disadvantages of 'invisible' braces.

It can cost hundreds or thousands of pounds to repair damage caused by scratching a car's paintwork. **Self-healing paint** is now used on some cars. This saves money on upkeep costs. The Sun's warmth or warm water allow the paint to reform after it has been scratched. The paint can be used on many hard surfaces and has been licensed for use on cars and mobile phones. It is more expensive than ordinary paint so it isn't used on low-cost items like toys. Its appearance is not suitable for some uses, such as painting rooms, and it is not available to use at home yet.

b Write down **two** advantages and **two** disadvantages of self-healing paint.

Superconductors

The National Grid transmits electricity across the UK. If the wires were made from superconductors instead of aluminium, the savings would be enormous. Electricity could be transmitted at low, safer voltages and less energy would be wasted heating the conductor and its surroundings. However, no superconductor works at room temperature.

Figure 2 Scratches on this car heal themselves

They must be cooled to very low temperatures. This makes superconductors very expensive to use and completely impractical for the National Grid. Superconductors are used to make very strong and compact electromagnets for MRI scanners and Maglev trains. They work even when their electricity supply is turned off, saving some energy.

Did you know ... ?
Maglev trains get their name from the term 'magnetic levitation'.

c Write down **two** advantages and **two** disadvantages of superconductors.

Figure 3 Superconductors only work at very cold temperatures so aluminium is used as the conductor in power lines in the National Grid instead

Chromic materials

Thermochromic and photochromic paints change colour according to the temperature and light intensity. They are available for use at home on different surfaces, but are expensive compared with traditional paints. There is a limited range of colours and they may need to be covered with a clear layer of varnish to protect the coatings. Many of these paints are toxic and difficult to use. Also, they can fade if exposed to high temperatures, some solvents or UV radiation. They are not water based so their use and disposal has a greater impact on the environment than some traditional paints. However, using these paints can give more interesting results than using traditional paints.

d Write down **two** advantages and **two** disadvantages of thermochromic and photochromic paints.

Examiner's tip
Questions on this topic will often be comprehension or data-analysis type questions, so make sure you read the question properly before you answer anything.

Summary questions

1. Copy and complete using these words:
   environment   expensive   money   variety

   Many modern materials are more .... than traditional materials. Many modern materials save .... by reducing the costs of repair or upkeep. Modern materials give people more .... in the way they use different products. Using modern materials may affect the .... more than traditional products.

2. Some hospitals are using special casts on patients after operations on their ankles and feet. The materials cost more but the patients don't need to stay in hospital as long. Write down **one** advantage and **one** disadvantage of this for the hospital and **one** advantage and **one** disadvantage for the patient.

3. Find out one other use of a modern material and write a paragraph explaining its advantages and disadvantages.

Key points
- New and exciting materials are being developed by research scientists for new technologies or to replace traditional materials.
- Many modern materials are often more expensive than traditional materials.
- Modern materials are often more specialised than traditional materials.
- Producing and disposing of modern materials may affect the environment more than traditional materials.

Summary answers

1. expensive, money, variety, environment

2. Hospital: reduced costs on patient care but more cost buying the cast in the first place

   Patient: less time is wasted making trips to the hospital but complications may not be spotted if there are fewer checkups.

3. For example, shrink wrapping allows food to be preserved for longer and saves costs but can be hard to remove and is not recyclable.

Answers to in-text questions

a These braces are more discrete and comfortable but they are expensive and need to be refitted every two weeks.

b The coating saves upkeep costs and is easy to use after application, but it is more expensive and can only be used by specialist companies.

c The electromagnet stays magnetised when the circuit is off; there is no energy wastage in a superconductor but superconductors can only be used at very low temperatures and can be expensive.

d They give a varied finish that people like. They can be used at home and on different surfaces but they are expensive and there is a limited range of colours.

Unit 3, Theme 2 – Making and improving products

Summary answers

1. **a** anode – positive; cathode – negative
   b Positive ions (cations) move toward the cathode. Negative ions (anions) move toward the anode.
2. **a** cathode
   b anode
   c They gain two electrons.
3. **a** Tin; resists corrosion from contents.
   b Nickel can cause allergies.
   c two from: electrocution, chemical burns, toxic electrolytes.
4. no, current, strong, e.g. Maglev trains or any other correct application.
5. **a** photochromic
   b It could warn you if you had spent too much time in the sun.
6. **a** Dirt will not stick to it.
   b It has a high melting point.
   c It is unreactive.

Unit 3, Theme 2 – Making and improving products

Summary questions

1. To perform electrolysis, two electrodes must be connected to a power source. The electrodes are placed in a liquid called an electrolyte.
   a What are the two electrodes called? Which is positive and which is negative?
   b Explain what happens in the electrolyte when the current is switched on.
2. During electroplating, metal ions are turned into solid metal.
   a Which electrode is the solid metal formed on?
   b Which electrode provides a source of metal ions?
   c What happens to Cu^{2+} ions to turn them into copper metal?
3. Many metals are electroplated to make them more useful.
   a What metal are cans of food electroplated with and why?
   b Why is nickel jewellery often plated with gold or silver?
   c Electroplaters must wear a lot of protective clothing at work. Name two hazards they protect themselves from at work.
4. Complete these sentences using your own words:
   Superconductors have .... resistance at very low temperatures. They are used to make electromagnets which stay magnetised when the .... is turned off. These electromagnets are used in MRI scanners because they are very .... for their size.
   One other use of superconductors is ....
5. A bracelet is made of a type of plastic that changes colour under UV light.
   a What is the name for this type of material?
   b How would the bracelet be useful?
6. Materials scientists are constantly trying to improve products.
   Teflon is a smart material. It is the trade name for a polymer called polytetrafluoroethene or PTFE. It has many applications as it is very slippery and unreactive.

A section of a Teflon® molecule

| Dirt will not stick to it   It has a high melting point   It is unreactive   It insulates   It prevents allergic reactions   It strengthens a structure |

   a Choose the reason why Teflon is used in clothing.
   b Choose the reason why Teflon is used as a coating for saucepans.
   c Choose the reason why Teflon is used to make containers for chemicals.

Practical suggestions

Practicals	AQA	k	📖
Electrolysis of copper sulfate solution using copper electrodes.	✓		
Investigate the factors that affect electrolysis of copper sulfate / electroplating of copper.	✓	✓	✓
Investigate smart materials – e.g. memory wire, pressure-sensitive resistance film.		✓	✓
Calibrate chromic strip to be used as a forehead thermometer.	✓		

Kerboodle resources

- Theme map: Improving products
- How Science Works: Wouldn't give a nickel for the whole process (12.2)
- WebQuest: Sportswear (12.4)
- Practical: Electroplating (12.2)
- Examination-style questions
- Answers to examination-style questions

Chapter 12 – Improving materials

Examination-style questions

1 Some metals need to be coated to make them more durable.

a Choose the correct word from each box to complete the sentences.

i Charged particles called .... are made by taking away or adding .... to atoms or molecules.

| atoms | electrons | ions |
| charges | protons | |

(2)

ii .... involves the movement of charged particles in an ....

| charging | combustion | electrolysis |
| air | an electrolyte | water |

(2)

iii One use for electroplating is to ....

| insulate |
| prevent allergic reactions |
| strengthen a structure |

(1)

b During electroplating, the article to be coated is made the cathode. Explain what happens at the cathode when the article is being coated. [H] (3)

c A company wanted to speed up the process of coating metals. The graph shows the results of one of their investigations.

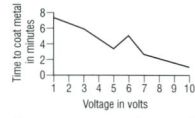

Voltage in volts

i Which voltage gave an anomalous result? (1)
ii Suggest what time the anomalous result should have been. (1)
iii Suggest what could have happened to cause the anomalous result. (1)
iv Write a conclusion based on the results in the graph. (1)

2 Opticians have been experimenting with different materials to make spectacles better.

a Smart materials have been used to make the frames. Describe the use of smart materials to make spectacle frames. (2)

b Chromic materials have been used to make the spectacle lenses. What is the advantage of making the lenses out of photochromic material? (2)

3 A medical company wanted to find the best superconductor to use in an MRI scanner.

a What is a superconductor? (2)

A multimeter is a device that can be used to measure the resistance of a device in an electrical circuit.

The diagram shows some of the apparatus used to find the best superconductor.

b What other variable would need to be measured? (1)

c *In this question you will be assessed on using good English, organising information clearly and using specialist terms where appropriate.*

Outline the method that the medical company could use to find the best superconductor in an MRI scan. (6)

4 Read the article about biosensors before answering the questions.

> **Smart wound dressings**
>
> Biosensors can measure the healing process of wounds. Scientists have made a wound dressing that can detect if anything goes wrong during the healing process.
>
> The sensors can detect the pH value indicating the acidity of the wound, and a protein present when the wound becomes inflamed. These things can be detected quickly without needing to take the wound dressing off and allowing infections in.

Evaluate the cost implications for the NHS of using these new biosensor wound dressings compared with traditional dressings (3)

Examination-style answers

1 a i ions
electrons (in that order) (2 marks)
ii electrolysis
electrolyte (in that order) (2 marks)
iii prevent allergic reactions (1 mark)

b Cathode is put in solution containing the coating metal.
Metal in solution is attracted to cathode.
$M^{n+} + ne^- \longrightarrow M$ (3 marks)

c i 6 (1 mark)
ii any number between 3.2 and 2.7 (1 mark)
iii Could have had the voltage too low OR could have timed it incorrectly. (1 mark)
iv The higher the voltage the quicker to coat. (1 mark)

2 a more flexible
will not break as easily (2 marks)

b Photochromic materials change colour in different light conditions. So can use lenses that darken in bright lights. (2 marks)

3 a material whose resistance becomes almost zero at low temperatures (2 marks)

b temperature (1 mark)

c Marks awarded for this answer will be determined by the Quality of Written Communication (QWC) as well as the standard of the scientific response.

There is a full and detailed description of the method the medical company could use. The answer is well structured with minimal repetition or irrelevant points. There is an accurate, fluent and clear expression of ideas with only minor errors in the use of technical terms, spelling, punctuation and grammar. (5–6 marks)

There is some description of the method the medical company could use with some omissions. The answer shows some attempt at structuring and the ideas are expressed with reasonable fluency and clarity. There are some errors in the use of technical terms, spelling, punctuation and grammar. (3–4 marks)

There is a brief description of the method the medical company could use. The answer is largely incomplete and may contain some valid points which are not clearly structured. It lacks fluency and/or clarity. It contains errors in the use of technical terms, spelling, punctuation and grammar. (1–2 marks)

No relevant content (0 marks)

Examples of points made in the response:
- Take resistance
- At lower and lower temperatures
- Repeat for accuracy
- Take at same temperatures for all superconductors for a fair test
- Keep same multimeter and thermometer for a fair test.
- The best would be the lowest resistance at the highest temperature.

4 Up to **two** advantages from:
- Will cost more than traditional wound dressings.
- Will save money on not having to treat infected wounds if dressing taken off.
- Only need medical attention if alarm triggered.
- Catching and treating problems early is cheaper than having to treat problems that have been left for some time.

Up to **two** disadvantages from:
- Will cost more than traditional wound dressings.
- Medical staff would have to be trained to use them.
- Problem with spreading infection from patient to patient if not cleaned properly between patients.
- Cost of recycling high to environment/economically.

(3 marks but at least 1 mark must be for an advantage and 1 mark for a disadvantage)

Unit 3, Theme 2 – Making and improving products

13.1 Selective breeding

Learning objectives

Students should learn:
- that selective breeding produces plants or animals with desired characteristics
- how an organism can be selectively bred
- that selective breeding reduces variation in a species.

Learning outcomes

Most students should be able to:
- name some characteristics that plants and animals are selectively bred for
- explain how selective breeding produces the required characteristics
- list some advantages and disadvantages of selective breeding.

Some students should also be able to:
- explain why selective breeding can cause genetic disorders in a species.

Specification link-up: Science B 3.5.2.3

- Explain how selective breeding of plants and animals involves selecting the parents with desired characteristics, crossing them, selecting from their offspring, and then repeating the process over several generations.

Within this context, candidates should be able to use scientific data and evidence to discuss, evaluate or suggest implications of the following:
- examples of risks associated with selective breeding …

Lesson structure

Starters

Strawberry plants – Ask students to sketch an ideal strawberry plant that would be selected by a farmer to breed. Students need to annotate the sketch to identify all the desired features. *(5 minutes)*

Desired characteristics – Ask students to work in pairs to make a list of the desired characteristics a farmer might select for in pigs, sheep and cows. To support students, provide them with a list of products that can be obtained from the three organisms. Students then match the products to the organism. To extend students, students should choose at least three products made from each organism, and state the part of the organism that is used to produce this product. *(10 minutes)*

Main

- Show students some images of different plants and animals. Discuss the characteristics of an idealised version of the organism. For example, a strawberry would ideally be bright red, large, very juicy and very sweet. A sheep would have thick, long wool.
- Introduce the idea of selective breeding. Farmers choose the 'best' animals or plants and breed them together. They then select the best offspring to breed again. This technique is repeated over many generations, until the farmer achieves the desired animal or plant.
- Students use this information to produce a cartoon strip that could be given to farmers, to explain how they could breed their animals or plants to produce the 'best' stock. The cartoon needs to show the selection for three generations. Students need to choose the species of animal or plant they wish to breed selectively and then need to decide on the desired characteristics.
- In small groups, students should use the information in the Student Book as a starting point for producing a table comparing the advantages and disadvantages of selective breeding. Through discussions within their group they should try to add other advantages and disadvantages.
- They then need to make a group decision, based on these findings, as to whether or not they would choose to breed plants or animals selectively. (Groups' opinions may differ on plants and animals.)
- The group's decisions can then be discussed as a class. Students may be aware of the health problems experienced by some breeds of dog or cat, which have been selectively bred over long periods of time. This piece of work can be extended using the practical activity in the Student Book.

Plenaries

Selective breeding definition – Ask students to write an explanation of what is meant by the term 'selective breeding'. *(5 minutes)*

Advantages and disadvantages of selective breeding – Ask students to produce a table comparing the advantages and disadvantages of selective breeding. To support students, provide them with some ideas they can sort into the advantages and disadvantages of selective breeding. To extend students, they should try to explain why selective breeding can produce genetic disorders. *(10 minutes)*

Support

- Students could be given a procedure for selectively breeding an animal and/or plant, which they could illustrate by drawing the characteristics the farmer is exploiting.

Extend

- Students could look into the problems caused by selective breeding, through the reduction in the gene pool of a species. The results could be fed back to the group, giving some everyday examples of these effects.

Science in context

Farmers selectively breed plants and animals to produce high yields. Florists and gardeners also selectively breed flowers and plants to produce high yields and to grow flowers for as long a season as possible. Breeders of pedigree or show animals selectively breed characteristics that appeal to prospective owners.

Chapter 13 – Selective breeding and genetic modification

Further teaching suggestions

Desired pet characteristics
Students select a pet or animal that they, or a member of their family, owns. What were the reasons for selecting this animal? What were the particular features of this species that are attractive? What traits would the students selectively breed for? Would breeding for this result in any health problems for future generations. Are there any potential health risks for this animal as it gets older?

Selective breeding programmes
Selective breeding programmes are used in zoos to try to protect animals from extinction. Students could discuss their feelings about zoos, and whether they think captive breeding programmes are ethical (e.g. some animals are transported around the country, or even to different countries, to reproduce).

Activity

Pedigree dogs
This task could be introduced by showing a clip from Crufts. Students should research pedigree dog breeding. Each student chooses a breed of dog and finds out which characteristics the dog is bred for, and any advantages and any health problems the breed may suffer from. This information can then be fed back to the class. Suitable breeds include greyhounds, German shepherds, dachshunds, labradors, springer spaniels, pugs and poodles.

Unit 3, Theme 1 – Making and improving products

13.1 Selective breeding

Learning objectives
- What is selective breeding?
- How are sheep selectively bred?
- How has selective breeding changed wheat into a more useful product?

links
For information about natural selection look back at 3.5 Evolution.

What is selective breeding?
To ensure that they maintain or improve their stock, farmers choose to breed from their best animals or plants. This is called **selective breeding**. It is something that has been carried out for thousands of years, ever since people first began to farm.

Farmers will select the animals they breed from, or the plants they grow, by characteristics that are of benefit. For example, they may choose:

- dairy cattle that produce lots of milk
- tomato plants that produce a high yield of tomatoes
- wheat that is resistant to a particular pest.

Selective breeding is one of the tools agricultural scientists have been using to try to produce enough food for everyone. This is becoming increasingly important as the world's population increases and people are living longer.

a What is meant by the term 'selective breeding'?

Selectively breeding sheep
Sheep farmers will select characteristics based on the produce they sell. Farmers who breed sheep for wool choose sheep that produce large, good quality fleeces. These can then be manufactured into wool products such as yarn or carpets. Farmers that raise sheep for meat, choose breeds and individuals that produce lean meat and a large number of offspring. Some farmers also raise sheep to produce cheese; they select individuals that have a high milk yield.

b How does the farmer choose which of the offspring to breed from?

c Why does it take several years of selective breeding to produce a flock of sheep with the desired characteristics?

Activity
Pedigree dogs
Pedigree dogs are selectively bred. They have the desired characteristics of their breed. However, many suffer from health problems as a result.

For example, the desire to have a sloping back with hind legs low to the ground causes hip problems in German shepherds, and the preference for a snub nose causes breathing problems in pugs. This is made worse when animals are 'in-bred'. This means closely related dogs, such as siblings, are mated. As a result of in-breeding, pedigree dogs have a much lower life expectancy than cross-breeds.

- Choose a species of pedigree dog.
- Research the desired characteristics of the breed, and any related health problems from which they suffer.
- Present your information as a leaflet or a poster that could be given to people researching which breed to choose as a new family pet or working dog.

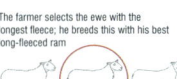
The farmer selects the ewe with the longest fleece; he breeds this with his best long-fleeced ram

The farmer chooses the best ewe and breeds again with his best ram

This process continues over many generations

Eventually all the sheep have the desired characteristic of a good-quality, long fleece

Figure 1 Selective breeding of sheep for long wool

Chapter 13 – Selective breeding and genetic engineering

Selectively breeding wheat
Wheat is a very important crop in the agricultural industry. It produces grain that is turned into flour, which has a wide number of uses, such as making bread. Selective breeding by farmers has changed the characteristics of wheat, as the table and Figure 2 show:

Wild wheat plants	Modern wheat plants
Ears are small and have few seeds	Ears are large and have many seeds
Stalks are brittle and ears often fall off	Stalks are stronger, so ears stay on
Ears ripen at different times	Ears ripen at the same time
Stalks grow to different heights	Stalks grow to the same height

d How do the features of modern wheat make it easier to harvest?

Many modern crops have also been selectively bred to have a high resistance to disease.

Problems of selective breeding
Selective breeding reduces the number of genes (the **gene pool**) from which a particular strain or variety of a species is created. It reduces variation. This means that, if a new disease arises, it might be that none of the organisms in that gene pool have the gene for resistance to this disease. This could result in a particular strain becoming extinct.

Wild wheat   Modern wheat

Figure 2 Effect of selective breeding of wheat

AQA Examiner's tip
Explaining selective breeding is often a three-mark question. The marks are for:
Select (the desired characteristics)
Breed (the best two organisms)
Repeat (over several generations)

Summary questions
1 Copy and complete the table with the characteristics a farmer may choose to breed selectively for:

Organism	Characteristics
Sheep	
Chicken	
Cow	
Wheat	
Apple tree	

2 What are the disadvantages of selective breeding?

3 A farmer wants to grow large, tasty strawberries. One species of strawberry is large and tasteless. Another species produces very small but very sweet and juicy strawberries. Draw a diagram to show how a farmer can selectively breed for the large and tasty characteristics.

Key points
- Selective breeding involves choosing the best organisms to breed to produce offspring with desired characteristics.
- Farmers selectively breed sheep for wool by breeding ewes and rams with the longest fleeces.
- Selectively bred wheat has larger ears with more seeds. The stalks are strong and grow to a similar height and the grain ripens at the same time. This makes the crop easier to harvest and produces higher yields.

Summary answers

1

Organism	Characteristics (examples)
Sheep	long wool; produces large number of offspring
Chicken	produces large number of eggs, of a good size
Cow	high milk production; large proportion of muscle for high meat production
Wheat	large ears; ears containing many seeds; strong stalks
Apple tree	high apple yield; attractive colour; sweet fruit

2 Selective breeding reduces variation in a species, which could result in useful characteristics being lost (such as resistance to disease). Some desired characteristics selected for can actually have a detrimental effect on their health and even reduce their life expectancy.

3 In the diagram students should show the largest strawberry from the large and tasteless plant and breed that with the sweetest strawberry from the small and juicy plant. From these offspring the largest and sweetest strawberries should be bred.

Answers to in-text questions

a Choosing to breed individuals with desired characteristics.

b The farmer chooses the animal with the desired characteristics – in this case the sheep with the longest and best-quality fleece.

c Sheep have to be chosen and bred over a number of generations to achieve the optimum desired characteristic. It takes time for the chosen offspring to mature to an age where they can reproduce, and produce their own offspring.

d The ears remain on the plant; the ears ripen at the same time; all plants grow to a similar height.

Unit 3, Theme 2 – Making and improving products

13.2 Genetically modified food

Learning objectives

Students should learn:
- that genetic engineering involves the transfer of 'foreign' genes into plants and animal embryos
- that genetic engineering produces plants and animals with desired characteristics
- that there are ethical issues associated with genetic engineering.

Learning outcomes

Most students should be able to:
- name some examples of genetically engineered plants
- explain simply how plants and animals are genetically engineered
- list some advantages and disadvantages of genetic engineering.

Some students should also be able to:
- evaluate the ethical arguments for and against genetic engineering.

Answers to in-text questions

a directly altering an organism's genes to produce desired characteristics
b It is much quicker, and more predictable.
c They are called foreign genes because they have come from another organism.
d It is insect resistant so insecticides do not need to be used during its cultivation. This saves money and means that less insecticide gets into the environment.

Support
- Students could be provided with a series of diagrams showing how corn was genetically engineered to produce Bt corn. Students should annotate the diagrams where applicable and rearrange them into the correct order.

Extend
- Students could write an evaluation of the potential costs to society, compared with the potential benefits, of allowing genetic engineering research to continue.

AQA Specification link-up: Science B 3.5.2.3

- Explain how genetic engineering involves the transfer of foreign genes into the cells of animals or plants at an early stage in their development so that they develop with desired characteristics.

Within this context, candidates should be able to use scientific data and evidence to discuss, evaluate or suggest implications of the following:
- examples of risk associated with genetic engineering.

Lesson structure

Starters

Genetic terminology – Ask students to place the following terms in order of size, from smallest to largest: gene, nucleus, chromosome, cell, organism. *(5 minutes)*

Student ideas – What do we mean by the term 'genetically modified'? Ask students in pairs, then as a class to discuss what they 'know', or have heard, about genetically modified products. To support students, provide them with some images of packaging, or brief headlines about GM foods to stimulate their discussion. To extend students, ask them to write a definition of what they think genetically modified means. *(10 minutes)*

Main

- Introduce the concept of genetic engineering – scientists insert a gene that codes for a desired characteristic into a cell's DNA. The organism will now display the desired characteristic and the cell will also pass the new instructions onto its offspring.
- Show students a variety of images of plants and animals that have been genetically engineered. The examples in the Student Book could be used as a starting point. Students should discuss how the characteristic is beneficial for the survival of the organism, or for commercial use.
- Using the flow diagram in the Student Book, discuss with students exactly how a plant can be genetically modified to display a desired characteristic. It would be useful to provide a cut-and-paste activity, asking students to sequence a similar genetic modification process, to reinforce this work once covered.
- Bt corn, which contains a toxin that kills insects, can also be used to illustrate how genetic engineering is carried out. In small groups, students could make plasticine models to show how this process was carried out, and then role-play this activity. They can use the flow diagram in the Student Book to support this activity. Alternatively, they could produce a flow diagram specific to this example, or a series of their own pictures to convey this information.
- Discuss the importance of using new scientific techniques to increase food production to feed the increasing population. Students could then complete the activity – answers are provided in the 'Activity' box opposite.
- Ask students if they can see any potential drawbacks of genetic modification – displaying a media headline can prompt discussion. Students should evaluate the benefits and potential drawbacks of using GM foods. Are there any risks associated with pollen from GM crops spreading into the environment? Is there a distinction to be made between modifying crops, and animals?

Plenaries

Genetic engineering definition – Provide students with the words of this definition (genetic engineering involves the transfer of foreign genes into an organism, to produce desired characteristics) in a muddled order. They need to rearrange the words into a sentence. *(5 minutes)*

Advantages and disadvantages of GM products – Ask students to produce a table comparing the advantages and disadvantages of GM products. To support students, provide them with some statements to sort into advantages and disadvantages. To extend students, ask them to support arguments for or against with reasoned scientific arguments. *(10 minutes)*

Chapter 13 – Selective breeding and genetic modification

Further teaching suggestions

Drought-resistant wheat debate
Students could write a campaign speech for a scientist who wishes to carry out research into a drought-resistant wheat plant. The speech is to be used at a public meeting where the issues are being discussed. During testing, the plant will be grown in fields surrounding a small town. The speech needs to list as many advantages of the research to society and attempt to allay public fears over the crop being grown. Students should then make a list of arguments the scientist might face from people attending the meeting, who are opposed to the testing.

Genetic engineering research
Search online news stories about genetic engineering such as 'glow in the dark' mice. For example, try www.bbc.co.uk. Discuss the advantages and disadvantages of each case.

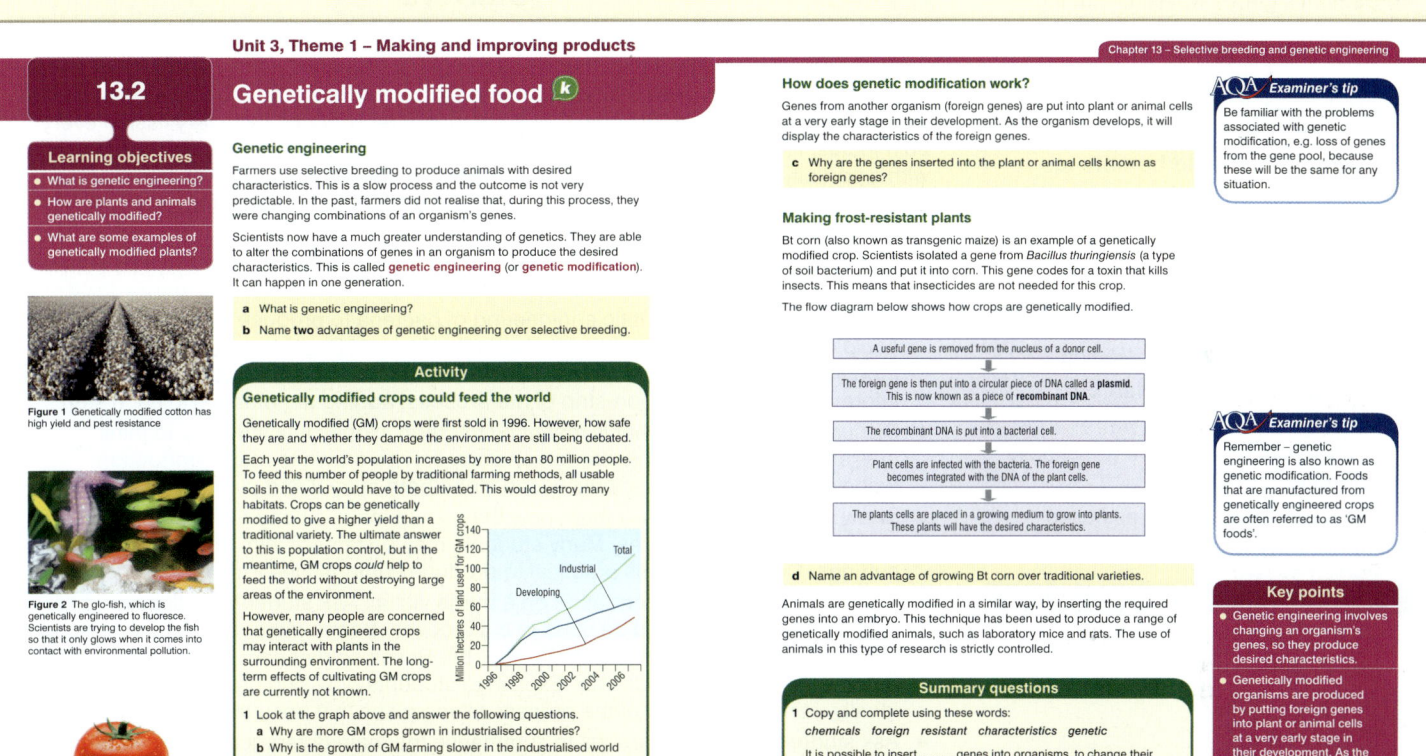

Activity

Genetically modified crops could feed the world

1. a Crops are genetically engineered in industrialised countries and are grown there in the experimental stages. Farmers in these countries therefore have access to this technology.
   b Many people have concerns over the safety of GM crops and the damaging effects it could have on wild plants. They have objected to the crops being grown and are choosing not to buy products containing GM ingredients. Many developing countries have large populations and want to grow GM crops to help feed them.
2. If all countries could grow bananas regardless of the climate, there would be no need for them to be exported from countries with hotter climates. This would mean that these countries would lose their export market, and therefore they would earn less money, damaging their economy.
3. If wild plants cross-pollinated with plants that did not produce seeds, many of their offspring would not produce seeds. This would mean that over a number of generations the wild species could die out – if no seeds are produced, no new plants can grow.

Summary answers

1. foreign, characteristics, genetic, resistant, chemicals
2. insect-resistant corn and high-yield/pest-resistant cotton
   glo-fish, and laboratory rats and mice
3. Otherwise the normal characteristics of the plant will already have developed.

Unit 3, Theme 2 – Making and improving products

13.3 Genetic engineering and cloning

Learning objectives

Students should learn:
- that genes can be inserted into bacteria to make them produce useful chemicals, including insulin and vaccines
- that a clone is a genetically identical copy of an organism
- that genetic engineering alters an organism's genes whereas cloning produces an exact copy of an organism's genes.

Learning outcomes

Most students should be able to:
- name some chemicals that can be produced by genetically engineered bacteria
- describe what a clone is
- describe the difference between genetic engineering and cloning.

Some students should also be able to:
- describe how bacteria are genetically engineered
- explain how tissue culture is used commercially to produce clones.

AQA Specification link-up: Science B 3.5.2.3

- Explain that cloning techniques involve laboratory processes to produce offspring that are genetically identical to the donor parent – tissue culture of plants and animal or human organs (e.g. skin and cartilage).
- Describe how human insulin is produced using genetically modified bacteria.

Lesson structure

Starters

What is a clone? – Ask students to draw a simple organism. Then draw what the organism would look like if they made a clone of the organism. *(5 minutes)*

Bacterial replication – Ask students to revise ideas on the number of bacteria that exist after several generations. Assume that a bacterium reproduces by binary fission every 20 minutes. How many bacteria exist after four hours? To support students, supply them with a partially completed results table. Students complete the table with the missing numbers. To extend students, ask them to display their data graphically – Number of bacteria (*y*-axis), Time / hours (*x*-axis). *(10 minutes)*

Main

- Introduce the idea of genetic engineering of bacteria to produce medicine. This could be done through the information in the Student Book on using genetically modified bacteria to produce insulin to treat diabetes. Students could summarise this information using a cartoon-strip of the process, adding annotations to each step.
- Discuss with the class what is meant by the term 'clone'. Ask students to think of examples of cloned objects – an analogy can be drawn to mass production techniques. Explain that scientists can clone both animal and plant tissues.
- Show students a plant. Ask them how they could make a copy of the plant without planting a seed it produces. Many students may be familiar with the idea of taking cuttings. Complete a class practical to produce cuttings (see Practical support).
- Provide students with information on a burns victim. Ask students to discuss what surgical techniques could be used to aid recovery from a major accident. A famous example to use would be Simon Weston, the British soldier who suffered extensive burns during the Falklands War. Explain to students how skin grafts are produced, ideally using a video clip of the process taking place in a hospital laboratory.
- Ask students to work in groups to summarise the main differences between genetic engineering and cloning. Their results should be a table similar to that in the Student Book.

Plenaries

Plant cuttings – Provide students with the steps required to take a plant cutting. They need to organise the steps into the correct order. *(5 minutes)*

Producing insulin – Ask students to produce a short summary of the steps involved in using bacteria to produce insulin. This could be displayed to the class in a range of forms – a poster, a cartoon strip, a flow diagram or as a role-play. To support students, provide them with access to the Student Book, to refer to as appropriate. To extend students, ask them to describe how the process could be adapted to produce a different chemical product. *(10 minutes)*

Support

- Provide students with a cut-and-paste activity for genetically engineering bacteria to produce insulin. Give students a series of diagrams that they need to rearrange into the correct order.

Extend

- Ask students to carry out a piece of research work into the production of medical products using genetically engineered organisms.

Further teaching suggestions

Leaflet

Students could produce a leaflet to try and convince people that genetic engineering is a valuable scientific tool. They should explain how genetic engineering is used to treat patients suffering from a range of diseases. Substances produced in this way include growth hormones, vaccines, factor VIII and antibiotics.

Chapter 13 – Selective breeding and genetic modification

Practical support

Propagation

Equipment and materials required: Herbs such as basil make ideal examples for students to use, sharp blade, agar plates, rooting hormones, pots, compost.

Details

The best place to take tissue samples is in between two leaf nodes, using a sharp blade. These can then be placed on agar plates containing rooting hormones. Once big enough, the plants can be transferred to pots containing compost.

Alternatively, students can take cuttings in a similar manner to a gardener. Remove a healthy shoot tip that is about 10 cm long. Trim off the lower leaves and cut cleanly beneath a leaf joint. Dip the cut end into rooting hormone, and plant directly into compost.

Safety: Wash hands after handling the rooting hormone. Take care with sharp blades.

Science in context

Scientists have to consider the ethical issues surrounding genetic engineering before creating a new variety of plant or animal or before cloning an organism. This is known as bioethics. There are many advantages of genetic engineering, both commercially and medically, but many people believe that interfering with an organism's genes is a 'step too far'.

Unit 3, Theme 1 – Making and improving products

13.3 Genetic engineering and cloning

Learning objectives
- How can bacteria be genetically engineered?
- What is a clone?
- What are the differences between genetic engineering and cloning?

Genetic engineering of bacteria

Bacteria can be genetically engineered to produce many useful chemicals. These include hormones, vaccines and antibiotics.

Genes that code for the production of the required chemical are inserted into the bacteria. The bacteria reproduce, producing millions of identical copies of the gene, and therefore the substance it codes for. The useful substances are then separated from the bacteria.

a Name **three** medical substances that can be produced by genetically engineered bacteria.

Insulin production

Bacteria have been genetically engineered to produce **insulin**. This hormone is used to control the blood glucose levels of many people suffering from **diabetes**. As bacteria reproduce very quickly, large amounts of insulin can be produced in a short period of time.

Bacteria are made to produce insulin by the following process:

1. Small circles of DNA called **plasmids** are found in bacteria. These are modified to include the piece of human DNA containing instructions for making insulin.
2. The plasmid is then inserted into the bacteria *Escherichia coli*. The bacteria now produce insulin.
3. The bacteria multiply, and produce large quantities of insulin in a fermenter.
4. Bacteria are then killed by heat (sterilisation), leaving behind the insulin.

b Why do bacteria reproduce better in a fermenter?

Cloning

When microorganisms reproduce, they produce genetically identical copies of themselves. These are called **clones**. This is exploited when bacteria are used to make useful chemicals.

c What is a clone?

Plant cloning

Taking a plant cutting is an example of cloning. Gardeners have been using this technique for centuries to produce more plants. Biotechnologists have developed this technique further, to help grow more food. Plant tissue culture allows the rapid production of many genetically identical plants. This is also known as **micropropagation** (see Figure 3).

Micropropagation offers several advantages over traditional planting techniques. Large numbers of plants can be produced, very quickly and need very little space. The plants are in a controlled environment and so are disease free.

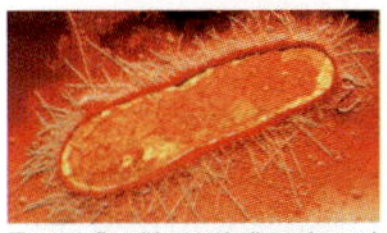

Figure 1 *E. coli* is genetically engineered to produce insulin

Figure 2 Bacteria are placed in a fermenter. This provides the perfect conditions for bacterial replication: warmth and nutrients.

Did you know …?
Many plants produce clones of themselves naturally. For example, strawberries send out runners, from which a new genetically identical plant will grow. This is known as asexual reproduction – meaning from one parent.

The main disadvantage of cloning plants is in the reduction of the gene pool. This can increase the risk of disease destroying a species.

Animal cloning

Human cells can be cloned in the laboratory and used for research into diseases and new drugs. For example, tissue culture is used to produce skin and cartilage.

This is carried out in the following way:

- Small pieces of a person's tissue are collected.
- These tissue fragments are transferred to an artificial, sterile environment.
- The cells continue to survive and function.
- When the tissue has grown to the desired size, it is replaced back into the person.

Tissue culture is used in skin grafting. New skin tissue is produced, which can be used to repair damage caused by burns or serious injury. As the new tissue is genetically identical to the donor material, the body will not recognise it as being 'foreign'. This minimises the risk of rejection.

Differences between genetic engineering and cloning

	Genetic engineering	Cloning
Genes present in organism	Produces a new set of genes – different from the original	Produces an exact copy of the genes present
Origin of genes	Uses 'foreign' genes, often transplanted from a different species	Uses genes copied from a member of the same species
Characteristics produced	Gives a new, desired characteristic to the organism	Produces a copy of the original characteristics

links
For more information on auxins (plant hormones) and their role in plant growth look back at 3.6 Plant growth.

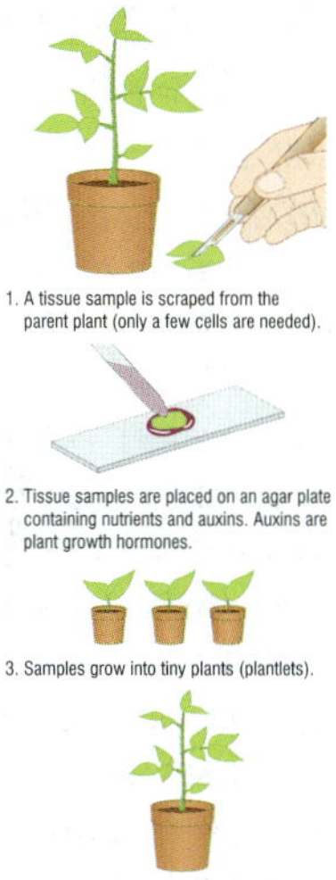

Summary questions

1 Match the following to make three sentences:

Cloned organisms	is a technique used to rapidly produce many identical plants
Micropropagation	produce many medicines including insulin
Genetically engineered bacteria	are genetically identical to their parent

2 Name **one** advantage and **one** disadvantage of cloning plants.

3 Draw a series of labelled diagrams to show how tissue culture can be used to produce many blueberry plants.

1. A tissue sample is scraped from the parent plant (only a few cells are needed).

2. Tissue samples are placed on an agar plate containing nutrients and auxins. Auxins are plant growth hormones.

3. Samples grow into tiny plants (plantlets).

4. Plantlets are planted in compost, and grown in a greenhouse where they develop into full-sized plants.

Figure 3 The process of micropropagation

Key points
- Genes can be inserted into bacteria to make them produce useful chemicals. These include insulin and vaccines.
- A clone is a genetically identical copy of an organism.
- Genetic engineering alters the genes of an organism; cloning produces exact copies of an organism.

Summary answers

1 Cloned organisms – are genetically identical to their parent
Micropropagation – is a technique used to produce rapidly many identical plants
Genetically engineered bacteria – produce many medicines including insulin

2 Advantages: plants produced very quickly, large numbers, grow in a small space, plants are disease-free
Disadvantages: reduced gene pool/genetic variation, increased susceptibility to disease

3 Tissue sample scraped from the blueberry plant ⟶ Tissue samples are placed on an agar plate containing nutrients and auxins ⟶ Samples grow into tiny plants ⟶ Small plants are planted in compost, and grown in a greenhouse where they develop into full sized blueberry plants.

Answers to in-text questions

a insulin, vaccines and antibiotics

b Has optimum conditions for growth – warmth and a source of nutrients.

c genetically identical copy of an organism

13.4 Ethics of genetic engineering and cloning

Unit 3, Theme 2 – Making and improving products

Learning objectives

Students should learn:
- that genetic research is highly controversial
- that a 'designer baby' develops from an embryo that has been created specifically for a desired characteristic
- that in gene replacement therapy a faulty gene can be replaced with a healthy gene.

Learning outcomes

Most students should be able to:
- state the benefits and potential drawbacks of genetic engineering and cloning
- describe what is meant by a 'designer baby'
- describe the process of gene replacement therapy.

Some students should also be able to:
- explain in detail the benefits and potential drawbacks of genetic engineering and cloning.

Support
- Provide students with a set of cards listing arguments for and against genetic research. Students should sort these cards into the appropriate category, to help form a reasoned opinion on the issues.

Students should be reminded that to succeed in an exam they must understand people's different viewpoints and be able to justify their beliefs using scientific reasons.

Extend
- Students could hold a debate – either in small groups, or as a whole class – into the ethics of cloning humans. What could the benefits and drawbacks to society be?

Specification link-up: Science B 3.5.2.3

- Understand that cloning techniques involve laboratory processes to produce offspring that are genetically identical to the donor parent – tissue culture of plants and animal or human organs (e.g. skin and cartilage).
- Explain how genetic engineering involves the transfer of 'foreign' genes into the cells of animals or plants at an early stage in their development so that they develop with desired characteristics.

Within this context, candidates should be able to use scientific data and evidence to discuss, evaluate or suggest implications of the following:
- the ethics of gene replacement therapy
- the economic, social and ethical issues concerning genetic engineering, genetically modified foods and 'designer babies'
- the ethics of genetic engineering compared to selective breeding.

Lesson structure

Starters

Genetics? – Students compile a list of areas of genetic research of which they are aware. This could include genetic engineering (often quoted in terms of GM foods), cloning, or mapping the human genome. *(5 minutes)*

Genetic engineering continuum – Attach a piece of string from one end of the classroom to the other labelled 'Strongly agree' at one end and 'Strongly disagree' at the other. Read out statements, some of which should be intentionally controversial, to which students should respond by placing themselves at positions along the continuum. *(10 minutes)*

Main

- Students, in groups, research the risks and benefits associated with genetic engineering. Students should be encouraged to explore their own beliefs associated with this subject area. Each group could be given one or two specific points to debate. Their discussions should then be summarised, and presented to the class.
- Possible benefits that could be discussed include:
  1. Reduction of genetic disorders in the gene pool
  2. Screening unborn babies for genetic disorders in early pregnancy or treating diseases
  3. Inserting embryonic cells into cancerous cells, to stop the cancer growing.
- Possible disadvantages that could be discussed include:
  1. Selecting the genetic make-up of future children
  2. Genetically engineered crops may spread into the wild, affecting native species
  3. Techniques, such as implanting an egg from one woman into the uterus of another, could become routine. Would it be possible to use less extreme techniques first?
- Revisit the concept of a clone. Discuss the advantages and disadvantages of cloning. More ideas for this task are included in the 'Activity' box.
- Explain what is meant by the term 'designer babies'. Discuss with students why some people feel that this is taking science a step too far. Time should be allowed for students to think deeply about this topic area – how might they feel if they were at risk of having a child with a genetic disease? If this could be avoided, is this morally acceptable?
- Explain what is involved in the technique of gene replacement therapy. Discuss the advantages and disadvantages of this technique.
- Ask students to discuss, as an overall area of research, how they feel about research into genetically engineered organisms, cloning and selective breeding. Which of these processes do they feel comfortable with? Which cause them concern? Students should be encouraged to back up their arguments using scientific evidence.

Plenaries

Advantages and disadvantages of genetic engineering – Ask students to name two advantages and two disadvantages of genetic engineering. *(5 minutes)*

Student definitions – Ask students to write definitions for 'gene replacement therapy' and 'designer baby'. To support students, provide them with key terms or phrases to be included in their definition. To extend students, they should provide a practical example of where each technique would be advantageous. *(10 minutes)*

Chapter 13 – Selective breeding and genetic modification

Activity

Would you like to be cloned?

This activity could begin by looking at the three photographs shown in the Student Book. These represent three views on cloning – harmful side effects to the animal (which lead to premature death), saving species from extinction and using your own cells to repair organs that are damaged. Students could debate the pros and cons of using cloning in each situation. Leading on from the debate, students need to focus on human cloning. They could carry out some research into possible advantages and disadvantages of this technique, using the internet.

Students should produce a report explaining whether or not they agree with the cloning of humans, using scientific reasoning to justify their view.

Possible advantages include – cloning organs, e.g. kidneys and bone marrow; some medical experts feel many life threatening diseases will become curable, e.g. diabetes, Parkinson's disease, and cystic fibrosis; scientists can use clones for research to understand how genes in cells can be switched off and on – this could lead to cures for many diseases; the reproduction of human skin has huge benefits for burn victims.

Possible disadvantages include – people believe that it is unethical; it is interfering with nature; it reduces genetic diversity; that there will be many health problems for the organism produced; people could design the characteristic or appearance of the children they wanted ('designer babies').

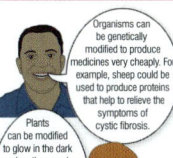
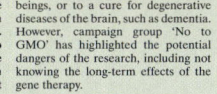

Further teaching suggestions

Dolly the sheep

Students could carry out research into Dolly the sheep. How was this procedure achieved? What are the potential advantages and disadvantages to society of animal cloning? When did Dolly die?

Answers to in-text questions

a Benefit: could boost human intelligence/cure dementia. Drawback: long-term effects are not known but there may be societal effect, for instance only the rich being able to afford the 'intelligence' gene for their children, creating a two-tier society.

b Arguments for: medicines produced (cheaply)/plants that glow when they need watering can reduce water usage/seedless fruits produced.
Arguments against: may result in wild plants not producing seeds/may reduce an animal's life span/may affect animal's quality of life.

c Advantage: embryos can be screened for genetic disorders.
Disadvantage: people may select characteristics, such as gender.

Summary answers

1 genetics, engineering, natural, animals

2 a IVF

  b a baby some of whose genetic characteristics have been selected

3 Arguments for include: Increase world food supply – pest or drought resistant crops, improved livestock (e.g. greater body mass/faster growing), provide medicines cheaply and in large quantities, seedless fruits. Arguments against include: Don't know long-term effects on an organism, could reduce animal's quality of life, unknown effect on wild organisms.

Unit 3, Theme 2 – Making and improving products

Summary answers

1.
   - Farmer chooses crops with the best characteristics.
   - These crops are then cross-pollinated.
   - Offspring grow.
   - These are then bred. This process is repeated for many years.
   - Farmer then chooses crops with the best characteristics.

2. **a** Stock will stay fresh for longer, so less stock would be wasted.
   b Some of the genes will have come from another organism.
   c Because the farmer will not have to wait for several generations for the change to occur.
   d colour, taste, rate of growth

3. **a** human donors – either alive, or dead
   b Organs are often difficult to source from human donors. Using animals would allow a limitless supply to be grown. It would also be possible to avoid problems of organ rejection in the recipient.
   c People feel this is unethical – it is not a natural process. The donor animal may experience suffering.

4. **a** circular ring of DNA
   b genetically identical copy
   c cheap, large amounts of insulin can be produced quickly

5. **a** To produce many genetically identical plants quickly.
   b large numbers of plants produced quickly, happens in a small space, produced in a disease free environment.
   c Reduces gene pool.

6.
   - A useful gene is removed from a donor nucleus.
   - The foreign gene is put into a plasmid.
   - The recombinant DNA put into a bacterial cell.
   - The bacteria reproduce producing lots of copies of the recombinant DNA.
   - Plant cells are infected with the bacteria.
   - The foreign gene becomes integrated into the plant's DNA.
   - The plants cells are grown into plants that display the desired characteristics.

Summary questions

1. The following sentences explain how farmers may selectively breed their crops. Rearrange them into the correct order:
   - Offspring grow.
   - Farmer chooses crops with the best characteristics.
   - Farmer then chooses crops with the best characteristics.
   - These crops are then cross-pollinated.
   - These are then bred. This process is repeated for many years.

2. The Flvr Svr tomato is genetically engineered to stay firm for longer.
   a Why is this an advantage for tomato sellers?
   b How are the genes in this tomato's nucleus different from those of a normal tomato?
   c Why is this process quicker than selective breeding?
   d What other features might a tomato grower choose to improve?

3. Some people require an organ transplant, due to damage or disease. Animals could be genetically modified, so they can provide organs for humans.
   a Where do organs for donation currently come from?
   b What are the arguments in favour of using animals in this way?
   c What are the arguments against using animals in this way?

4. Bacteria can be genetically engineered to produce insulin.
   a What is a plasmid?
   b What is a clone?
   c Name two advantages of producing insulin using bacteria.

5. **a** What is the technique of micropropagation used for?
   b Name two advantages of this technique
   c Name one disadvantage of this technique.

6. The following sentences describe how a plant can be genetically modified to display a specific characteristic. Rearrange the sentences into the correct order:
   - Plant cells are infected with the bacteria.
   - The foreign gene becomes integrated into the plant's DNA.
   - A useful gene is removed from a donor nucleus.
   - The bacteria reproduce producing lots of copies of the recombinant DNA.
   - The recombinant DNA is put into a bacterial cell.
   - The foreign gene is put into a plasmid.
   - The plants cells are grown into plants that display the desired characteristics.

Practical suggestions

Practicals	AQA	k	📖
Grow new plants from tissue cultures.	✓		✓

Kerboodle resources

- Video: Dairy Farmer (13.1)
- Support: Selective breeding and genetic engineering (13.1. and 13.2)
- Viewpoint: Genetically modified crops (13.2)
- Viewpoint: Genetically modified crops (13.2)
- Animation: Genetic modification (13.3)
- Support: You're never alone with a clone (13.3)
- Viewpoint: Designer babies (13.4)
- Examination-style questions
- Answers to examination-style questions
- Test yourself: Improving products

Chapter 13 – Selective breeding and genetic engineering

Examination-style questions

1. Genetic engineers have been working to improve organisms. The table shows possible outcomes of genetic modification.

   Tick the boxes to show which of these are advantages, disadvantages or neither.

Outcome	Advantage	Disadvantage	Neither
Could spread and destroy native species			
Glow in the dark pigs			
Herbicide-resistant wheat			
Possible introduction of cancer			
Removal of genetic disorders			

   (5)

2. Farmers have been using selective breeding for centuries.
   a Put the sentences in the correct order to describe the process of selective breeding of sheep.
      A Farmer repeats over several generations
      B Farmer chooses the best offspring
      C Farmer chooses sheep that have lots of wool
      D Farmer breeds the sheep (3)
   b Suggest another quality of sheep that a farmer might want to improve. (1)
   c Give a disadvantage of selectively breeding sheep. (1)

3. *In this question you will be assessed on using good English, organising information clearly and using specialist terms where appropriate.*
   a Human insulin can be made using genetically modified bacteria. Explain how they have done this. (6)
   b Before human insulin was produced through genetic engineering, diabetics used to be treated with insulin from pigs. Suggest **one** disadvantage of this treatment. (1)

4. Greyhound dogs have been selectively bred to run very fast in races. The table shows the average time for a greyhound to run 2 km in different years.

Year	1920	1930	1940	1950	1960
Time in seconds	161.00	160.02	157.88	152.25	149.86
Year	1970	1980	1990	2000	
Time in seconds	145.32	142.35	138.26	136.85	

   a Suggest the average time to run 2 km in 1935. (1)
   b In which year might it have taken 139.28 seconds? (1)
   c Describe an experiment that greyhound racers could do to collect results for this year. (3)
   d Calculate the percentage decrease in the overall time taken to run 2 km. (4)

5. A couple who have a son with a blood disease researched different treatments. They read this newspaper article on 'designer babies'.

   Designer babies could save money for the NHS

   Children who are very ill with incurable diseases might require a lot of medication for long periods of time. That could cost the NHS as much as £50,000 a year and nearly £2 million in a lifetime.

   It has been estimated that the cost of producing a designer baby through IVF treatment is £20,000.

   Dr Edwards explained how a designer baby is chosen. 'Tests are conducted on the mother's embryos to find one that does not have their brother or sister's inherited disease and is a suitable genetic match. It is then implanted in the mother's womb to create a baby from which bone marrow or cells can be transplanted into their ill sibling.'

   a Suggest why the couple might be against having a designer baby. (1)
   b If the couple's son lived until he was 20 years old, what could be the saving for the NHS if he could be treated using a 'designer' sibling? (4)
   c What could be the long-term implications of being able to select the characteristics of a baby before it is born? (2)

6. Selective breeding and genetic modification are both methods used to change the characteristics of organisms.

   Sort these advantages out into those about selective breeding, those about genetic modification and those that apply to both.
   • Bigger crops   • Easy for farmer to do
   • Get desired characteristics   • Quick

   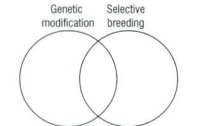

   (2)

Examination-style answers

1.

Outcome	Advantage	Disadvantage	Neither
Could spread and destroy native species		✓	
Glow in the dark pigs			✓
Herbicide resistant wheat	✓		
Possible introduction of cancer		✓	
Removal of genetic disorders	✓		

(5 marks)

2. a C D A B (All 4 correct = 3 marks, 3 correct = 2 marks, 1 or 2 correct = 1 mark)
   b meat (1 mark)
   c loss of other characteristics (1 mark)

3. a Marks awarded for this answer will be determined by the Quality of Written Communication (QWC) as well as the standard of the scientific response.

   There is a clear, balanced and detailed explanation of how human insulin can be made using genetically modified bacteria. The answer is well structured with minimal repetition or irrelevant points. There is an accurate, fluent and clear expression of ideas with only minor errors in the use of technical terms, spelling, punctuation and grammar. (5–6 marks)

   There is some explanation of how human insulin can be made using genetically modified bacteria, with some omissions. The answer shows some attempt at structuring and the ideas are expressed with reasonable fluency and clarity. There are some errors in the use of technical terms, spelling, punctuation and grammar. (3–4 marks)

   There is a brief explanation of how human insulin can be made using genetically modified bacteria. The answer is largely incomplete and may contain some valid points which are not clearly structured. It lacks fluency and/or clarity. It contains errors in the use of technical terms, spelling, punctuation and grammar. (1–2 marks)

   No relevant content (0 marks)

   Examples of points made in the response:
   - (Plasmid) is (circular) bacterial DNA
   - Cut open by enzymes
   - Mixed with human insulin gene
   - Bacteria produces human insulin
   - All copies of bacteria also make human insulin
   - Human insulin then removed from bacteria using enzymes.

   b Body rejects pig insulin
   OR
   Pig insulin not the same as human insulin
   OR
   Religious / ethical objections, e.g. vegetarian (1 mark)

4. a any number between 158 and 159 (1 mark)
   b any year between 1981 and 1989 (1 mark)
   c Time dog running 2 km
      Repeat OR try with different dogs
      Take an average (3 marks)
   d 161 − 136.85
      = 24.16
      24.15 ÷ 161 × 100
      = 15% (4 marks)

5. a ethical / religious reasons (1 mark)
   b 20 × 50 000
      = 1 000 000
      1 000 000 − 20 000
      = £980 000 (both unit and amount needed for the mark) (4 marks)
   c Any **two** from:
      May create a superhuman race
      Could get rid of all genetic diseases
      Would be able to choose gender, hair colour or other characteristics
      May be prejudices against those who are not genetically enhanced
      Less biodiversity. (2 marks)

6.

   Genetic modification   Selective breeding

   (2 marks)

Unit 3, Theme 3

Improving our environment

AQA Specification link-up: Science B 3.5.3

We all need to be aware of the effects that making and using products in everyday life can have on the world around us. Scientists are constantly looking for ways to lessen the damage we do by improving the ways we make, use and dispose of products.

Energy is expensive and should not be wasted. Energy consultants need to understand how heat is transferred. Home owners need to show how energy efficient their house is before they can sell it.

Environmental scientists realise that pollution-free air in the home is important for our health and well being.

In this theme there are three contexts:
- 3.5.3.1 Environmental concerns when making and using products
- 3.5.3.2 Saving energy in the home
- 3.5.3.3 Controlling pollution in the home

Unit 3, Theme 3
Improving our environment

In Unit 3, Theme 3 you will work in the following contexts, covered in Chapters 14 and 15:

Environmental concerns when making and using products

Manufacturing new products is not always good for the environment. The more materials are processed, the more energy is used to produce them. Generating this energy usually produces greenhouse gases, which are contributing to climate change. Other types of production can also harm the environment. Growing enough food to feed everyone requires the use of fertilisers. These can be washed into rivers and end up killing fish through a process called eutrophication.

Getting rid of old products can be as damaging as making new ones. Landfill sites all over the country are filling up as we try to find new ways to dispose of waste. To help with this problem, materials scientists are constantly trying to invent new forms of biodegradable packaging.

Saving energy in the home
Heat transfers

The kitchen is a room of extremes of temperature. Fresh food is stored at low temperatures in fridges or frozen at sub-zero temperatures in freezers. We boil water in kettles and cook food at much hotter temperatures. How can we do this as efficiently as possible in the same room? In different buildings, temperatures are controlled by controlling heat transfers so that the temperature inside is comfortable whether there is ice outside or blazing sunshine.

Saving money on our heating bills

Heating bills can be very expensive for some houses, but neighbouring houses may have very different heating bills. Modern homes are built to be well insulated due to our building regulations but if you have an older house what can you do to reduce your bills? Some owners spend large amounts of money on changes that are not as cost-effective as a quicker simpler change. How can you tell if a change is worth doing? Householders need to know if the amount of energy saved is significant and how big the saving will be.

Controlling pollution in the home
Pollution in the home

We are used to pollution being a problem caused by traffic and power stations. It is surprising to realise there can be pollution inside our homes as well. Dust, moulds, spores and pollen can all build up inside homes to levels that cause ill effects in people.

Household hazards

We use many products in our homes to keep them clean and free from germs. Some products have been used for years but other newer products are less familiar. The packaging carries a lot of (hazard) information about how to store or use a product safely.

The silent killer

Homes use boilers to heat water and for central heating. Most of the time, we take them for granted until something goes wrong. A faulty boiler may just mean a cold shower in the morning but in some cases faulty boilers can result in deaths.

Environmental concerns when making and using new products

Activity:
- A chocolate bar can be used here, with the class examining its fate once eaten. Students could plan an experiment to find out the amount of energy generated by a chocolate bar wrapper if it was incinerated. Students can also explain why performing the practical might be too hazardous, because of the fumes given off by additives and pigments in the wrapper.
- This context also offers a good opportunity for a longer term investigation, looking into the degradable properties of various supermarkets' carrier bags. Students could leave some bags in daylight, bury some in string bags, and leave some indoors as a control group. Leave at least three weeks for there to be any observable difference.

Possible misconceptions

Students often forget about conservation of mass when thinking about biodegradation. Where do the molecules end up?

Saving energy in the home

Heat transfers affect our lives daily, from the clothes we wear and the homes we live in to cooking and storing the food we eat. Efforts to control energy transfers surround students in their homes and schools.

Activity:
- Students review their knowledge of the arrangements of particles in solids, liquids and gases for discussion of conduction and convection.
- To prepare for learning about thermal radiation, students review the electromagnetic spectrum. They consider uses of radiant heaters in the home, e.g. bar heaters, toasters, grills.

Possible misconceptions

Common misconceptions to consider are:
- Particles in a solid expand as they get hotter (they actually vibrate more, so effectively take up more space but their size does not change).
- Temperature and energy are the same (an object has more energy when it is at a higher temperature but energy is also linked to the type of material and the size of the object. Small hot objects may have less energy than larger cooler objects).
- Objects at freezing point have no energy (all objects at temperatures above absolute zero, −273°C, have some energy and so their particles vibrate but cooler objects have less energy).
- Black surfaces attract energy (they are better at absorbing it, but they do not attract it).

Activity:
- Students prepare a graffiti wall of energy saving tips that can be referred to throughout these lessons on energy transfers.

Possible misconceptions

Remind students that energy travels from a hotter place to a cooler place – a home gets cooler because the energy is transferred to the surroundings, not because cold is gained from the surroundings.

Chapter 14 – Environmental concerns when making and using products

Possible misconceptions

Students may believe that if there are no symptoms, the pollution is not present. Also, students may believe that symptoms are directly linked to pollutants when in fact these may have no effect. Remind students that you cannot always taste, smell or see pollutants.

Activity:

- Students review simple chemical reactions including oxidation reactions.

Possible misconceptions

Students may not realise that faulty gas appliances cause problems in very well sealed homes because the air cannot be drawn in from outside.

The problems of carbon monoxide apply to all fossil fuel burners as well as car and motor engines (e.g. when a vehicle is turned on in a garage with the door closed or a heater is used in an enclosed space).

Activity:

- Students list the household products that they use or see in their homes, advertised on TV, or in shops, with their different uses. You may be able to show an advert for different cleaning products. Students discuss any special precautions linked to these products, e.g. special instructions or packaging. If available, hand round a selection of packets or adverts for different products to see what they claim to do and any special instructions.

Possible misconceptions

Many familiar products such as bleach can be very harmful to people. Students do not always assess the risk of familiar products compared with unfamiliar products correctly and this is why hazard warning symbols are helpful. They may also fall for advertising slogans and believe branded products are more effective. This can be linked, in their minds, to the strength of the product.

Controlling pollution in the home

Activity:

- In small groups, students decide on a definition for airborne pollutant and the types of particle that they consider are pollutants. Compare these definitions and then students use these ideas to list activities in the home that could increase the level of airborne pollutants and possible ways of detecting these and the effects on people in the building.

Unit 3, Theme 3 – Improving our environment

14.1 Producing greenhouse gases

Learning objectives

Students should learn:

- that there are three main greenhouse gases
- that these gases are produced when products are made and used
- that greenhouse gases cause global warming.

Learning outcomes

Most students should be able to:

- state the three main greenhouse gases
- explain that these gases may cause global warming
- state how each of these gases is emitted.

Some students should also be able to:

- explain how greenhouse gases are created when products are made and used.

Science in context

This is very topical as students will be aware of trends in the weather. Ensure the distinction between climate (long term) and weather (short term) is clearly understood. Students are likely to learn about these issues in geography, PSHE and other subjects too, as well as media coverage.

Support

- Concentrate on ensuring students have a sound grasp of the main ideas. They may well be confused by random statistics or ideas that have stuck in their minds from encounters with this topic outside the lesson. Aim to discover misconceptions as you begin the topic.

Extend

- Discuss the evidence that scientists have for the levels of greenhouse gases present in the atmosphere in previous years. Direct measurements have been made in the last couple of centuries or so, but to investigate times before then, indirect methods like analysing ice cores have been used. How reliable are these? Students could research this.

AQA Specification link-up Science B 3.5.3.1

- Describe the main ways in which making and using products may result in increased emissions of natural greenhouse gases into the atmosphere causing global warming including:
  - carbon dioxide from the combustion of fossil fuels in vehicles and power stations
  - methane from decomposition of rubbish in landfill sites and various forms of agriculture
  - nitrous oxide from vehicle exhausts and power stations and as a result of increased use of nitrogen-based fertilisers.

Within this context, candidates should be able to use scientific data and evidence to discuss, evaluate or suggest implications of the following:

- changes to the composition of … air as a consequence of industrial activity.

Lesson structure

Starters

What's in the molecule? – Remind students about the composition of molecules. Support students by prompting them to say that carbon dioxide contains carbon and oxygen for example. This should help students understand why methane is a product of anaerobic respiration. You could also ask them to match a list of compounds if they are provided with a list of elements each compound contains. *(5 minutes)*

My ideas on greenhouse gases – Students rank some or all of the following (or similar) statements (e.g. as agree/neutral/disagree). Discuss their responses and use this as an opportunity to correct some misconceptions. For example:
'I can name some greenhouse gases.'
'I know how some greenhouse gases are produced.'
'Only human actions cause greenhouse gases.'
'Human actions have increased the amount of greenhouse gases.'
'Greenhouse gases are always harmful.' *(10 minutes)*

Main

- This subject is very topical and there will be many misconceptions. You could use a graffiti wall for students to add ideas and to identify areas of concern.
- The internet is an excellent resource for researching environmental issues, but you have to choose sites with care as they may be biased or contain factual inaccuracies.
- Students are very likely to know carbon dioxide is a greenhouse gas, but are less likely to be aware that methane and nitrous oxide are also greenhouse gases.
- Students research each gas, perhaps in groups of three. They produce a brief PowerPoint slide show showing the main ways that each gas is produced and its impact.
- Discuss the terms:
  carbon footprint (how much carbon dioxide results from our activities).
  carbon taxes (activities producing high levels of carbon dioxide may be taxed at a higher rate as a discouragement).
  carbon offsetting (rather than change their behaviour, some people carry on with normal activities, but pay money to an organisation who pledges to plant trees. As they grow, the trees absorb carbon dioxide).

Plenaries

What I have learned – Students write down two facts about carbon dioxide, methane and nitrous oxide. *(5 minutes)*

Spiderchart – Extend students by getting them to prepare a spider chart linking some or all of these terms: greenhouse gas, global warming, carbon dioxide, atmosphere, fossil fuel (you could also include nitrous oxide, methane, bacteria, car exhausts). Support students by getting them to complete these (or similar) sentences using these words: greenhouse gas, global warming, atmosphere, fossil fuels.
Carbon dioxide is a ………………………. Carbon dioxide is produced when …………………. burn. Carbon dioxide is increasing in the …………. Carbon dioxide absorbs energy causing …………. *(10 minutes)*

Further teaching suggestions

Burning a fuel and greenhouse gases

Demonstrate that burning a fuel gives off carbon dioxide (test with limewater, which will turn cloudy) as well as water vapour (another greenhouse gas). Wear eye protection as limewater is an irritant.

Rotting vegetation and greenhouse gases

Show students a test tube, set up a few days in advance, containing vegetation decaying under water. This is one way methane is produced.

Did you know ... ?

Carbon footprints can be calculated for organisations such as schools as well as for individuals. Knowing your carbon footprint can help you change your behaviour, and comparing the carbon footprint for individuals in different countries can be very revealing.

Unit 3, Theme 3 – Improving our environment

14.1 Producing greenhouse gases

Learning objectives
- What are greenhouse gases?
- How are greenhouse gases produced?

We use many different products in our daily lives compared with earlier generations. The quantity of different products has increased dramatically. We buy goods more often and throw more away. More people live on Earth now than ever before. This all adds up to a big impact on the environment worldwide. Scientists are now looking for more environmentally friendly ways to:

- manufacture products
- reduce their impact when in use
- reduce their effect on the environment when they are thrown away.

Scientists are very concerned about three gases in particular: **carbon dioxide**, **methane** and **nitrous oxide**. These gases are produced in large quantities and are examples of **greenhouse gases**. Our actions have increased the concentration of these gases found in our atmosphere. The atmosphere traps more of the Sun's energy when the concentration of greenhouse gases increases. This effect is called **global warming** and it is expected to change our climate in future. There is some evidence to suggest this is already taking place.

a What are the main greenhouse gases?
b What is global warming?

Figure 1 Burning fossil fuels produces carbon dioxide and other greenhouse gases

Carbon dioxide

Carbon dioxide is the greenhouse gas produced in largest quantities. The quantity produced when **fuels** are burned has increased greatly in the last 70 years. The **fossil fuels** – coal, oil and gas – have taken millions of years to form. Fossil fuels are burned in increasing amounts in vehicles and in power stations to generate electricity. Fossil fuels are also used for heating and cooking. Smaller quantities of fuels such as wood and **biofuels** are also burned, and they produce carbon dioxide too.

c How does burning fossil fuels contribute to global warming?

Methane

Methane is 20 times more effective at trapping heat than carbon dioxide per molecule. It is the next most important overall contributor to global warming after carbon dioxide. Methane is released into the atmosphere as part of the carbon cycle in several ways:

- from the action of bacteria when plants and other organic materials decay with very little air present – this can happen in swamps, rice paddy fields and hydroelectric schemes
- when organic material rots in a landfill site or compost heap
- from coal mining activities and other industrial processes
- from bacteria in the guts of animals, including farm animals.

Figure 2 Bacteria in paddy fields release methane, a greenhouse gas

d Write down **four** ways that methane can be produced.

Activity

Greenhouse gases from different countries

Find out how the amount of each of the three main greenhouse gases produced by different countries varies.

Present your data to the class either as a poster or PowerPoint presentation.

Nitrous oxide

Nitrous oxide is about 310 times more effective at absorbing energy than carbon dioxide per molecule. It is produced naturally, but increasing amounts are produced from agriculture.

Intensive farming methods use nitrogen-based fertilisers to increase crop yields. If large amounts of fertiliser are used, denitrification can take place. **Denitrification** is when bacteria convert nitrates into nitrous oxide if there is little oxygen present in the soil. Using animal waste in large quantities as a fertiliser also produces nitrous oxide.

Nitrous oxide is also made when fuels are burned at very high temperatures, as in car engines. Nitrogen is a relatively unreactive gas, but in these conditions it does react with oxygen. However, catalytic converters in cars can convert nitrous oxides in exhaust gases into harmless nitrogen gas.

links
For more information on the effects of greenhouse gases see 14.2 The effects of greenhouse gases.

Figure 3 Intensive farming contributes to global warming

e Write down **two** ways nitrous oxides are produced in large quantities.

Summary questions

1 Here are three of the main greenhouse gases:
carbon dioxide   methane   nitrous oxide
  a Put the gases in order of the quantities found in the atmosphere, starting with the most abundant.
  b Put the gases in order of their 'warming effect', starting with the gas that has the biggest 'warming effect' per molecule.
2 Explain why putting waste on a compost heap can produce the same greenhouse gases as sending it to a landfill site.
3 Prepare a leaflet summarising the main ways that each of the three greenhouse gases in question 1 are produced.

Did you know ... ?

You can calculate how much your lifestyle and actions contribute to the main greenhouse gases – this is called your carbon footprint. Carbon footprint calculators can be found online.

Key points

- Greenhouse gases include carbon dioxide, methane and nitrous oxide.
- Carbon dioxide is produced when fuels are burned.
- Methane is mainly produced when bacteria break down organic matter in the absence of oxygen.
- Nitrous oxide is produced by nitrogen-based fertilisers from intensive farming and by car engines.

Summary answers

1 a carbon dioxide, methane, nitrous oxide
  b nitrous oxide, methane, carbon dioxide

2 In both cases the same rotting process takes place and methane is produced.

3 Students should elaborate on these main ideas:
  Carbon dioxide is produced when fuels are burned.
  Methane is mainly produced when bacteria break down organic matter in the absence of oxygen.
  Nitrous oxide is produced from intensive farming and by car engines when it is released in exhaust gases.

Answers to in-text questions

a carbon dioxide, methane, nitrous oxide
b the warming of the Earth's atmosphere as a result of a greater concentration of greenhouse gases
c Fossil fuels have trapped carbon for millions of years. Now we are burning them far quicker than they are forming, an excess of carbon dioxide is produced in the atmosphere.
d bacteria in animal guts, coal mining, organic material rotting in landfill, or rotting when very little air is present.
e When burning fuel in car engines and when large amounts of nitrogenous fertilsers are used.

Unit 3, Theme 3 – Improving our environment

14.2 The effects of greenhouse gases

Learning objectives

Students should learn:
- what is meant by global warming
- what the Kyoto Protocol is and what it includes
- why the Kyoto Protocol is needed.

Learning outcomes

Most students should be able to:
- describe how the greenhouse effect takes place
- state what a treaty is and why it is needed
- state why the Kyoto Protocol is needed.

Some students should also be able to:
- explain in detail how global warming occurs
- explain what is included in the Kyoto Protocol and what it is intended to achieve.

AQA Specification link-up Science B 3.5.3.1

- Explain how increased greenhouse gases absorb more long-wave radiation from the Earth and therefore more heat is retained in the atmosphere.
- Know about international agreements such as the Kyoto agreement on climate change to achieve the stabilisation of the dangerous gases carbon dioxide, methane and nitrous oxide.

Lesson structure

Starters

Global warming and electromagnetic radiation – Students sketch the electromagnetic spectrum, showing wavelength/frequency/energy changes. Point to different members of the spectrum and students identify members with longer/shorter wavelengths, etc. *(5 minutes)*

My ideas on global warming – Students rank some or all of the following (or similar) statements (e.g. as agree/neutral/disagree). 'I know how global warming is caused', 'I know what the effects of global warming are', 'Global warming and climate change are the same thing', 'My actions can reduce global warming', 'Every country affects global warming'. Extend students by getting them to discuss their responses in pairs and to identify misconceptions. Support students by explaining common misconceptions after asking for a display of hands for each response. *(10 minutes)*

Main

- The NASA website includes video clips linked to the Kyoto Protocol.
- Check students understand the terms greenhouse gases, global warming, climate change.
- Check they understand how global warming occurs. Both the Sun and the Earth radiate electromagnetic radiation.
  - The Sun is hot so it radiates light and UV, some of which passes through our atmosphere.
  - The Earth absorbs the visible and UV radiation, warming up.
  - The Earth is cooler than the Sun so it radiates infrared, radiating more as it heats up.
  - Our atmosphere allows some infrared to escape and absorbs then re-radiates the rest back to Earth.
  - When the energy absorbed by the Earth matches the energy emitted through the atmosphere, the Earth stays at a reasonably constant temperature.
  - Increasing amounts of greenhouse gases in our atmosphere mean less infrared from the Earth can escape but the same amount of radiation from the Sun can pass through.
  - The atmosphere is gradually warming, fractions of a degree per decade. This is global warming, and these changes could cause climate change.
- Some issues for discussion:
  - Global warming happens naturally in cycles, e.g. the Sun is more active in cycles of 10–20 years so it radiates more energy. How can we find out how much global warming is due to human activity and how much is due to natural events? The Earth has been warmer during some periods in its past.
  - Indirect information provides evidence of past temperatures, but how reliable is it? It relies on some assumptions as well as some sound scientific basis, e.g. analysis of gases in ice cores, looking at writings and records from the past, analysing tree rings from ancient trees.
  - It is only recently that we have been able to make detailed, accurate scientific records. How well calibrated were the instruments used in the past?
  - How much should we trust computer predictions and simulations using assumptions that may not be correct?
  - Should developing countries be allowed to emit higher levels of greenhouse gases? How fair is carbon trading?

Plenaries

Let's make Kyoto work – Students think up a slogan to encourage people to stick to the targets from the Kyoto agreement. *(5 minutes)*

Global warming – Extend students by asking them to write out the stages in global warming using a flow chart or diagram. Support students by asking them either to put stages that you give to them in order, or to add labels to a diagram you provide. *(10 minutes)*

Support

- Use flow charts and diagrams to explain the main points. The process causing global warming need not be covered in detail, but be careful of making a simplistic explanation that causes more misconceptions.

Extend

- There is a wide range of ethical issues to discuss here. Encourage students to carry out independent research with a clear bibliography. Detailed debates or presentations will add greatly to the topic.

Chapter 14 – Environmental concerns when making and using products

Further teaching suggestions

Finding more about global warming

There are a wide range of sites that explain material clearly for students. The UK government has also developed websites to explain its specific actions after Kyoto.

Science in context

Discuss the Kyoto agreement more closely, e.g. should we include carbon emissions on goods we use purchased in different countries? How much can an individual do compared with a company?

Did you know ... ?

The Kyoto Protocol is a potentially important step in the fight against irreversible climate change. It is a legal agreement that is being implemented now and whose impact is likely to affect this generation of students in their future lives.

Unit 3, Theme 3 – Improving our environment

14.2 The effects of greenhouse gases

Learning objectives
- Why does global warming occur?
- What is the Kyoto protocol?
- Why do we need the Kyoto protocol?

The Sun is so hot it radiates short wavelength light and ultraviolet radiation. Our **atmosphere** reflects some of the Sun's radiation, allowing about two-thirds to pass through. The Earth absorbs this energy, and warms up. As Earth warms, it emits more radiation as long wavelength (infrared) radiation. If the amount of energy absorbed matches the energy emitted, the temperature of Earth stays constant.

a What happens to the Sun's radiation when it reaches our atmosphere?

Some gases in the atmosphere are effective at absorbing infrared radiation radiated by the Earth. They re-radiate it in all directions so some is directed back to Earth. This raises the temperature in our atmosphere higher than it would be otherwise. This is called the **greenhouse effect** and is one reason why the Earth can sustain life. Gases like carbon dioxide, methane and nitrous oxides are called **greenhouse gases**.

The balance of gases in our atmosphere has changed over the last century. Burning fossil fuels releases carbon that was trapped in the Earth to the atmosphere as carbon dioxide. The proportion of gases like carbon dioxide, methane and nitrous oxides has increased in our atmosphere. Gradually our atmosphere is warming up. This increase in the average temperature of the Earth's atmosphere is called **global warming**.

Figure 1 Global warming occurs when more radiation is absorbed than is emitted.

b What do greenhouse gases do to infrared radiation?

c What is meant by global warming?

Global warming

Global warming is worrying because scientists believe it is changing our climate. No one knows how much the temperatures will rise, or what the effects will be. Some effects could be:
- there may be more droughts
- sea levels may rise
- polar ice caps and glaciers may melt
- there may be more hurricanes and storms.

These effects are called **climate change** and could have a big impact on our way of life.

d What are the main effects of climate change?

One country on its own cannot control global warming. The UK only contributes about 2 per cent of the world's greenhouse gases. Gases from all countries spread throughout the atmosphere. If a country is producing less greenhouse gases, it may be less able to trade with other countries. That is because it is manufacturing fewer products or having to raise costs as a result of controlling emissions. If all countries agree to reduce activities producing greenhouse gases, we may be able to limit climate change.

The Kyoto agreement

The **United Nations** is a group of industrialised nations including the European Union. In 1997 many of its representatives met in Kyoto, Japan. Some of them signed an international agreement called the **Kyoto Protocol**. It set targets for these countries to reduce their greenhouse gas emissions by about 5 per cent of their 1990 levels over the years 2008–12. It was a legally binding document, forcing them to monitor their emissions, meet the targets and report back. This is an important first step, but some countries did not sign and others did not get the Kyoto protocol ratified (approved) by their governments.

links
For information on producing greenhouse gases look back at 14.1 Producing greenhouse gases.

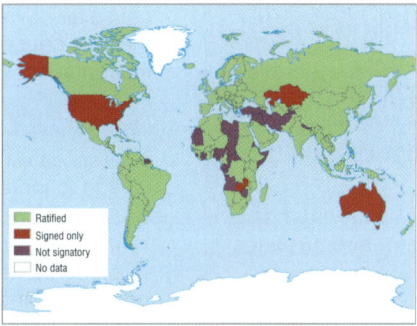

Figure 2 Many countries signed the Kyoto Protocol but some did not

e What was the purpose of the Kyoto Protocol?

Did you know ... ?

The UK is reducing its emissions by making homes more energy efficient, producing more energy from renewable sources, setting limits on business emissions and raising taxes on landfill sites to encourage recycling.

Summary questions

1 Copy and complete using these words:
global warming   shorter wavelength   longer wavelength   climate change

The Sun radiates .... radiation and the Earth emits .... radiation.
.... is when temperatures in our atmosphere increase.
.... may cause more droughts, hurricanes and rising sea levels.

2 Write a sentence that explains the difference between greenhouse effect, global warming and climate change.

3 Explain why one country cannot tackle climate change on its own.

4 Find out how the Kyoto Protocol has affected different countries in the last ten years.

Key points
- The greenhouse effect is when energy is retained in the atmosphere by greenhouse gases.
- Global warming is the increase in the average temperature of our atmosphere.
- Global warming will cause climate change.
- The Kyoto Protocol is a legally binding agreement.
- Many countries in the world agreed to cut their emissions of greenhouse gases.

Summary answers

1 shorter wavelength, longer wavelength, global warming, climate change

2 The greenhouse effect traps energy in the atmosphere so it warms up (global warming) and this causes climate change such as droughts.

3 Many countries contribute to the problem so the impact of one country may be tiny; the country will put itself at an economic disadvantage if it stops producing goods and services.

4 For example, the UK is on target for its greenhouse gas reductions and should improve its target (partly helped by the global recession).

Answers to in-text questions

a Some is reflected and some passes through the atmosphere and is absorbed by Earth.

b absorb and re-emit the radiation

c Global warming is the increase in temperature of the atmosphere as a result of the greenhouse effect.

d droughts, rising sea levels, melting ice caps and glaciers, more hurricanes and storms

e To set legally binding targets for most nations on Earth to reduce or limit greenhouse gas emissions.

Unit 3, Theme 3 – Improving our environment

14.3 Threats to the countryside

Learning objectives

Students should learn:
- that intensive farming uses a range of chemicals to increase crop production
- that eutrophication occurs when fertilisers pollute water courses
- that indicator species are used by scientists to monitor environmental pollution levels.

Learning outcomes

Most students should be able to:
- describe the role of fertilisers, herbicides and pesticides in intensive farming
- describe the process of eutrophication
- name some examples of indicator species
- describe how indicator species are used to monitor pollution levels.

Some students should also be able to:
- explain in detail how living organisms can be used to monitor air and water pollution.

Specification link-up Science B 3.5.3.1

- Describe how leaching of artificial fertilisers, pesticides and herbicides causes pollution in lakes and rivers (eutrophication).
- Explain how indicator species may also be used to monitor changes in pollution levels:
  a water pollution – bloodworm, water louse, sludge worm, rat-tailed maggot
  b air pollution – lichen.
- Use data from indicator species to evaluate levels of pollution.

Within this context, candidates should be able to use scientific data and evidence to discuss, evaluate or suggest implications of the following:
- changes to the composition of … water as a consequence of industrial activity.
- use data from indicator species to evaluate levels of pollution.

Lesson structure

Starters

Chemical additives – Ask students to make a list of chemicals that farmers can use to increase their crop production (yield). *(5 minutes)*

Air and water pollution – Ask students in small groups to discuss and list factors that can cause air or water pollution. To support students, provide them with anagrams of factors that can cause pollution. For example, power stations, industry, fertilisers. To extend students, ask them to say how each factor may harm the environment. *(10 minutes)*

Main

- Discuss with students what is meant by intensive farming. What chemicals do farmers add to increase their crop yield? Discuss herbicides, pesticides and fertilisers.
- Show students a photo of a lake covered in algae with dead fish. How do they think this has occurred? Students could offer suggestions based on prior knowledge – few students will initially appreciate that fertilisers are the cause. Provide students with annotations of the flow diagram in the Student Book. They need to arrange these into the correct order and draw illustrations to represent what is occurring. This could form a poster activity for the class.
- Explain what is meant by the term 'indicator species'. Introduce the use of lichen in detecting air pollution, and sludge worms in detecting sewage pollution in water. If conditions allow, students could complete a short survey of the presence (or otherwise) of lichen in the school grounds. Students should discuss how they could record samples to gain a good estimation of the total number of lichen present.
- Provide students with data showing the presence of lichen at different distances from a power station or industrial area. Students should complete a data analysis exercise using this information.
- Discuss briefly how waste water is treated and released back into water courses. This could raise some concerns amongst students, particularly if they have seen images in the media of sewage leaks. Extend the discussion to describe how environmental scientists might monitor the quality of water, through the use of indicator species.
- Provide students with data on the presence of invertebrates in water courses. Supply information on the location of clean water, a source of pollution, and an area of diluted pollution. Also supply information on the number and species of populations present at each location. Students should complete a data analysis exercise using this information.

Support

- Students could be given a sequence of pictures to sort, to explain how eutrophication occurs.

Extend

- Students could be given data to analyse the financial rewards of intensive farming techniques, e.g. yield per acre. This can be weighed up against the potential ecological and ethical drawbacks.

Plenaries

True or false –
- Pesticides kill weeds. [F]
- Sludge worms are an indicator of clean water. [F]
- Lichen can be used to estimate levels of air pollution. [T]
- Fertilisers can run off into lakes and cause eutrophication. [T]
- Fertilisers promote healthy plant growth. [T] *(5 minutes)*

Eutrophication – Ask students to write a series of sentences, or draw a flow diagram, explaining what happens during eutrophication. To support students, provide them with a series of diagrams that they can annotate. To extend students, ask them to suggest other ways farmers can add nutrients to soil, or prevent nutrients from being removed from the soil. *(10 minutes)*

Chapter 14 – Environmental concerns when making and using products

Further teaching suggestions

How can pesticides harm wildlife?
Students can research how pesticides can be damaging to the wildlife population or human health. Students begin by considering a simple food chain, e.g. lettuce → rabbit → fox. What are the numbers of organisms involved at each stage of the food chain? What would happen to the rabbits /foxes if some of the lettuces had a toxic chemical sprayed on them?

This discussion can then be expanded to look at the spraying of DDT. It is important for students to be aware why DDT was used as an insecticide and its benefits in terms of controlling malaria. The drawbacks of using DDT should be discussed.

River sampling
Students could visit a river and take samples of the invertebrates present. Students could identify the invertebrate species using classification keys. From this they can deduce the cleanliness of the water source.

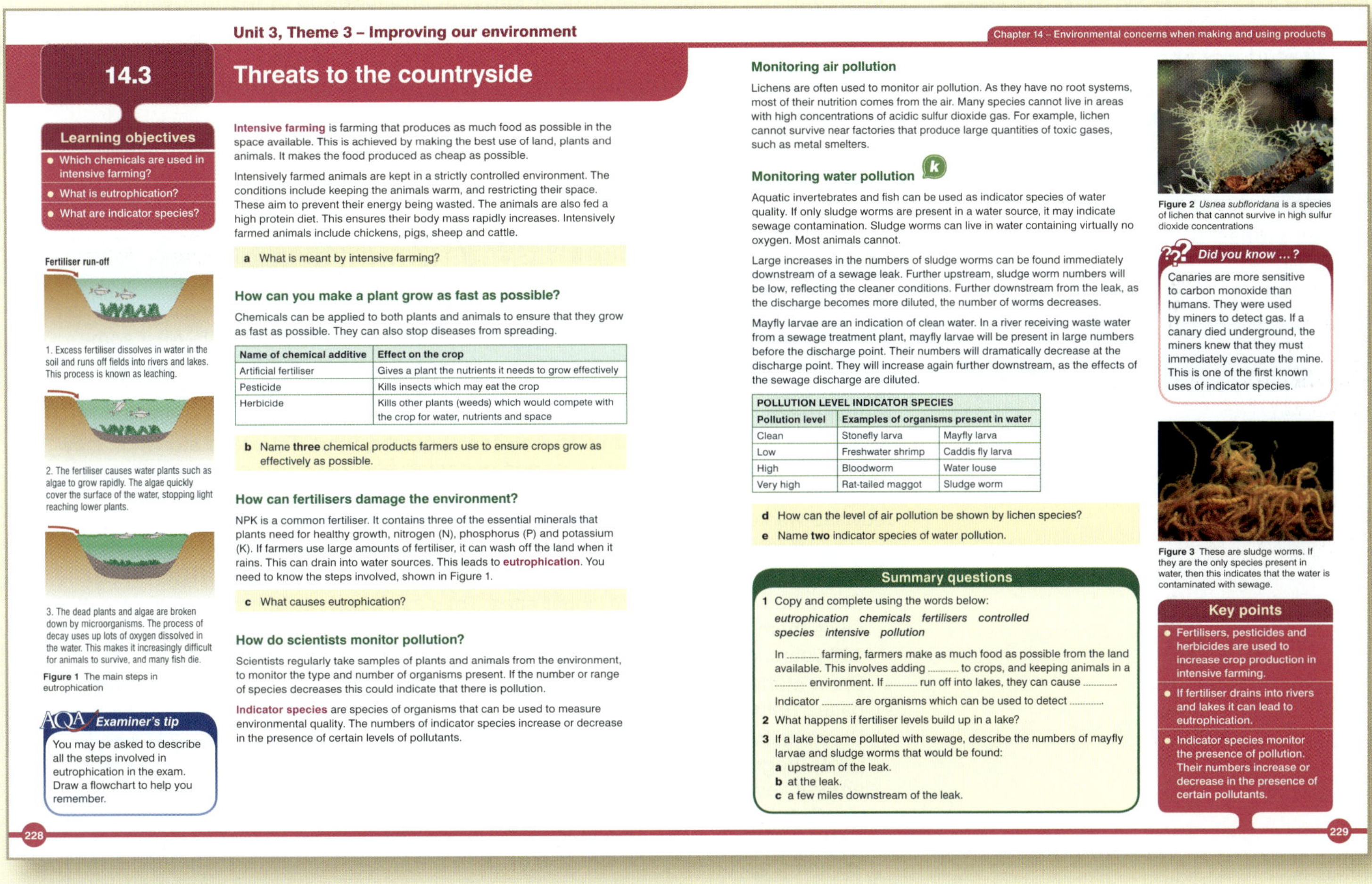

Unit 3, Theme 3 – Improving our environment

14.3 Threats to the countryside

Learning objectives
- Which chemicals are used in intensive farming?
- What is eutrophication?
- What are indicator species?

Fertiliser run-off

1. Excess fertiliser dissolves in water in the soil and runs off fields into rivers and lakes. This process is known as leaching.

2. The fertiliser causes water plants such as algae to grow rapidly. The algae quickly cover the surface of the water, stopping light reaching lower plants.

3. The dead plants and algae are broken down by microorganisms. The process of decay uses up lots of oxygen dissolved in the water. This makes it increasingly difficult for animals to survive, and many fish die.

Figure 1 The main steps in eutrophication

AQA Examiner's tip
You may be asked to describe all the steps involved in eutrophication in the exam. Draw a flowchart to help you remember.

Intensive farming is farming that produces as much food as possible in the space available. This is achieved by making the best use of land, plants and animals. It makes the food produced as cheap as possible.

Intensively farmed animals are kept in a strictly controlled environment. The conditions include keeping the animals warm, and restricting their space. These aim to prevent their energy being wasted. The animals are also fed a high protein diet. This ensures their body mass rapidly increases. Intensively farmed animals include chickens, pigs, sheep and cattle.

a What is meant by intensive farming?

How can you make a plant grow as fast as possible?
Chemicals can be applied to both plants and animals to ensure they grow as fast as possible. They can also stop diseases from spreading.

Name of chemical additive	Effect on the crop
Artificial fertiliser	Gives a plant the nutrients it needs to grow effectively
Pesticide	Kills insects which may eat the crop
Herbicide	Kills other plants (weeds) which would compete with the crop for water, nutrients and space

b Name **three** chemical products farmers use to ensure crops grow as effectively as possible.

How can fertilisers damage the environment?
NPK is a common fertiliser. It contains three of the essential minerals that plants need for healthy growth, nitrogen (N), phosphorus (P) and potassium (K). If farmers use large amounts of fertiliser, it can wash off the land when it rains. This can drain into water sources. This leads to **eutrophication**. You need to know the steps involved, shown in Figure 1.

c What causes eutrophication?

How do scientists monitor pollution?
Scientists regularly take samples of plants and animals from the environment, to monitor the type and number of organisms present. If the number or range of species decreases this could indicate that there is pollution.

Indicator species are species of organisms that can be used to measure environmental quality. The numbers of indicator species increase or decrease in the presence of certain levels of pollutants.

Monitoring air pollution
Lichens are often used to monitor air pollution. As they have no root systems, most of their nutrition comes from the air. Many species cannot live in areas with high concentrations of acidic sulfur dioxide gas. For example, lichen cannot survive near factories that produce large quantities of toxic gases, such as metal smelters.

Monitoring water pollution
Aquatic invertebrates and fish can be used as indicator species of water quality. If only sludge worms are present in a water source, it may indicate sewage contamination. Sludge worms can live in water containing virtually no oxygen. Most animals cannot.

Large increases in the numbers of sludge worms can be found immediately downstream of a sewage leak. Further upstream, sludge worm numbers will be low, reflecting the cleaner conditions. Further downstream from the leak, as the discharge becomes more diluted, the number of worms decreases.

Mayfly larvae are an indication of clean water. In a river receiving waste water from a sewage treatment plant, mayfly larvae will be present in large numbers before the discharge point. Their numbers will dramatically decrease at the discharge point. They will increase again further downstream, as the effects of the sewage discharge are diluted.

POLLUTION LEVEL INDICATOR SPECIES		
Pollution level	Examples of organisms present in water	
Clean	Stonefly larva	Mayfly larva
Low	Freshwater shrimp	Caddis fly larva
High	Bloodworm	Water louse
Very high	Rat-tailed maggot	Sludge worm

d How can the level of air pollution be shown by lichen species?
e Name **two** indicator species of water pollution.

Summary questions
1. Copy and complete using the words below:
eutrophication chemicals fertilisers controlled species intensive pollution

In _____ farming, farmers make as much food as possible from the land available. This involves adding _____ to crops, and keeping animals in a _____ environment. If _____ run off into lakes, they can cause _____.
Indicator _____ are organisms which can be used to detect _____.

2. What happens if fertiliser levels build up in a lake?
3. If a lake became polluted with sewage, describe the numbers of mayfly larvae and sludge worms that would be found:
   a upstream of the leak.
   b at the leak.
   c a few miles downstream of the leak.

Figure 2 *Usnea subfloridana* is a species of lichen that cannot survive in high sulfur dioxide concentrations

Did you know … ?
Canaries are more sensitive to carbon monoxide than humans. They were used by miners to detect gas. If a canary died underground, the miners knew that they must immediately evacuate the mine. This is one of the first known uses of indicator species.

Figure 3 These are sludge worms. If they are the only species present in water, then this indicates that the water is contaminated with sewage

Key points
- Fertilisers, pesticides and herbicides are used to increase crop production in intensive farming.
- If fertiliser drains into rivers and lakes it can lead to eutrophication.
- Indicator species monitor the presence of pollution. Their numbers increase or decrease in the presence of certain pollutants.

Summary answers

1. intensive, chemicals, controlled, fertilisers, eutrophication, species, pollution

2. Fertiliser causes algal blooms, which prevent light reaching plants at the bottom of the lake. When these die they are decomposed by microorganisms. This removes oxygen from the water, resulting in the death of fish.

3. **a** large numbers of mayfly larvae
   b large numbers of sludge worms / no mayfly larvae
   c large numbers of mayfly larvae

Answers to in-text questions

a producing as much food as possible, by making the best use of land, plants and animals
b artificial fertilisers, pesticides, and herbicides
c fertiliser
d Numbers of lichen and lichen species decrease if sulfur dioxide is present.
e two from: stonefly larvae, mayfly larvae, freshwater shrimps, caddis fly larvae, bloodworms, water louses, rat-tailed maggots, sludge worms

Unit 3, Theme 3 – Improving our environment

14.4 Disposing of our waste

Learning objectives

Students should learn:
- that landfills and incinerators are not sustainable
- the advantages and disadvantages of biopolymers
- that photo-degrading polymers and oxy-degrading polymers will degrade relatively quickly. **[HT only]**

Learning outcomes

Most students should be able to:
- list the relative advantages and disadvantages of landfill sites and waste incineration
- describe the advantages and disadvantages of biopolymers.

Some students should also be able to:
- explain how degradation-accelerating methods work. **[HT only]**

Support
- To support students, focus on defining what is and is not biodegradable/recyclable.

Extend
- Using internet resources such as government statistic sites, students can research and compare the amount of waste produced and recycled by other countries. Students could also research populations and calculate figures for waste per capita.

AQA Specification link-up Science B 3.5.3.1

- Describe the methods of degrading plastics:
  a photo-degradable – those that degrade after prolonged exposure to sunlight **[HT only]**
  b oxo-degradable – an additive helps to break down the plastic allowing access by microbes. **[HT only]**
- Know that water-soluble plastics such as polyvinyl alcohol (PVOH) and ethylene vinyl alcohol (EVOH) can be used for plastic films for packaging and shopping bags. **[HT only]**

Within this context, candidates should be able to use scientific data and evidence to discuss, evaluate or suggest implications of the following:
- the environmental impact of landfill sites for the disposal of waste materials including plastics
- the advantages and disadvantages of using plants to make plastics
- disposal of waste material by incinerating or recycling
- the advantages and disadvantages of using biodegradable products in landfill.

Lesson structure

Starters

What do you think? – Ask students what they think happens to rubbish. After some discussion in pairs and small groups, students feed back their ideas to the class. *(5 minutes)*

How much? – Present students with the bar chart from the Student Book. In pairs, they write down as many statements as they can about the chart. To extend students, they could estimate the masses of waste and quantitatively describe the patterns they see. To support students, ask leading questions, like 'What has happened to the amount of non-recycled waste since 1983?' *(10 minutes)*

Main

- Show students pictures of landfill sites. They can match up items with how long it takes them to biodegrade (i.e. paper – 2–4 weeks, orange peel – 4–5 weeks, tin can – 100 years, glass jar – unknown).
- Show students video clips of polylactic acid (PLA) production or the use of corn-based polymers.
- Students form four equal 'expert groups' to research oxo-degradation **[HT only]**, photo-degradation **[HT only]**, and plant polymers. Each group has 20 minutes to answer key questions such as 'What does it involve?', 'What are the advantages and disadvantages?', etc. After 20 minutes, students then form groups of four, one from each expert group. Groups now have a further 20 minutes to produce a poster on 'cleaning up landfills'.
- Students can use the internet to research UK supermarkets for materials they use for carrier bags. Different supermarkets have different approaches. Which supermarket carrier bag degrades the fastest? Which has the least environmental impact?
- Students work in small groups to redesign the packaging of several common items. They might suggest alternative materials perhaps and give reasons for their suggestions. Groups will attempt to reach the goal of 25% reduction in packaging materials currently used and explore alternative materials. Students must indicate which parts of the packaging now include accelerated-degradation materials and why.

Plenaries

Pictionary – Students must describe oxo-degradation **[HT only]**, photo-degradation **[HT only]**, landfill site and biopolymers using only pictures. *(5 minutes)*

Mind-mapping – Students produce a concept map linking ideas from the lesson to a central theme of waste disposal. To support students, provide words and themes, asking them to group key words under each theme. To extend students, require them to write on each link in their concept map, explaining what the link between those two ideas is. *(10 minutes)*

Chapter 14 – Environmental concerns when making and using products

Further teaching suggestions

Rubbish log

In preparation for the lesson, students could keep a rubbish log for 24 hours, recording everything they throw away. The class results could be compiled and analysed. Yearly estimates for different classes of rubbish could be calculated.

Unit 3, Theme 3 – Improving our environment

14.4 Disposing of our waste

Learning objectives
- How do we dispose of our waste?
- How can waste be disposed of sustainably?

Did you know …?
It has been estimated that every year each person in the UK on average throws away seven times their own body mass in waste.

New products can increase our quality of life. Modern materials can have more useful properties than traditional materials. The industries involved in making these new products create jobs and lots of money. Nearly every new product that is made will need to be disposed of sooner or later.

Dispose? Degrade? Recycle?

Most of the UK's rubbish ends up in **landfill sites**. Landfill is simply dumping waste into huge holes in the ground and covering it with soil. As our waste output increases, this process has become more complicated. Landfill areas are now often lined first, to prevent waste leaking into the environment. Complex systems remove or capture the liquids and gases seeping from the waste.

The UK is running out of landfill sites near the places where waste is produced. It isn't sustainable to continue using landfill sites. The materials that end up in landfill sites are just buried and lost. They are not reused. However, landfills are cheap.

Incineration is a way of disposing of waste by burning it. It can be more useful than landfill because it saves space and energy. After incineration, the waste is reduced to ashes, which take up a lot less space than landfills. Also, the energy released from burning the waste can be used.

The disadvantage of incinerating waste is that toxic chemicals are produced. Some of these are gases. They must be filtered out of the incinerator's emissions. The ashes that are produced are highly toxic and need to go to specialised landfills.

Some materials don't need to be processed like this because they break down naturally (**biodegrade**). They can break down because bacteria, fungi and other **decomposers** can feed on them. Most food waste is biodegradable, as are materials that come from living organisms. Synthetic materials are often not biodegradable. Biodegraded material can often be used again. For example, compost from food scraps and grass cuttings can be used as a fertiliser. However, materials only biodegrade in the right conditions. Decomposers need air, warmth and moisture to stay alive. Also, the waste products of some decomposers can be dangerous. For example, the flammable gas methane could be produced. If methane builds up inside a sealed landfill, there is a risk of explosion.

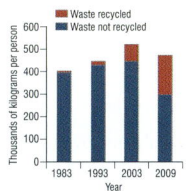

Figure 1 Recycling of waste has increased tremendously within your lifetime
Source: Department for Environment, Food and Rural Affairs

Examples of biodegradable items	Examples of non-biodegradable items
Vegetable peelings	Foil wrappers
Paper and card	Some plastic packaging
Garden waste	Glass

To avoid having to use landfill or incineration we can **recycle** or re-use as much as possible. By recycling, we can avoid having to extract the raw materials needed to make new products. However, we also create more energy demands by sorting, transporting and processing the recyclable materials. If recycling isn't performed carefully, it can use as much energy as producing materials from scratch and still cause pollution.

Figure 2 15 million tonnes of waste still go to UK landfills every year

a What types of microbes help to biodegrade materials?

Making better polymers [Higher]

Plastic bags are made from polymers that are manufactured from products of crude oil. As well as the problems of getting rid of them, using lots of plastic bags uses up crude oil that could be used as fuel. Even though plastics make up a small percentage of the mass of waste, they take up most of the space. Reducing the volume of plastics on landfills could drastically reduce the volume of waste.

One solution to this problem is making polymers that can degrade (break down) more easily. Shopping bags and packaging can now be made from Polyvinyl Alcohol (PVOH) and Ethylene Vinyl Alcohol (EVOH). These materials are water soluble and slowly break down in the environment. However, they need to be stored carefully as humidity shortens their shelf life.

b How can PVOH and EVOH help reduce the size of landfills?

Biopolymers

Like EVOH and PVOH, biopolymers behave in a similar way to polymers from crude oil. They can be used to make shopping bags and other packaging materials. They are made from plant **biomass**, such as corn or potatoes. PLA (Polylactic Acid) is a type of biopolymer made from potato starch. It can biodegrade and doesn't use up crude oil.

Biopolymers are not perfect, however. They often have lower melting points than conventional polymers and can be less durable. Some also need specific conditions and special equipment to biodegrade them. As well as this, farmland has to be set aside to grow biopolymer crops instead of food crops.

c What is PLA and what is it made of?

Photo-degradation and oxo-degradation

Additives can be used when making plastic helping them biodegrade faster:

Oxo-degradable plastics contain an additive that helps the plastic break down. This allows microbes to digest the plastic. Oxo-degradable plastic bags break down thousands of times faster than conventional ones.

Photo-degradable plastics break down when they are exposed to light. The plastic rings holding your drinks cans together have been made from photodegradable LDPE since the 1990s.

Figure 3 A degradable plastic bag

Summary questions

1 Use the information in this spread to produce a summary table:

Activity	Description	Advantages	Disadvantages
Landfills			
Incineration			
PVOH and EVOH			

2 List the advantages and disadvantages of oxo-degradable plastics. [H]

3 Why is photo-degradable plastic a good choice for six-pack rings? [H]

Key points
- Disposing of products in landfills and incinerators damages the environment.
- Polymers like EVOH and PVOH are used in packaging because they are water soluble and will biodegrade.
- Using biopolymers, photo-degradable and oxo-degradable materials speeds up biodegradation. [H]

Summary answers

1 **Landfills**: Description: burying waste. Advantage: cheap. Disadvantage: toxic waste can bleed into soil, takes up a lot of space, energy not reclaimed
Incineration: Description: burning waste. Advantage: less land, energy can be recycled. Disadvantage: toxic wastes and gases.
PVOH and EVOH: Description: degradable polymers. Advantage: water soluble, degrade. Disadvantage: need to be kept dry when stored.

2 Advantage: break down much faster
Disadvantage: made from fossil fuels

3 'Useful' life of the product is spent indoors in the dark, so plastic only degrades after disposal (if not buried beneath other waste).

Answers to in-text questions

a bacteria and fungi (decomposers)
b If polymers can degrade, plastic packaging will take up less space.
c PLA is polylactic acid. It is made from potato starch.

Unit 3, Theme 3 – Improving our environment

Summary answers

1

Name of gas	How it is produced	Its impact on global warming
Carbon dioxide	When fossil fuels are burned	Most impact – found in large quantities
Methane	From rotting vegetation; in animal guts; from mines	Medium impact – more effective at trapping heat than CO_2 but is not present is such large quantities
Nitrous oxide	From catalytic car engines; from intensive farming methods	Small impact – very effective at trapping heat but found in small quantities

2 A FALSE. Greenhouse gases have always been in the atmosphere.
B TRUE
C FALSE. Global warming is the increase in global temperatures; climate change is the associated changes in weather, e.g. increasing storms and drought.

3 a To add nutrients to the soil.

b
- Excess fertiliser dissolves in water in the soil.
- Fertiliser runs off fields into rivers.
- Fertiliser causes water plants such as algae to grow rapidly.
- Algae quickly cover the surface of the lake, stopping light reaching lower plants.
- Many plants die.
- The dead plants and algae are broken down by bacteria.
- Oxygen is used up by decay-causing bacteria.
- Many fish die.

c eutrophication

4 a PVOH and EVOH are water soluble so will degrade when disposed of. This will reduce the mass and volume of landfills.

b Incineration – takes up less space, reclaims some of the energy, but produces toxic gases and other waste. Recycling – avoids having to obtain the raw materials again, can be pointless if not done efficiently.

5 photo-degrading plastics, which break down in the sun. oxo-degrading plastics, which contain additives that help bacteria break the bags down. **[HT only]**

6 a indicator species

b The number of different species could be counted at different distances from the town. The more lichen growing, the less sulfur dioxide.

Unit 3, Theme 3 – Improving our environment

Summary questions

1 Complete this table:

Name of gas	How it is produced	Its impact on global warming
	When fossil fuels are burned	
Methane		
		Small impact – very effective at trapping heat but found in small quantities

2 Which one of the statements below is true? Correct the statements that are not true.
A Greenhouse gases have only been in the atmosphere for the last century.
B The Kyoto Protocol is a legal agreement for countries to reduce their output of greenhouse gases.
C Global warming and climate change are the same thing.

3 a Why do farmers use fertilisers?
b Rearrange the following sentences to explain how fertiliser can get into a river and what the effects might be.
- Many fish die.
- Fertiliser causes water plants such as algae to grow rapidly.
- Many plants die.
- Excess fertiliser dissolves in water in the soil.
- Fertiliser runs off fields into rivers.
- Oxygen is used up by decay-causing bacteria.
- The dead plants and algae are broken down by bacteria.
- Algae quickly cover the surface of the lake, stopping light reaching lower plants.

c What is this process called?

4 Billions of plastic shopping bags are given away by supermarkets every year. Most of them end up on landfills.
a Describe how PVOH and EVOH can help reduce the environmental impact of shopping bags.
b Describe two alternatives to sending the shopping bags to landfills. Give the advantages and disadvantages of each.

5 In addition to biodegradability we can also speed up the degradation of plastics in other ways. Explain two other methods. **[HT]**

6 Lichen can be used to detect the level of sulfur dioxide in the atmosphere.

Sulfur dioxide dissolves in rainwater to form acid rain and is released from some factories and power stations.
a Give the name for an organism used to detect levels of pollution.
b Describe how lichen can be used to show how levels of sulfur dioxide in the atmosphere change with distance from a town.

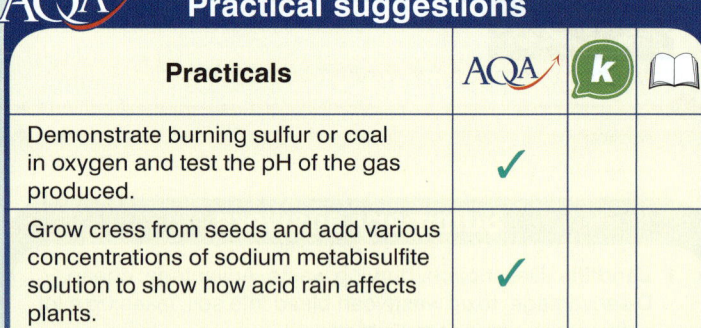

Practical suggestions

Practicals	AQA	k	📖
Demonstrate burning sulfur or coal in oxygen and test the pH of the gas produced.	✓		
Grow cress from seeds and add various concentrations of sodium metabisulfite solution to show how acid rain affects plants.	✓		
Investigate the effects of SO_2 on growth of cress seedlings.	✓		

Kerboodle resources

- Theme map: Improving our environment
- How Science Works: Governments get hot under collar about global warming (14.1)
- Simulation: Indicator species used to monitor water pollution (14.3)
- Bump up your grade: What a load of rubbish! (14.4)
- Examination-style questions
- Answers to examination-style questions

Chapter 14 – Environmental concerns when making and using products

Examination-style questions

Environmental scientists have been looking into the processes that release harmful gases into the atmosphere. Choose chemicals from the box to answer the questions.

| Ammonia | Carbon dioxide | Helium |
| Methane | Nitrous oxide | |

1.
   a. Which **two** gases are released from the combustion of fossil fuels? (2)
   b. Which gas is produced from decomposition of rubbish from landfill sites? (1)
   c. Which gas is produced due to the use of fertilisers? (1)
   d. Give the chemical formula for methane. (1)

2. Greenhouse gases in the atmosphere are causing problems for the planet.
   a. Explain how greenhouse gases are causing global warming. (2)
   b. Suggest **one** problem caused by global warming. (2)

3. Many things can cause pollution. Put the things that cause pollution under the correct heading in the table.

Individual people	Industrial processes	Intensive farming	Natural

- Artificial fertilisers
- Bacteria that break down organic matter in wetlands
- Cars
- Extraction of metals from their ores
- Herbicides
- Litter (6)

4. Plastics made from plants are becoming more widely used. Read the newspaper article below then answer the questions.

Biodegradable plastic – a revolutionary packaging

The applications of plant-based plastics are on the increase. One firm has developed food trays made from tapioca starch obtained from a plant called cassava. This material can withstand temperatures from −40°C to 220°C. It can also be coloured and printed. The advantage is that when this type of plastic is buried after use, or put into a composter, it breaks down naturally into water and carbon dioxide. The disadvantage is that it costs more than plastic made from crude oil. The cost of a plant-based plastic drinks bottle is up to 10% higher than regular plastic bottles.

   a. Why is it an advantage that food trays can withstand temperatures from −40°C to 220°C? (1)
   b. In this question you will be assessed on using good English, organising information clearly and using specialist terms where appropriate.
   Discuss the advantages and disadvantages of using plants to make plastics. (6)
   c. If an oil-based plastic drinks bottle costs 79p, up to how much could a plant-based bottle cost? (4)

5. Farmers want to grow a high yield of crops to increase profit.
   a. Which general product could a farmer put on his fields to replace nutrients so the plants will grow better? (1)
   b. Explain how this product can cause fish to die in rivers and lakes. (3)

Examination-style answers

1.
   a. carbon dioxide, nitrous oxide (answers in either order) (2 marks)
   b. methane (1 mark)
   c. nitrous oxide (1 mark)
   d. CH_4 (1 mark)

2.
   a. Gases absorb more long-wave radiation from the Earth. (1 mark)
   So heat is kept in the atmosphere. (1 mark)
   b. Any sensible problem and explanation, e.g.
   - ice caps melting causes flooding / loss of habitat
   - Increased amount / intensity hurricanes / adverse weather are caused by climate change
   - water shortage causes increased desertification

   (1 mark for the problem and 1 mark for the explanation)

3.
Individual people	Industrial processes	Intensive farming	Natural
Cars Litter	Extraction of metals from their ores	Artificial fertilisers Herbicides	Bacteria that break down organic matter in wetlands

(6 marks = 1 mark for each cause of pollution under the correct heading)

4.
   a. Can be frozen or heated up (1 mark)
   b. Marks awarded for this answer will be determined by the Quality of Written Communication (QWC) as well as the standard of the scientific response.

   There is a clear, balanced and detailed response discussing the advantages and disadvantages of using plants to make plastics. The answer is well structured with minimal repetition or irrelevant points. There is an accurate, fluent and clear expression of ideas with only minor errors in the use of technical terms, spelling, punctuation and grammar. (5–6 marks)

   There is some discussion of the advantages and disadvantages of using plants to make plastics, with some omissions. The answer shows some attempt at structuring and the ideas are expressed with reasonable fluency and clarity. There are some errors in the use of technical terms, spelling, punctuation and grammar. (3–4 marks)

   There is a brief discussion of the advantages and disadvantages of using plants to make plastics. The answer is largely incomplete and may contain some valid points which are not clearly structured. It lacks fluency and/or clarity. It contains errors in the use of technical terms, spelling, punctuation and grammar. (1–2 marks)

   No relevant content (0 marks)

   Examples of points made in the response:
   - Plants are renewable
   - Plant-made plastics are biodegradable
   - Need a big area of land to grow plants on
   - May use fertilisers, herbicides and pesticides on plants
   - These can cause eutrophication
   - Non-biodegradable plastics are cheaper
   - Non-biodegradable plastics use up land in landfill
   - The breakdown of non-biodegradable plastics can produce toxic materials
   - These toxic materials can leak into the environment.

   c. 10% of 79 OR 79 ÷ 10
   = 7.9
   79 + 7.9
   = 86.9 OR 87 pence (4 marks)

5.
   a. fertiliser (1 mark)
   b. Leach into rivers and lakes.
   Causes algae to over grow.
   Takes all the oxygen out of the water. (3 marks)

Unit 3, Theme 3 – Improving our environment

15.1 Conduction

Learning objectives

Students should learn:
- what is meant by heat conduction
- how to increase or reduce the rate of heat conduction
- how to control the rate of heat conduction in the home.

Learning outcomes

Most students should be able to:
- state that conduction takes place in solids
- give examples of good heat conductors and bad heat conductors (good insulators)
- describe ways unwanted heat transfers by conduction are reduced in the home.

Some students should also be able to:
- explain how conduction takes place using the particle model
- explain how unwanted heat transfers by conduction are reduced in the home.

Support

- Students find an explanation of conduction using particles hard to understand. Concentrate on ensuring students can identify materials that are good or poor conductors and how they are used to increase or decrease heat transfers.

Extend

- Explain how the structure of the material affects how well heat is transferred. Good thermal conductors like metals contain a lattice of particles. Some electrons are not strongly bonded to any particular particle. These 'free' electrons move rapidly through the conductor when it is heated transferring the thermal energy rapidly.
- Ask students to explain why gases are such poor heat conductors using the particle theory (the molecules do not interact with each other to any great extent).

AQA Specification link-up Science B 3.5.3.2

- Describe how heat is transferred by conduction … in the home.
- Describe ways of minimising heat loss in the home (e.g. insulation, double glazing …).

Lesson structure

Starters

Comparing conduction – Hand round samples of different materials, e.g. metal, wood and plastic spoons. Students decide which feels coolest. The objects are at room temperature so heat travels from your hand to each object, as your hand is hotter. The metal spoon conducts the heat quickest so it feels coldest. *(5 minutes)*

Atomic arrangements – Students review the different arrangements of solids, liquids and gases. Extend students by asking them to sketch the arrangement, stating if particles can change places or move (discuss vibrations in fixed places). They can modify their diagram to show what happens as a solid warms up. To support students, only ask them to sketch the correct diagrams, identifying differences and similarities. *(10 minutes)*

Main

- Show an animation of what happens at a molecular level. Particles (atoms or molecules) vibrate in fixed positions passing energy to their neighbours.
- Particles vibrate more when hotter, so their separation increases and the solid expands. (In simplistic terms, the particles do not expand, but they effectively occupy more space because the size of their vibrations is larger.)
- Solids, liquids and gases all conduct heat. Solids are better conductors because the particles are closer together so the energy transfer is more efficient.
- The amount of conduction depends on the material and its shape.
- Demonstrate conduction by placing rods made from different materials (e.g. different metals, glass, wood and plastic) in beakers of hot water and measuring how quickly heat travels along the rod. Attach drawing pins to the rod using Vaseline or wax and see which ones fall off first.
- Wrap containers of hot liquid in different materials to see which combination reduces the rate of heat loss. You can investigate using several layers of the same material or different types of material. These practical suggestions are both good opportunities to practise a range of controlled assessment skills.

Plenaries

Simulating conduction – Stand students in a line, so they look at the back of the student in front. The student at the back of the line touches the student in front on the shoulder, then starts to walk on the spot. As soon as one student feels a touch, they touch the person in front then walk on the spot. Each walking student represents a vibrating particle. Touching shoulders represents energy passing from particle to particle. This is a fairly slow method of heat transfer. If appropriate, explain that good thermal conductors have electrons free to move within the lattice of particles. These move fast so the energy travels much quicker. Represent this energy transfer using paper balls thrown from the back of the line towards the front. *(5 minutes)*

Conduction at home – Students identify materials that are good or bad heat conductors. Extend students by asking them to list examples in the room (or their homes) explaining how they decrease heat transfer by conduction (e.g. double glazing, curtains, carpets). If there is time, the students rank these uses in order of importance. Then the students can list other types of insulation, including sketching their macro-structure with 'air pockets', e.g. loft insulation, hot water cylinder jackets, etc. Support students by suggesting places in a typical room where heat losses may occur and ask them to suggest ways that the conduction can be reduced (e.g. by adding carpets, curtains). *(10 minutes)*

Chapter 15 – Our environment at home

Further teaching suggestions

Clothing and heat loss
Investigate the materials used to make winter clothing or bedding. Identify features that reduce conduction.

Did you know …?
Layers of dead skin covering the soles of our feet are poor heat conductors, so walking quickly over hot coals means heat does not reach the nerves. People who exfoliate their skin beforehand can suffer burns!

Science in context
In recent years, the materials used to make buildings have been chosen to reduce or increase heat transfer by conduction. New building regulations now include strict controls on the conductivity of materials used, to reduce the energy losses.

Unit 3, Theme 3 – Improving our environment

15.1 Conduction

Learning objectives
- What is conduction?
- How can we change the amount of conduction?
- Where do we make use of conduction in the home?

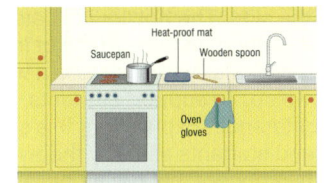

Figure 1 Conductors of heat energy help us cook things evenly. Insulators of heat energy stop hot things causing damage.

Heat always travels from hotter places to cooler places. In solids, heat is transferred by **conduction**. The heat is transferred more quickly if:
- there is a large temperature difference
- the object has a large cross-section
- the object is made from a good heat **conductor**.

a Write down **three** things that affect how quickly heat is transferred.
b A hot cup of tea and a glass of tap water are left at room temperature. Explain which one loses or gains heat more quickly.

We cook food at high temperatures and store it at low temperatures. Fridges and freezers are lined with insulating material. This slows down heat transfers from the room into the fridge or freezer. **Insulators** are materials such as cloth, wood or plastic, which do not conduct heat well.

Oven gloves are made from padded cloth. We use them to take hot food out of the oven. The cloth, with air trapped between its fibres, is an insulator that slows down the heat transfer from the hot dish to our hand.

Many saucepans are made from metal. Metals are good heat conductors. Heat spreads quickly and evenly through a metal saucepan when it is on a hot hob. This means the food heats up quickly.

c Explain which of these are good heat conductors: copper, plastic, aluminium, wood, carpet, iron.
d Why does wrapping fish and chips in newspaper help to keep them hot?

In a solid, particles vibrate in fixed positions. Particles vibrate more in the part of a solid that is heated. The bigger vibrations pass energy onto neighbouring particles which start to vibrate more. This way, energy passes through the solid, warming up parts further away from the heat source.

Materials like metals are good heat conductors. Their particles are in fixed positions, but some electrons are free to move in the solid. When the metal is heated, the **free electrons** spread the heat quickly through the material.

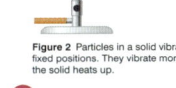

Figure 2 Particles in a solid vibrate in fixed positions. They vibrate more when the solid heats up.

Most non-metals are insulators. They do not have free electrons so conduction is slow. Materials with trapped air pockets make good insulators. Examples of these include breeze blocks, double glazing, duvets and padded clothing. Particles in a gas do not pass energy on to each other easily.

e Drawing our curtains helps to keep the home warmer. Explain why.

Activity
Modelling conduction
To visualise why conduction happens best in a solid, line up in a row with a few of your friends and pretend you are all particles. Link elbows and, starting at one end (the hot end) one person starts moving. Soon you will find that everyone in the row cannot help but move. Next, hold hands or wrists to show particles in a liquid. When the person at the hot end moves now, their movement is not transferred as readily.

- How could you pretend to be particles in a gas? And, how will the movement be transferred this time?

Most heat is wasted in our homes through the walls, windows, floors and roofs as these have large surface areas. We can reduce these heat transfers in many ways.
- Many homes are built using breeze blocks because trapped air bubbles in the concrete reduce conduction.
- Many homes have cavity walls. There is an inner wall and an outer wall separated by a few centimetres. The trapped layer of air between the walls reduces conduction. Some people fill this with foam for extra insulation.
- Double glazed windows have two layers of glass with air trapped between them to reduce conduction.
- Carpets and rugs reduce conduction through floors.
- Loft insulation is a thick layer of padding made from fibreglass strands. It is rolled between the roof joists and reduces conduction through the ceiling.

f Why are heat losses from the home bigger in winter?
g Explain why carpets are useful for reducing heat losses in the home.

Summary questions
1 Copy and complete using the words below:
  conductor electrons insulator nucleus vibrate
  If part of an object is heated, its nuclei will .... more, passing energy to a neighbouring ....
  Conduction is quicker in metals because some .... are free to move.
  Copper is a good heat ....
  Polystyrene is a good ....
2 Explain how conduction is reduced in these cases:
  a A child wears several layers of clothes on a cold day.
  b A house with a thatched roof stays warm in winter and cool in summer.
3 Explain why metal is used to increase heat transfers in these cases:
  a Metal pipes carry cooling fluid through a fridge.
  b A baking tray is made from metal.

Figure 3 Air bubbles in breeze blocks help to reduce conduction

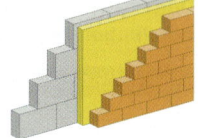

Figure 4 Cavity walls in modern homes help to insulate them

Did you know …?
People can walk safely over hot coals. The skin on our feet and the ash on top of the coal conduct heat very slowly. This allows people to walk quickly over the hot coals that are at about 500 °C.

Key points
- Conduction is when heat is transferred between neighbouring particles.
- Metals are good heat conductors.
- Non-metals and gases are poor heat conductors, or good insulators.

Summary answers
1 vibrate, nucleus, electrons, conductor, insulator
2 **a** The clothes are insulating and each extra layer traps an extra layer of air which is also insulating.
  b Thatched roofs are thick and made from straw which is an insulator. Air is trapped in pockets between the layers of straw.
3 **a** The pipes are metal to conduct heat quickly from inside of the fridge to the cooling fluid.
  b The tray conducts heat from the hot oven so the food cooks evenly.

Answers to in-text questions
a temperature difference, cross-section, material
b Hot tea loses heat quickest – large temperature difference.
c Copper, aluminium, iron – all metals
d Newspaper is a good insulator.
e It traps a layer of air between the curtain and window, which is insulating. The curtain is cloth which is a good insulator.
f The temperature difference between the home and outside is greater.
g Floors have a large surface area so the heat loss would be great. The carpet has trapped air pockets, and is made from an insulator.

Unit 3, Theme 3 – Improving our environment

15.2 Convection

Learning objectives

Students should learn:
- what is meant by convection
- how to increase or reduce the rate of convection
- how to control the rate of convection in the home.

Learning outcomes

Most students should be able to:
- state that convection takes place in liquids and gases (fluids)
- explain how unwanted heat transfers by convection are reduced in the home
- state how changes in density set up a convection current.

Some students should also be able to:
- explain in detail how changes in density affect convection and set up a convection current.

AQA Specification link-up Science B 3.5.3.2
- Describe how heat is transferred by … convection … in the home.

Lesson structure

Starters

Trick question – Ask: what weighs more – a kilogram of lead or a kilogram of feathers? In fact both weigh the same but lead is denser than feathers. This confuses many students. Explain that an understanding of density will be needed to understand convection. *(5 minutes)*

Hot water floats – Demonstrate hot water is less dense than cooler water. Half fill a 1 litre beaker with cold water. Colour about 400 ml of very hot water with a small amount of potassium permanganate. Very carefully tilt the beaker of cold water and pour the hot water into the large beaker down the side. The hot liquid floats on top of the cold water. (Practise this first. A piece of paper resting on the surface of the cold liquid may help). It does work, and can be very effective. Make sure there is a white background so the difference in colour of liquids shows up. To support students, invite them to touch the beaker and feel the temperature difference. Explain that this shows that warm liquids float on top of cool liquids. To extend students ask why the two liquids do not mix. Students should link this idea to the concept that hot air balloons float. *(10 minutes)*

Main

- Contextualise the concept of convection, e.g. considering the 'Science in context' box opposite.
- Show an animation of convection taking place. At a molecular level, warmer particles (atoms or molecules) vibrate more when hotter, so their separation increases and the volume of warm fluid (gas or liquid) that they take up expands. Clear up the misconception that the particles do not expand.
- Warmer fluid becomes less dense than cooler fluid (density = mass/volume) so it rises. This happens in liquids and gases.
- Convection is when particles carry thermal energy from a hotter place to a cooler place as they move.
- Convection currents are a special case of convection. Convection is quick and efficient when circulating currents are set up if a fluid is heated from the bottom or cooled from the top.
- Set up a circus of experiments to demonstrate convection:
  – Suspend a spiral of foil above a candle flame. It spins as the hot gases rise.
  – Place a crystal of potassium manganate(VII) in two beakers containing cold water. Heat one beaker gently from the bottom. The purple dye spreads much more quickly in the beaker heated from the base as a convection current is set up. CLEAPSS Hazcard 81 Potassium manganate(VII) – oxidising and harmful.
  – Place an ice cube containing food dye in a beaker of cold water. The cold coloured melt water sinks to the bottom of the beaker as a convection current is set up.
  – Explain these effects using the concept of density.
- Investigate the design of houses identifying features that reduce/increase convection, e.g. chimneys increase convection; draught excluders reduce it. Students identify features in their own homes for homework.

Plenaries

Convection or conduction – Students list similarities and differences between the two methods of heat transfer. To support students, give them a selection of ideas to compare, e.g. particles are involved; it takes place in solids; particles change places; quicker method of heat transfer, etc. *(5 minutes)*

Simulating convection – One part of the room represents a source of heat – provide small pieces of paper to represent energy. The furthest part of the room from the source of heat represents the coolest place. Students near the source of heat pick up paper (gaining energy), carrying it to the coolest part of the room. They deposit the paper (losing energy), and move back to the source of heat to collect more. This model has limitations but shows how circulating particles transfer energy through convection currents. Students should be able to identify differences between this model and the model of conduction. *(10 minutes)*

Support
- Students find an explanation of convection using particles hard to understand. Ensure students can identify situations where convection takes place. Coloured labelled diagrams help to reinforce teaching on convection currents.

Extend
- Discuss convection in terms of particle arrangements. Ask students to explain why it cannot take place in solids.

Science in context

Many features of buildings and appliances are designed to reduce or increase heat transfers by convection, e.g. new building regulations are designed to ensure modern homes do not have draughts in windows or doors.

Unit 3, Theme 3 – Improving our environment

15.2 Convection

Learning objectives
- What is convection?
- How can we change the rate of convection?
- Where do we make use of convection in the home?

  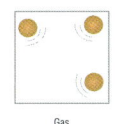

Solid — Liquid — Gas

Figure 1 Particles are in fixed positions in solids, but can change places in liquids and gases

Liquids and gases are both **fluids**. Heat is transferred by **convection** in fluids. The particles in the fluid move, carrying energy with them.

When a fluid is warm, it is less **dense** than when it is cool. When a fluid is heated, particles gain energy and vibrate more, moving further apart and so the fluid takes up more space. The warmer fluid expands and becomes less dense. Warm water takes up more space than the same mass of cool water. This is why warm water rises and the cooler, denser water sinks. The same is true for gases such as air.

a What is meant by a fluid?

b How can you tell that water is more dense than cooking oil?

Figure 3 A wind is felt when cool air is sucked into a bonfire to replace the hot smoke

Figure 2 Hot air in this balloon is less dense than the surrounding air, so it floats

Heat is lost by convection through gaps in badly fitting window frames and doors.

Hot smoke is less dense than surrounding air so it rises up a chimney. Heat is lost up the chimney because of convection.

Fan ovens use fans to move hot air inside an oven so the food cooks evenly. Without a fan, food at the top of the oven cooks quicker because convection means the hot air rises.

The burners in a gas oven are at the base, so convection helps the heat to spread more evenly.

c Explain why hot air balloons rise.

d Where is the coolest place in a fridge?

Heat spreads quickly if a liquid or gas is heated from the base, or cooled from the top. This sets up convection currents.

The heating element is fitted at the base of a kettle. When the kettle is on, it heats up the water next to it. The warm water rises and cool water sinks to take its place. The heat circulates and spreads quickly through the water.

Some fridges have an ice box at the top of the fridge. Cooler air near the ice box sinks and warmer air rises to take its place, transferring heat away from the fridge.

e Where should you heat something if you want to set up a convection current?

f Explain how convection currents help an ice cube to cool a drink quickly.

We can reduce unwanted convection by blocking up gaps or creating small pockets of air.

People often block up old fireplaces or install draught excluders. These are strips of foam that block up gaps in window and door frames. Closing doors and windows also stops heat losses by convection.

Because of convection, warm air rises, so a lot of heat is lost through the roof of a house. Loft insulation stops the warm air reaching the loft and escaping through the roof.

g Write down five ways to reduce convection in the home.

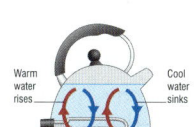

Figure 4 Convection currents spread heat evenly through the water

Did you know...?
Molten rock inside the Earth is constantly moving due to convection currents. These currents move tectonic plates, causing earthquakes and volcanoes.

links
For information of plate tectonics look back at 1.5 Changes in the Earth's surface.

Summary questions

1 Copy and complete using the words below:
convection fluid rise sink density
Convection is when heat is transferred through a ....
The particles move, carrying energy when .... takes place.
The .... of a fluid changes if its temperature changes.
Hot fluids will .... and cool fluids will .... if a convection current is set up.

2 Explain how you could reduce convection in these cases:
a Keeping a drink hot.
b Stopping convection through a letter box.

3 Explain how convection currents help a radiator heat a room.

Key points
- Convection occurs when particles move, carrying heat with them.
- Convection takes place in liquids and gases.
- Convection currents spread heat in fluids when they are heated from the base or cooled from the top.

Summary answers

1 fluid, convection, density, rise, sink

2 **a** Use a lid, use a thermos mug.
  b Cover the letter box.

3 Air above the radiator gets warm and rises; cooler air sinks to take its place.

Answers to in-text questions

a Fluids are liquids or gases.
b Oil will float on top of water.
c The hot air inside them is less dense than the surrounding air.
d at the bottom
e at the bottom
f The cooler water from the ice sinks and warmer liquid rises to be cooled by the ice cube.
g Block up gaps or old fireplaces or install draught excluders; close doors especially those leading to stairwells and windows; install loft insulation.

Unit 3, Theme 3 – Improving our environment

15.3 Radiation

Learning objectives

Students should learn:
- what is meant by radiation
- how to increase or reduce the rate of radiation
- how to control the rate of radiation in the home.

Learning outcomes

Most students should be able to:
- state that heat is radiated from the surfaces of hot objects
- describe how to control heat transfers by radiation in the home
- explain how changes in colour affect the amount of radiation.

Some students should also be able to:
- explain in detail ways in which heat transfers by radiation can be controlled in the home.

AQA Specification link-up Science B 3.5.3.2
- Describe how heat is transferred by … radiation in the home.

Lesson structure

Starters

Electromagnetic spectrum – Students list the members of the electromagnetic spectrum. Explain that thermal radiation is infrared radiation. *(5 minutes)*

Radiation at home – Students list devices at home that radiate energy (e.g. radio radiates sound, lamp radiates light and cooker radiates heat). Students use these examples to give their definition of radiation (e.g. an object gives out energy but no part of the object moves, etc.). Support students by giving them a list of objects (radio, hot cup of tea, etc.) and a list of different types of energy and ask them to say what type of energy is radiated by each object. Use this to explain the concept of radiation. *(10 minutes)*

Main

- Many students think black surfaces attract heat. This is wrong. Black surfaces absorb heat well – but they do not *attract* it.
- Objects at a steady temperature absorb and emit the same amounts of radiation. Objects warming up absorb more heat than they emit, etc.
- Black objects absorb and emit more radiation in a given time than white objects at the same temperature. (See 'Science in context' box opposite.)
- Set up a circus of experiments to demonstrate thermal radiation. These are a simple selection of ideas:
  - Use wax to attach drawing pins to two sheets of thin metal (e.g. foil pie dishes). Paint the back of one sheet black and leave the other sheet shiny. Place each sheet about 10 cm from a Bunsen burner flame with the drawing pin side facing away from the flame. The drawing pin falls off the blackened sheet first as this sheet absorbs the heat more efficiently.
  - Pour boiling water into two identical containers, one painted black on the outside and one that is shiny. Place thermometers in each beaker to investigate which emits more radiation and cools quickest.
  - A Leslie cube is a metal cube filled with boiling water. The different faces are different colours. Use a temperature probe about 2–5 cm away to compare the heat radiated by the different coloured surfaces.
  - Have beakers of water at different temperatures. Students rank these without touching them (by holding the back of their hand near the surface of the beaker, they can compare the amount of radiation emitted). Hot objects emit heat at a faster rate than cooler objects.
- Combine the ideas from 15.1, 15.2 and 15.3. Set students the task of designing a container to keep a drink as hot as possible, or an ice cube from melting for as long as possible. Students should aim to reduce conduction, convection and radiation.

Plenaries

'Splat' – Write a list of about 12 key words on the board (e.g. conduction, convection, radiation, solid, liquid, gas, black, white, insulator, etc.). Two students stand by the board and another student defines one of these words. The student who 'splats' (puts their hand over) the correct word stays at the board and the questioner takes the place of the other student. Someone else in the class asks the next question. *(5 minutes)*

Different heat transfers – Students write down differences and similarities between radiation and conduction and/or convection. To support students, give them key words or phrases to classify as similarities or differences. *(10 minutes)*

Support
- Pieces of foil and black paper are useful visual aids. Remind students that white, shiny surfaces reflect heat – this helps some students to remember that white, shiny surfaces do not absorb heat well.

Extend
- Remind students that no particles are involved in radiation – just waves of energy. Also explain that radiation happens at the surface of the object.

Science in context

The insides of fridges are normally white – this reduces heat radiation. The insides of ovens are normally black – this increases the amount of radiation emitted.

Unit 3, Theme 3 – Improving our environment

15.3 Radiation

Learning objectives
- What is radiation?
- How can we change the amount of radiation absorbed by a surface?
- Where do we make use of radiation in the home?

Many buildings in hot countries are painted white. This reduces heat transfers into the house by **radiation**, keeping the building cool.

Radiation that heats things up is also called **infrared radiation**. This radiation forms part of the electromagnetic spectrum. The energy is transferred in waves from the surface of a hot object. The waves travel at the speed of light.

Hotter objects emit (give out) more radiation in a given time from their surface than cooler objects. When radiation is emitted, objects cool down. A hot drink emits more radiation faster than a lukewarm drink, cooling down more quickly.

Radiation is absorbed by objects that are cooler than their surroundings. The objects heat up. This is why an ice cube warms up and melts if it is left at room temperature.

Figure 1 Painting your house white can keep you cool in sunny weather

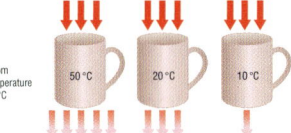

Figure 2 Objects emit or absorb radiation depending on their temperature relative to their surroundings

a What is another name for radiation that heats up objects?
b Explain which of these emits more radiation? A hot drink in a white mug or a cold drink in a white mug?

Black matt surfaces are good at absorbing radiation. Solar panels on house roofs are coloured black to absorb the Sun's radiation effectively. White shiny surfaces reflect radiation and do not absorb radiation well.

All objects emit radiation if they are hotter than their surroundings. Black matt surfaces emit radiation faster than white shiny surfaces. A hot drink in a black mug emits radiation and cools down more quickly than the same drink in a white mug.

c Which colour is best at absorbing and emitting radiation?
d Explain why a refrigerated lorry painted white is more efficient to run than a dark-coloured lorry.

Some things are coloured black to increase heat transfers by radiation.
- Ovens cook food at high temperatures, so they must emit radiation easily. This is why they are black inside.
- Semiconductor chips in computers get extremely hot. They are attached to heat sinks which are painted black to increase the rate of heat loss by radiation and help keep them cool.

Figure 3 Black solar panels absorb the Sun's energy efficiently

Some things are white or shiny to reduce heat transfers by radiation.
- If food is cooking too quickly, it is often covered with a layer of foil. This reflects some radiation and reduces the speed it cooks at.
- Some people fit sheets of foil behind their radiators. This reflects radiation back into the room, stopping it from passing through the wall.
- Survival blankets are shiny to reflect back the body's heat from a casualty.

e Why are the inside of fridges white?
f The insides of many saucepans are black. How does this help cook food?

Figure 5 Survival blankets save lives by stopping casualties from losing heat to their surroundings

Figure 4 Black cooling fans inside equipment increase heat losses by radiation

Activity

Modelling energy transfer

Line up with a few of your friends again. This time you need several balls (or pieces of paper) to represent energy. For conduction, one ball is passed down the line. For convection, each person picks up a ball and then moves to the back of the line to show convection current. For radiation, one person shines a torch at the person at the end of the row, missing out all the people (particles) in-between.

Summary questions

1 Copy and complete using the words below:
  black infrared radiation surface white
  Radiation is the transfer of energy by .... rays.
  It happens at the .... of objects.
  Hot objects emit more .... than cool objects.
  .... objects emit more radiation than .... objects.
2 Why does a black car heat up quicker than a white car on a hot day?
3 Electric heaters have a curved shiny backing behind the heating element. Explain how this helps direct more heat into the room.
4 Thermos flasks reduce heat transfers by conduction, convection and radiation. Find a diagram of a thermos flask and label it to show how these heat transfers are prevented.

Key points
- Radiation is emitted when infrared waves transfer heat from the surface of objects.
- Hot objects radiate energy more quickly than cooler objects.
- Black matt objects radiate more energy in a given time than white shiny objects.

Summary answers

1 infrared, surface, radiation, black, white
2 It absorbs more radiation.
3 The shiny backing reflects heat forwards so more heat is directed out of the front of the heater.
4 The diagram should show:
  an inner flask with silvered sides to reduce radiation
  vacuum between the walls of the inner flask to reduce conduction
  inner flask resting on plastic supports to reduce conduction
  lid to reduce convection; made of plastic to reduce conduction.

Answers to in-text questions

a infrared
b The hot drink as hot objects emit more radiation than cool objects.
c black
d Heat transfers by radiation are reduced.
e To reduce radiation.
f Black is a good emitter of radiation.

Unit 3, Theme 3 – Improving our environment

15.4 Will you save money?

Learning objectives

Students should learn:
- how to reduce energy losses in the home
- how to calculate payback time
- the difference between efficiency and cost effectiveness.

Learning outcomes

Most students should be able to:
- explain different methods of reducing heat losses in the home
- calculate payback time for different measures
- state what is meant by efficiency, cost effectiveness and payback time
- explain the difference between efficiency and cost effectiveness.

Some students should also be able to:
- use payback time to assess a suitable energy-saving measure for different situations.

Support
- Payback calculations are tricky unless the answer is in whole years. Students should practise converting fractions of years into months.

Extend
- Students should consider wider issues, e.g. which methods used to heat a home are most efficient and other advantages or disadvantages; should inefficient equipment be banned or is it more wasteful to discard functioning but less efficient equipment? How realistic are the estimates on Energy Performance Certificates? Should all homes be equally efficient?

AQA Specification link-up Science B 3.5.3.2
- Describe ways of minimising heat loss in the home (e.g. insulation, double glazing, hot water tank jackets, thermostatic controls, draught excluders).
- Explain the term 'payback time' in relation to installing energy-saving measures.
- Explain the difference between efficiency and cost-effectiveness.

Within this context, candidates should be able to use scientific data and evidence to discuss, evaluate or suggest implications of the following:
- the efficiency and cost-effectiveness of methods used to reduce domestic energy consumption.

Lesson structure

Starters

Review heat transfers – Students identify key features of conduction, convection and radiation. Students also identify materials used in buildings that are insulators or heat conductors. *(5 minutes)*

How is my home insulated? – Students brainstorm different heat saving measures installed in their homes. To extend students, ask them to list the reasons why some homes use different measures, e.g. old buildings do not have cavity walls. They may also think of cost, ease of installation and different features of properties. To support students, help them to identify which of these measures are found in most homes and which are less common by counting how many students have each measure installed. *(10 minutes)*

Main

- A walk around school identifying energy-saving measures can be useful (or set this as a task for homework in the previous lesson).
- Identify features built into the design of the property and which measures were added later.
- Students will know some rooms are warmer than others – ask them to identify the features that affect this.
- Use the internet to see how several of the energy-saving websites put this theory into practise.
- Provide information about the costs and projected savings of measures such as installing insulation and double glazing. This can come from leaflets or websites. Students do payback calculations and discuss their findings. Ensure students consider situations when a longer payback may be worthwhile.
- Discuss whether a householder should install a new boiler. New models have improved technology that is much more efficient than older boilers. However, their initial cost means that in some situations they are not cost effective. Information comparing the efficiency and cost of boilers is widely available. Include the use of thermostatic controls in a house. How many thermostats in a three-bedroom house? Where should they be located? Would they be good value for emoney? Consider payback time.
- If time allows, students could prepare a poster or leaflet either explaining one aspect of saving energy in the home, or to inform people of the ways that heat can be lost.

Plenaries

Save heat – Students think of a slogan to encourage householders either to install extra energy-saving measures, or to use their homes more efficiently. *(5 minutes)*

Mind map – Students produce a mind map with three key areas: energy-saving measures; payback time and its calculation; cost effectiveness and efficiency. If there is time, they should link in the ideas of types of heat transfer. To support students, provide them with key words and a layout to complete with information they have learnt in the lesson. *(10 minutes)*

Chapter 15 – Our environment at home

Further teaching suggestions

Energy-saving in the home

Students research current costs of installing different energy-saving measures for different buildings using secondary sources. They could also look at varying providers and find the actual cost of fully insulating an older house.

Did you know …?

Building regulations mean that houses being built now have to achieve higher standards of insulation than ever before.

Science in context

All homes are heated and insulated in some way. Students may be aware of the increasing costs of fuel and may have heard the term 'fuel poverty'.

Unit 3, Theme 3 – Improving our environment

15.4 Will you save money?

Learning objectives
- How can heat losses be reduced in our homes?
- What is payback time?
- What is the difference between efficiency and cost-effectiveness?

An energy performance certificate is needed if you are selling a house. This certificate shows how well energy is saved at the moment, and suggests improvements.

We have already seen some of these improvements that reduce heat losses, such as:

- draught excluders fitted round gaps in doors and windows
- hot water tank jackets to insulate the hot water tank so water stays hot for hours
- loft insulation to reduce heat losses through the roof
- double glazing to reduce heat losses through windows
- cavity wall insulation to reduce heat losses through walls. This is a layer of foam which fills the gap between a house's cavity walls.

Having effective thermostatic controls on the heating system is also important.

Figure 1 The amount of heat lost from different houses depends on many factors

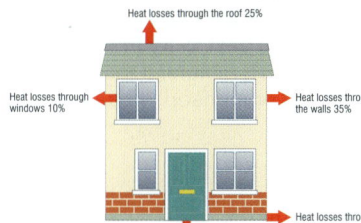

Figure 2 This shows the amount of heat lost from different parts of a house

a Explain why energy performance certificates produced at different times for the same house could be different.

Identical houses in the same street have different heating bills as different energy-saving measures will be installed. The same house has different heating bills if the way the equipment is used changes. If people leave the heating on for longer, or set it at a higher temperature, their bills will be higher. Thermostats turn the heating on or off automatically to keep rooms at a constant temperature. They can be fitted to radiators so the temperature in each room can be controlled.

Payback time

Energy-saving measures cost money to install. The time taken for the savings you make to cover the cost of installation is called the **payback time**. It is usually measured in years.

Figure 3 Insulating a hot water tank saves money

payback time = cost of installation ÷ annual savings

The table shows the installation cost and annual savings for several different measures.

Energy-saving method	Cost of installation (£)	Annual saving (£)
Draught proofing	100	50
Hot water tank lagging	12	36
Double glazing	4500	135
Loft insulation	250	150
Cavity wall insulation	500	115

Using this information, the payback time for lagging a hot water tank is: 12 ÷ 36 = 0.33 years (or 4 months).

b Calculate the payback time for each of the measures in the table.

Efficiency versus cost-effectiveness

Some energy-saving measures give small annual savings. If the measure is cheap and easy to carry out, it can still be cost-effective. This means that your savings are greater than the costs. A person renting a flat for a couple of years would save money if they used energy-saving light bulbs, fitted lagging over the hot water tank and used draught proofing.

Some measures would only be cost-effective if you plan to live in a house for many years. Double glazing and cavity wall insulation save a lot of money each year, but the initial cost is high. However, double glazing has other benefits: it can improve the house's appearance and cut down on noise.

Some changes reduce heating bills by improving efficiency. A boiler heats the water used in a central heating system and provides you with hot water. The efficiency of a boiler can vary between 55 and 90 per cent. Using a more efficient boiler will save heating costs by about £150 per year. However, new boilers may only last 8–10 years. If the payback time is longer than that, or the owner plans to move, this may not be cost-effective.

c Explain why using energy-efficient light bulbs is usually cost-effective and efficient.

Summary questions

1 Copy and complete using the words below:
installing loft insulation savings using a thermostat

Payback time is the time taken for the .... to match the installation costs.

.... is one way to reduce heat losses from a building.

.... is a way to use central heating more efficiently.

2 Which two energy-saving measures should be installed in these situations?
a A student renting a flat for a year.
b A couple living in a flat for two years.
c A family living in a house for 15 years.

3 Write a paragraph explaining several ways that heating bills can be reduced without spending money.

Figure 4 Some boilers are more efficient than others

AQA Examiner's tip

There has to be a certain amount of maths in each exam paper, so payback time is one topic that you may find comes up regularly. Show all your working and do not forget the units.

Did you know …?

Grants from local councils help householders pay to install loft insulation. This makes it more cost-effective.

Key points
- Payback time = cost of installation ÷ annual savings.
- Efficient equipment is not always cost-effective to install.
- Most methods to reduce heat losses in buildings increase the insulation installed.

Summary answers

1 savings, insulation, using a thermostat

2 **a** E.g. the student could provide energy efficient light bulbs, draught proofing, hot water lagging as some of these items could be taken with them when they leave.

 b E.g. energy-efficient light bulbs, draught proofing, hot water lagging (N.B. no loft if they live in a flat)

 c E.g. as before but also loft insulation, thermostats, possibly cavity wall insulation. Double glazing is rarely cost effective unless a long-term view is taken.

3 e.g. Turning heating down using thermostat/controls on boiler; having heating on for less time using a timer switch; keeping doors and windows closed and curtains drawn when it is cold.

Answers to in-text questions

a Different energy-saving measures could be installed.

b 2 years, 4 months, 33 years and 4 months, 1 year and 8 months, 4 years and 4 months

c Energy-efficient bulbs use less electricity to produce the same amount of light. They are cheap so their payback time in electricity savings is a couple of years. Some users may take the bulbs if they move home, replacing them with the original less efficient bulbs.

Unit 3, Theme 3 – Improving our environment

15.5 U-values

Learning objectives

Students should learn:
- what is meant by a U-value
- the impact of different U-values on a building.

Learning outcomes

Most students should be able to:
- state what is meant by a U-value
- compare heat losses through materials with different U-values
- explain how materials with different U-values are used in the home.

Some students should also be able to:
- evaluate the implications of using materials with different U-values.

AQA Specification link-up Science B 3.5.3.2

- Know that the U-value is the measure of the rate of heat loss through a material.

Within this context, candidates should be able to use scientific data and evidence to discuss, evaluate or suggest implications of the following:
- the efficiency and cost-effectiveness of methods used to reduce domestic energy consumption
- U-values of different types of material.

Lesson structure

Starters

Where is most heat lost? – Show thermograms for different buildings. Students point out where most heat is lost. To support students, ask them to list the places and identify the different materials involved. To extend students, ask them to give reasons why the heat loss varies in these places and link this heat loss to thickness and type of materials used. *(5 minutes)*

Comparing materials – Place temperature-sensitive strips on a selection of beakers made from different materials, e.g. glass, polystyrene, metal and a beaker wrapped in bubble wrap. Remind students that the strips are all the same colour because the materials are all the same temperature. Fill the cups with hot water (e.g. 40°C). The strips take different times to change colour, but all become the same colour eventually. Explain what a U-value is and ask students to rank the cups in order of U-value, with the slowest change matching the lowest U-value. It is best to try this experiment first, as you want the colour changes to spread over about a minute. If the colour change is too quick, use cooler water. Before you carry out the comparison, students should predict the rank order. *(10 minutes)*

Main

- Introduce the idea of U-values, linking this to the type of materials used. Explain that the U-value is per unit area and the U-value will vary if different thicknesses of materials are used. This is why government guidance for loft insulation has recommended greater thicknesses than in the past.
- If possible, hand round a selection of different materials for students to handle, or add labels around the room showing the typical U-value of each material. You can explain to more able students that the U-value is linked to a property of materials called thermal conductivity, as well as its thickness.
- Give students access to websites aimed at builders and developers to investigate more about the practical aspects of U-values, for example the implications of different U-values for the homeowner and how materials with different U-values combine if used adjacent to each other. U-value calculators are also included, which can be interesting for higher attaining students.
- Students prepare a short PowerPoint presentation explaining what a U-value is, giving some typical values, and linking this to the idea of energy efficient homes.
- As an alternative, Students prepare an annotated diagram showing a thermogram of a building and labelling the different materials with typical U-values. Additional information includes an indication of the impact each part of the building has on the overall heat loss (e.g. walls may have a low U-value, but a large surface area).

Plenaries

Statements – In pairs, students make statements comparing energy transfers and the U-values, e.g. the wall has the lowest U-value so heat travels slower through the wall compared to the window. They correct each other if misleading statements are made. To support students, ask them to say if statements given to them by you are true or false. *(5 minutes)*

Presentation – Students show their presentations on U-values. *(10 minutes)*

Support

- Students can handle different material samples, preferably with each U-value already marked on it. Samples should have the same cross-sectional area if possible, but can be different thicknesses. Use different samples to explain how under the same conditions, heat travels more quickly if the U-value is higher. Students should be able to rank the samples by heat loss, or choose the best sample for specific situations, e.g. which sample lets least heat through in a given time?

Extend

- Introduce the equation:
  heat loss = U-value × area × temperature difference
  Use this to compare the heat losses in different shaped buildings, and the impact of different weather conditions.

Further teaching suggestions

Modelling heat transfers at home

You may already have a small model of a house, but otherwise in small groups students can build a model of a house using a cardboard box (e.g. a small shoe box) using different materials to insulate each wall. Attach temperature-sensitive strips in different positions on each wall, or use a temperature probe. Place a source of heat inside the house and monitor which strips change colour first. The source of heat could be a filament bulb, or a large beaker of hot water. Candles are not suitable due to the risk of fire. Students should consider other factors that may affect their conclusions (e.g. the position of the source of heat, the impact of convection, etc.).

Science in context

New building regulations mean new buildings must achieve energy efficient standards, including acceptable U-values for different materials. Existing buildings must be upgraded to achieve specified energy performance targets in the future. This issue is very topical. Many websites – often aimed at builders or homeowners – explain the issues and science very well. The topic is likely to crop up when people move homes, or in news stories as the regulations are introduced more widely.

Unit 3, Theme 3 – Improving our environment

Chapter 15 – Our environment at home

15.5 U-values

Learning objectives
- What is a U-value?
- What does a U-value tell you about a material?
- How are materials with different U-values used to control heat losses?

Figure 1 The colours of a thermogram compare the heat loss from parts of a building

Figure 2 Some building materials are better heat conductors than others

All buildings waste heat through walls, roofs, windows and floors. Different building materials lose heat faster than others. Homes lose heat more quickly if walls and roofs are not insulated.

U-values compare the rate of heat loss through different materials. If a material has a high U-value, heat passes through it quickly. Good conductors, such as metals, have a high U-value. Good insulators of heat, such as foam, have a low U-value.

a What do U-values tell us about a material?

b Heat passes more quickly through windows than insulated walls. Does glass have a higher or lower U-value than an insulated wall?

U-values and building

Builders try to choose materials with a low U-value. This helps the building stay at a comfortable temperature. If it is cold outside, a low U-value reduces the rate of heat losses. More heat stays inside the building. In hot countries, heat travels into a building from outside more slowly if its materials have low U-values. The inside stays cooler.

A building using materials with low U-values needs less energy to stay at a comfortable temperature. Less energy is needed to heat the home in cold weather. As well as that, less energy is needed for air-conditioning and cooling fans in hot weather. The owner spends less money on energy bills, and there is a smaller impact on the environment. The building is more comfortable to live in too.

Figure 3 Materials with a low U-value are good for buildings in hot and cold countries

c Why are materials with low U-values good for homes in hot and cold climates?

d Write down two benefits of using materials with low U-values.

U-values of different materials

How quickly heat is lost from a building depends on the building materials, surface area and the temperature difference.

The table shows typical U-values of different building materials. How quickly heat is lost depends on the temperature difference and area of the material, so U-values are measured in $W/(m^2°C)$.

Place	Type of material	U-value in $W/(m^2°C)$
Outer wall	22 cm solid brick	2.2
Outer wall	28 cm brick-block cavity – insulated	0.6
Ground floor	solid concrete	0.8
Ground floor	suspended – timber	0.7
Roof	pitched with felt, 100 mm insulation	0.3
Roof	flat, 25 mm insulation	0.9
Window	metal frame, single glazed	5.8
Window	upvc frame, double glazed – 20 mm gap	2.7

Walls have the largest surface areas, so the materials used for walls have a big impact on heat losses. Many buildings have two walls with a cavity (gap) between to reduce the U-value, and slow down heat losses. The building's running costs will be much lower if the walls are well insulated.

On a frosty day, the temperature difference between the outside and inside of a building can be 25 °C. This is why heat losses and heating bills are much higher in a cold winter.

e Use information from the table. Which material should a builder use for: (i) the outer wall, (ii) the ground floor, (iii) the roof, (iv) the window frames?

f It is more expensive to build a home using cavity walls. Explain why this extra cost is worth paying.

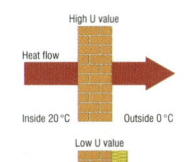

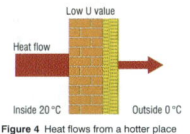

Figure 4 Heat flows from a hotter place to a cooler place. It passes more slowly through materials with low U-values.

Summary questions

1 Copy and complete using the words below:
   cooler   reduces   slower
   Heat flows from a hotter place to a _____ place.
   The heat flow is _____ if the material has a lower U-value.
   Insulating the wall reduces heat losses because it _____ the wall's U-value.

2 Why is less heat lost between rooms inside a house than from walls on the outside of the home?

3 Explain how a builder can use U-values to design an energy-efficient home. Do low U-values mean that the home is energy-efficient?

4 This equation describes the rate of heat loss through a wall:
   heat loss = U-value × area × temperature difference.
   Explain why heat losses are greater through windows (U-value $5 W/(m^2°C)$) compared with heat losses through a wall (U-value $1 W/(m^2°C)$). Double glazed windows have a U-value of $2.9 W/(m^2°C)$. Explain how fitting double glazing will affect the heat losses.

Did you know ... ?

Building regulations used to state the highest U-values allowed for building materials. Now, new homes will be carbon neutral, so materials, boilers and air-tightness must be designed to improve the building's energy efficiency.

Key points

- U-values measure how quickly heat travels through a material.
- Low U-values mean heat is lost more slowly through the material.
- Most buildings use materials with low U-values.

Summary answers

1 cooler, slower, reduces

2 There is a smaller temperature difference between rooms inside the house.

3 Choosing materials with a low U-value means that there are fewer heat losses from the home so heating bills are lower, and the energy bills are lower overall. However, heating is only part of the energy usage in a home so lighting, whether windows are open or closed, the choice of heating methods and whether the heating is on a timer or thermostat have an impact. (Other reasonable arguments should be given credit.)

4 The heat loss through windows is (5 times) greater as the U value is (5 times) higher.
Double glazing reduces heat losses through windows by about half as the U value is about half the value of single glazed windows.

Answers to in-text questions

a The U-value tells us the rate of heat loss through a material.

b higher

c These materials stop heat travelling into, as well as out of, the building.

d reduced energy bills, reduced impact on environment, more comfortable to live in

e i brick/block cavity;   ii suspended timber;
   iii pitched felt;   iv UPVC frame

f The energy savings over the home's lifetime will pay for the additional cost when building the home. Also the home will be more comfortable (there is also a legal requirement to build energy efficient homes).

Unit 3, Theme 3 – Improving our environment

15.6 Pollution in the home

Learning objectives

Students should learn:
- how dust, mould, spores, pollen, smoke and fumes from household products contribute to poor air quality indoors
- that indoor air quality can be improved by dusting, vacuuming, killing mould and using extractor fans.

Learning outcomes

Most students should be able to:
- name and describe some household air pollutants
- name some ways to improve air quality indoors.

Some students should also be able to:
- describe in detail the harm that particular household pollutants can cause.

AQA Specification link-up Science B 3.5.3.3

- Name some of the common pollutants in homes (dust, mould and spores, pollen, smoke, fumes from household products).
- Name some of the common symptoms of exposure to high indoor pollution levels (asthma, headaches, tiredness, dizziness, nausea, itchy nose, sore throat).

Within this context, candidates should be able to use scientific data and evidence to discuss, evaluate or suggest implications of the following:
- methods of reducing pollution in the home including the use of less toxic products.

Lesson structure

Starters

Atishoo! – In small groups, students list as many substances as possible that make them sneeze or cough. How many of them are found in the home? *(5 minutes)*

Scale activity – Present students with an elecronmicrograph of a dust mite. Include a scale on the picture. Students use the scale to calculate the actual size of the mite. To extend students, they can convert this into mm. To support students, focus on describing its features. Where do they think it lives? *(10 minutes)*

Main

- Use pictures and diagrams to explain to the students how an asthma attack occurs.
- As a numeracy activity, students could use demographic data to estimate how many people in their school might have asthma. If this is scaled down to the size of the class, is it still accurate? Why might it not be? Are students' homes or the school in a large city or the countryside?
- Using the information from the Student Book and their own research, students can produce a leaflet for sufferers of asthma, explaining some of the triggers that could set off an asthma attack.
- Students can set up an investigation into particulates in the air. By leaving dust traps (see 'Practical support') in different rooms, they can compare the relative particulate levels in different parts of the school. Provided with a map, students could annotate it with hot-spots for asthma sufferers to avoid.
- Students can create an advertisement for a new product that improves indoor air quality (i.e. a bathroom extractor fan, fungicide, anti-static cleaner). Their advertisement has to explain scientifically how their product reduces pollutant levels and why the pollutant is harmful in the first place.

Plenaries

Tweet – Students must write about a type of indoor air pollution using 140 characters or less. To extend students, ask them to include a picture to go with their tweet. To support students, provide key words for each type of pollutant. *(5 minutes)*

Revision questions – In pairs, students write sets of revision questions to ask other members of their table. *(10 minutes)*

Support

- For each pollutant, focus on summarising where it is and what it does. Use photographs of mould, pollen, soot, etc. to help reinforce the learning.

Extend

- Show students advertisements for air ionisers and invite them to suggest how they might work. Students can write a flow chart explaining the process.

Further teaching suggestions

Dust mites

Students can find electron micrographs of dust mites from the internet and create a display explaining where they have come from.

Practical support

Investigating particulates

Equipment and materials required: Slides, microscopes, 1 × 1 mm graph paper, clear adhesive tape.

Details

Cut out a 1 × 1 cm square of graph paper and stick it onto a slide with tape. Avoid fingerprints. Fix a second piece of tape to the slide, sticky side up, using pieces of tape either side of it. The end result will be a sticky grid in the centre of the slide.

These dust traps can be left in various places around the school (although care should be taken with glassware). After a week, the grids can be examined under a microscope. Each dust trap can be given a score out of 100 according to how many squares have particulates in them.

15.6 Pollution in the home

Learning objectives
- What substances can pollute air in the home?
- What are the symptoms of exposure to high levels of indoor pollution?
- How can air in the home be cleaned?

Whether it's from car exhausts, factory chimneys or power stations, we often think of air pollution as something that happens outdoors. A lot of **air pollutants** are released into the outside environment, but many stay within homes. These substances can cause **respiratory illness**, infections and, in extreme cases, even death.

Respiratory illness is the second biggest killer in the UK (heart disease is the biggest). It affects over 8 million people. There are over 40 different types of respiratory disease. The most common include **asthma**, **hay fever** and **emphysema**.

The build up of pollutants in the home can be caused by our efforts to be more energy efficient. With air conditioners and climate control, there's often little reason to open windows. This results in air being recycled again and again – so any pollutants can't escape. Air pollution includes dust, moulds and spores (e.g. from damp walls), smoke (e.g. from cigarettes), building materials, boilers, pollen and fumes from household products.

Figure 1 One in every 11 children in the UK is being treated for asthma

a Name two respiratory illnesses.
b Why can pollutants build up in homes and offices?

Indoor air pollution affects particular parts of our homes:

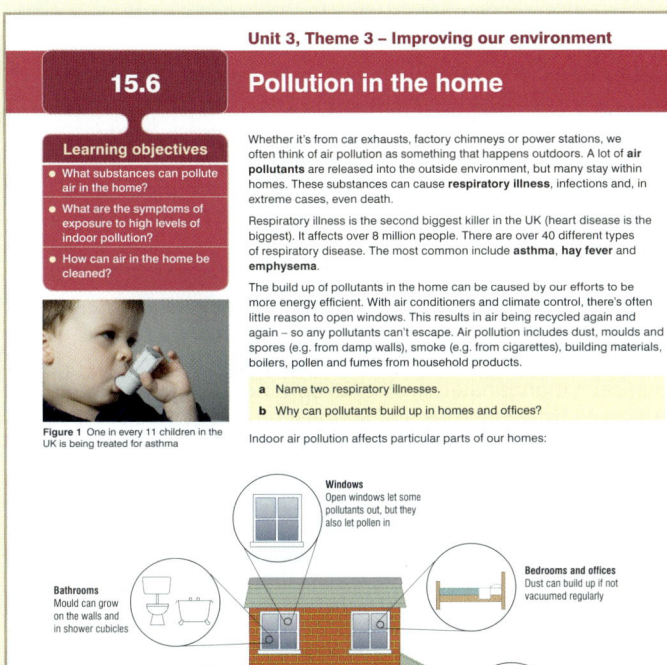

Figure 2 Where is the pollution in your home?

Mould and spores

Mould is a type of fungus that grows well in damp conditions. It reproduces by releasing millions of **spores** into the air. These spores can cause **allergies**. They can also produce poisonous chemicals called **mycotoxins**.

Mould can be killed by specialist anti-fungal bathroom cleaners. Opening windows or using an extractor fan in the bathroom is the best way to prevent mould growing in your home because it reduces humidity.

Dust

Dust is actually made mainly of you. It's formed from the dead flakes of skin cells you shed all the time, as well as fibres from clothing and furnishings. Dust can trigger allergies, itching and asthma attacks. It can also provide food for **dust mites**. Every home has dust mites living in it but if there are too many of them their waste can also cause allergic reactions. The static electricity on a TV screen can attract dust easily, so anti-static wipes and cleaners are a good way to keep dust at bay.

c How can dust mites be harmful?

Soot and smoke

Soot and smoke are both made up of microscopic carbon particles that can permanently damage the **alveoli** in your lungs. This condition is called emphysema. The main symptom is shortness of breath. Very few homes in the UK now use wood fires. Gas-burning fires hardly produce any soot. However, passive smoking remains an issue in some homes.

Fuels and cleaning products

These can cause chemical damage to the respiratory system and even poison you. They should always be used in well-ventilated areas. Signs of respiratory damage include tiredness, nausea and headaches.

Carbon monoxide

Carbon monoxide is a gas that can be given off by faulty boilers. Carbon monoxide poisoning can kill.

Pollen

Pollen is a reproductive cell released by plants. Like mould spores, it can cause allergic reactions. Pollen allergies are called hay fever and affect around 20 per cent of people in the UK. Different people are allergic to different types of pollen. Keeping windows closed on hot days can stop pollen getting into your home. **Antihistamines** can be taken to reduce the allergic reaction.

d What causes hay fever?

Figure 3 Dust mites live off your dead skin cells

links
For more information about carbon monoxide poisoning see 15.8 The silent killer.

Summary questions

1 Summarise the information on this page as a table with the headings; 'Area in home', 'Type of pollution', and 'Harm caused'.
2 Name three types of indoor pollution caused by living organisms.
3 Describe four ways to reduce the effects of indoor pollution.

Key points
- Closed air systems in homes can lead to a build up of indoor pollutants.
- Indoor air pollutants include dust, mould and spores, smoke, fumes from household products, carbon monoxide and pollen.
- Some of the symptoms of exposure to high levels of indoor pollution are asthma, headaches, tiredness, dizziness, nausea, an itchy nose and/or a sore throat.

Summary answers

1

Area in home	Type of pollution	Harm caused
Bathroom	Mould and spores	Allergies, toxins
Bedrooms and offices	Dust, dust mites	Allergies, asthma attacks
Fireplaces	Soot and smoke	Emphysema
Garages	Fuels and cleaning products	Respiratory damage, poisoning
Boilers	Carbon monoxide	Poisoning
Windows	Pollen	Hay fever

2 dust mites, mould, pollen

3 extractor fans, close windows when the pollen count is high, antistatic cloth helps pick up dust, hoovering to remove dust, avoid wood fires, anti-fungal bathroom sprays

Answers to in-text questions

a **two** from: asthma, hay fever, emphysema
b Air conditioning systems don't let air escape, so pollutants build up.
c The waste products of dust mites can cause allergic reactions.
d pollen

Unit 3, Theme 3 – Improving our environment

15.7 Household hazards

Learning objectives

Students should learn:
- that many household products are potentially dangerous
- that hazard symbols alert us to the dangers so we can take proper precautions.

Learning outcomes

Most students should be able to:
- match hazard symbols to their meanings
- describe precautions to take when dealing with particular hazards.

Some students should be able to:
- link specific products to the hazards associated with them and explain how they can be used safely.

AQA Specification link-up Science B 3.5.3.3
- Interpret hazard labels on household products.
- State the risks associated with these hazards, and know ways of minimising these risks.

Within this context, candidates should be able to use scientific data and evidence to discuss, evaluate or suggest implications of the following:
- the hazards and risks caused by using household products.

Lesson structure

Starters

Match-up – Students connect pictures of hazard symbols to the names of the hazards. *(5 minutes)*

How dangerous? – Present students with a selection of domestic products, either bottles or photographs. Students score each one out of five for how dangerous they think it is. To extend students, ask them to explain their decisions for each product. To support students, ask them to divide them into 'Safe' and 'Dangerous'. *(10 minutes)*

Main

- Use a presentation with pictures of hazard symbols to review students' understanding of their meanings.
- Demonstrate some hazardous substances to the class – concentrated acid with sucrose, glycerol with potassium manganate(VII), ethanol burning. Students observe the reaction, explain how it could be dangerous and suggest hazard symbols to put on the chemicals.
- Provided with a shopping list of products, e.g. bleach, oven cleaner, etc., students can research their ingredients, producing a detailed breakdown of why particular products are hazardous. Alternatively, students could perform a product audit of their home before the lesson and come prepared with a list of materials that have hazard symbols on them.
- Students can research alternative herbal cleaning recipes on the internet.
- Students could plan an experiment to compare the efficacy of conventional and 'natural' cleaning agents on a range of stains. This is an excellent opportunity to develop a range of controlled assessment skills in planning and then carrying out the investigation if you have the time.

Plenaries

Slogan – In pairs, students write a slogan to help people remember to read product labels carefully. *(5 minutes)*

Design a symbol – Some domestic chemicals give off poisonous gases when mixed together (i.e. chlorine-based bleach and acids). Students can design their own hazard symbols for such chemicals. To extend students, ask them to come up with their own hazards. To support students, present them with a scenario they must produce a hazard symbol for. *(10 minutes)*

Support
- Students could design their own safety posters to put up at home.

Extend
- To extend students, they could use COSHH student safety sheets when assessing the hazards and risks of the chemicals discussed in the lesson.

Further teaching suggestions

Allergies and common household objects

After reading about the 'toxic sofas' in the Student Book, students could research other products that have turned out to be harmful. For instance, a number of cases of iPhone allergy have been reported.

Practical support

Acid and sugar demonstration

Equipment and materials required: Eye protection, fume cupboard, beaker (100 cm^3), 50 g sucrose, 20 cm^3 concentrated sulfuric acid, bucket of water, plastic bag.

Details
Wear eye protection throughout. In a fume cupboard, add about 50 g sucrose to a 100 cm^3 beaker. Clamp the beaker. Carefully add the concentrated sulfuric acid to the sugar. The sugar will turn yellow, then brown. After a minute it will blacken, and a spongy mass of carbon will rise up the beaker, releasing steam and sulfur dioxide fumes. Explain to students that the acid is removing the hydrogen and oxygen from the carbon in the sugar as water, point out that their body tissues contain a great deal of water. Take care when disposing of the resulting lump of carbon. Drop it into a bucket of water then wrap the carbon in a plastic bag before putting in the waste bin.

Safety: Wear eye protection and nitrile gloves. CLEAPSS Hazcard 98A Sulfuric acid – corrosive.

Glycerol and potassium manganate(VII) demonstration

Equipment and materials required: Eye protection, fume cupboard, tin lid, potassium manganate(VII) crystals (approx. 10 g), glycerol (approx. 20 cm^3), pipette.

Details
Pour some potassium manganate(VII) crystals onto the tin lid and pipette glycerol over the top. The reaction can take more than a minute to begin, giving enough time to describe to students how oxidising agents can cause flammable materials to spontaneously burst into flame. The glycerol should burst into flame and burn for a minute or two.

Safety: Wear eye protection and use a safety screen when demonstrating. CLEAPSS Hazcard 81 Potassium manganate(VII) – oxidising and harmful. CLEAPSS Hazcard 37 Glycerol.

Burning ethanol demonstration

Equipment and materials required: Eye protection, tin lid, safety mat, ethanol, matches.

Details
Pour a little ethanol (20 cm^3) into a tin lid on top of a safety mat. Move the bottle away to a safe distance and ignite the ethanol. Do this once with the lights on (preferably in a bright place so students realise how difficult it is to see) and with the lights off. Draw students' attention to how quiet and easy it is to miss the flame.

Safety: CLEAPSS Hazcard 40A Ethanol – highly flammable and harmful.

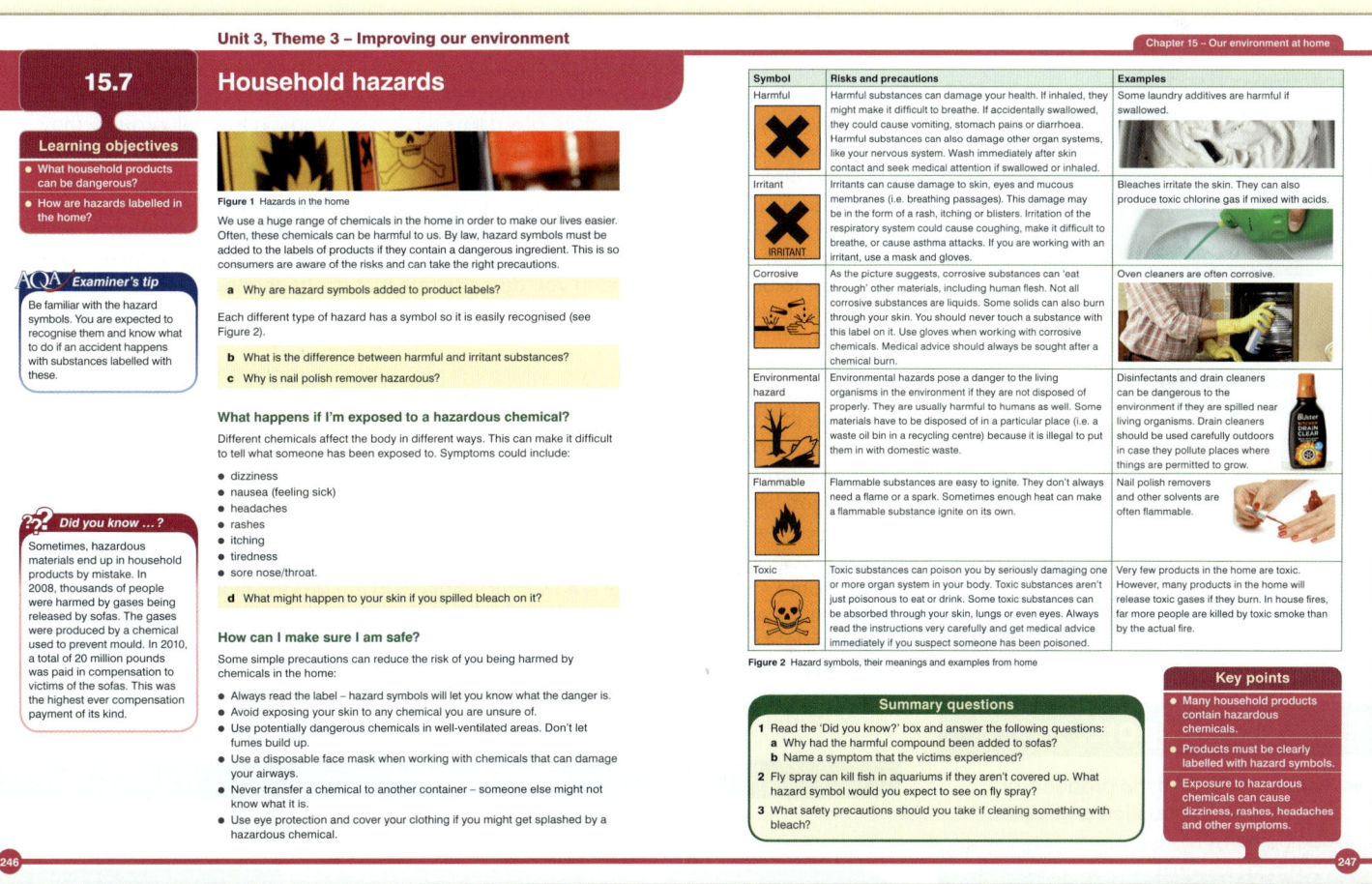

Summary answers

1. **a** To prevent mould growing.
   b chemical burns, eczema, skin cancer
2. environmental hazard
3. gloves, mask if working close up

Answers to in-text questions

a So that consumers know what the dangers are and can take the right precautions.
b Harmful: damage to health
Irritant: specific damage to skin and mucus membranes
c flammable
d Bleach can cause damage to skin. This damage may be in the form of a rash, itching or blisters.

Unit 3, Theme 3 – Improving our environment

15.8 The silent killer

Learning objectives

Students should learn:
- how boilers need a good air supply in order for complete combustion to take place and to work efficiently
- that incomplete combustion produces carbon monoxide and carbon
- that carbon monoxide replaces oxygen in haemoglobin, which can cause death.

Learning outcomes

Most students should be able to:
- explain why a boiler needs a good air supply to work efficiently, including a word equation to show the reaction
- state that incomplete combustion happens when not enough oxygen is present, and write a word equation for incomplete combustion
- describe the symptoms of carbon monoxide poisoning.

Some students should also be able to:
- explain in detail the effects of carbon monoxide poisoning on the body.

AQA Specification link-up Section B 3.5.3.3
- Explain why domestic boilers need an adequate supply of air to work efficiently.
- Explain how incomplete combustion of fuels used in domestic boilers results in lower energy output and the formation of toxic combustion products (carbon monoxide and soot).

Within this context, candidates should be able to use scientific data and evidence to discuss, evaluate or suggest implications of the following:
- the importance of ventilation in the home.

Lesson structure

Starters

Combustion recall – Students rearrange an equation for complete combustion. To extend students, give them the chemical formulae as well and ask them to balance it. *(5 minutes)*

'Whoosh bottle' – Demonstrate combustion of ethanol with a 'whoosh bottle', then repeat to demonstrate that the ethanol won't burn properly when there is less oxygen available. (Details in 'Practical support'.) *(10 minutes)*

Main

- Provide students with molecular models of oxygen and methane. They can show complete combustion by rearranging them into carbon dioxide and water.
- Present students with a picture of a domestic boiler. Working in small groups, they can write a flow-chart explaining how it works. Draw attention to the importance of good air flow in the boiler.
- Students can perform an investigation into incomplete combustion by timing how long it takes a Bunsen burner to boil a fixed volume of water with its air hole open by different widths (see 'Practical support'). As well as measuring the time taken for the water to boil, students watch you demonstrate that sooty deposits can be seen when the air hole is closed. Results can be presented as line graphs of size of opening versus time taken to boil water. This can be used to develop a range of controlled assessment skills.
- Use a simplified picture of haemoglobin to show oxygen molecules attaching to it. State that CO binds to the haemoglobin about 230 times as tightly and use this to explain why CO prevents the blood carrying oxygen.
- Show students pictures of hyperbaric chambers, explaining that the high pressure of oxygen helps force CO out.

Plenaries

Symptoms – Students make a flow chart explaining the progression of carbon monoxide poisoning symptoms. To support students, provide them with statements to sequence into the right order. To extend students, include information about the concentration of CO in the blood at different stages. *(5 minutes)*

Safety poster – Students working in pairs can produce a simple yet informative safety poster warning about the dangers of incomplete combustion. *(10 minutes)*

Answers to in-text questions
a carbon monoxide, carbon dioxide, water and carbon
b It is colourless, odourless and tasteless.
c That incomplete combustion is taking place.

Support
- Focus on recall of the products of incomplete combustion and how it can be harmful.

Extend
- Given information on the specific capacity of water, students performing the Bunsen investigation can calculate the amount of energy needed to make water boil, and thus the rate of energy transfer during each experiment. This could also be converted into watts (W).

Further teaching suggestions

CO Detectors
Students can discuss why carbon monoxide detectors are designed to monitor concentration over time, rather than just concentration. (This is to reduce false alarms from common sources of CO in the home, such as cigarette smoke.)

Practical support

Whoosh bottle

Equipment and materials required: Eye protection, large water cooler bottle (empty), matches, splint attached to a metre rule, denatured ethanol, 400 cm³ beaker

Details

Pour about 5 cm³ ethanol into the water cooler bottle, swirling it around and tilting the bottle so that it can cover all surfaces. Then pour out the excess ethanol into the beaker.

Ensure all students are at least three metres away from the bottle and turn off the lights. Wearing eye protection, light the splint (attached to the end of the metre rule) and carefully put it into the bottle. The resulting combustion reaction will produce a loud whooshing noise as flame emerges from the neck of the bottle. Point out the water that has formed inside the bottle. When the students ask you to repeat it, follow the same steps as the first time, making sure that the bottle has cooled and all previous flames are extinguished, swirling the ethanol around and pouring out the excess. When the ethanol fails to ignite for a second time, ask students why they think this has happened.

Safety: Follow CLEAPSS Supplementary Risk Assessment SRA 06. CLEAPSS Hazcard 40A Ethanol – highly flammable and harmful.

Bunsen burners and incomplete combustion

Equipment and materials required: Eye protection, safety mat, tripod, gauze, 400 cm³ beaker, water, thermometer, Bunsen burner, ruler, stopclock.

Details

Students boil the water from cold (or increase the temperature by a certain amount, such as 20 °C), timing how long it takes. This can be repeated with the Bunsen burner's air hole open at different widths, as measured with the ruler. Results can be collected as 'width' versus 'time taken to boil'. Students should notice a clear relationship between the width of the air hole and how quickly the water boils. (CAUTION: care must be taken to ensure materials cool down sufficiently between trials).

Safety: Take care with hot water and hot equipment.

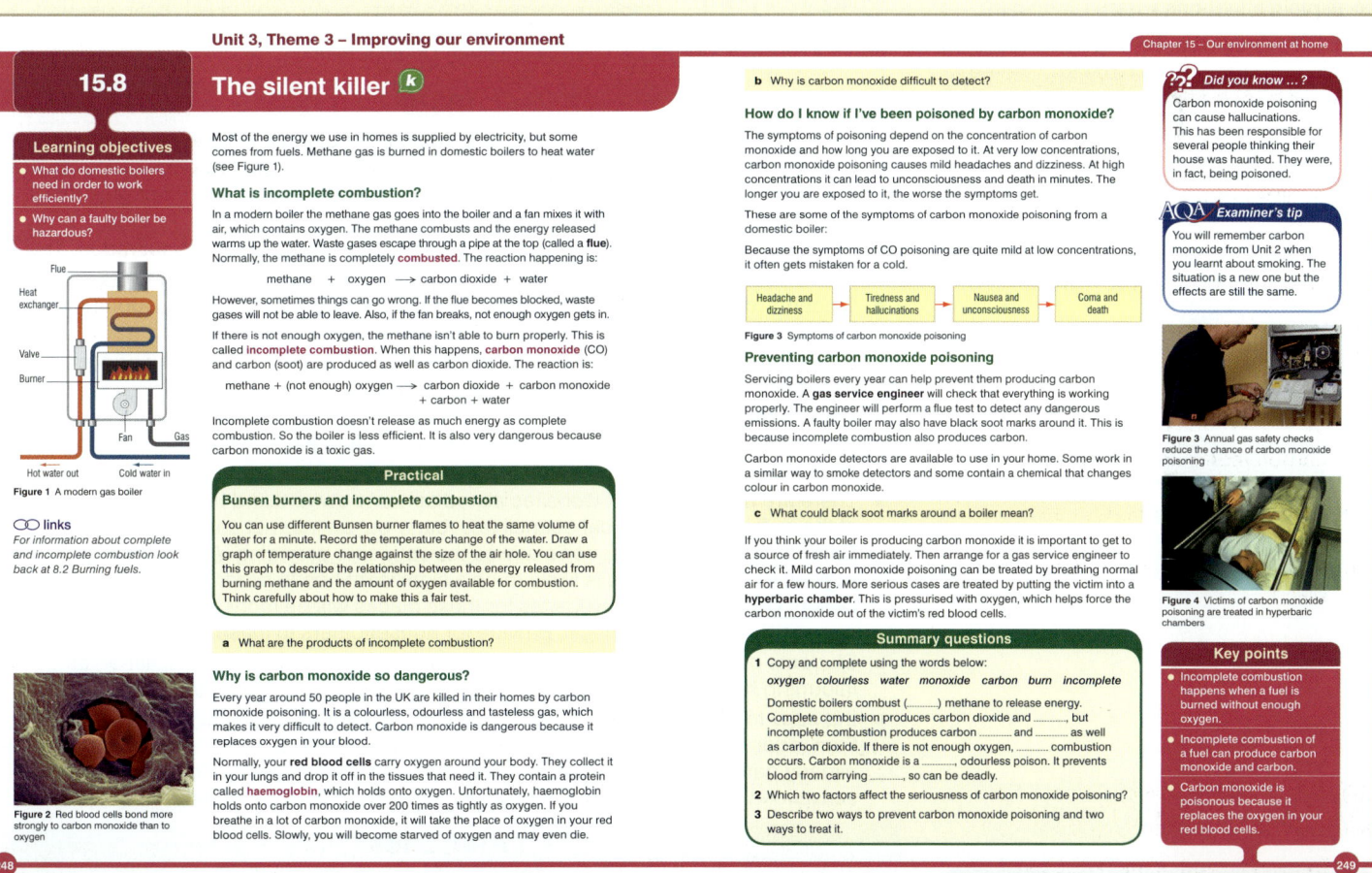

Summary answers

1. burn, thermal, water, monoxide, carbon, incomplete, colourless, oxygen
2. length of exposure
   concentration of gas
3. Prevention: Annual boiler safety checks by a gas service engineer, using a carbon monoxide detector.
   Treatment: Plenty of fresh air, hyperbaric chamber in severe cases.

Did you know …?

In 2005, a 23-year-old woman had recently had a new boiler installed. The boiler was defective and when she took a shower, the house filled with carbon monoxide. She was found later terrified and unable to breathe, claiming to have seen a ghost. There are other examples of this phenomenon that students can investigate.

Unit 3, Theme 3 – Improving our environment

15.9 Radon gas

Learning objectives

Students should learn:
- that radon is a radioactive gas
- how radon can be a health risk
- how to reduce the impact of radon gas.

Learning outcomes

Most students should be able to:
- state radon is radioactive gas and can cause cancer
- explain how to reduce the build up of radon in homes.

Some students should also be able to:
- explain situations where radon can build up.

AQA Specification link-up Science B 3.5.3.3

- Know that radon is a radioactive gas and is a cause of cancer.
- Understand that if rocks and soil beneath the home contain large concentrations of radium or uranium, radon may become a pollutant.

Within this context, candidates should be able to use scientific data and evidence to discuss, evaluate or suggest implications of the following:
- the importance of ventilation in the home
- the dangers of radon gas in the home.

Lesson structure

Starters

What is radon? – Ask students to find radon on the periodic table and predict some of its properties; it is the heaviest of the inert gases so chemically unreactive. It is one of the heavier elements and this is a key reason why it is radioactive. *(5 minutes)*

What gas is there? – Students consider ways to detect something in the air, e.g. how could you tell if there is smoke in the air? Room spray? etc., explain that this lesson is about a gas that cannot be seen or smelled. *(10 minutes)*

Main

- Review the structure of atoms and the properties of alpha and beta particles and gamma rays. Radon emits alpha particles, alpha particles are not penetrating, so are not damaging outside the body, but they are ionising – so they are damaging inside the body.
- Either show a short video clip, or hand round a short news article about radon. This is a topical subject, but reports can be alarmist so choose the article with care.
- Students identify the properties of radon gas (radioactive, colourless, odourless, seeps from rocks such as uranium – uranium ore is surprisingly widespread).
- Students identify the problems caused by radon (increased risk of lung cancer, especially for smokers) and the problems detecting it.
- Discuss the reasons why radon is a problem in buildings only (it builds up in enclosed spaces but disperses in the open air) and move on to discuss methods used to stop the build up of radon in homes.
- There are many excellent websites aimed at home owners and businesses. Look at the website for your local council (e.g. carry out an internet search for radon Oxfordshire) to see the specific advice the council is providing. Action areas are regions where radon gas is likely to be a significant health hazard (e.g. Cornwall, Devon, Cotswolds) – the websites for these places will be more informative and you can see the help that councils provide for householders in these regions.
- Students prepare a poster summarising the main points about radon, aimed at informing householders about risks, and what they can do about this.
- It is easy to over state the dangers – our lifestyle choices put us at far more risk than radon. Radon is a risk in some areas, especially for smokers whose lungs are already damaged and if exposure takes place over many years. However, there is sensationalist journalism, e.g. regarding negligible risks from granite work tops, etc.

Plenaries

Check it out – Students make up a slogan to encourage people in high risk areas to test their homes for radon. *(5 minutes)*

Is there radon near you? – Ask students to look at the radon map of the UK and identify if their area is at high risk for radon. Extend students by providing a geology map for the UK. Ask students to look for links between the two maps of the rock types and the risk of radon gas (e.g. look for regions where granite is common). *(10 minutes)*

Support
- Concentrate on the effects of radon (damage to lung cells when it is breathed in) rather than the process of ionisation of molecules in the cells.

Extend
- Use COSHH safety sheets or refer to specific regulations for employers to examine different ways of reducing risks from radon.

Further teaching suggestions

Comparing risks

Students obtain and use data on risks of radon exposure with other life style choices that increase cancer risks.

Unit 3, Theme 3 – Improving our environment

Chapter 15 – Our environment at home

15.9 Radon gas

Learning objectives
- What is radon?
- Why is radon a health hazard?
- How can we reduce the risks of radon building up in homes?

Figure 1 Granite rocks naturally produce radon gas

Lung cancer is a killer. Smoking is the most damaging thing you can do to your health, increasing your risk of lung cancer enormously. Second-hand smoke can also cause cancers in non-smokers. The next most common cause of lung cancer is radon gas.

Radon gas is a radioactive gas, produced naturally from uranium in rocks. It is colourless and odourless, and seeps from the ground and into our homes. It is not usually a big health risk to most people. Since cells in a smoker's lungs are already damaged, the radon can be more harmful, making lung cancer more likely.

a Where does radon gas come from?
b Why is radon gas a health hazard?

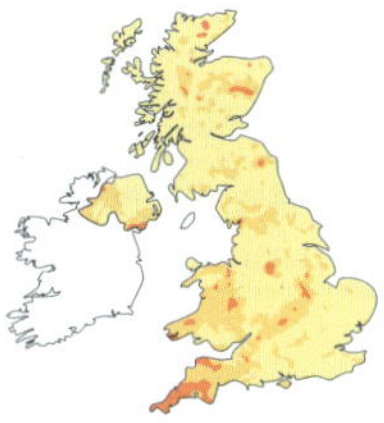

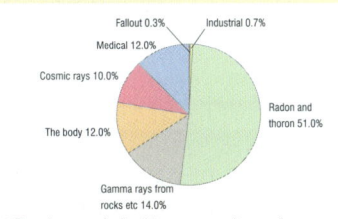

Figure 2 Radon occurs in different amounts across the UK. The darker the colour on the map, the higher the level of radon.

Based on information supplied by The Department for Environment, Food and Rural Affairs.

Figure 3 The main sources of radioactivity we are exposed to every day

We are naturally exposed to radioactivity all the time (see Figure 3). This includes radioactivity from building materials, soil, cosmic rays and food. The background radioactivity is usually too low to be a health risk. The amount of radon gas in your home depends on where you live. Some parts of the country have higher levels than others. If the rocks and soil below your home contain large amounts of uranium or plutonium, it is likely more radon gas will seep inside.

This is a problem if the gas is allowed to build up inside the home. Over many years, the radon can damage cells in a person's lungs. Some people will develop lung cancer as a result. If the house is well ventilated, the radon gas cannot build up. This means that the health risk is very low.

c Where does radioactivity come from naturally?
d How can you prevent radon gas becoming a problem in the home?

Reducing radon levels

The amount of radon in a building can be tested using small detectors. Most homes in the UK have a radon level of 20 Bq per cubic metre. If the levels are above 200 Bq per cubic metre, the health risks are greater and action should be taken.

It is possible to reduce the level of radon in buildings. However, this can be too expensive unless radon levels are high. The risks of radon are very small in the short term, but increase over long periods of time. Reducing radon levels is important if people are smokers or exposed over long times.

In some areas of the country, it can be hard to sell a home that may have high radon levels unless action has been taken to reduce the levels. Businesses may need to control their radon levels to protect the health of their workers.

The three main ways to reduce radon levels are to:
- seal the floor so radon cannot seep in from the ground underneath
- increase the ventilation so the radon cannot build up inside
- use an extractor fan under the building to pump the radon away before it enters the building.

The best method depends on the level of radon and type of building.

e Write down three ways to reduce the radon level in a building.
f Explain three reasons why some people do not take action to keep radon levels below a certain level.

Figure 4 Radon tests help people decide whether to reduce radon levels in their home

links
For information on radioactivity look back at 11.7 What is radioactivity?

Summary questions

1 Complete the following sentences using these words:
   cancer   gas   uranium   ventilation
   Radon is a radioactive _____ that can cause _____.
   Radon seeps from rocks that contain _____.
   You can reduce radon levels by increasing _____ in a building.

2 Explain two reasons why radon is not a problem for most people.

3 You have received a letter saying that the radon levels in your home are above 200 Bq per cubic metre. Explain what you can do to reduce the levels of radon in your home.

Key points
- Radon is a radioactive gas that can cause cancer.
- Radon seeps from rocks under buildings that contain uranium and plutonium.
- Radon levels can be reduced by ventilating the building.

Summary answers

1 gas, cancer, uranium, ventilation

2 Only certain rock types emit significant amounts of radon; non-smokers are much less affected by radon than smokers.

3 Increase the ventilation, e.g. using an extractor fan or seal cracks in the floor that allow radon to enter or install a pump to pump away the radon. (Students may sensibly say open windows, etc. – this is the right idea but not effective enough if the radon levels are high.)

Answers to in-text questions

a the ground, rocks
b It damages cells in the lungs and can cause lung cancer.
c Building materials, cosmic rays, rocks, food, etc. (Students may also refer to the pie chart.)
d Don't allow it to build up, e.g. stop it coming in by sealing, pump it out or ventilate the home well.
e Seal the floor, increase ventilation, install a pump under the floor.
f They may not realise it is a problem as you cannot see or smell it; installing the pumps, etc. is expensive and disruptive; they may not believe it is a risk as the effects take many years to develop.

Unit 3, Theme 3 – Improving our environment

Summary answers

1. **a** **three** from: nuclei vibrate in fixed positions, particles that are heated vibrate more, the vibration is passed to the neighbouring atoms, free electrons carry heat (energy) through metals

   b any **three** of these or other correct answers:
   carpets made from wool
   in lofts using fibre glass insulation
   double glazing / cavity walls, layer of trapped air
   cavity wall insulation, foam between the walls

2. radiation, conduction, convection, convection

3.

	Best colour	Best type of surface
Absorbing heat radiation	black	dull
Emitting heat radiation	black	dull
Reflecting heat radiation	white	shiny

4. **a** Installing loft insulation will reduce heating bills. The payback time is less than two years and so most people will still be living in the same home.

   b There may be a large cost; some people may move out before the original cost is recouped; it is messy to install as it is fibreglass.

   c The lowest U value materials mean the lowest heat losses through the walls/windows. Extra initial costs will be repaid over several months/years in lower heating bills.

5. **a** near windows – pollen, hayfever

   b offices and bedrooms – dust, allergies from dust mites

   c bathrooms – mould, allergies from spores

6. **a** i C
   ii B
   iii E

   b any two from:
   asthma
   headaches
   tiredness
   dizziness
   nausea
   itchy nose
   sore throat

   c any one from: extractor fan to expel air from house; close windows when the pollen count is high to stop pollen coming in house; antistatic cloth helps pick up dust; hoovering removes dust; anti-fungal bathroom sprays stop mould growing.

7. **a** petrol – (highly) flammable

   b bleach – irritant

   c oven cleaner – corrosive

   d rat killer – toxic

8. **a** boiler

   b incomplete combustion

   c It stops the blood carrying oxygen.

9. **a** The risks of lung cancer are very low in non-smokers even if they are exposed to radon, and cancers take many years to develop. Radon is colourless and odourless.

   b The measures may be expensive, messy or time consuming to install or require professional installation. People may not realise the risks, and may not want potential house purchasers worrying about radon exposure.

Unit 3, Theme 3 – Improving our environment

Summary questions

1. **a** Explain how heat is transferred in solids by conduction.
   b Write down **three** places where conduction is reduced in a building. What materials are used as insulators?

2. Complete these sentences using the words below. You can use the words more than once:
   conduction   convection   radiation

   The inside of a fridge is white to reflect .... and reduce heat transfers.
   A double glazed window has a layer of air trapped between two panes of glass. This reduces ....
   Warm air from a radiator rises and circulates round a room. This is ....
   Draught proofing reduces heat losses by ....

3. Complete the gaps in this table using some or all of these words:
   black   white   shiny   dull

	Best colour	Best type of surface
Absorbing radiation		
Emitting radiation		
Reflecting radiation		

4. **a** Explain **two** reasons why many people will save money by installing loft insulation. It costs £250 to insulate a typical home and the savings are likely to be £150 per year.
   b Give **two** reasons why some people would not install loft insulation.
   c Explain why a builder should always aim to use materials with the lowest U-value, even if these may be more expensive initially.

5. Name some pollutants you might find in the following parts of a home, and describe how they can harm you.
   a Near windows
   b Offices and bedrooms
   c Bathrooms

6. The labelled diagram shows many pollutants in the home.

   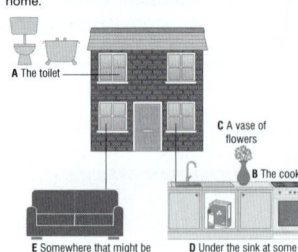

   A The toilet
   C A vase of flowers
   B The cooker
   E Somewhere that might be dusty, such as under the sofa
   D Under the sink at some household products

   a For each pollutant in the list, suggest a letter from the diagram that might be responsible.
   i Pollen
   ii Smoke
   iii Dust
   b Name **two** common symptoms of exposure to high indoor pollution levels.
   c Suggest and explain a simple method of reducing the level of indoor pollution.

7. Suggest which hazard symbol should appear on containers of the following substances.
   a Petrol
   b Bleach
   c Oven cleaner
   d Rat killer

8. Carbon monoxide poisoning kills more than 50 people the UK every year.
   a What household item can produce carbon dioxide if isn't working properly?
   b What is the name of a common chemical reaction that can produce carbon monoxide?
   c Why is carbon monoxide poisonous?

9. Radon is a radioactive gas that seeps from rocks. It builds up in homes and can cause an increased risk of lung cancer for people living in some houses.
   a Explain why people may not be aware of the risks of radon gas.
   b Explain why people may be unwilling to install measures that prevent the build up of radon gas.

AQA Practical suggestions

Practicals	AQA		📖
Demonstrate the production of solid particles by incomplete combustion using a Bunsen burner yellow flame or a candle flame to heat a boiling tube of cold water.		✓	✓

Kerboodle resources

- Support: The importance of fresh aid indoors (15.8)
- Practical: Heat transfer (15.3)
- Interactive activity: Energy loss (15.4)
- Revision podcast: Insulating the home
- Examination-style questions
- Answers to examination-style questions
- Test yourself: Improving our home environment

Chapter 15 – Our environment at home

Examination-style questions

Match the type of heat transfer to its definition.

Heat transfer	Definition
Conduction	An electromagnetic wave
Convection	Heat energy is passed to neighbouring particles
Radiation	Occurs because of the movement of particles in liquids and gases

(2)

The table shows three different ways to insulate walls.

Method of insulation	Cost in £	Annual saving in £	Payback time in years	CO_2 saved per year in kg
Solid wall	10500		26.25	2100
Cavity wall	250	115		610
Thermal wall paint		25	4	75

a Complete the table. *(3)*
b In this question you will be assessed on using good English, organising information clearly and using specialist terms where appropriate.
Write a conclusion based on the data in the table, suggesting with reasons which method of insulation is the best. *(6)*

The 'European Agreement concerning the International Carriage of Dangerous Goods by Road' is in charge of making sure hazardous substances are transported safely. They have several classes of materials.
- Class 1 Explosive substances and articles
- Class 2 Gases
- Class 3 Flammable liquids
- Class 4.2 Substances liable to spontaneous combustion
- Class 5.1 Oxidising substances
- Class 6.1 Toxic substances
- Class 7 Radioactive material
- Class 8 Corrosive substances

a What class would be given to a crate of rat poison carrying this symbol? *(1)*

b What class would be given to a tanker of liquid pentane carrying this symbol? *(1)*

c What class would be given to a lorry containing waste material from a nuclear power station? *(1)*
d What class would be given to a lorry carrying bottles of hydrochloric acid with this symbol? *(1)*

4 Choose the correct word from each box to finish the sentences.
a When fitting a domestic boiler, care has to be taken to ensure a good supply of ....

> air   electricity   water

(1)

b Otherwise .... combustion occurs.

> full   incomplete   radioactive

(1)

c This leads to the formation of .... products including ....

> infectious   penetrating   toxic
> carbon monoxide   methane
> nitrous oxide

(2)

5 Use the equation to answer the questions.

$$\text{efficiency} = \frac{\text{useful power out}}{\text{total power in}}$$

a How efficient is a new wall-mounted gas boiler that uses 28.00 kW and converts 25.62 kW into useful energy for heating? *(2)*
b How efficient is a similar boiler that sits on the floor and transfers 30.60 kW and wastes 3.65 kW of energy? *(4)*
c Choose a reason why the second boiler is not as efficient as the first.

| Energy transfer through the floor | Energy transfer through the walls | Needs more power to run |

(1)

Examination-style answers

1

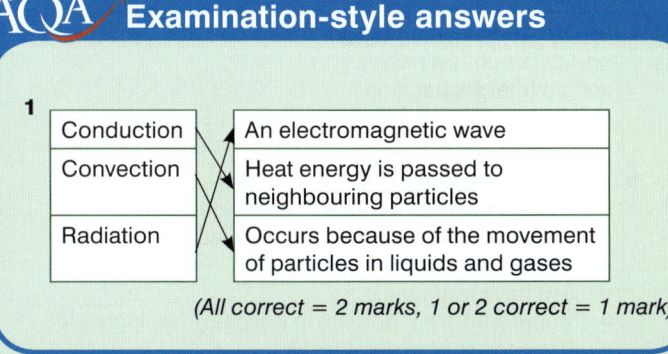

(All correct = 2 marks, 1 or 2 correct = 1 mark)

2 a

Method of insulation	Cost in £	Annual saving in £	Payback time in years
Solid wall	10500	400	26.25
Cavity wall	250	115	2.17
Thermal wall paint	100	25	4

(3 marks)

b Marks awarded for this answer will be determined by the Quality of Written Communication (QWC) as well as the standard of the scientific response.

There is a clear, balanced and detailed conclusion based on the data in the table suggesting with reasons which method of insulation is the best. The answer is well structured with minimal repetition or irrelevant points. There is an accurate, fluent and clear expression of ideas with only minor errors in the use of technical terms, spelling, punctuation and grammar. *(5–6 marks)*

There is some conclusion based on the data in the table suggesting with reasons which method of insulation is the best, with some omissions. The answer shows some attempt at structuring and the ideas are expressed with reasonable fluency and clarity. There are some errors in the use of technical terms, spelling, punctuation and grammar. *(3–4 marks)*

There is a brief conclusion based on the data in the table suggesting with reasons which method of insulation is the best. The answer is largely incomplete and may contain some valid points which are not clearly structured. It lacks fluency and/or clarity. It contains errors in the use of technical terms, spelling, punctuation and grammar. *(1–2 marks)*

No relevant content *(0 marks)*

Examples of points made in the response:
- Solid wall insulation saves the most carbon dioxide
- Carbon dioxide is a greenhouse gas
- So saving CO_2 reduces global warming
- Thermal wall paint is the cheapest to install
- Cavity wall insulation takes just over 2 years to save the money spent on its installation.

3 a 6.1 *(1 mark)*
  b 3 *(1 mark)*
  c 7 *(1 mark)*
  d 8 *(1 mark)*

4 a air *(1 mark)*
  b incomplete *(1 mark)*
  c toxic *(1 mark)*
    carbon monoxide *(1 mark)*

5 a $25.62 \div 28$
     91.5% or $(25.62 \div 28) \times 100$ *(1 mark)*
  b $30.60 - 3.65$ *(1 mark)*
     $= 26.95$ *(1 mark)*
     $26.95 \div 30.60$ *(1 mark)*
     $= 88.07\%$ or $(26.95 \div 30.60) \times 100$ *(1 mark)*
  c Loses heat through the floor. *(1 mark)*

Unit 3 – Making my world a better place

AQA Examination-style answers

1 a Scotland
17 – 5 OR 17.5 – 5
= 12 OR 12.5 *(2 marks)*

b 1987 / 1988 / 1989 *(1 mark)*

c i 369 570 ÷ 100 000 = 3.7
3.7 × 9   OR 3.7 × 10
33.3 – 37 *(2 marks)*

ii increase *(1 mark)*

d Any **three** from:
- The amount of alcohol consumed by other European women did not change much OR went up a little then back down.
- The amount of alcohol consumed by Scottish women stayed the same until about 1975.
- The amount of alcohol consumed by Scottish women increased.
- Scottish women consumed far less alcohol than other European women but now drink far more.
- There are many other factors involved in liver sclerosis so not really able to say. *(3 marks)*

2 a Marks awarded for this answer will be determined by the Quality of Written Communication (QWC) as well as the standard of the scientific response.

There is a clear, balanced and detailed description of the stages that follow once a lymphocyte encounters a pathogen. The answer is well structured with minimal repetition or irrelevant points. There is an accurate, fluent and clear expression of ideas with only minor errors in the use of technical terms, spelling, punctuation and grammar. *(5–6 marks)*

There is some description of the stages that follow once a lymphocyte encounters a pathogen, with some omissions. The answer shows some attempt at structuring and the ideas are expressed with reasonable fluency and clarity. There are some errors in the use of technical terms, spelling, punctuation and grammar. *(3–4 marks)*

There is some description of the stages that follow once a lymphocyte encounters a pathogen. The answer is largely incomplete and may contain some valid points which are not clearly structured. It lacks fluency and/or clarity. It contains errors in the use of technical terms, spelling, punctuation and grammar. *(1–2 marks)*

No relevant content *(0 marks)*

Examples of points made in the response:
- Lymphocyte creates an antibody
- Antibody sticks to pathogen
- Phagocyte notices antibody on pathogen
- Phagocyte engulfs pathogen
- Digestive enzymes inside the phagocyte attack the pathogen and break it down
- Pathogen is destroyed.

b reproduce rapidly
produce toxins *(2 marks)*

AQA Examination-style questions

1 Drinking alcohol increases the risk of liver sclerosis.
The graph shows the number of women that died from sclerosis of the liver in different parts of Europe from 1955 to 2000.

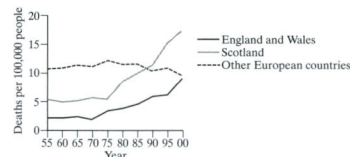

a Which part of Europe had the biggest change over the time period and by how much did it change? *(2)*

b In what year were the number of deaths from sclerosis of the liver the same in Europe and Scotland? *(1)*

c i Leeds in the north of England had a female population of 369 570 in 2000. Estimate how many of them would have died that year from sclerosis of the liver. *(2)*

ii How do you expect this number to change in the future? *(1)*

d Suggest **three** things that you can say about the drinking habits of other European women when compared with Scottish women from 1955 to 2000. *(3)*

2 *In this question you will be assessed on using good English, organising information clearly and using specialist terms where appropriate.*
Our white blood cells defend us against microbes entering our bodies.

a Describe the stages that follow once a lymphocyte encounters a pathogen. *(6)*

b Explain why a small number of bacteria that get through a cut in your skin into your body can make you feel ill. *(2)*

3 Some drugs can be used to improve the quality of our lives.

a Match the medical drug to its use.

Medicinal drug	Use
Aspirin	Kills bacteria
Paracetamol	Anti-inflammatory
Antibiotic	Painkiller

(2)

b Give **two** reasons why doctors are worried about the over-prescription of antibiotics. *(2)*

4 Many household products have hazard symbols on their packaging.

a Match the hazard symbol to its meaning.

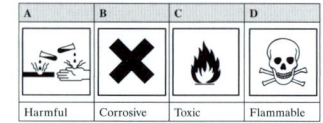

A	B	C	D
Harmful	Corrosive	Toxic	Flammable

(4)

b Another danger in the home is radon gas.

i Why might there be radon gas in a home? *(1)*

ii Suggest **one** reason why radon gas can be dangerous. *(1)*

3 a aspirin – anti-inflammatory
paracetamol – painkiller
antibiotic – kills bacteria
(3 correct = 2 marks, 2 correct = 1 mark, 1 correct = 0 marks)

b Any **two** reasons from:
- Bacteria become resistant to antibiotics.
- NHS have to spend more money researching new antibiotics.
- NHS have more patients.
- People who are ill and are in hospital have increased chance of contracting disease.
- Increased cost to NHS in prescribing different antibiotics. *(2 marks)*

4 a A: Corrosive
B: Harmful
C: Flammable
D: Toxic *(4 marks)*

b i Rocks or soil beneath home can contain radium OR uranium.
ii Radioactive gas can cause cancer. *(2 marks)*

Unit 3 – Making my world a better place

End of Unit 3 questions

Materials scientists are constantly looking for new products to make our lives easier.
a Thermochromic materials change colour according to their temperature.
   Suggest a reason for using thermochromic materials in each of these cases.
   i   A baby's bottle (1)
   ii A t-shirt (1)
   iii A bottle of milk (1)
   iv A ceramic hob (1)
b Give the name for a material that changes colour according to how much light there is. (1)

Some scientists want to use cloning techniques to make new organs for humans. The new organs will cut down waiting times and hopefully reduce the numbers of organs that are rejected by their new host.
Cloning is done by removing the nucleus from a donor egg, and replacing it with the DNA from the organism to be cloned.
a What is a clone? (1)
b Suggest **two** reasons why some people might be against cloning organs in this way? (2)

Lichen is an indicator species that is very sensitive to sulfur dioxide, which causes acid rain. There are many different varieties of lichen. There are those that are very hardy nearest the town centre. They are green and crusty. Further from the town more types of lichens can grow that look leafy and shrubby.
The table shows the number of different varieties found growing on walls at different distances from the town centre.

Distance from town centre in km	Number of different varieties of lichen found
0	0
2	15
4	13
6	15
8	25
10	27
12	46
14	46
16	65

a Explain at which distance from the town there is an odd result. (2)
b Write and explain a conclusion based on the results. (2)

Bump up your grades

The important thing here is to practise comprehension questions and look at describing patterns in graphs and tables. If a student can extract information from a body of text, it could make all the difference. Also, make sure that your students show their working out, you would be surprised to see the number of students who just write a number. If it is incorrect, they get no marks. Practise, practise, practise.

Grade C students may be required to compare various types of medicine including vaccines and antibiotics. Also, a description of the harmful effects of tobacco and alcohol is expected at this level. A description of the different parts of the body's defence system is needed along with knowledge of the characteristics of the three types of ionising radiation. Describing the need for electrolysis and the steps involved in selective breeding would put someone over the C/D borderline. And again, lots of practice is needed with the equations.

Grade A students need to be able to describe the steps involved in genetic modification. Electrolysis questions involving knowledge of the processes at the cathode and anode are at this level.

An F grade student would need to suggest how tobacco and alcohol could harm the body, they should be able to name parts of the body's defence system and name some hazards around the home. Knowledge of the various hazard symbols would be an advantage.

---

5 a i So you can see if it is too hot.
   ii Fashion / look good
   iii So you can see if it has become too hot.
   iv So you can tell if it has become hot enough OR cold enough to touch. (4 marks)
  b photochromic (1 mark)

6 a a genetically identical copy of a donor parent (1 mark)
  b any **two** from:
   - Less biodiversity
   - Against ethics / religion
   - No human rights for clone
   - Taking life of person that could have been born from egg (2 marks)

7 a 2 km because the rest of the results go in a pattern apart from this one. (2 marks)
  b greater variety of lichens found further away from the town centre, therefore more sulfur dioxide nearer the town centre (2 marks)

Controlled Assessment

Planning and risk assessing

Learning objectives

Students should learn:
- how to undertake a centre assessed piece of work.
- how to plan and risk assess an investigation successfully.

Learning outcomes

All students should be able to:
- start a centre assessed task
- plan the steps in an investigation
- carry out a risk assessment.

Specification link-up
- 3.6.2 Plan an investigation
- 3.6.3 Assess and manage risk when carrying out practical work

Lesson structure

Starter

Make a plan – Ask students to write a step by step plan for a simple task at home, for example making a cup of tea or getting washed and dressed in the morning. They can then risk assess the activity. Did they get everything in the correct order? *(10 minutes)*

Main

- Explain to students how their Controlled Assessment will operate and be marked.
- Students can study an experiment that is new to them or that is familiar from a previous year, for example making calcium carbonate from calcium chloride and sodium carbonate. This gives a number of different risks and also encourages the use of student safety sheets for stage 3 students. A new investigation gives students the chance to carry out research. The use of research to provide a basis for an investigation will be needed in the Controlled Assessment.
- Differentiation is by outcome. All students should be provided with the main headings for a risk assessment. They can highlight all of the risks and assess the experiment.
- Stage 1 students may only identify the main chemicals and some of the glassware. Stage 2 students should be able to identify the risks and suggest control measures. Stage 3 students should be able to provide detailed risks and control measures and should be encouraged to use Hazcards but read them carefully so that they identify the correct calcium salts.

Plenary

Question and Answer – Ask about the risks that have been identified and the control measures to check understanding and comprehensiveness. Refer back to the information found on Hazcards. *(10 minutes)*

Science in context

In many lines of work, employees are expected to follow Standard Operating Procedures or SOPs. SOPs are useful in that they allow for reproducible results across different groups/organisations carrying out the same investigation, even in different countries. The employee is expected to risk assess their own work and use information such as CLEAPSS or COSHH to enable them to carry out experiments safely without causing themselves or others harm. Failure to adhere to these could result in injury and/or dismissal.

Support

- Students may only identify a few of the risks. When writing a plan, some students may need a writing fame including headings such as 'First I will', 'Then I will'. They often find the bullet pointed list of instructions easier.

Extend

- Students should be able to write a detailed risk assessment identifying all risks and describing their control measures.

The role of the teacher in the centre assessed unit

The centre assessed unit contributes 25% of the total GCSE. Students will have studied a range of topics throughout the GCSE. The centre assessed unit may draw from any part of the course. You will be offered three possible investigations. You can do all of them or just one or two but only one can be submitted.

The centre assessed unit involves researching, planning and carrying out an investigation safely, taking measurements, making conclusions and evaluating the investigation. It is a good idea to practise some of the skills beforehand so that when students carry out the centre assessed investigation, they are familiar with all of the different components.

The unit is assessed at three different stages, each being hierarchical. A template of the requirements will be provided as a mark scheme. Stage 1 is the simplest and requires the least amount of detail and skill. Stage 3 is the most complex and requires detail, more complex ideas and scientific reasoning.

Planning and risk assessing

Planning the investigation

Students are expected to write a plan for their investigation which must be in the future tense or written as a list of instructions. At stage 1, the student should state the aim of their investigation. The plan can be quite simple and may contain errors such as steps being in the wrong order. For example, in an investigation looking at the effect of different fertilisers on the growth of a plant, if the seeds are added to the soil before the fertiliser and the soil is stirred, the seeds may be too deep in the soil to grow. At stage 2 the aim should be described and the plan should be more coherent and should contain enough detail for another person to carry it out. For example, it should include *what* measurements will be carried out and *how*. Also include a prediction of the possible relationship between the variables under investigation. At stage 3, the aim of the experiment should be explained. For example, if students are looking at the effect of different types of antacids, this would help doctors to prescribe the most effective one for different patients. They should suggest a quantitative relationship between variables if possible. The plan should be detailed and should include all steps in the correct order and specify amounts, volumes and concentrations of chemicals. This should all be completed independently. The teacher needs to include annotation to show the level of guidance that has been given.

Risk assessing the investigation

All investigations, however simple, contain risks. It is important that students realise these risks before carrying out the investigation. At stage 1, the students need simply to identify the main risks. For example, if investigating antacids, these would be tested on acid and would involve some glassware. At stage 2, the student should identify the risks and how to control them. This could include common sense details such as wearing goggles and gloves. At stage 3, the student needs to identify all of the relevant risks and control measures. The hazards need to be identified specifically, for example the names of any chemicals used. The risks should also be correct for the chemicals being used. For example, the acid used would be an irritant (not corrosive). The control measures taken should be based on scientific reasoning. This should all be completed independently. The teacher needs to include annotation to show the level of guidance given.

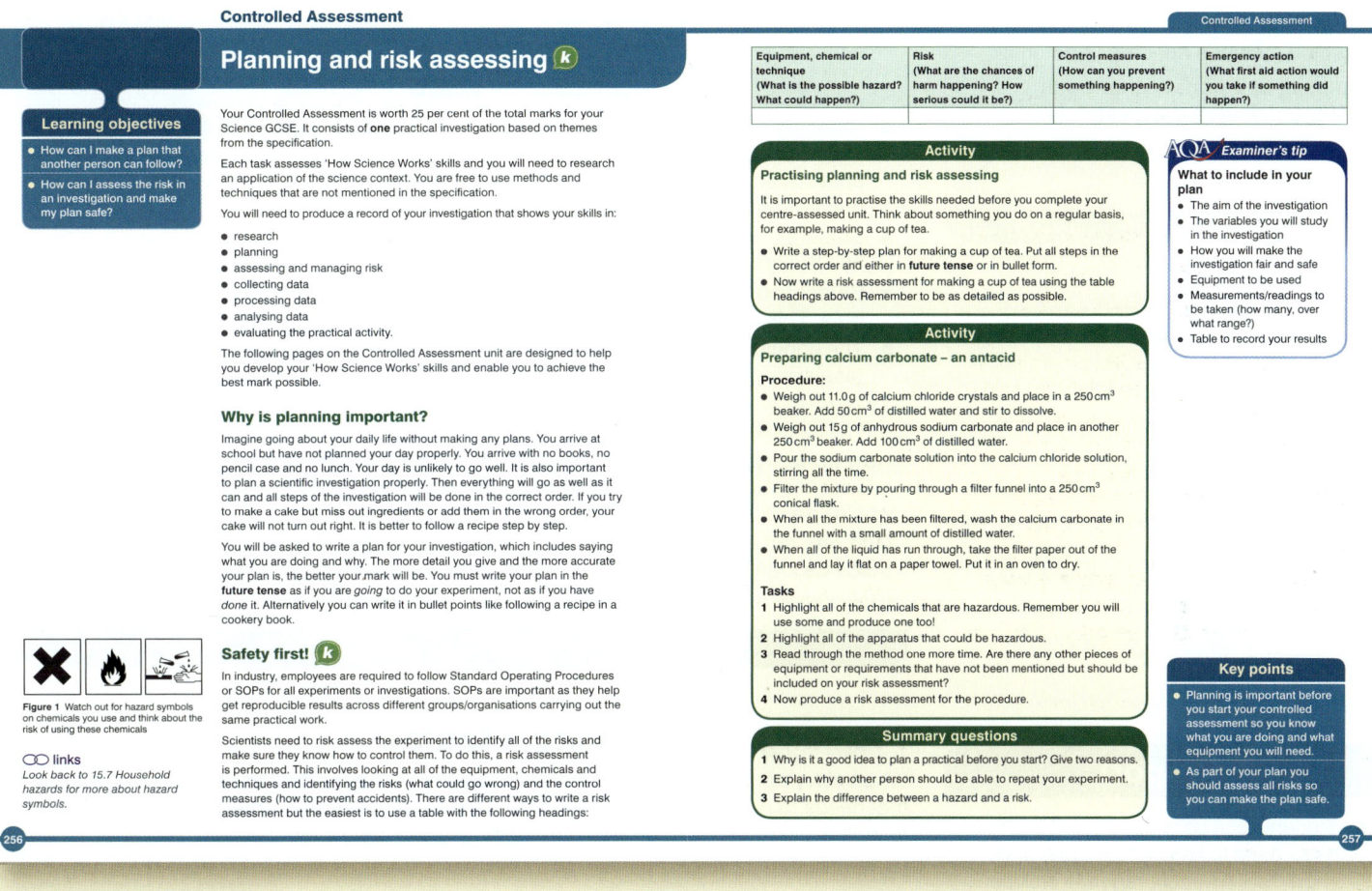

Practical support

Preparing an antacid

Material and equipment required: Spatula, calcium chloride crystals, 3 × 250 cm^3 beaker, distilled water, glass rod, anhydrous sodium carbonate, measuring cylinder (100 cm^3), filter funnel, filter paper, 250 cm^3 conical flask, wash bottle, paper towel, eye protection, access to electric balance and oven.

Details

The experiment to make calcium carbonate can actually be carried out and used as an extension activity to lead into an investigation into the effectiveness of antacids. Students can use their own product and compare it to other known brands.

Safety: Wear eye protection. CLEAPSS Hazcard 47A Hydrochloric acid – corrosive (Student Safety Sheet 36). CLEAPSS Hazcard 19A Calcium chloride – irritant (Student Safety Sheet 33). CLEAPSS Hazcard 95A Sodium carbonate – irritant (Student Safety Sheet 20). CLEAPSS Hazcard 19B Calcium carbonate (Student Safety Sheet 33). Note: students should not be given access to CLEAPSS Hazcards.

Summary answers

1 To be make sure that all of the equipment and chemicals are prepared and ready to use; to be aware of the risks; to know how to prevent accidents happening; if they do, to know the first aid treatment that would be needed.

2 To be able to follow the instructions to verify the findings.

3 A hazard is something that can cause us harm whereas a risk is the chance of the hazard actually causing harm.

Controlled Assessment

Collecting and recording evidence

Learning objectives

Students should learn:
- how to plan a table to record data collected in an experiment
- how to present data as a graph
- why it is necessary to carry out repeat measurements.

Learning outcomes

All students should be able to:
- plan and draw a table of results, which includes headings and units
- plot an appropriate graph with suitable scales, labels and units
- decide when repeats are necessary and why.

AQA Specification link-up
- 3.6.4 Collect primary and secondary data
- 3.6.5 Select and process primary and secondary data

Lesson structure

Starters

Make a table – Give students some data from an experiment. They should organise the data into a suitable table with headings and units. *(5 minutes)*

Complete a table – Give students a table of results that is incomplete, for example missing headings and units and with no column for averages. Students alter the table to make it complete and meaningful. *(5 minutes)*

Main

- Students carry out an experiment to look at the effectiveness of different antacids. They can use the calcium carbonate they made in a previous lesson or can use antacid remedies. They should decide what measurements they will record, the number of repeats and the headings and units needed.
- Students draw a graph to present their findings, deciding the type of graph needed and the appropriate scale. They should ensure they label the axes and include units.
- Students can carry out the 'Practise planning a table' activity and plot a graph of the data. This allows for a line graph to be drawn.
- The conclusion and evaluation can be carried out as part of the next lesson (analaysing and evaluating evidence).

Plenaries

How do we record results? – If carrying out the investigation, through question and answer, discuss how students have recorded their results and what type of graph they would use. *(5 minutes)*

Odd one out – Show a table of results, such as the 'Spot the odd one out' task. Students identify the anomalous results and discuss the type of graph they would plot. *(10 minutes)*

Support
- Students can be given a blank grid that helps them to lay out their table. They should still be encouraged to include their own headings and units, although there may be some omissions or errors. When plotting a graph, they may require some help choosing an appropriate graph and scale.

Extend
- Students should decide the appropriate number of decimal places for their results, should construct the most suitable graph and include lines of best fit where appropriate.

Planning and drawing a table

When carrying out an experiment or investigation, it is important to think about the data that needs to be collected and how this needs to be presented. To enable another person to look at the results and understand them fully, the table should include headings for all measurements to show what the numbers mean and units should be included.

At stage 1, students may construct a simple table but some of the headings and units may be missing or unclear. At stage 2, a table with more than two columns should be constructed and the headings and units should be clear. Additionally at stage 3, the data recorded should be given to the appropriate number of decimal places. For example, if measuring volume using a burette, one decimal place would be expected and this should be consistent for all of the results collected.

Why repeat an experiment?

It is not always necessary to repeat an experiment, for example if carrying out flame tests it is very likely that the flame colour will be the same each time. However, in most cases where numerical data are being recorded, repeats are necessary to check precision and improve accuracy. Students are required to recognise when repeats should be carried out at stage 2 and at stage 3, they should explain why repeats were necessary. If repeats were not necessary or appropriate, students can still gain credit at stage 2 and 3 but in both cases they must explain why. Students should understand the term anomalous result in order to discuss why it is necessary to repeat an experiment and discard anomalous results from a set of repeat readings before calculating the mean.

Controlled Assessment

Which graph to draw?

A very easy way of analysing the data collected is to draw a graph. Students are required to draw graphs which are appropriate to the data they have collected. If the data are categoric then a bar chart is sufficient. However, at the higher stages, if it is possible to plot a line graph and produce a line of best fit, then this must be done to gain credit. This requires students to investigate an independent variable that is continuous. At stage 3, the student should independently choose the type of graph and the scale to use, which should be appropriate. If a line of best fit has been drawn then it should be accurate and drawn with a sharp line rather than dot to dot, sketched or drawn with a thick pen or pencil. Remind students that the line of best fit is not always a straight line!

Practical support

Material and equipment required: Calcium carbonate made in the practical preparing antacid (details on p.257) indigestion tablets; mortar and pestle; Universal Indicator solution or other acid/base indicator with dropping pipette, 0.1 mol/dm³ hydrochloric acid; range of measuring cyclinders, beakers, conical flasks, burette and stand, tile, funnel, water, eye protection.

Details
As this is an open-ended planning exercise students suggestions will vary. A common plan will involve mixing a fixed mass of different indigestion tablets in distilled water, adding an acid–base indicator to each and titrating against dilute hydrochloric acid to determine the volume of acid that can be neutralised by the alkaline solution of each indigestion remedy. Check student plans for safety before any practical work commences and allow time for trial runs so students can find out the best quantities to use to get reasonable results.

Safety: Check students' plans for health and safety before practical work begins. Wear eye protection. Make sure students use dropping pipettes sensibly. CLEAPSS Hazcard 19A (Student Safety Sheet 36) Calcium chloride – irritant. CLEAPSS Hazcard 95A (Student Safety Sheet 33) Sodium carbonate – irritant. CLEAPSS Hazcard 47A (Student Safety Sheet 20) Hydrochloric acid – corrosive. CLEAPSS Hazcard 19B (Student Safety Sheet 33) Calcium carbonate.

Controlled Assessment

Collecting and recording evidence

Learning objectives
- How can I record and present my data from an investigation in a suitable table?
- Why should I repeat my measurements or observations?
- What is the most appropriate type of graph to use to present my data?

Planning a table of results

Imagine trying to do your homework with all of your books, pens and pencils scattered all over your desk. You are unlikely to find everything you need and you may even lose your homework amongst all of the chaos. When you carry out an experiment, it is a good idea to plan your results table before you start so that, when you record your results, they are well ordered and you can find everything.

The first step in planning a results table is to decide what you are going to measure in the investigation. For example, if you are boiling some water in a beaker and you want to measure the temperature, you would need to record the temperature over a certain time interval. You then need to decide how many times you are going to do the experiment and include enough columns in your table. If you have repeated your experiment, it is a good idea to include enough columns for each test result and a column for means (averages). The final step is to make sure that you have included headings and units.

The more organised your table is, the easier it is for you to record your results, analyse your findings and for someone else to follow what you have done as well.

Activity
Practise planning a table

Imagine that you are going to carry out an experiment to find the best material to use to make a saucepan. You are going to boil 100 cm³ of water in four different materials until the water reaches 100 °C. You are going to do this for glass, copper, steel and aluminium.

Plan a table to record your results. Think about:
- what you need to record
- the headings and units you need to include
- how many times you will do the experiment
- whether you need to include a column for the means.

Why repeat measurements?

It is always good practice to repeat measurements to check your findings and to see if your results are repeatable and precise. The more times you can repeat an experiment, the more accurate your findings are likely to be. By repeating, you can check whether the results agree or are close enough to show that they are precise. For example, if you time how long it takes to boil 50 cm³ of water and find that the first time it takes 38 seconds, the second time it takes 42 seconds and the third time it takes 1 minute and 14 seconds you should be able to spot an **anomalous result**. There is always a little bit of variation in any repeat measurement. However, this is a result that does not fit the pattern and usually happens because a mistake has occurred. Which is the anomalous result in the set of three repeat readings above?

AQA Examiner's tip
Remember that:
A categoric variable is described by words. It has certain fixed values. Examples are 'the type of metal' or 'the location of the water sample'.
A continuous variable can have any numerical value. So anything measured is a continuous variable. Examples are 'time', 'temperature' or 'distance'.

It is always a good idea to repeat measurements if you find anomalous results so that all of your results in a repeat set are close together (precise). If you haven't got time to repeat an anomalous reading it can be discarded – identify it, then don't take it into account when calculating the mean (average).

Presenting data in a graph

Depending on the data you have recorded, you can present your data in different forms. A graph is an easy way of analysing your findings very quickly.

A **bar chart** is used for categoric data; for example, showing the density of different metals.

A **line graph** is used for continuous data; for example, showing how temperature changes over time. If you can, you should draw a line of best fit which is a line that goes through the middle of the distribution of all the points. (Don't forget that a line of best fit can be a curve or a straight line.)

The **independent variable** is the variable that you choose to vary during the investigation. This is plotted on the x (horizontal)-axis. The **dependent variable** is the one that you measure to judge the effect of varying the independent variable. This is plotted on the y (vertical)-axis.

Choosing an appropriate scale for your graph is very important. The scale should allow you to clearly see your graph. You should try to use as much of the graph paper as possible but choose a scale that is easy to read. For example, do not choose a scale with 10 squares for each 30 units. (Imagine trying to plot a value such as 17.4 units on that scale!) You should also make sure that your scale is linear. This means that each square on the graph represents the same quantity. For example, if you are plotting a graph of temperature against time, each square for time should represent the same number of minutes or seconds.

Activity
Spot the odd one out

Some students have carried out an experiment to look at acidity in water taken from four different reservoirs. They have recorded how much water from each reservoir was needed to neutralise 25 cm³ of sodium hydroxide. Their results table is shown below.

Reservoir	Volume of water needed to neutralise 25 cm³ of sodium hydroxide (cm³)			
	Test 1	Test 2	Test 3	Mean
Matlock Dye Works	28.1	28.6	29.3	36.3
Butterley Reservoir	62.5	44.2	60.2	38.8
Carsington Water	20.2	21.7	22.4	30.5
Ratcliffe on Soar Power Station	14.3	52.5	51.6	39.5

1 Which results are anomalous? How do you know?
2 What would be the best way to present these findings in graph form?
3 Draw a graph of the results. Remember to use appropriate scales, label your axes and include units.

Practical
Investigating the effectiveness of antacids

Carry out an experiment to look at the effectiveness of two antacids. Before you start, plan a results table to record your results. How could you present your findings as a graph?

Summary questions

1 Why is it important to include headings and units in a results table?
2 Why might you repeat certain measurements within your experiment?
3 What type of graph would be used to display the results of the following investigations?
  a How does temperature affect the speed of a tennis ball?
  b Which type of supermarket carrier bag is the strongest?
4 Look at the table above. The student tested a fifth sample of water from a canal. The results of the three tests were 18.4, 18.8 and 18.6 cm³. What is the mean of these three readings?

Key points
- When planning an investigation you should also produce a table to record your data in.
- Repeat any measurements or observations a number of times to improve the accuracy of your results.
- A bar chart is used for categoric data.
- A line graph is used for continuous data.

Activity

Answers to 'Spot the odd one out'

1 Test 2 for Butterley Reservoir and test 1 for Ratcliffe on Soar Power Station are anomalous because they do not fit the pattern since they are not close to the other results.
2 A bar chart would be the best way to present the findings.
3 Reservoir name on the x axis, volume of water in cm³ on the y axis. An axis break could be used so that the y axis scale is from 30.0 to 40.0.

Summary answers

1 Headings and units should be included to make it clear to yourself and another person what the data have been collected. The units give the numbers' meaning and allow you to compare two sets of results fairly.
2 You might repeat a measurement that is anomalous, i.e. it does not fit the pattern of the other data.
3 a line graph
  b bar chart
4 18.6 cm³

Controlled Assessment

Analysing and evaluating evidence

AQA Specification link-up:
- 3.6.6 Analyse and interpret primary and secondary data

Learning objectives

Students should learn:
- to analyse data displayed on a graph, spotting any patterns
- to suggest improvements to the method used in their investigation.

Learning outcomes

All students should be able to:
- use detailed scientific knowledge and understanding to explain conclusions drawn from the data
- give detailed reasoning when explaining how to improve their investigation, including appropriate technical vocabulary.

Lesson structure

Starter

Spot the pattern – Give students some sketch graphs with axes labelled and a variety of different lines drawn, e.g. showing positive linear, negative linear and direct proportionality. Ask the students to identify the relationship between the variables. To support students, start the sentence for them and let them complete the relationship, e.g. 'The greater the load on the spring ...'. To extend students also give them a more complex graph to comment on, e.g. rate of enzyme activity against temperature. *(10 minutes)*

Main

- Go over the answers to the starter together, using the graphs provided in the Student Book as general examples of the specific relationships between variables that they have been identifying.
- Show an example of a complex relationship as well as a sketch graph where there is no obvious link between the two variables.
- Stress the need to refer back to predictions made at the start of an investigation and how important it is to explain the analysis of your graph by applying their scientific knowledge and understanding.
- When going on to look at the skills of evaluation, there is an excellent opportunity to revise all the technical vocabulary that a good grasp of 'How science works' requires. Ask the students to produce their own glossary of words such as accuracy, precision, resolution, repeatability, reproducibility, validity, range, interval, independent variable, dependent variable, control variable.
- In evaluating their investigation students should also think about errors in practical work – random (e.g. human error, which can be scattered around the true value) or systematic (e.g. zero error on a balance, which will be consistently high or low).
- Carry out the activity 'Practising evaluation' in which students criticise the scenario presented to them. Higher attaining students could be asked to produce a plan that would yield more repeatable and reproducible data from the investigation.

Plenary

How to improve – Using Summary question 2, ask students to produce a checklist of 'tick-box' statements that could be used at the end of any evaluation. To support students you could provide key words to include. To extend students provide them with the Level 3 criteria from the specification and ask them to justify how using their checklist would enable a student to achieve the highest level. *(10 minutes)*

Support

- Provide students with a writing frame to get them started on their conclusions and evaluations, e.g. 'The pattern I can see on my graph is ....', 'I think this pattern happened because ....', etc.

Extend

- Students should use secondary sources of data to help with their conclusion. They should make sure the secondary source is from a reliable source and valid for what they are investigating.

Drawing conclusions

The patterns and relationships observed in data represent the behaviour of the variables in an investigation. However, it is necessary to look at patterns and relationships between variables with the limitations of the data in mind in order to draw conclusions.

Evaluating

In evaluating a whole investigation the repeatability, reproducibility and validity of the data obtained must be considered.

Repeatability refers to the precision of the data collected. The students should comment on the range within each set of repeated measurements – these should be consistent and any significant variation should be explained in terms of possible errors, with suggested improvements included.

Reproducibility refers to the consistency of your data compared to data collected by other groups doing the same investigation. This comparison might be with others in the class or from secondary sources.

Validity takes into account the factors of repeatability and reproducibility. Validity also considers whether your investigation actually gathered the data needed to answer the question under investigation. When considering this, students should evaluate the fairness of the tests carried out and suggest improvements to their method. The validity of conclusions drawn from data will also take into account the limitations of the data collected and suggest further work that could be carried out.

In the actual controlled assessment task this data processing, analysis and evaluation stage requires a high level of teacher control. The guidance states that students must work individually under direct supervision to write up their findings, analyse their own and the secondary data and present their evaluations and conclusions. It is at this stage that teachers should present students with class or teacher obtained data for them to comment on the reproducibility and effectiveness of their own work.

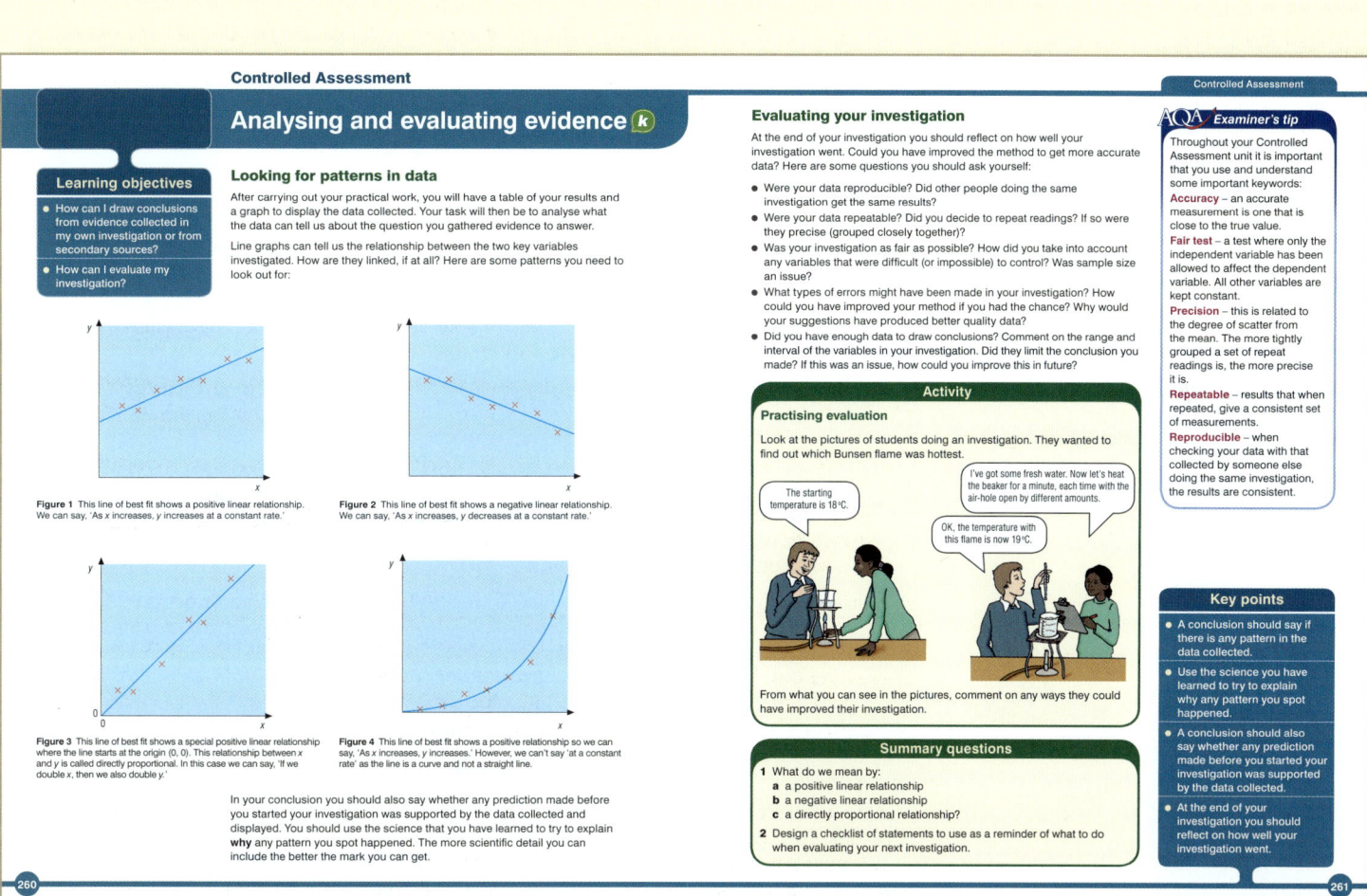

Further teaching suggestions

Room for improvement!

Role-play an incompetent scientist in a demonstration, asking students to list each error you make whilst performing an investigation. Go over the list of errors asking for the explicit reasons why they were poor practice.

Summary answers

1 a a relationship in which the dependent variable increases as the independent variable decreases, at a constant rate

   b a relationship in which the dependent variable decreases as the independent variable increases, at a constant rate

   c a relationship in which the dependent variable doubles as the independent variable doubles

2 student evaluation checklist to cover aspects of repeatability, reproducibility and validity

Text © James Hayward, Jo Locke, Nicky Thomas 2011
Original illustrations © Nelson Thornes Ltd 2011

The right of James Hayward, Jo Locke and Nicky Thomas to be identified as authors of this work has been asserted by them in accordance with the Copyright, Designs and Patents Act 1988.

All rights reserved. No part of this publication may be reproduced or transmitted in any form or by any means, electronic or mechanical, including photocopy, recording or any information storage and retrieval system, without permission in writing from the publisher or under licence from the Copyright Licensing Agency Limited, of Saffron House, 6–10 Kirby Street, London, EC1N 8TS.

Any person who commits any unauthorised act in relation to this publication may be liable to criminal prosecution and civil claims for damages.

Published in 2011 by:
Nelson Thornes Ltd
Delta Place
27 Bath Road
CHELTENHAM
GL53 7TH
United Kingdom

11 12 13 14 15 / 10 9 8 7 6 5 4 3 2 1

A catalogue record for this book is available from the British Library

ISBN 978 1 4085 0836 7

Cover photograph: Getty Images/Tanya Constantine

Illustrations by GreenGate Publishing, with additional artwork by Wearset

Page make-up by GreenGate Publishing

Printed and bound in Spain by GraphyCems